Probability, Statistics, and Reliability for Engineers and Scientists

Second Edition

Bilal M. Ayyub
Richard H. McCuen

CHAPMAN & HALL/CRC

A CRC Press Company

Boca Raton London New York Washington, D.C.

Library of Congress Cataloging-in-Publication Data

Ayyub, Bilal M.
Probability, statistics, and reliability for engineers and scientists / by Bilal M. Ayyub and Richard H. McCuen.—2nd ed.
p. ; cm.
Revised ed. of: Probability, statistics, & reliability for engineers. c1997.
Includes index.
ISBN 1-58488-286-7 (alk. paper)
1. Engineering mathematics. 2. Reliability (Engineering). I. McCuen, Richard H. II. Ayyub, Bilal M. Probability, statistics, & reliability for engineers. III. Title.

TA330 .A99 2002
519.5′02462—dc21 2002276810

Direct all inquiries to CRC Press LLC, 2000 N.W. Corporate Blvd., Boca Raton, Florida 33431.

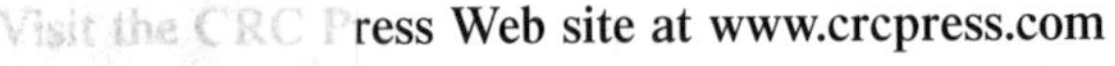

Visit the CRC Press Web site at www.crcpress.com

International Standard Book Number 1-58488-286-7
Library of Congress Card Number 2002276810
Printed in the United States of America 1 2 3 4 5 6 7 8 9 0
Printed on acid-free paper

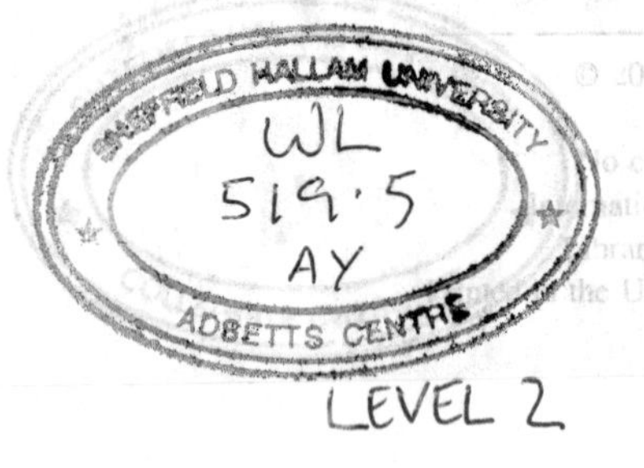

Dedication

To my wife, Deena, and our four children, Omar, Rami, Samar, and Ziad.

Bilal M. Ayyub

Preface

In preparing this book, we strove to achieve the following educational objectives: (1) introduce probability, statistics, reliability, and risk methods to students and practicing professionals in engineering and the sciences; (2) emphasize the practical use of these methods; and (3) establish the limitations, advantages, and disadvantages of the methods. The book was developed with an emphasis on solving real-world technological problems that engineers and scientists are asked to solve as part of their professional responsibilities.

Upon graduation, engineering and science majors will need to be able to collect, analyze, interpret, and properly apply vast arrays of data as part of their responsibilities in technological environments. Graduates must have a solid academic foundation in methods of data analysis and synthesis, as the analysis and synthesis of complex systems are common tasks that confront even entry-level professionals. The education of entry-level professionals must, therefore, include the topics covered in this book, presented in such a way that students come to recognize the importance and relevance of the material to complex problems for which they will be responsible as professionals.

This book is intended to be used in courses that are the first formal introduction to the subjects of probability, statistics, reliability, and risk analysis. Rather than taking either a theoretical or an applied approach to the topics, we have tried to strike a balance. The underlying theory, especially the assumptions that are central to the methods, is presented, but then the proper application of the theory is presented through realistic examples, often using actual data. Every attempt is made to show that methods of data analysis are not independent of each other. Instead, we show that real-world problem solving often involves applying many of the methods presented in different chapters. Data analysis should be viewed as a continuum rather than the isolated application of one method. The fact that this book is separated into 16 chapters should not suggest that the material should be viewed as 16 unrelated topics. To effectively work in a technological environment requires the attitude that methods in probability, statistics, reliability, and risk form a continuum of concepts, and we have tried to instill this philosophy throughout the book.

Problems that are commonly encountered by engineers and scientists require decision making under conditions of uncertainty. The uncertainty can be in the definition of a problem, the available information, the alternative solution methodologies and their results, and the random nature of the solution outcomes. Studies show that in the future engineers and scientists will need to solve more complex design problems with decisions made under conditions of limited resources, thus necessitating increased reliance on the proper treatment of uncertainty. Therefore, this book is intended to better prepare future engineers and scientists, as well as assist practicing professionals, in understanding the fundamentals of probability, statistics, reliability, and risk methods, especially their applications, limitations, and potentials.

An aspect of probability and statistics that is most difficult to grasp is the concept of sampling variation. In the practice of engineering and science, only one sample of data is generally available. It is important to recognize that the statistical results would be somewhat different if a different sample had been collected, even if that sample was equally likely to have occurred. Simulation is a means of demonstrating the sample-to-sample, or sampling, variation that can be expected. For this reason, we have incorporated a section on simulation at the end of Chapters 1 to 15. Simulation is one way of generating a better appreciation for sampling variation that is inherent in statistical problems; however, omitting the sections on simulation does not diminish a reader's understanding of the other sections or chapters. We show that simulation has wide application as a modeling tool. Its application enables users of the book to assess the effects of violations in theoretical assumptions of statistical methods. In a sense, simulation is the modeling tool by which sensitivity analyses can be performed. From assessing the power of a statistical test when applied to a certain type of problem to identifying the limitation of a particular sample size, computer simulation is an important tool in all topics of probabilistic, statistical, reliability, risk, and uncertainty analysis.

We have developed this book with a dual use in mind: as a self-learning guidebook and as a required textbook for a course. In either case, the text has been designed to achieve important educational objectives, as discussed in Chapter 1.

The 16 chapters of the book cover the following subjects: (1) an introduction to the text that covers uncertainty types and decision analysis; (2) graphical analysis of data and the computation of important characteristics of sample measurements and basic statistical characteristics; (3) the fundamentals of probability; (4) an introduction to discrete random variables, their distributions, and their application; (5) a discussion of widely used continuous probability distributions and their application in simulation experiments; (6) the joint behavior of random variables and the probabilistic characteristics of functions of random variables; (7) a formal presentation of Monte Carlo simulation; (8) statistical analyses that include parameter estimation and identification of the sampling distribution of random variables; (9) an introduction to hypothesis tests, their uses in decision making, and commonly used tests; (10) an overview of experimental design procedures for both single- and two-factor analyses of variance; (11) confidence-interval estimation, sample-size determination, and probability-model selection; (12) an introduction to a variety of topics related to the use of correlation and regression analyses; (13) curve fitting or nonlinear model development using regression analysis; (14) an introduction reliability analysis of components with application; (15) system reliability, risk, and decision analysis; and (16) the use of Bayesian methods in engineering.

The book was designed for an introductory course in probability, statistics, and reliability with an emphasis on applications. In developing the book, a set of educational outcomes as detailed in Chapter 1 motivated the structure and content of this text. Ultimately, serious readers will find the content of the book to be very useful in problem solving and decision making, especially where practical applications are of interest.

In each chapter of the book, computational examples are given in the individual sections of the chapter, with more detailed engineering applications given in a concluding section. Also, each chapter includes a set of exercise problems that cover the materials of the chapter. The problems were carefully designed to meet the needs of instructors in assigning homework and the readers in practicing the fundamental concepts.

The book can be covered in one or two semesters depending on the level of a course or the time allocated for topics covered in the book. The chapter sequence can be followed as a recommended sequence; however, if necessary, instructors can choose a subset of the chapters for courses that do not permit complete coverage of all chapters or for courses that cannot follow the order presented. After completing Chapters 1, 2, and 3, readers will have sufficient background to follow and understand the materials in Chapters 4 and 5 followed by Chapter 6 and 7. Afterwards, readers can follow any of the following tracks of chapters: Chapter 8 to 13, Chapters 14 and 16, and Chapter 16 according to the sequence indicated below, which illustrates possible sequences of these chapters in terms of their interdependencies:

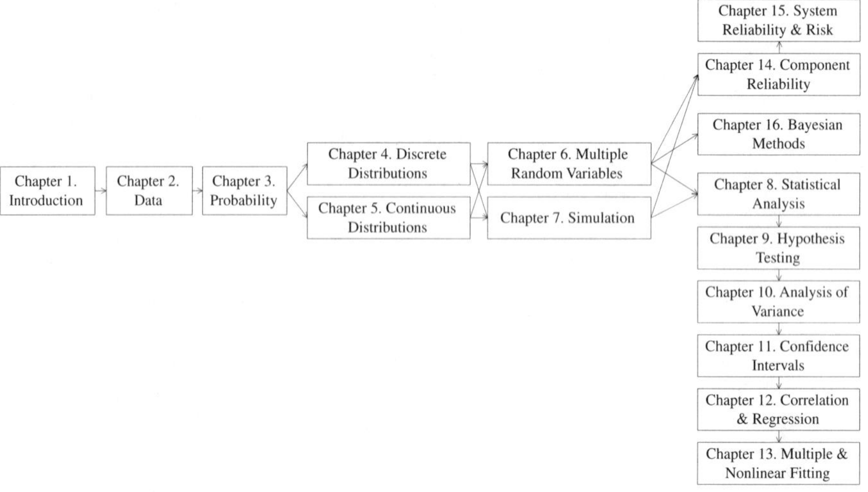

The book also provides problem statements for four simulation projects or case studies at the end of the chapters. These projects are developed in stages in the chapters. Students and readers can use these projects to perform analyses that should enhance their understanding of concepts presented in the chapters. Sample projects can be downloaded or viewed at http://ctsm.umd.edu under "simulation projects." These projects were designed based on many years of experience in teaching probability, statistics, and reliability at the University of Maryland and were found to be effective in communicating these concepts and enhancing the understanding of students.

Acknowledgments

This book was developed over several years and draws on our experiences in teaching courses on probability, statistics, and reliability. Drafts of most sections of the book were tested in several courses at the University of Maryland for many years before its publication, starting four years before publication of the first edition. This testing period has proven to be a very valuable tool in establishing its contents and final format and structure.

We would like to acknowledge all of the students who participated in the testing of the book for their comments, input, and suggestions. The students who took courses on computational methods in civil engineering during the semesters from 1992 to 1997 contributed to this endeavor, as did students who took courses on probability and statistics during the semesters from 1997 to 2001. Their feedback was very helpful and greatly contributed to the final product. Also, we would like to acknowledge the instructors who used the book during this testing period: Drs. Ibrahim Assakkaf, P. Johnson, and R. Muhanna. The assistance of Dr. Ru-Jen Chao, Dr. Maguid Hassan, and Nizar Al-Rihani in critically reviewing the problems at the end of the chapters and developing some of the example problems and problem solutions is gratefully acknowledged. Reviewers' comments that were provided to us by the publishers were used to improve the book to meet the needs of instructors and enhance the educational process. We acknowledge the publisher and the reviewers for these comments. We also acknowledge the efforts of the editorial staff, artists, and designers at CRC Press LLC for the production of the book.

We would like to invite users of the book, including both readers and instructors, to send us any comments on the book by e-mail at **ba@umd.edu**. These comments will be used in developing future editions of the book. Also, they are invited to visit the website http://ctsm.umd.edu/CRCPress for additional applications, project materials, sample projects, and an errata.

Bilal M. Ayyub
Richard H. McCuen
2002

The Authors

Bilal M. Ayyub, Professor of Civil and Environmental Engineering at the University of Maryland (College Park) and the General Director of the Center for Technology and Systems Management, is a researcher and consultant in the areas of structural engineering, systems engineering, uncertainty modeling and analysis, reliability and risk analysis, and applications related to civil, marine, and mechanical systems. He completed his B.S. degree in civil engineering in 1980 and completed both the M.S. (1981) and Ph.D. (1983) in civil engineering at the Georgia Institute of Technology. He has performed research projects funded by the U.S. National Science Foundation, Coast Guard, Navy, Army Corps of Engineers, Maryland State Highway Administration, American Society of Mechanical Engineers, and several engineering companies. Dr. Ayyub has served the engineering community in various capacities through societies that include ASNE, ASCE, ASME, SNAME, IEEE, and NAFIPS. He is a fellow of ASCE, ASME, and SNAME and is a life member of ASNE and USNI. He chaired the ASCE Committee on the Reliability of Offshore Structures, and currently chairs the SNAME panel on design philosophy and the ASNE journal committee. He was the General Chairman of the first, second, and third International Symposia on Uncertainty Modeling and Analysis held in 1990, 1993 and 1995 and the NAFIPS annual conference in 1995. He is a registered Professional Engineer (PE) with the State of Maryland and is listed in *Who's Who in America* and *Who's Who in the World.*

Dr. Ayyub is the author and co-author of over 300 publications in journals and conference proceedings, and reports. He was the recipient of the ASNE "Jimmie" Hamilton Award for the best papers in the *Naval Engineers Journal* in 1985 and 1992. Also, he was the recipient of the ASCE "Outstanding Research-Oriented Paper" in the *Journal of Water Resources Planning and Management* for 1987, the ASCE Edmund Friedman Award in 1989, the NAFIPS K.S. Fu Award for distinguished service, and the Walter L. Huber Research Prize of ASCE in 1997.

His publications include *Elicitation of Expert Opinions for Uncertainty and Risks* (CRC Press, 2001; author); *Uncertainty Modeling and Analysis in Civil Engineering* (CRC Press, 1998; editor); *Uncertainty Modeling in Vibration, Control, and Fuzzy Analysis of Structural Systems* (World Scientific, 1997; edited with A. Guran, and A. Haldar); *Uncertainty Analysis in Engineering and the Sciences: Fuzzy Logic, Statistics, and Neural Network Approach* (Kluwer Academic, 1997; edited with M.M. Gupta); *Uncertainty Modeling in Finite Element, Fatigue, and Stability of Systems* (World Scientific, 1997; edited with A. Haldar and A. Guran); *Numerical Methods for Engineers* (Prentice Hall, 1996; authored with R. McCuen); *Uncertainty Modeling and Analysis: Theory and Applications* (North-Holland/Elsevier, 1994; edited with M.M. Gupta); and *Analysis and Management of Uncertainty: Theory and Applications* (North-Holland/Elsevier, 1992; edited with M.M. Gupta).

Richard H. McCuen, Professor of Civil Engineering at the University of Maryland at College Park, received degrees from Carnegie Mellon University and the Georgia Institute of Technology. He received the Icko Iben Award from the American Water Resource Association and was co-recipient of the 1988 Outstanding Research Award from the American Society of Civil Engineers Water Resources Planning and Management Division. Topics in statistical hydrology and stormwater management are his primary research interest.

He is the author of 17 books and over 200 professional papers, including *Modeling Hydrologic Change* (CRC Press, 2002); *Hydrologic Analysis and Design, Second Edition* (Prentice Hall, 1998); *The Elements of Academic Research* (ASCE Press, 1996); *Estimating Debris Volumes for Flood Control* (Lighthouse Publications, 1996; with T.V. Hromadka); and *Dynamic Communication for Engineers* (ASCE Press, 1993; with P. Johnson and C. Davis).

Table of Contents

Chapter 1
Introduction

TABLE OF CONTENTS

1.1. INTRODUCTION

The processes of engineering analysis and design can be systematically performed within a systems framework. Generally, an engineering project can be modeled to include a segment of its environment that interacts significantly with it to define an engineering system. The boundaries of the system are drawn based on the goals and characteristics of the project, the class of performances (including failures) under consideration, and the objectives of the analysis.

The first step in engineering problem solving is to define the architecture of the system. The definition can be based on observations at different system levels that are established based on the analysis goals of a project. The observations can be about the different elements (or components) of the system, interactions among these elements, and the expected behavior of the system. Each level of knowledge that is obtained about an engineering problem defines a system to represent the project. As additional levels of knowledge are added to previous ones, higher epistemological levels of system definition and description are possible which, taken together, form a hierarchy of the system descriptions.

An epistemological hierarchy of systems requires a generalized treatment of uncertainty in both the architecture of the system and the collected information. This treatment can be based, in part, on probability and statistical concepts, as well as other related tools. Therefore, engineering systems should be viewed with an understanding of the knowledge content of collected information, including its associated uncertainties. Also, the user should understand the limitations of prediction models that are due to the inherent insufficiency of models resulting from the assumptions used to develop them. The uncertainty dimension in the analysis of engineering systems can result in valuable insight or information that is necessary to make rational decisions. Such a decision process considers the available information or knowledge, decision choices, alternative decision outcomes, and associated uncertainties. These aspects of decision analysis are discussed in Section 1.1.1.

The need for a proper treatment of uncertainty in engineering systems as well as the advancement of computer technology and its availability to practicing engineers have impacted engineering education in many aspects. In order to prepare engineering students and practicing engineers for current and future challenges, an introduction to and an emphasis on computer-based methods are necessary. Currently, practical engineering problems demand the use of computers to obtain solutions in a timely manner and with an acceptable level of accuracy.

The objective of this book is to introduce the fundamentals of probability, statistics, and reliability to engineering and science students and practicing engineers and scientists in order to deal with uncertainty in engineering systems. The practical aspects of the use of these methods is emphasized throughout the book, with practical applications provided in each chapter. Although the book was developed with an emphasis on engineering, science, and technological problems, these methods can be used to solve problems in many other fields. The methods are provided with varying levels of detail and emphasis. A critical presentation of these methods is provided to enhance the reader's understanding.

Probability, statistics, and reliability methods can be used for hand or computer-based computations. The importance of these methods in engineering can be effectively demonstrated in cases dealing with complex problems where analytical solutions cannot be obtained or hand calculations cannot be made. In this book, we use common engineering and science problems to demonstrate the computational procedures. The examples were intentionally selected with traceable solutions so that readers can reproduce them. It is helpful, but not necessary, for the reader to be familiar with the fundamentals of a computer language.

The use of any computational method without a proper understanding of its limitations and shortcomings can have serious consequences. Before using a method, users should become well versed with these methods in terms of the computational details and their limitations, shortcomings, and strengths. The methods presented in the book, especially the statistical and reliability methods, should not be used without a complete understanding of the underlying theory and knowledge of computational methods or procedures.

1.1.1. Decision Making in Engineering and Science

Engineering and science are disciplines in which their members are often in decision-making positions. To make the best decisions, engineers and scientists must be aware of and fully understand alternative solution procedures. This need has become more important in professional practice as the problems that engineers and scientists must address have become more important to society and as the design methods have become more complex. In reference to decisions, the term *best* is multifaceted. For example, engineers need accurate solutions that are both unbiased and have high precision. The solution must be cost effective and have minimal environmental consequences. The adopted solution should improve the community and be aesthetically appealing. These are just some of the criteria used in making engineering decisions.

When engineers are given a design problem, they typically approach the solution using a systematic procedure. One formulation of this process is as follows: (1) identify the problem, (2) state the objectives, (3) develop alternative solutions, (4) evaluate the alternatives, and (5) implement the best alternative. In this process, engineers should (1) consider the uncertainties associated with the alternative solutions, (2) assess the array of possible outcomes of each alternative with their associated uncertainties, and (3) evaluate all data and prediction models used in the analyses, also with their associated uncertainties. Probability, statistics, and reliability methods can be used to help ensure that each of these tasks is properly handled.

Decision problems can be classified as either single- or multiple-objective problems. For example, modeling that uses the principle of least squares (as used in Chapter 12) to define a curve fit to data is considered to involve a single objective. A multiple-objective problem may seek to minimize the total expected cost while both maximizing safety and minimizing environmental damage. Decision analysis requires that all objectives are clearly stated. Also, the term *expected* in the example objectives means "on the average," which implies the need to properly model the associated uncertainties. For cases of multiple objectives, the objectives must all be stated in the same units. Also, weights that reflect the importance of the objectives and that can be used to combine the objectives must be assigned. Then, the problem can be formulated in a structure that is suitable for attaining the decision objectives.

The development of alternative solutions is the step that is especially critical to success in meeting project objectives. A team of engineers must have a sound technical background and a broad understanding of alternative design methods. To properly evaluate each alternative, a complete understanding of the technical basis of the design method is required, then alternatives must be evaluated and a selection made.

Computers have increased the number of alternative solution procedures that are available to engineers and scientists. Whereas the engineers of past generations were limited to graphical, analytical, and simple empirical methods, engineers can now consider numerical and simulation methods in their design work. The increased complexity of the numerical and simulation methods is believed to improve the accuracy of the solutions, and, because these methods are easily implemented with a computer, the increased effort over the less detailed methods is minimal. Thus, the increase in design effort is more than offset by the expected improvement in accuracy. To achieve this benefit, however, an engineer must fully understand the more complex methods so that they can be properly applied.

1.1.2. Expected Educational Outcomes

While a textbook can be the principal medium for an individual working independently to gain an understanding of a body of knowledge, books such as this one are typically used as part of a course on probability, statistics, and reliability methods. This book has been designed to be used in either capacity, as a self-learning guidebook or as a required textbook for a course. In either case, the text has been designed to produce the following educational outcomes:

1. The reader of the book should be able to (a) collect data on a problem and describe the data using graphical and descriptive measures; (b) develop a probabilistic model for the problem; (c) perform probability operations and evaluations; (d) use discrete and continuous random variables to model some aspects of the problem; (e) assess the joint behavior of random variables with associated correlations; (f) evaluate the probabilistic characteristics of functions of random variables; (g) perform statistical analyses of the data, such as histogram development, univariate analysis, probability model selection, analysis of variance, hypothesis testing, parameter estimation, confidence-interval estimation, and selection of sample sizes; (h) perform correlation and regression analyses for fitting a curve or a model to data; (i) perform random-number and random-variable generation for the Monte Carlo simulation of selected random variables; (j) evaluate the reliability of a component of a system or the reliability of the entire system; (k) perform risk analyses and risk-based decision making; and (l) utilize Bayesian methods for updating probabilities or statistical measures.
2. The reader should become familiar with the use of simulation methods, which are essential for solving complex problems in real-world applications.
3. The reader should be able to select from alternative methods the one method that is most appropriate for a specific problem and state reasons for the selection.
4. The reader should be able to formulate algorithms to solve problems.
5. The reader should understand the limitations of each method, especially the conditions under which they may fail to provide a solution with acceptable accuracy.
6. The reader will become familiar with engineering applications of the various probability, statistical, reliability, risk, and Bayesian methods.

In developing the topics presented in the book, these expected educational outcomes motivated the structure and content of this text. Ultimately, the serious reader will find the material very useful in engineering problem solving and decision making.

1.2. TYPES OF UNCERTAINTY

The analysis of a system often involves the development of a model of the system. The model can be viewed as an abstraction of some aspects of the system. In performing this abstraction, an analyst or engineer must decide which aspects of the system to include and which to leave out. Figure 1-1 shows uncertainties in these aspects that can make model development difficult. Also, depending on the state of knowledge about the system and the background of the analyst or engineer, other aspects of the system might not be known, thus increasing the overall uncertainty of the system. In these three categories (i.e., abstracted, nonabstracted, and unknown or unknowable aspects of the system), several types of uncertainty can be present. Figure 1-1 provides examples of uncertainties within each category. This book covers some of these uncertainty types that can be dealt with using probability, statistics, reliability, and Bayesian methods (in Chapters 3 to 16). This section provides a brief description of these uncertainty types. Uncertainty can be viewed as a category within ignorance as shown in Figure 1-2. The state of ignorance for a person or society can be unintentional or deliberate due to an erroneous cognition state and not knowing relevant information, or ignoring information and deliberate inattention to something for various reasons such as limited resources or cultural opposition, respectively. The latter type is a state of conscious ignorance that is not intentional, and, once recognized, evolutionary species try to correct for that state for survival reasons with varying levels of success. The former ignorance type belongs to the blind ignorance category; therefore, ignoring means that someone can either unconsciously or deliberately refuse to acknowledge or regard or leave out an account or consideration for relevant information. These two states should be treated in developing a hierarchal breakdown of ignorance. Ignorance can be viewed to have a hierarchal classification based on its sources and nature as shown in Figure 1-2. Ignorance can be classified into two types: blind ignorance (also called *meta-ignorance*) and conscious ignorance (also called *reflective ignorance*).

Blind ignorance includes not knowing relevant know-how, objects-related information, and relevant propositions that can be justified. The unknowable knowledge can be defined as knowledge that cannot be attained by humans based on current evolutionary progressions, or cannot be attained at all due to human limitations, or can only be attained through quantum leaps by humans. Blind ignorance also includes irrelevant knowledge that can be of two types: (1) relevant knowledge that is dismissed as irrelevant or ignored, and (2) irrelevant knowledge that is believed to be relevant through nonreliable or weak justification or as a result of *ignoratio elenchi*. The irrelevance type can be due to untopicality, taboo, and undecidability. Untopicality can be attributed to intuitions of experts that could not be negotiated with others in terms of cognitive relevance. Taboo is due to socially reinforced irrelevance. Issues that people must not know, deal with, inquire about, or investigate define the domain of taboo. The undecidability type deals with issues that cannot be designated true or false because they are considered insoluble, solutions that are not verifiable, or a result of *ignoratio elenchi*. A third component of blind ignorance is fallacy, which can be defined as erroneous beliefs due to misleading notions.

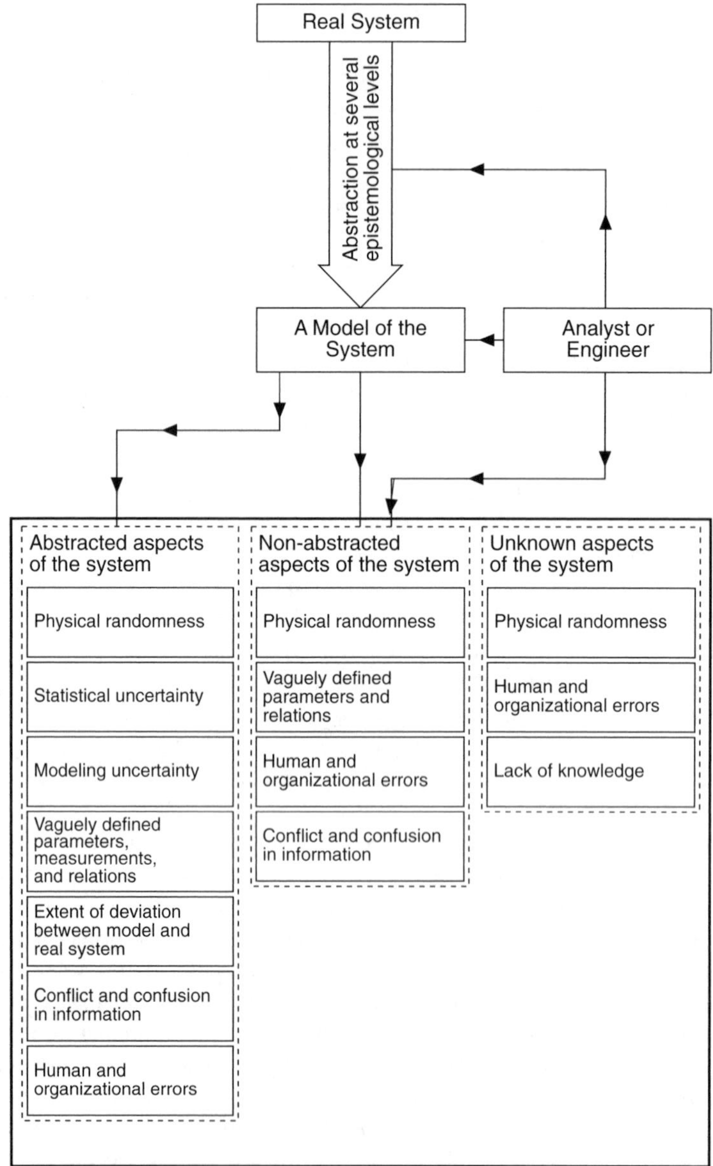

FIGURE 1-1 Uncertainty types for systems.

The philosopher Kurt Gödel (1906–1978) showed that a logical system could not be both consistent and complete and could not prove itself complete without proving itself inconsistent and vice versa. Also, he showed that problems exist that cannot be solved by any set of rules or procedures; instead, for these problems one must

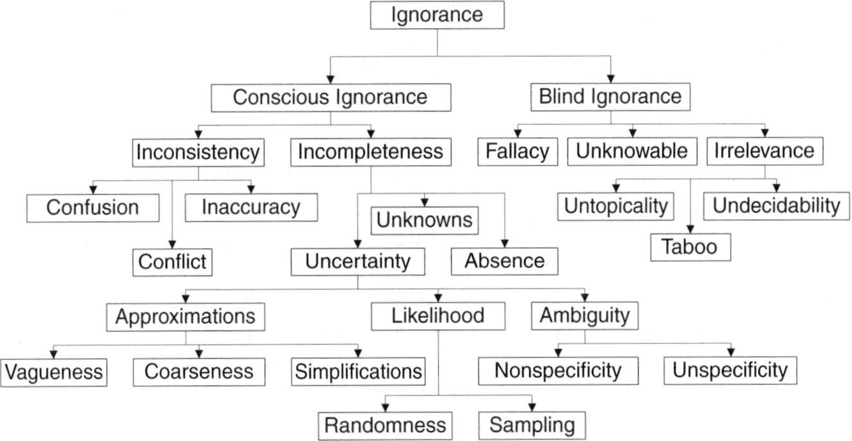

FIGURE 1-2 Ignorance types.

always extend the set of axioms. This philosophical view of logic can be used as a basis for classifying the conscious ignorance into inconsistency and incompleteness.

Inconsistency in knowledge can be attributed to distorted information as a result of inaccuracy, conflict, contradiction, and/or confusion, as shown in Figure 1-2. Inconsistency can result from numerical assignments and substitutions that are wrong, conflicting, or biased, producing confusion, conflict, or inaccuracy, respectively. The confusion and conflict result from in-kind inconsistent assignments and substitutions, whereas inaccuracy results from a level bias or error in these assignments and substitutions.

Incompleteness is defined as incomplete knowledge and can be considered to consist of (1) absence and unknowns as incompleteness in kind, and (2) uncertainty. The unknowns or unknown knowledge can be viewed in evolutionary epistemology as the difference between the becoming knowledge state and current knowledge state. The knowledge absence component can lead to one of these scenarios: (1) no action and working without the knowledge, (2) unintentionally acquiring irrelevant knowledge leading to blind ignorance, or (3) acquiring relevant knowledge that can have various uncertainties and levels. The fourth possible scenario of deliberately acquiring irrelevant knowledge is not listed, as it is not realistic.

Uncertainty can be defined as knowledge incompleteness due to inherent deficiencies with acquired knowledge. Uncertainty can be classified based on its sources into three types: ambiguity, approximations, and likelihood. The ambiguity comes from the possibility of having multiple outcomes for processes or systems. Recognition of some of the possible outcomes creates uncertainty. The recognized outcomes might constitute only a partial list of all possible outcomes, leading to unspecificity. In this context, unspecificity results from outcomes or assignments that are not completely defined. The incorrect definition of outcomes (i.e., error in defining outcomes) can be called *nonspecificity*. In this context, nonspecificity results from outcomes or assignments that are improperly defined. The unspecificity is a form of knowledge absence and can be treated similarly to the absence category under incompleteness. The nonspecificity can be viewed as a state of blind ignorance.

The human mind has the ability to perform approximations through reduction and generalizations (i.e., induction and deduction, respectively) when developing knowledge. The process of approximation can involve the use of vague semantics in language, approximate reasoning, and dealing with complexity by emphasizing relevance. Approximations can be viewed to include vagueness, coarseness, and simplification. Vagueness results from the non-crisp nature of belonging and nonbelonging of elements to a set or a notion of interest, whereas coarseness results from approximating a crisp set by subsets of an underlying partition of the set's universe that would bound the crisp set of interest. Simplifications are assumptions made to make problems and solutions tractable.

The likelihood can be defined in the context of chance, odds, and gambling. Likelihood has the primary components of randomness and sampling. Randomness stems from the nonpredictability of outcomes. Engineers and scientists commonly use samples to characterize populations, thus the last type.

In subsequent sections, the classification provided in Figure 1-1 is discussed in detail, because uncertainties in engineering systems are considered to be mainly attributed to ambiguity and vagueness in defining the variables or parameters of the systems and their relations. The ambiguity component is generally due to noncognitive (or aleatory) sources, which include: (1) physical randomness, (2) statistical uncertainty due to the use of limited information to estimate the characteristics of these parameters, and (3) model uncertainties that are due to simplifying assumptions in analytical and prediction models, simplified methods, and idealized representations of real performances. The vagueness-related uncertainty is due to cognitive (or epistemic) sources, such as: (1) the definition of certain parameters — for example, structural performance (failure or survival), quality, deterioration, skill and experience of construction workers and engineers, environmental impact of projects, conditions of existing structures; (2) other human factors; and (3) defining the interrelationships among the variables or parameters of the problems, especially for complex systems.

1.2.1. Noncognitive or Aleatory Types

Engineers and researchers currently deal with the ambiguity types of uncertainty in predicting the behavior and designing engineering systems using the theories of probability (Chapters 3 to 6) and statistics (Chapters 7 to 13). Probability distributions are used to model system parameters that are uncertain. Probabilistic methods that include reliability methods, probabilistic engineering mechanics, stochastic finite-element methods, reliability-based design formats, random vibration, and other methods were developed and have been used for this purpose. In this treatment, however, a realization of the presence of a cognitive type of uncertainty was established. Subjective probabilities that are based on mathematics used for the frequency type of probability are used to deal with this uncertainty. Uniform and triangular probability distributions are sometimes used to model this type of uncertainty. The Bayesian techniques have also been used to deal with gaining information about these parameters (see Chapter 16). The underlying distributions and probabilities can, therefore, be updated. Regardless of the nature of the gained information, whether it is cognitive or noncognitive, the same mathematical assumptions and tools are used in these approaches.

1.2.2. Cognitive or Epistemic Types

The cognitive types of uncertainty arise from mind-based abstractions of reality. These abstractions are, therefore, subjective and lack crispness or precision. This vagueness is distinct from ambiguity in source and natural properties. The axioms of probability and statistics are limiting for the proper modeling and analysis of this type and may not be completely relevant or completely applicable. The vagueness type of uncertainty can be modeled using the theories of fuzzy sets and possibility. In engineering, the theory has proven to be a useful tool in solving problems that involve the vagueness type of uncertainty. For example, the theory has been successfully used in (1) strength assessment of existing structures and other structural engineering applications; (2) risk analysis and assessment in engineering; (3) analysis of construction failures, scheduling of construction activities, safety assessment of construction activities, decisions during construction, and tender evaluation; (4) impact assessment of engineering projects on the quality of wildlife habitat; (5) planning of river basins; (6) control of engineering systems; (7) computer vision; and (8) optimization based on *soft* constraints. This type of uncertainty is considered to be beyond the scope of this book.

1.3. INTRODUCTION TO SIMULATION

A central theme of this book is data analysis. Generally, an engineer or scientist works with data measured from real systems. For example, transportation engineers frequently use traffic counts at intersections or accident data for various configurations of control signals in designing roadways. Environmental engineers collect water-quality data, and their analyses of the data are used to decide the type of water-quality treatment that is needed. This book emphasizes the analysis of such real-world data. Unfortunately, real-world data may be too limited for reliable decision making. Real-world data may not cover extreme situations that are important in design. For example, streamflow records often do not include the extreme floods that are important in assessing flood risk. For these reasons, engineers often use simulated data to help make decisions. Simulation is formally covered in depth in Chapter 7; however, it is introduced in all of the earlier chapters to show its usefulness as a supplement to the traditional approaches to data analysis.

1.3.1. Simulation by Coin Flipping

The basics of simulation are easily understood using an example based on the flip of a single coin. Assume that, at a particular location on a river, the water quality is considered good about 50% of the time and unacceptable 50% of the time. Also, assume that an environmental engineer tests the quality of the water once a day and the quality on one day is independent of the quality on the previous day. This situation is much like the flip of a coin in that probabilities (i.e., likelihood estimates) are 0.5, and the outcome of the flip of the coin on any one trial is independent of the outcome from the previous flip. Let's also assume that the environmental engineer only has data from the past 2 weeks, and the sample of 14 values is inadequate to make a decision about the problem. Specifically, the engineer knows that polluted water on two consecutive days is not damaging to the aquatic life, but damage occurs if the water is of unacceptable quality for three or more consecutive days. The actual data for the 14 days were as follows:

AAUAUUAUUAAAUA

in which A indicates acceptable water quality and U unacceptable water quality. Thus, in the engineer's data, the water was of acceptable quality on 57% of the days and no instances of aquatic damage occurred. Should the engineer, therefore, believe that aquatic damage will not occur in the future? Of course not! Because the engineer needs to make a decision based on the probability of aquatic damage in the future, the engineer decides to simulate the process and use the simulated probability to make the decision. Flipping a coin 56 times produced the following 8-week sequence:

HHTHTHH HHHTHTH TTHTTTH THHHTHT

THTTTTH HTHHHTH THTHTTH HHTTHTH

where H is heads and T is tails. If a tail is considered an unacceptable water quality, then two occurrences of unacceptable quality happened in 8 weeks, which represents once every 4 weeks or 13 times a year. Of course, for decision making, the engineer would want to use a longer sequence of simulated data. This example demonstrates the use of simulation in a real-world problem in order to establish a better understanding of the problem, its consequences, and possible solutions.

1.3.2. Generation of Random Numbers

In the example of the previous section, the random occurrence of events was simulated using the flip of a coin. For any one event, only one of the two possible outcomes can occur, H or T, with each having the same likelihood of occurrence. Graphically, this appears as:

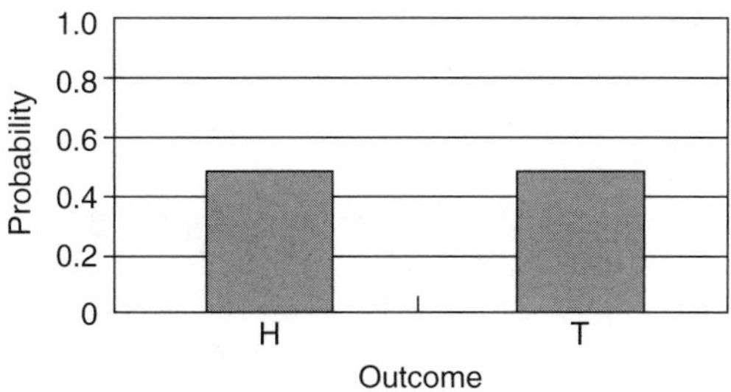

If two coins were flipped and the number of heads counted, one of three possible outcomes could occur: 0, 1, or 2. Letting X be the number of heads, the probability $P(x)$ can be graphed as:

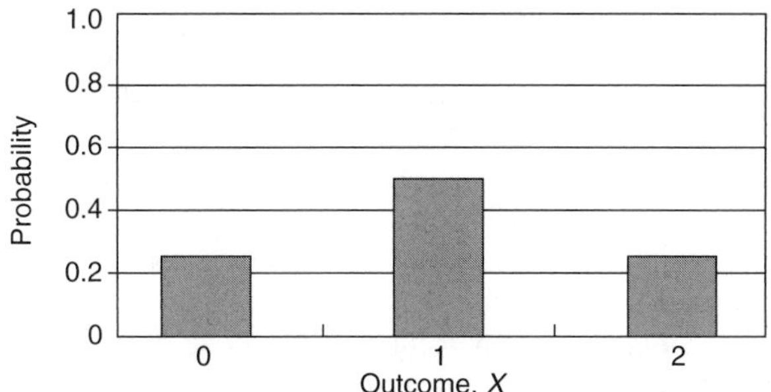

Values of 0, 1, and 2 would result from these sequences: 0 for the pair (T, T), 1 for the pairs (H, T) or (T, H), and 2 for the pair (H, H).

If random events were generated with the roll of a single die, one of six outcomes is possible. If the die is fair, each outcome y is equally likely, so each would have a probability of 1/6. Graphically, this appears as:

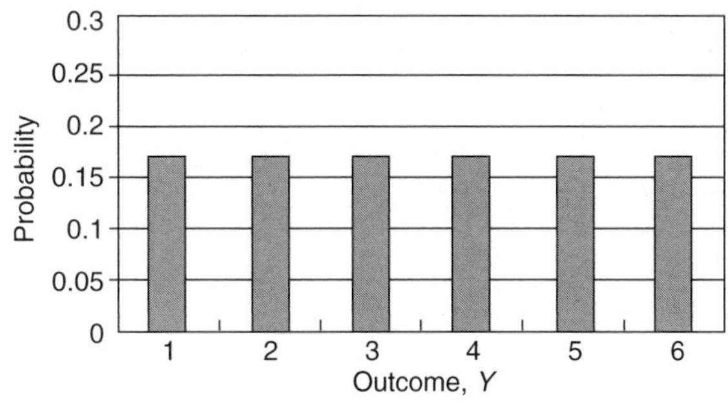

If a pair of dice were rolled simultaneously, the probability of the sum of dots from the two die (Z) would appear graphically as:

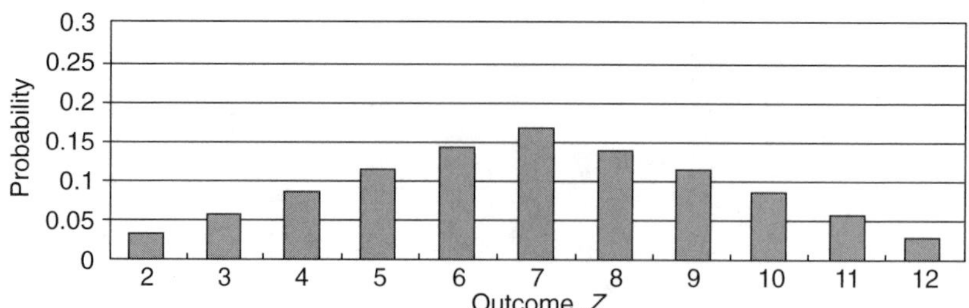

It is left to the reader to show how this would occur.

In each of these cases, only integer values were possible. That is, for example, if a single die is rolled, a value of 3.5 is not possible. Values from the flip of a coin or roll of a die are considered discrete. In many engineering cases, this is desirable, such as when the engineer is interested in the number of traffic fatalities, the number of floods per decade, the number of earthquakes above an intensity of 6.0 per century, or the number of drums of a hazardous waste.

In many other cases, engineers need values over a continuum, such as the stopping distance of cars, the magnitude of a flood, the compression strength of concrete, or the weight of fertilizer used per acre. Values on a continuum can be generated with a spinner (such as those included in some board games) placed over a 360° protractor. The arrow can be spun and the angle read, with possible values from 0 to 360°. The precision at which the values are recorded would depend on the accuracy of the spinner. Values of the angle (A) to the nearest 5° might be possible for a crude apparatus, while values to eight digits might be possible with an electronic apparatus. Thus, the probability of the outcome of a spin is 1/360 and could be graphed as:

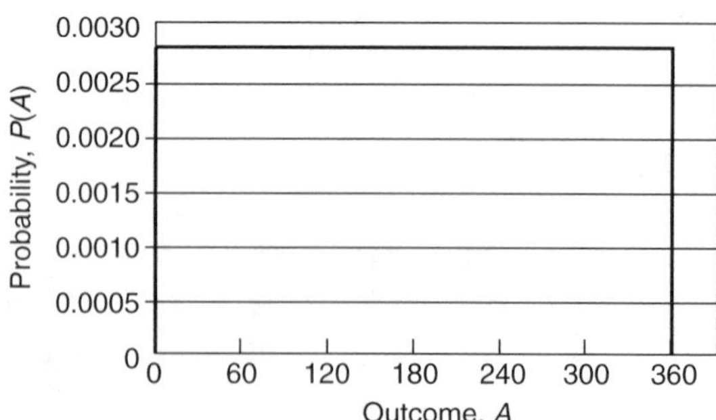

It should be obvious that we could transform the angle A to a new variable B that would take on values from 0 to 1 using the transformation $B = A/360$. This would appear as:

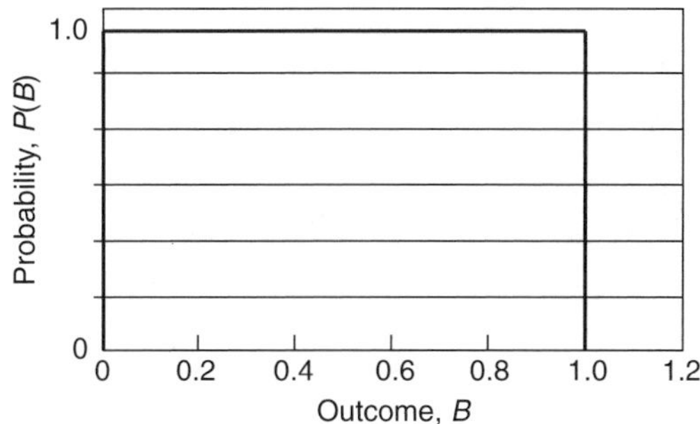

This distribution from 0 to 1 is very useful and is used frequently.

1.3.3. Computer Generation of Random Numbers

A central element in simulation is a random-number generator. In practice, computer packages are commonly used to generate the random numbers used in simulation; however, it is important to understand that these random numbers are generated from a deterministic (well-defined and predictable) process and are more correctly called *pseudo* random numbers. Because the random numbers are derived from a deterministic process, it is important to understand the limitations of these generators.

Random-number generators produce numbers that have specific statistical characteristics. Obviously, if the generated numbers are truly random, an underlying population exists that can be represented by a known probability function using figures as in Section 1.3.2. A single die is the most obvious example of a random-number generator. If we rolled a single die many times, we could tabulate a frequency histogram. If the die is a fair die, we could expect the sample histogram to consist of six bars with almost equal heights. Of course, the histogram for the population would consist of six bars of equal height, as shown previously. Other random-number generators would produce random numbers having different distributions, and when a computerized random-number generator is used it is important to know the underlying population. In this section, the midsquare method and the *rand* function method for random-number generation are described for illustration purposes. Other methods are described in Chapter 7.

1.3.3.1. Midsquare Method

The midsquare method, one of the simplest but least reliable methods of generating random numbers, is provided for demonstration purposes and is not suitable for use in simulation in practice. However, it is a method that illustrates problems associated with deterministic procedures. The general procedure is as follows:

1. Select at random a four-digit number, which is referred to as the *seed*.
2. Square the number and write the square as an eight-digit number using preceding (lead) zeros if necessary.

3. Use the four digits in the middle as the new random number.
4. Repeat steps 2 and 3 to generate as many numbers as necessary.

As an example, consider the seed number of 2189. This value produces the following sequence of four-digit numbers:

04791721
62678889
46076944
00591361
34963569
92833225
69422224

At some point one of these numbers must recur, which will begin the same sequence that occurred on the first pass. For example, if the four-digit number of 3500 occurred, the following sequence would result:

12250000
06250000
06250000
06250000

Such a random-number sequence does not pass statistical tests for randomness. While the procedure could be used for very small samples or five-digit numbers could be used to produce ten-digit squares, the midsquare method has serious flaws that limit its usefulness. However, it is useful for introducing the concept of random-number generation.

The midsquare method cannot be used for actual simulation studies, but numerous methods that are more reliable are available. Other more reliable methods for generating random numbers are discussed in Chapter 7.

1.3.3.2. The rand *Function*

Computer programming languages, including program commands in calculators and functions available in spreadsheets, have the *rand* function or an equivalent function for generating random numbers. For example, a spreadsheet in Microsoft Excel™ can be used to demonstrate the use of the *rand* function. To use the function, the reader should type the following function, including the equal sign and the parentheses, in a cell of a spreadsheet to obtain a random number:

random number = *rand*( )

When the return key is pressed, a random value is generated and placed in the cell. The *rand* function returns a uniformly distributed random number greater than or equal to 0 and less than or equal to 1 in a cell. A new random number is returned every time the worksheet is calculated. Worksheets that include many *rand* function uses might become busy reevaluating the *rand* functions every time the worksheet is recalculated. A solution to this problem is to turn off the worksheet calculation or to copy all the *rand* function cells and paste them as values only at the same locations, therefore retaining the numerical values without the *rand* functions. The following list of eight numbers was generated using the *rand* function of Excel:

0.106102761
0.358171156
0.810248935
0.666115294
0.129626466
0.272854219
0.912612626
0.979930229

It can be observed that all the numbers in this list are in the range of zero to one. Readers should try to generate their own random numbers using both a calculator and spreadsheet program.

1.3.4. Transformation of Random Variables

One example of a transformation was shown in Section 1.3.2. In that case, a value on a continuous scale from 0 to 360 was transformed to a continuous scale from 0 to 1. Transformation is quite frequently necessary and is discussed in detail in Section 7.4; however, the rudiments are discussed herein and used in Chapters 2 to 6 and 8 to 13.

Assume that a die rather than a coin is available. How could we generate flips of a coin with the die? This is easily accomplished by transforming the value of the die to the value of the coin. When the die is rolled, the occurrence of a 1, 2, or 3 would represent a head, while a value of 4, 5, or 6 would represent a tail. Graphically, this transformation appears as:

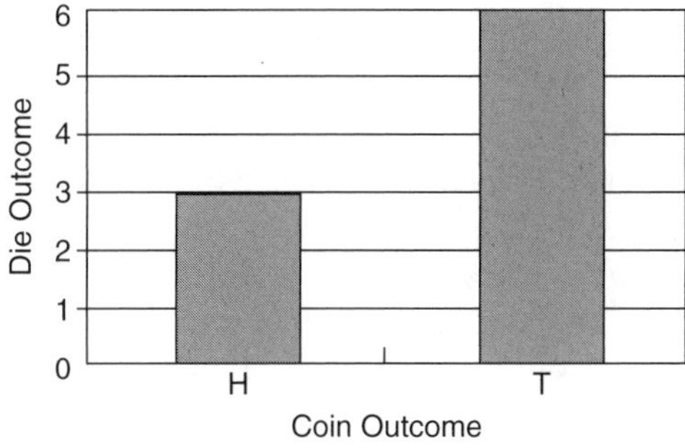

We enter along the ordinate with the outcome of the roll of a die and move horizontally. If the line for the head (H) is intersected, then we assume a head occurred. If we move horizontally and the line for the tail is first intersected, then we assume a tail occurred. How could we simulate the flip of an unbalanced coin for which the probability of a head is 1/3 rather than 1/2? The reader should construct a transformation graph for this case (for the moment, we will call these graphs, where one value is transformed to another, *transformation graphs*).

The transformation graph for the case of the 0-to-360 scale transformed to a 0-to-1 scale is:

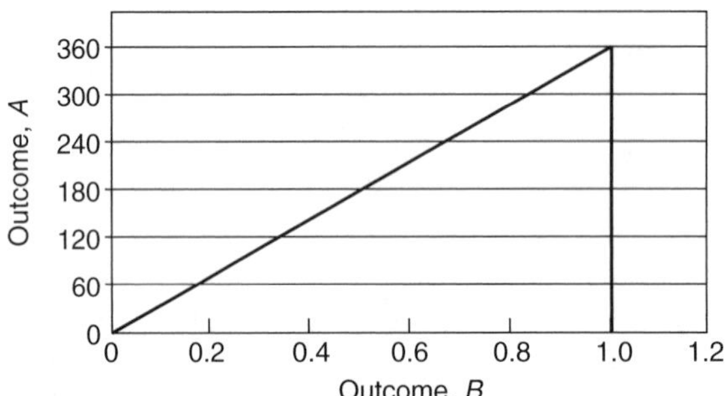

The value of B could be transformed to simulate the outcome (Y) roll of a die using the following transformation graph:

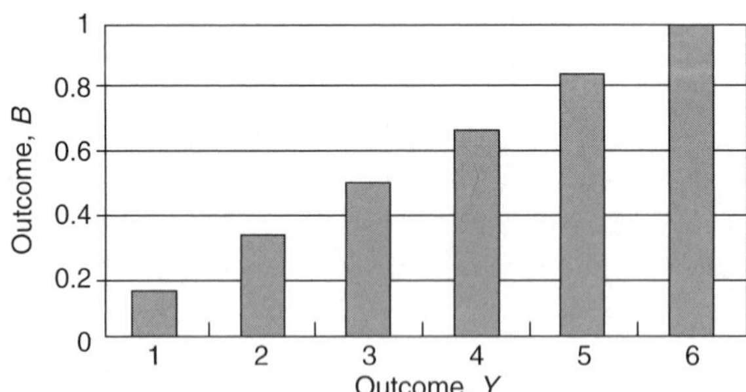

Entering along the ordinate with the value of B, move horizontally until one of the vertical lines is intersected. The transformed variable Y is then read from the abscissa. For example, if B equals 0.42, then Y is 3.

Transformation graphs such as these can be used in simulation studies. Transformation models can be used instead of graphs and are widely used in simulation studies. They are central to the discussions of simulation throughout this book.

Example 1-1: Generation of Wave Heights Using Linear Transformation

The height of waves in an ocean at some point in time and over a spatial area is random in nature. To illustrate the generation of the height of a wave, we will use a linear transformation model due to its simplicity. It should be noted that a linear model is not realistic, and other models are available to more accurately represent wave heights. An example linear transformation model is used to generate random values of wave height (h) in the range of 0 to 5 meters. The linear model transformation can be expressed as:

$$h = u(b - a) + a \tag{1-1}$$

where h is the wave height, u is a random number, a is the lower limit on wave height of 0, and b is the upper limit on wave height of 5 meters. The random number u can be generated using the *rand* function in each simulation trial or cycle (i). Table 1-1 show example results of this generation of wave heights.

TABLE 1-1 Uniform Generated Numbers (u) and Wave Heights (h)

Random number (u)	Wave height (h)
0.121593361	0.607967
0.140811553	0.704058
0.913545411	4.567727
0.617829368	3.089147
0.680623817	3.403119
0.708733255	3.543666
0.539608737	2.698044
0.706784267	3.533921
0.200539197	1.002696
0.91497701	4.574885

1.4. PROBLEMS

1-1. Select an engineering system for which you can define different levels of abstractions on the system with different abstraction aspects. Identify the abstracted aspects of the system, the nonabstracted aspects of the system, and the unknown aspects of the system. What are the uncertainty types for the selected system? Describe the uncertainty types using examples.

1-2. Show the development of the probability graph $P(Z)$ vs. Z given in Section 1.3.2.

1-3. Show the transformation graph that transforms the roll of a die (W) to the value of the flip of an unfair coin (C) for which the probability of a tail is 5/6.

1-4. Assume the value of X is random and can take on values from 0 to 1. Assume the following transformation graph relates X to the annual number of fatal accidents (N) at an intersection:

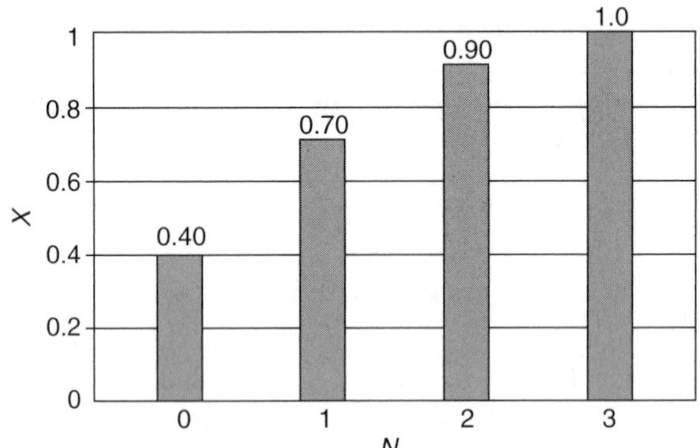

A random-number generator yields the following six values of X: 0.37, 0.82, 0.64, 0.25, 0.02, 0.94. How many fatal accidents occurred in each of the 6 years?

1-5. The probabilities of the largest magnitude of an earthquake in any decade are as follows:

	Magnitude					
	3 to 4	4 to 5	5 to 6	6 to 7	7 to 8	8 to 9
Probability	0.78	0.13	0.04	0.03	0.01	0.01

or

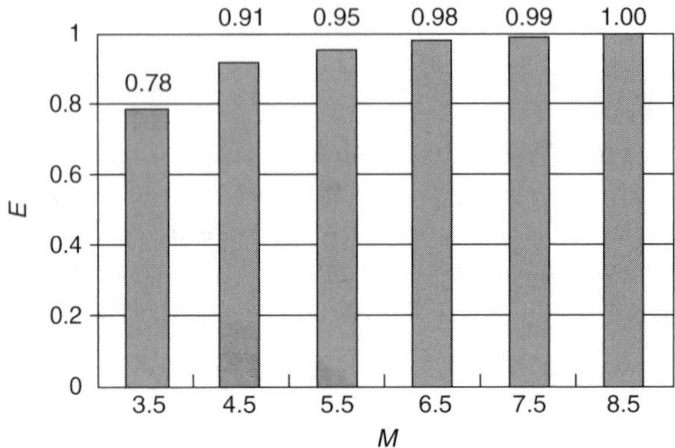

Construct a transformation graph that could transform a random number (E) over the range from 0 to 1 to the magnitude of an earthquake (M), where M takes on the center value for each interval of the magnitude. For each of the following values of E, find the simulated value of M:

$$E = \{0.27, 0.62, 0.13, 0.49, 0.96, 0.06, 0.84\}$$

1-6. Use the midsquare method to generate 10 random numbers using a seed of 3456.

1-7. Use the midsquare method to generate 10 random numbers using a seed of 8371.

1-8. Use the midsquare method to generate 20 random numbers using a seed of 9658.

1-9. Use the midsquare method to generate 20 random numbers using a seed of 2468.

1-10. Use the *rand* function in a spreadsheet to randomly generate 20 random numbers. Repeat the process for another 20 random numbers. Is the first set of random numbers the same as the second set?

1-11. Use the *rand* function in a spreadsheet to randomly generate 20 random numbers, and use linear transformation to produce random values in the range [2,6].

1-12. Use the *rand* function in a spreadsheet to randomly generate 100 random numbers, and use linear transformation to produce random values in the range [22,132].

1-13. Use the *rand* function in a spreadsheet to randomly generate 100 random numbers, and use linear transformation to produce random values in the range [–5,5].

1-14. Use the *rand* function in a spreadsheet to randomly generate 100 random numbers, and use linear transformation to produce random values in the range [–15,325].

1.5. SIMULATION PROJECTS

This section provides problem statements for four simulation projects or case studies. These projects are developed further in subsequent chapters where appropriate. Students and readers can use these projects to perform analyses that should enhance their understanding of the concepts presented in the chapters. Sample projects can be downloaded or viewed at http://ctsm.umd.edu under simulation projects. After completing the analyses, a report should be written. The report should have a professional structure and style with all concepts utilized clearly explained. The report should include a title page, an executive summary, a table of contents, and various sections devoted to each analysis. Example sections include assessing the statistical characteristics of the generated random variable, computing exceedence probabilities, and performing parametric analysis. Each

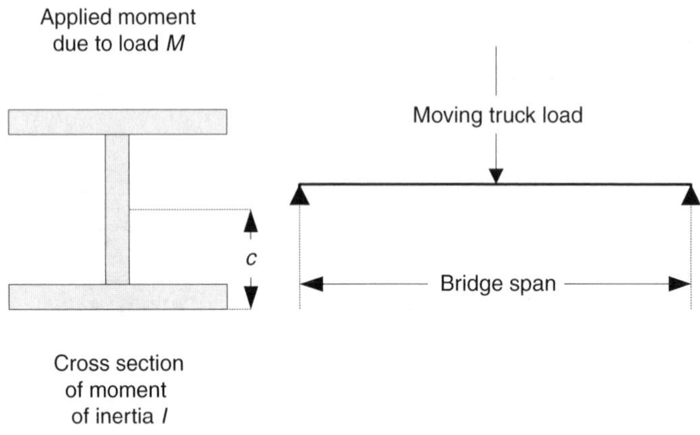

FIGURE 1-3 Steel beam in a bridge.

section should begin with a descriptive heading; the format used for headings in this book can be used as an example of headings. Most reports will include an introduction section, a conclusions section, and a sufficient number of intermediate sections to cover the analyses. All data should be placed in tables and graphical presentations as needed. The format used in this book for tables and figures should be followed. Note that all tables and figures are numbered, with references in the text to the table or figure indicated by the number. Large tables or cases involving a large number of figures should be placed in appendices to the report. Students should work in teams and set up a website with all project details, reports, and related links.

1.5.1. Structural Beam Study

The case study introduced in this section and used in subsequent chapters can be viewed as an example project. The objective of this study is to investigate the stress at the extreme fibers of a structural steel beam used in a bridge (shown in Figure 1-3). The stress at the extreme fibers can be computed as:

$$\sigma = \frac{Mc}{I} \leq f_y \tag{1-2}$$

where σ is the computed stress, M is the applied moment on a section due to external loads, c is the distance from the neutral axis to the extreme fibers, I is the centroidal moment of inertia of the cross section, and f_y is the yield stress of the material of the beam. The variables M, c, I, and f_y are called *basic random variables*. The meaning of a random variable will be introduced in Chapter 3. Table 1-2 shows the parameters

TABLE 1-2 Linear Transformation Parameters for the Beam Failure Analysis Case Study

Random variable	Mid-interval	a	b	$b-a$
c	10	9.5	10.5	1
M	3500	1000	6000	5000
I	1000	900	1100	200
f_y	50	45	55	10

needed for performing linear transformation according to Equation 1-1. Using linear transformation for the four random variables, generated values can be obtained based on the *rand* function in a spreadsheet. The results of 20 simulation trials are provided in Table 1-3. Table 1-3 is structured in three parts. Table 1-3A lists the 20 generated random numbers. It should be noted that each random variable requires its own set or stream of generated random numbers in order to produce random variables that are not related. Table 1-3B transform the generated random numbers into random variables using a linear transformation process as defined in Equation 1-1 and according to the parameters defined in Table 1-2. Table 1-3C uses Equation 1-2 to compute the stress (σ) and compares this computed stress to the generated yield stress of the material. Failure is defined in this case study as reaching the first yield of the materials at the extreme fibers. The last column in Table 1-3C was developed based on this comparison. In this case study, the moment M due to the load on the beam was assumed to show a wide range of zero to 6000 in order to demonstrate the possible failure state of the beam. In real structures, the occurrence of failure is not as frequent as demonstrated in this case study. Table 1-3C shows five failures in 20 simulation trials (i.e., 25% of the cases). The reliability of this steel beam is defined as 100% minus 25%, or 75%.

1.5.2. Stream Erosion Study

The erosion of streams can significantly increase the cost of treating water for water supply, damage the habitat of fish, and cause in-stream structures, such as coffer dams, to fail. Thus, prediction of stream erosion rates is necessary. Empirical models are widely used for making predictions. Table C-1 (see Appendix C) includes a set of data for 62 small streams. The data include three predictor variables: the mean soil particle diameter, the slope of the stream channel, and the discharge rate of the water in the stream per unit foot of width. The erosion rate (Y) is measured as a weight per unit time per unit foot of width.

Using the data of Table C-1, find the minimum and maximum values of Y. Then generate 50 values of Y using the *rand* function. Transform each of the 50 random values (u_i) to values of Y using the following transformation model:

$$Y_i = \min + u_i(\max - \min) \tag{1-3}$$

TABLE 1-3A Generated Stresses in a Beam, Yield Stresses, and Failure Analysis: Generated Random Numbers

Simulation trial i	Generated random number (u) for c	Generated random number (u) for M	Generated random number (u) for I	Generated random number (u) for f_y
1	0.265683	0.444767	0.534083	0.824389
2	0.112984	0.795123	0.142713	0.313709
3	0.130852	0.040925	0.610206	0.119275
4	0.540974	0.91155	0.388728	0.562588
5	0.986392	0.043964	0.444948	0.302641
6	0.599745	0.419549	0.067631	0.42764
7	0.109176	0.599827	0.778881	0.12306
8	0.582111	0.173442	0.07341	0.682096
9	0.444096	0.049815	0.479772	0.960011
10	0.20698	0.680786	0.711584	0.88489
11	0.96046	0.625776	0.266105	0.622753
12	0.166624	0.240704	0.653093	0.398477
13	0.897585	0.709163	0.22793	0.647695
14	0.310298	0.945637	0.317049	0.307006
15	0.931585	0.553377	0.007468	0.36257
16	0.340714	0.634572	0.858001	0.556425
17	0.473103	0.987604	0.376823	0.011421
18	0.119418	0.84858	0.733391	0.651616
19	0.769009	0.941	0.060353	0.513985
20	0.652001	0.413037	0.68325	0.807705

TABLE 1-3B Generated Stresses in a Beam, Yield Stresses, and Failure Analysis: Generated Random Variables

Simulation trial i	Generated c	Generated M	Generated I	Generated f_y
1	9.765683	3223.834	1006.817	53.24389
2	9.612984	4975.614	928.5426	48.13709
3	9.630852	1204.623	1022.041	46.19275
4	10.04097	5557.749	977.7457	50.62588
5	10.48639	1219.822	988.9896	48.02641
6	10.09974	3097.744	913.5263	49.2764
7	9.609176	3999.133	1055.776	46.2306
8	10.08211	1867.208	914.6821	51.82096
9	9.944096	1249.075	995.9544	54.60011
10	9.70698	4403.928	1042.317	53.8489
11	10.46046	4128.882	953.2209	51.22753
12	9.666624	2203.521	1030.619	48.98477
13	10.39759	4545.813	945.5859	51.47695
14	9.810298	5728.184	963.4097	48.07006
15	10.43158	3766.887	901.4936	48.6257
16	9.840714	4172.859	1071.6	50.56425
17	9.973103	5938.021	975.3646	45.11421
18	9.619418	5242.902	1046.678	51.51616
19	10.26901	5704.998	912.0707	50.13985
20	10.152	3065.187	1036.65	53.07705

TABLE 1-3C Generated Stresses in a Beam, Yield Stresses, and Failure Analysis: Computed Stress, Failure Analysis, and Reliability Assessment

Simulation trial i	Generated f_y	Computed stress (σ) according to Equation 1-1	Did failure occur?
1	53.24389	31.26978	No
2	48.13709	51.51137	Yes
3	46.19275	11.35135	No
4	50.62588	57.07539	Yes
5	48.02641	12.93394	No
6	49.2764	34.24797	No
7	46.2306	36.39822	No
8	51.82096	20.58136	No
9	54.60011	12.47138	No
10	53.8489	41.01329	No
11	51.22753	45.30954	No
12	48.98477	20.66779	No
13	51.47695	49.98539	No
14	48.07006	58.32948	Yes
15	48.6257	43.58834	No
16	50.56425	38.32018	No
17	45.11421	60.71627	Yes
18	51.51616	48.18449	No
19	50.13985	64.23261	Yes
20	53.07705	30.01764	No

1.5.3. Traffic Estimation Study

Traffic engineers need estimates of the number of vehicles on roads in a local community. These estimates are used to set the timing of traffic lights and for the design of roadways. Table C-2 contains a set of data for a median-sized city that is separated into 46 zones. The variable of interest (Y) is the number of daily work trips made in a zone. This variable is some function of the zone population, the number of dwellings in the zone, and the number of vehicles in the zone.

Using the data of Table C-2, find the minimum and maximum values of Y. Then generate 25 values of Y using the *rand* function. Transform each of the 25 random values (u_i) to values of Y using the following transformation model:

$$Y_i = \min + u_i(\max - \min) \tag{1-4}$$

1.5.4. Water Evaporation Study

Evaporation of water stored in a reservoir is a significant source of water loss. This loss reduces the amount of water available for power generation, irrigation, and recreation. It may also reduce the aesthetics of the site. Therefore, predictions of evaporation are made prior to constructing a dam so that evaporation losses can be accounted for in the design of the dam.

Table C-3 contains a set of data from an existing dam. The daily evaporation rate is the variable (Y) to be predicted. The predictor variables are the mean daily temperature, the mean daily relative humidity, and the mean daily wind speed, which are variables thought to influence the predicted value of lake evaporation. Find the minimum and maximum values of Y from Table C-3. Generate 40 values of Y using the *rand* function. Transform the 40 random values (u_i) to values of Y using the following transformation model:

$$Y_i = \min + u_i(\max - \min) \tag{1-5}$$

Chapter 2
Data Description and Treatment

TABLE OF CONTENTS

2.1. INTRODUCTION

It is common in engineering to deal with certain types of dispersion and uncertainty by collecting data and information about some variables that are needed to solve a problem of interest. The data can then be used to establish some understanding about the relations among the different variables. After collecting the data, it is necessary to utilize techniques for describing, treating, and analyzing them. The objective of this chapter is to introduce common techniques to describe and summarize data.

2.2. CLASSIFICATION OF DATA

Graphical analyses are widely used in engineering for both making initial assessments of data characteristics and presenting results to be used by engineers. There are a number of ways of graphically presenting data, each having advantages for specific types of data. Therefore, before discussing the graphical methods, it is necessary to review methods for classifying data.

Data can be measured on one of four scales: nominal, ordinal, interval, and ratio. The four scales are given in order of numerical value. Variables defined on one scale can be reduced to a lower scale of measurement but cannot be described on a higher scale of measurement; however, when a variable is transformed to a lower scale of measurement, there is an accompanying loss of information.

In addition to the scale of measurement, data can be classified based on their dimensionality. The dimensionality is a measure of the number of axes needed to present the data graphically. Graphical analyses are usually limited to use with one-, two-, and three-dimensional data.

2.2.1. Nominal Scale

The nominal scale of measurement is at the lowest level, because there is no order to the data. Measurements consist of simply identifying the sample as belonging to one of several categories. Nominal measurement scales are both discrete and qualitative; however, numbers may be assigned to the categories for the purpose of coding.

Frequently used examples of variables measured on a nominal scale include: (1) gender: female or male; (2) political affiliation: Republican, Democrat, Independent, or other; or (3) college major: engineering, sciences, physical education, or other. Engineering data are sometimes provided using a nominal scale — for example: (1) project failed or did not fail; (2) fatal and nonfatal accidents; or (3) land use, such as urban, rural, forest, institutional, commercial, or other.

2.2.2. Ordinal Scale

The ordinal scale of measurement is considered to be a higher scale than the nominal scale because it has the added property that there is order among the groups; however, the magnitude of the differences between groups is not meaningful. For example, military ranks are measured on an ordinal scale. The major is above the sergeant and the sergeant is above the private, but we cannot say that a major is two or three times higher than a sergeant.

Variables of interest in engineering that are measured on an ordinal scale include the infiltration potential of soil texture classes and hazard classifications for dam design (high, moderate, or low hazard). Soils are classified into one of several categories, such as sand, sandy loam, clay loam, and clay. In this respect, soil texture is measured on a nominal scale; however, if we consider the infiltration potential of the soil, then we can put the soil textures in order according to the infiltration potential, high to low.

2.2.3. Interval Scale

The interval scale of measurement has the characteristics of the ordinal scale, in addition to having a meaningfulness in the separation between any two numbers on the scale. Temperature is defined on the interval scale. We recognize that a difference in temperature of 5°C is less than a difference of 10°C. Values on an interval scale may be treated with arithmetic operators. For example, the mean value of a set of test grades requires addition and division.

Engineering data are frequently recorded on an interval scale. The yield strength of steel, the compression strength of concrete, and the shear stress of soil are variables measured on an interval scale. The annual number of traffic fatalities and the number of lost worker-hours on construction sites due to accidents are also engineering variables recorded on an interval scale.

2.2.4. Ratio Scale

The ratio scale represents the highest level of measurement. In addition to the characteristics of the interval scale, the ratio scale has a true zero point as its origin, not like the interval scale, for which the zero point is set by some standard. For example, in the interval scale the zero point for temperature (°C) is set at the point where water freezes. However, it could have been either set at the point where water boils or based on some other substance.

The standard deviation, which is discussed in Section 2.5.2, is measured on a ratio scale. The zero point is that for which there is no variation. The coefficient of variation and many dimensionless ratios such as the Mach and Reynolds numbers are measured on a ratio scale.

2.2.5. Dimensionality of Data

The dimensionality of data was defined as the number of axes needed to represent the data. Tabular data with one value per classification is an example of one-dimensional data. For example, if a transportation engineer gives the number of fatal traffic accidents for each state in 1991, the variable is described using one-dimensional, interval-scale data. It can be represented in tabular form as a function of the state. It could also be represented pictorially with the number of fatalities as the ordinate (vertical axis) and the state as the abscissa (horizontal axis). In this case, the ordinate is on an interval scale, while the abscissa is on a nominal scale; when presented this way, it appears as a two-dimensional graph.

Two-dimensional plots are very common in engineering. For example, the solution of a dependent variable y of a differential equation could be plotted as a function of the independent variable x. In this case, both variables are expressed on an interval scale. As another example, the corrosion rate of structural steel as a function of the length of time that the steel has been exposed to the corrosive environment is a two-dimensional plot. If we have data for different types of steel (carbon steel, copper steel, and weathered steel), the three relationships can be presented on the same graph, with the steel types identified. In this case, steel type is a nominal variable, so the two-dimensional plot includes two variables on interval scales and one variable on a nominal scale.

2.3. GRAPHICAL DESCRIPTION OF DATA

The first step in data analysis is often a graphical study of the characteristics of the data sample. Depending on the objectives of the analysis and the nature of the problem under consideration, one or more of the following graphical descriptors are commonly used: area charts, pie charts, bar charts, column charts, scatter diagrams, line charts, combinations, and three-dimensional charts. The selection of a graphical descriptor type should be based on (1) its intended readers, (2) the type of data, (3) the dimensionality of the problem, and (4) the ability of the graphical descriptor to emphasize certain characteristics or relations of the parameters of the problem. In this section, examples are used to illustrate the different types.

2.3.1. Area Charts

Area charts are useful for three-dimensional data that include both nominal and interval-independent variables, with the value of the dependent variable measured on an interval scale and cumulated over all values of the nominal variable. The independent variable measured on an interval scale is shown on the abscissa. At any value along the abscissa, the values of the dependent variable are cumulated over the independent variable measured on the nominal scale.

Example 2-1: Area Chart for Traffic Analysis

A traffic engineer is interested in analyzing the traffic at an intersection. A vehicle approaching the intersection can proceed in one of the following directions: straight, left turn, or right turn (U-turns are assumed to be illegal). The vehicles were counted at the intersection and classified according to direction. The counts were established

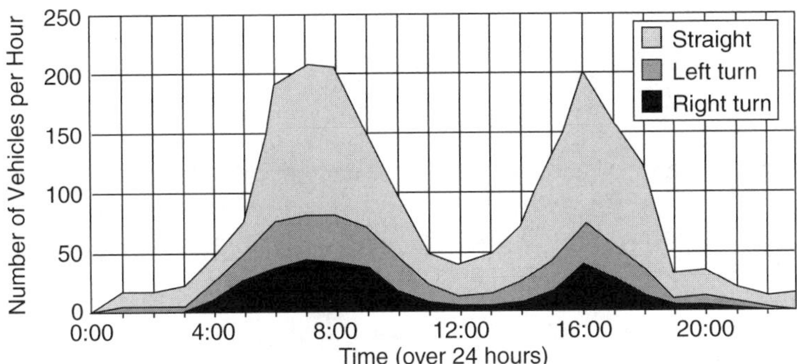

FIGURE 2-1 Traffic at an intersection.

for 24 hours in a typical business day. The results are shown in Figure 2-1 in the form of an area chart. The area chart shows that at 8:00 approximately 40 vehicles made a right turn, approximately 75 vehicles made either a right or left turn, and a total of approximately 200 vehicles went through the intersection.

2.3.2. Pie Charts

Pie charts are commonly used to graphically present data recorded as fractions, percentages, or proportions. The 100% of the circular pie chart is separated into pie slices based on fractions or percentages. The variable presented in a pie chart is measured on an interval scale.

Example 2-2: Pie Chart for Shipment Breakdown

A shipping company facilitates for customers the transfer of any item from any location to another within the United States. The company transfers 25, 30, and 45% of the items by air, ground, and sea transportation, respectively. In this case, the variable is the form of transportation, with values recorded as a percentage of all transportation. The breakdown of these items by shipping method is shown in Figure 2-2 in the form of a pie chart.

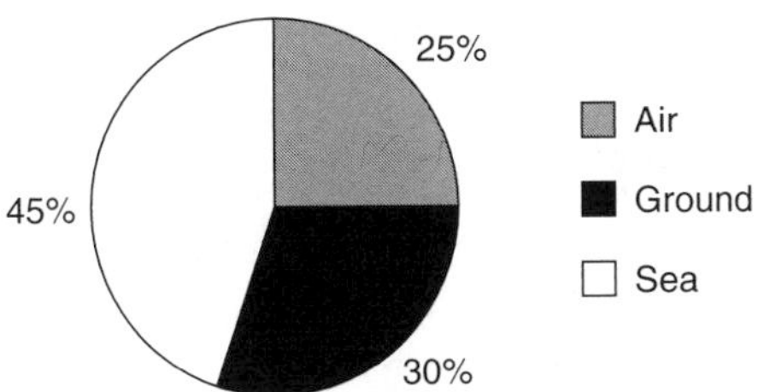

FIGURE 2-2 Shipping methods.

2.3.3. Bar Charts

Bar charts are also useful for data recorded on an interval scale with one or more independent variables recorded on nominal or ordinal scales. The dependent variable can be a magnitude or a fraction. Both one- and two-dimensional bar charts can be used, with a dimension that is based on the number of independent variables.

Example 2-3: Bar Chart for Reinforcing Steel Production

A reinforcing-steel manufacturer provides steel of three different yield strengths: 40, 50, and 60 ksi. The production manager would like to keep track of production for both reinforcing-steel type and the four quarters of a production year. Thus, the amount of steel produced is the dependent variable. The two independent variables are the steel type and the yearly quarters. The dependent variable is on an interval scale, while the independent variables are on an ordinal and nominal scale, respectively. A bar chart, in this case, can meet the requirements of the manager as shown in Figure 2-3. In this case, the bars are shown vertically because the descriptors can be easily included on the abscissa and as a side note.

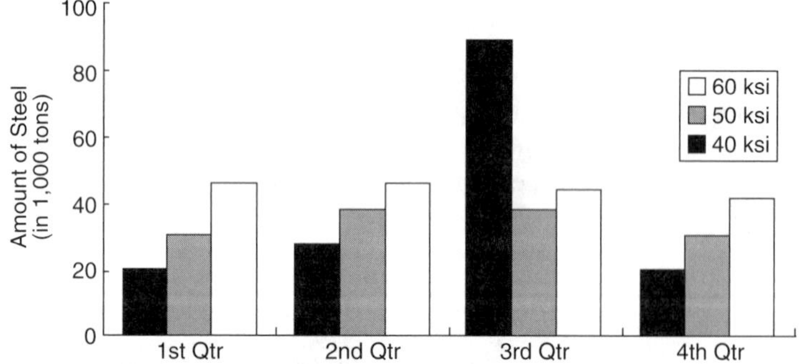

FIGURE 2-3 Reinforcing steel production by yield strength and quarter.

Example 2-4: Bar Chart for Capacity of Desalination Plants

In this case, the dependent variable is the capacity (in million gallons per day, mgd) of water from desalination plants worldwide. The independent variable is the process used in the desalination. Figure 2-4 shows the distribution. The bar chart, in this example, is shown sideways to facilitate presenting the descriptors of the six processes.

2.3.4. Column Charts

Column charts are very similar to bar charts but with the added constraint that the dependent variable is expressed as a percentage (or fraction) of a total. In this case, one of the independent variables is used for the abscissa, while the dependent variable is shown as percentages (or fractions) of the second independent variable. The dependent variable is shown as the ordinate.

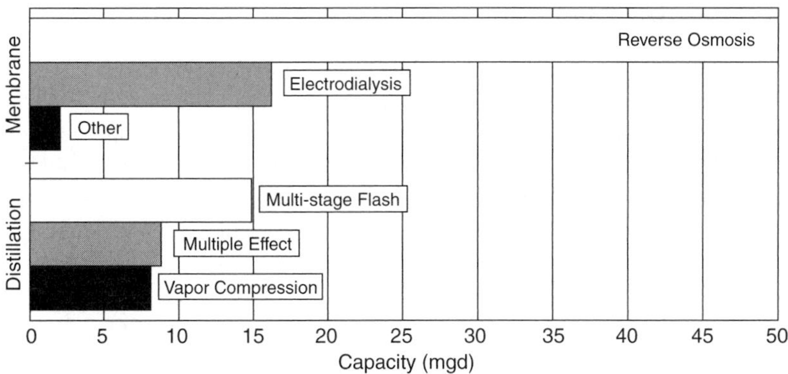

FIGURE 2-4 Worldwide desalination capacity (mgd) for available processes.

Example 2-5: Column Chart for Reinforcing Steel Production

An alternative method for displaying the reinforcing steel production of Example 2-3 is by column charts. Figures 2-5A and B show example charts. Note that in Figure 2-5A, steel production is displayed as a percentage of the total steel produced for each quarter.

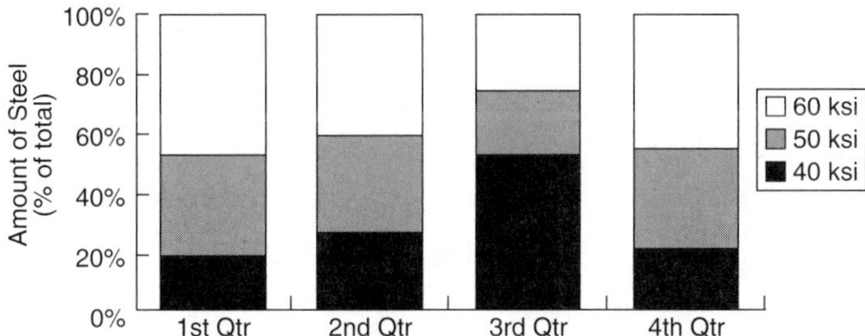

FIGURE 2-5A Reinforcing steel production (as percentage) by yield strength and quarter.

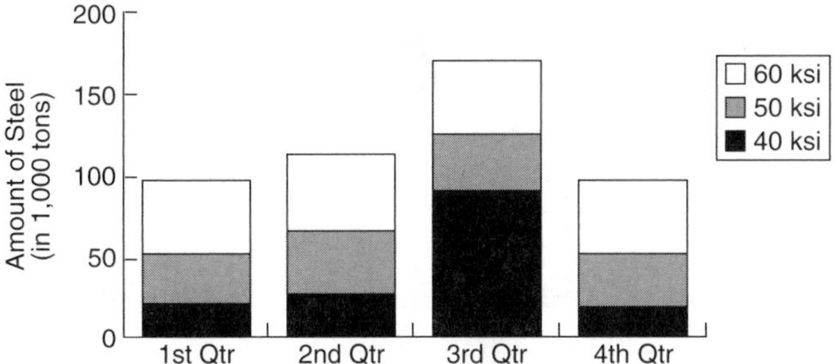

FIGURE 2-5B Reinforcing steel production by yield strength and quarter.

2.3.5. Scatter Diagrams

When both the independent and dependent variables are measured on interval or ratio scales, the data are best presented with scatter plots. Usually, the variable to be predicted (dependent variable) is shown on the ordinate and the independent variable on the abscissa. The ranges for the variables on axes are based on the minimum and maximum values of the measured data, possible values for the variables, or values that may be expected to occur in the future.

Example 2-6: Scatter Diagram for Yield Strength and Carbon Content

The yield strength of steel depends on several factors, such as the carbon content. An experimental program was designed to understand this relationship. Figure 2-6 shows the measured values. Because we are interested in the yield strength as a function of the carbon content, the yield strength is placed on the ordinate. The limits of the plot were set by the values of the data. An alternative plot with a range of 0 to 200 ksi and 0 to 0.1% carbon content could also be used to display the data. It is evident from this figure that under similar test conditions the yield strength increases as the carbon content increases.

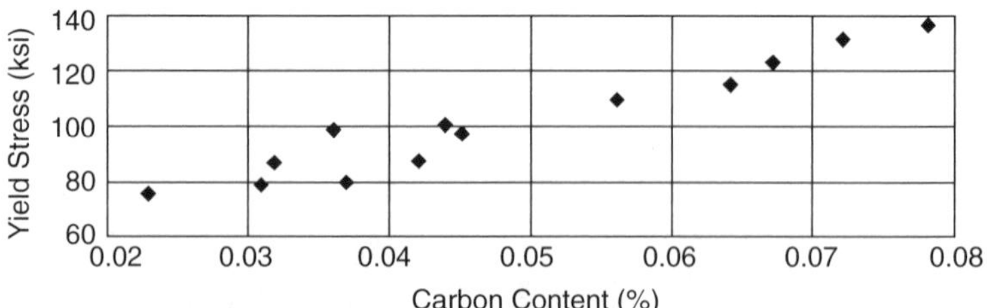

FIGURE 2-6 Yield strength and carbon content.

2.3.6. Line Graphs

Line graphs are used to illustrate mathematical equations. With both variables measured on interval or ratio scales, the variable to be predicted is usually shown as the ordinate. Line graphs are frequently used for design work when the design method is relatively simple.

Example 2-7: Line Chart for Peak Discharge Rates

For one region of the state of Maryland, peak discharge rates (Q in ft^3/sec, cfs) can be estimated as a function of drainage area (A in mi^2) by the following equations:

$$Q_2 = 55.1A^{0.672} \tag{2-1a}$$

$$Q_{10} = 172A^{0.667} \tag{2-1b}$$

$$Q_{100} = 548A^{0.662} \tag{2-1c}$$

in which the subscripts on Q reflect the return frequency (T, in years) of the storm. Thus, Q_{100} is the 100-year peak discharge. Designs with these equations are usually limited to drainage areas of less than 100 mi². Figure 2-7 shows the peak discharge for drainage areas up to 100 mi². The two-dimensional line chart shows the dependent variable on the ordinate and one independent variable as the abscissa, with the second independent variable in the legend taking three values. All three variables are measured on interval scales.

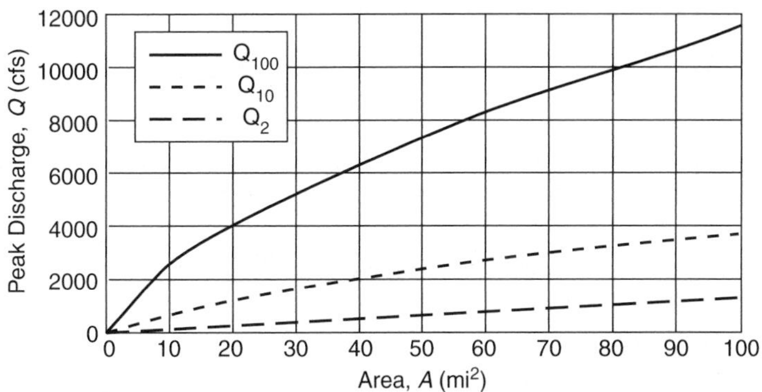

FIGURE 2-7 Peak discharge rate vs. drainage area and return period.

Example 2-8: Line Chart for Yield Strength and Carbon Content

To establish the relationship between yield strength and carbon content, two independent laboratories were requested to perform similar tests on similar sets of specimens. The results from the two laboratories are shown in Figure 2-8. The length of the lines connecting the points reflects the magnitude of the variations that are reported based on measurements made at different laboratories. The figure displays the differences in results from the two laboratories in order to help an analyst assess the significance of these differences.

2.3.7. Combination Charts

In combination charts, two or more of the previously discussed graphical methods are used to present data. For example, a line graph and bar chart can be combined in the same plot. A combination chart that includes both a scatter plot and a line graph is also commonly used to present experimental data and theoretical (or fitted) prediction equations.

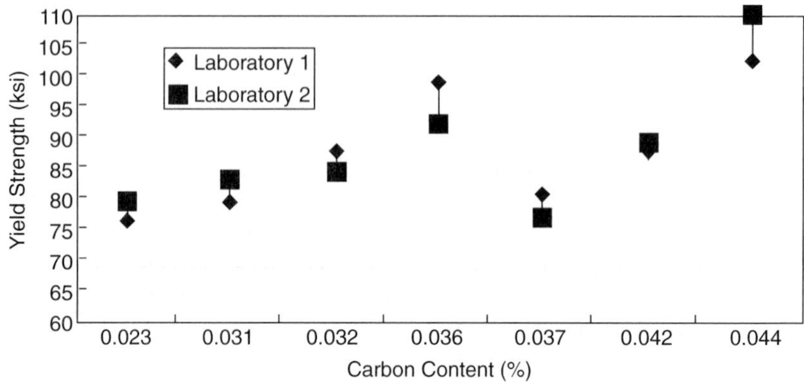

FIGURE 2-8 Line charts for yield strength and carbon content.

Example 2-9: Operation of a Marine Vessel

The annual number of operation hours of a high-speed Coast Guard patrol boat is of interest. The annual number of hours was recorded for 10 years. The measured data, shown in Figure 2-9 as vertical bars, represent sample information. The analytical model used to fit the data is shown in the figure as a solid line representing an assumed frequency model for the population. The combination chart in this case provides an effective presentation tool.

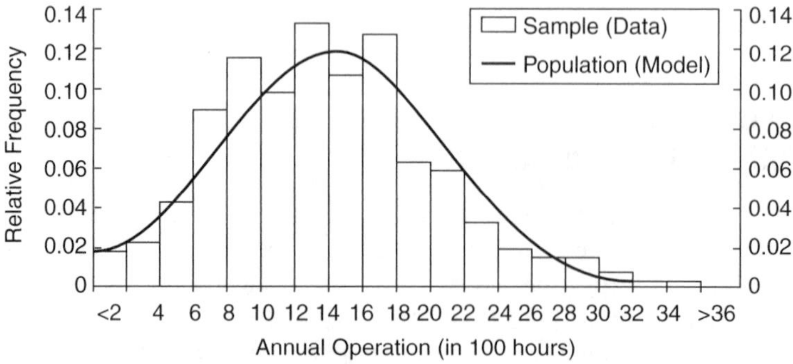

FIGURE 2-9 Combination chart for operational profile.

Example 2-10: Sample Distribution and Probability Function

A combination graph is useful in comparing sample measurements and an assumed probability function. Measurements of daily evaporation (in./day) were collected for each day in a month. A histogram of 30 values is shown in Figure 2-10. An increment of 0.02 in./day is used to form this histogram. The measured evaporation ranges

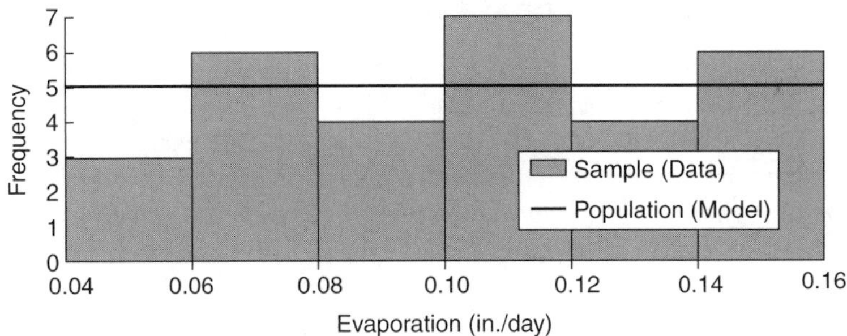

FIGURE 2-10 Combination chart for evaporation data.

from 0.046 to 0.157 in./day; thus, lower and upper bounds of 0.04 and 0.16 in./day are used. Based on the figure, it is reasonable to assume that each increment on the abscissa of the graph is equally likely. Therefore, a uniform frequency of 5, which results from 30 measurements divided by 6 intervals, can be used as the model.

Example 2-11: Corrosion Penetration vs. Exposure Time

Measurements of corrosion were made yearly on a set of 10 steel specimens. The measured data are plotted in the combination graph of Figure 2-11. A power model was fitted to the data (see Chapter 13 for the definition of a power model) and is also shown in Figure 2-11. The fitted power model is:

$$\text{Penetration} = 28.075t^{0.39943} \tag{2-2}$$

where t is the time exposure. The combination chart of Figure 2-11 is useful because it shows both the measured data and a graph that can be used to make predictions. The scatter of the measured points about the fitted line indicates the accuracy of the model.

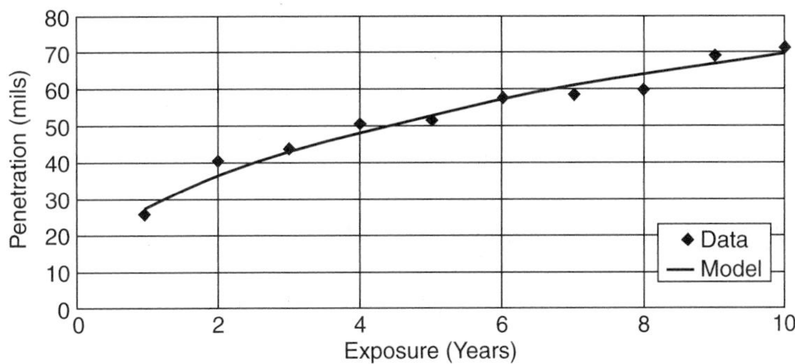

FIGURE 2-11 Combination chart for corrosion prediction.

2.3.8. Three-Dimensional Charts

Any of the charts described in the previous examples can be displayed in three dimensions (three-dimensional); however, three-dimensional charts are commonly used to describe the relationships among three variables. For example, Figure 2-12A shows a three-dimensional pie chart for the data of Example 2-2. Another example is shown in Figure 2-12B, in which the speed (in mph) and number of vehicles per hour passing through an intersection are shown at different times over a 24-hr period.

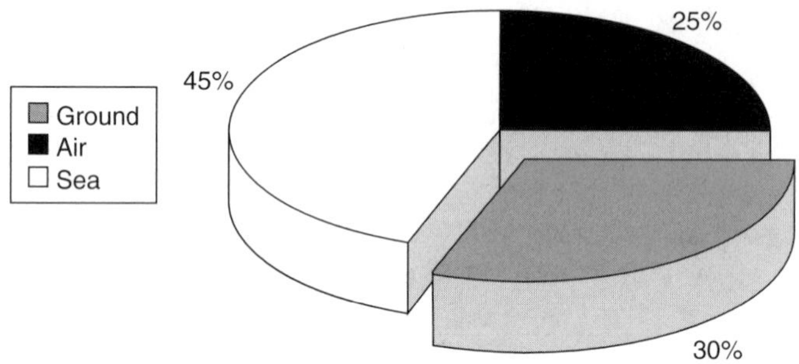

FIGURE 2-12A Three-dimensional pie chart.

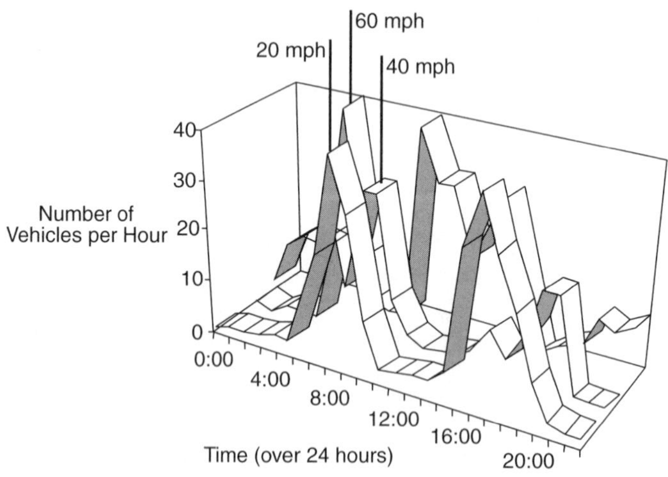

FIGURE 2-12B Three-dimensional surface chart.

2.4. HISTOGRAMS AND FREQUENCY DIAGRAMS

Histograms and frequency diagrams are special types of charts that are commonly used to display and describe data. They are developed from the data for some variables of interest. A histogram is a plot (or a tabulation) of the number of data points (count or frequency) vs. selected intervals or values for a parameter. A frequency diagram (or frequency histogram) is a plot (or tabulation) of the frequency of occurrence vs. selected intervals or values of the parameter. The relative frequency can be viewed as a fraction of the total sample that occurs in an interval; therefore, it is called a *fraction histogram.* The number of intervals (k) can be subjectively selected depending on the sample size (n). The number of intervals can be approximately determined as:

$$k = 1 + 3.3\log_{10}(n) \tag{2-3}$$

Also, the number of intervals can depend on the level of dispersion in the data. The relative frequency diagrams can be derived from the histogram by dividing the number of data points that correspond to each interval by the sample size. In practice, it is common to try several numbers of intervals or interval sizes and subjectively assess the appropriateness of the resulting histograms for the purpose of selecting one of them to best represent the data. A histogram that has small intervals will unnecessarily capture the sampling variability and *noise* in the data, whereas a histogram that has large intervals will not properly show the overall variability in the data. As a limiting case, a histogram with very small intervals will have one sampled value in an interval with many intervals that are empty (i.e., without data in them). On the other hand, a histogram with very large intervals will have all sampled values in one interval (i.e., it does not show the variability in the data). The meaning of these diagrams and their usefulness and development are illustrated in the following example.

Example 2-12: Grades of Students

Students are always interested in the frequency histogram of test scores. This can be developed by determining the number of test scores in various groups, such as the letter grades or intervals of 10 points. If a test is given to 50 students and the numbers of scores are tabulated for each grade level (i.e., A, B, C, D, and F), a histogram can be plotted. If the number of students receiving grades of A, B, C, D, and F were 5, 11, 18, 10, and 6, respectively, a graph of the number of students vs. the grade level indicates that the grades have a bell-shaped plot. The resulting histogram and frequency diagrams are shown in Figures 2-13A and 2-13B. Alternatively, the frequency of grades could be computed for intervals, such as 0–10, 11–20, 21–30, ..., 91–100, and the frequency could be plotted against these intervals.

The effectiveness of a graphical analysis in identifying the true shape of the frequency diagram depends on the sample size and the intervals selected to plot the abscissa. For small samples, it is difficult to separate the data into groups that provide a reliable indication of the frequency of occurrence. With small samples, the configuration of the histogram may be very different for a small change in the intervals selected for the abscissa; for example, the grade distribution might appear very different if the intervals 0–15, 16–25, 26–35, ..., 76–85, 86–100 are used instead of the 10-point intervals described previously.

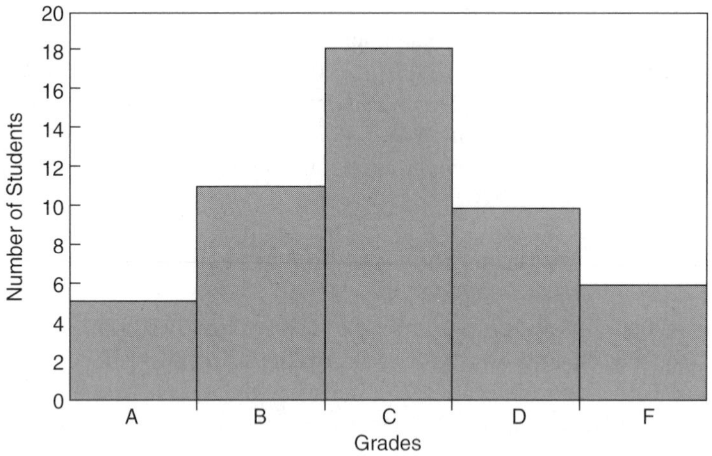

FIGURE 2-13A Histogram of grades.

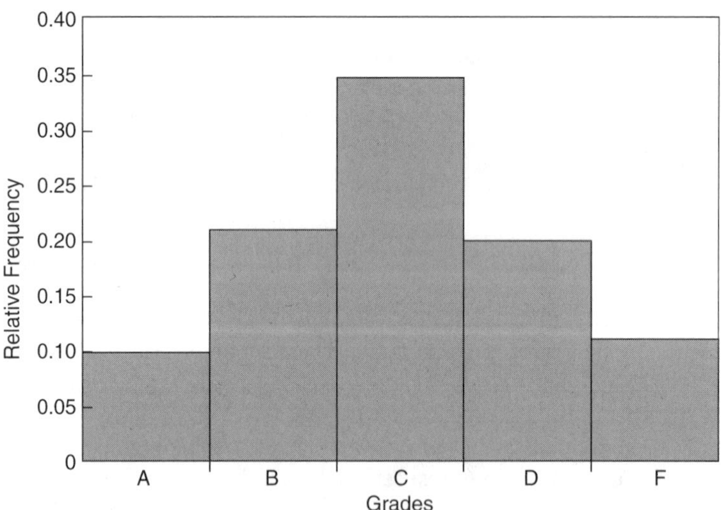

FIGURE 2-13B Relative frequency histogram of grades.

Example 2-13: Thickness Measurement of Corroded Steel Plates

Exposing steel to a corrosive environment, such as in the case of a steel bridge spanning a waterway or a cargo ship making voyages regularly, leads to loss of thickness of structural components. A corroded steel plate was measured at 20 locations and produced the following measurements in mm: 7.807, 8.886, 8.694, 8.185, 9.235, 8.526, 6.890, 8.953, 6.284, 6.533, 8.953, 8.112, 7.372, 9.640, 7.344,

8.837, 8.900, 9.048, 7.253, and 8.588. The minimum and maximum values are 6.284 and 9.640, respectively. These extreme values and examination of the data can be used to select a suitable constant-interval size. As an example, a size of 0.500 mm was selected, starting with a thickness of 6.000 mm and incrementally increasing the thickness to 10.000 mm. Table 2-1 shows these intervals (or bins), the counts in each bin, and the relative frequency. The relative frequency equals the count in the table divided by 20. The results are shown in Figures 2-14A and 2-14B.

TABLE 2-1 Frequency and Fraction Histogram of Thickness Measurements

Interval or bin for x	Frequency	Relative frequency
$x \leq 6$	0	0
$6 < x \leq 6.5$	1	0.05
$6.5 < x \leq 7$	2	0.10
$7 < x \leq 7.5$	3	0.15
$7.5 < x \leq 8$	1	0.05
$8 < x \leq 8.5$	2	0.10
$8.5 < x \leq 9$	8	0.40
$9 < x \leq 9.5$	2	0.10
$9.5 < x \leq 10$	1	0.05

2.5. DESCRIPTIVE MEASURES

In engineering it is sometimes desirable to characterize some data by certain descriptive measures. These measures, which take numerical values, can be easily communicated to others and quantify the main characteristics of the data.

Most data analyses include the following three descriptive measures at the fundamental level:

1. Central tendency measures
2. Dispersion measures
3. Percentile measures

In this section, an introductory description of these measures is provided. A formal discussion of these measures is provided in subsequent chapters. Additionally, box-and-whisker plots are introduced as a graphical means of presenting these measures.

2.5.1. Central Tendency Measures

A very important descriptor of data is the central tendency measure. The following three types can be used:

1. Average value
2. Median value
3. Mode value

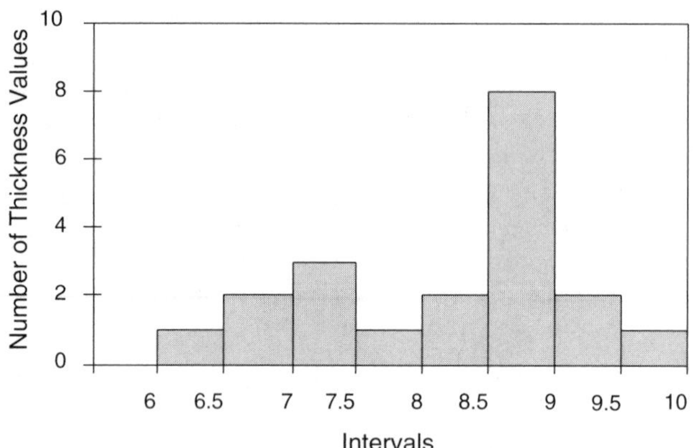

FIGURE 2-14A Frequency histogram of plate thickness.

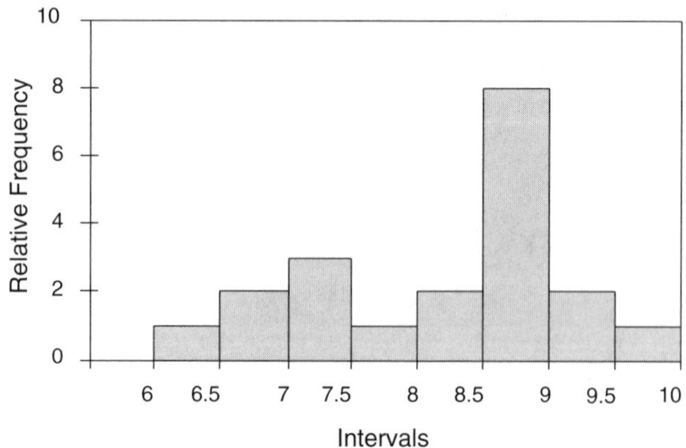

FIGURE 2-14B Relative frequency histogram of plate thickness.

The average value is the most commonly used central tendency descriptor. For n observations, if all observations are given equal weight, the average value is given by:

$$\bar{X} = \frac{1}{n}\sum_{i=1}^{n} x_i \tag{2-4}$$

where x_i is a sample point, and $i = 1, 2, \ldots, n$.

The median value x_m is defined as the point that divides the data into two equal parts; that is, 50% of the data are above x_m and 50% are below x_m. The median value can be determined by ranking the n values in the sample in decreasing order, 1 to n. If n is an odd number, the median is the value with a rank of $(n + 1)/2$. If n is an even number, the median equals the average of the two middle values — that is, those with ranks $n/2$ and $(n/2) + 1$.

The mode value x_d is defined as the point or points with the highest frequency or relative frequency of occurrence that can be observed within regions or ranges of x. This point can be determined with the aid of the frequency histogram.

Although these measures convey certain information about the underlying sample, they do not completely characterize the underlying variables. Two variables can have the same mean, but different histograms, thus measures of central tendency cannot fully characterize the data. Other characteristics are also important and necessary.

Example 2-14: Mean Value of Grades

Consider the grades of the students discussed in Example 2-12. Assume the following grade points (levels) that correspond to the letter grades:

$$\begin{aligned} A &= 4 \\ B &= 3 \\ C &= 2 \\ D &= 1 \\ F &= 0 \end{aligned}$$

Therefore, the average grade point of the class can be computed as follows:

$$\bar{X} = \frac{4+4+4+4+4+3+\ldots+3+2+\ldots+1+\ldots+1+0+\ldots+0}{50} = 1.98 \qquad (2\text{-}5)$$

In this equation, there are 5 of the 4 value, 11 of the 3 value, 18 of the 2 value, 10 of the 1 value, and 6 of the 0 value. Therefore, this equation can be written as:

$$\bar{X} = \frac{5\times4+11\times3+18\times2+10\times1+6\times0}{50} = 1.98 \qquad (2\text{-}6)$$

The average value indicates that on the average the class is at the C level. By inspecting the frequency histogram in Figure 2-13B, the median value is also C, as 32% of the grades are in the A and B levels and 32% of the grades in the D and F levels. Therefore, the C grade divides the grades into two equal percentages. This method of finding the median is proper for values on an ordinal scale, but not on a continuous scale. Again, the C grade is the mode, because it has the highest frequency of 36%. In this example, the average, median, and mode values are the same value. However, in general, these values can be different.

2.5.2. Dispersion Measures

The dispersion measures describe the level of scatter in the data about the central tendency location. The most commonly used measure is the variance and other quantities that are derived from it. For n observations in a sample that are given equal weight, the variance (S^2) is given by:

$$S^2 = \frac{1}{n-1}\sum_{i=1}^{n}(x_i - \bar{X})^2 \tag{2-7}$$

The units of the variance are the square of the units of the variable x; for example, if the variable is measured in pounds per square inch (psi), the variance has units of $(\text{psi})^2$. Computationally, the variance of a sample can be determined using the following alternative equation:

$$S^2 = \frac{1}{n-1}\left[\sum_{i=1}^{n} x_i^2 - \frac{1}{n}\left(\sum_{n=i}^{n} x_i\right)^2\right] \tag{2-8}$$

Equation 2-8 provides the same answer as Equation 2-7 when computations are made using an appropriate number of significant digits. Equations 2-7 and 2-8 provide an estimate of the variance that is an average of the squared difference between the x values and their average value. In these equations, $(n - 1)$ is used to compute the average deviation, instead of (n), in order to obtain an unbiased estimate of the variance as discussed in Chapter 8.

Two commonly used derived measures based on the variance are the standard deviation and the coefficient of variation. By definition, the standard deviation (S) is the square root of the variance as follows:

$$S = \sqrt{\frac{1}{n-1}\left[\sum_{i=1}^{n} x_i^2 - \frac{1}{n}\left(\sum_{n=i}^{n} x_i\right)^2\right]} \tag{2-9}$$

It has the same units as both the underlying variable and the central tendency measures; therefore, it is a useful descriptor of the dispersion or spread of a sample of data.

The coefficient of variation (COV, or δ) is a normalized quantity based on the standard deviation and the mean; therefore, it is dimensionless. The COV is defined as:

$$\text{COV} = \frac{S}{\bar{X}} \tag{2-10}$$

It is also used as an expression of the standard deviation in the form of a percent of the average value. For example, consider $\bar{X}$ and S to be 50 and 20, respectively;

therefore, COV(X) = 0.4 or 40%. In this case, the standard deviation is 40% of the average value.

Example 2-15: Dispersion Measures of Concrete Strength

A sample of five tests was taken to determine the compression strength (in ksi) of concrete. Test results are 2.5, 3.5, 2.2, 3.2, and 2.9 ksi. Compute the variance, standard deviation, and coefficient of variation of concrete strength. The mean value of concrete strength $\overline{X}$ is given by:

$$\overline{X} = \frac{2.5 + 3.5 + 2.2 + 3.2 + 2.9}{5} = 2.86 \text{ ksi} \tag{2-11}$$

The variance of concrete strength is computed using Equation 2-8 as follows:

$$S^2 = \frac{2.5^2 + 3.5^2 + 2.2^2 + 3.2^2 + 2.9^2 - \dfrac{(2.5 + 3.5 + 2.2 + 3.2 + 2.9)^2}{5}}{5 - 1} = 0.273 \text{ ksi}^2 \tag{2-12}$$

Therefore, the standard deviation is given by:

$$S = \sqrt{0.273} = 0.52249 \text{ ksi} \tag{2-13}$$

The coefficient of variation can be computed as follows:

$$\delta \text{ or } \mathrm{COV}(X) = \frac{S}{\overline{X}} = \frac{0.52249}{2.86} = 0.183 \tag{2-14}$$

The relatively large coefficient of variation, 18.3%, suggests that the average value is not reliable and that additional measurements might be needed. If, with additional measurements, the coefficient of variation remains large, relatively large factors of safety should be used for projects that use the concrete.

2.5.3. Percentiles

A p percentile value (x_p) for a parameter or variable based on a sample is the value of the parameter such that p% of the data is less than or equal to x_p. On the basis of this definition, the median value is considered to be the 50 percentile value. It is common in engineering to be interested in the 10, 25, 50, 75, and 90 percentile values for a variable.

Example 2-16: Operation of a Marine Vessel

The annual number of operation hours of a high-speed Coast Guard patrol boat was discussed in Example 2-9. The annual number of hours was recorded for 10 years. An analytical model used to fit the data is shown in Figure 2-9 as a solid line representing an assumed frequency model for the population. The model is also shown in Figure 2-15. Also shown in this figure is the 25 percentile value, which is 702 hr. The 75, 90, and 10 percentile values are 1850, 2244, and 578 hr, respectively. The calculation of the percentile values, in this example, requires knowledge of the equation of the model and probabilistic analysis. The model used in Figure 2-15 is called the *normal probability density function*. The area under this model is a measure of likelihood or probability. The total area under the model is 1. Therefore, the 25 percentile value is the x value such that the area under the model up to this value is 0.25. This value is 702 hr, as shown in Figure 2-15.

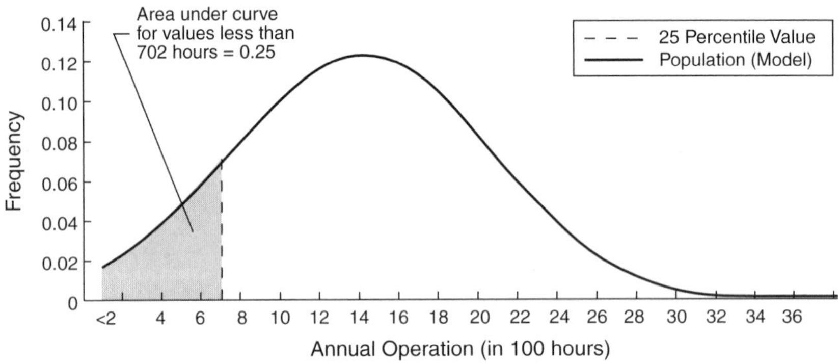

FIGURE 2-15 Percentile values for the operational profile.

2.5.4. Box-and-Whisker Plots

A box-and-whisker plot is a graphical method for showing the distribution of sampled data, including the central tendency (mean and median), dispersion, percentiles (i.e., 10, 25, 75, and 90 percentiles), and the extremes (minimum and maximum). Additionally, it can be used to show the bias about the standard value and, if the figure includes multiple plots for comparison, the relative sample size.

To construct a box-and-whisker plot, the following characteristics of a data set must be computed:

1. Mean and median of the sample
2. Minimum and maximum of the sample
3. 90, 75, 25, and 10 percentile values

The plot consists of a box, the upper and lower boundaries of which define the 75 and 25 percentiles, and upper and lower whiskers, which extend from the ends of the box to the extremes, as shown in Figure 2-16. At the 90 and 10 percentiles, bars that are one half of the width of the box are placed perpendicular to the whiskers. The mean and median are indicated by solid and dashed lines, respectively, that are the full width of the box.

Figure 2-16 shows the box-and-whisker plot for the maximum daily ozone concentration. The mean and median values are 59 and 52 ppb (parts per billion), respectively. The 10, 25, 75, and 90 percentile points are 24, 36, 79, and 97 ppb, respectively.

If a figure includes more than one box-and-whisker plot and the samples from which each plot is derived are of different sizes, then the width of the box can be used to indicate the sample size, with the width of the box increasing as the sample size increases. Figure 2-17 shows box-and-whisker plots of samples of a toxic

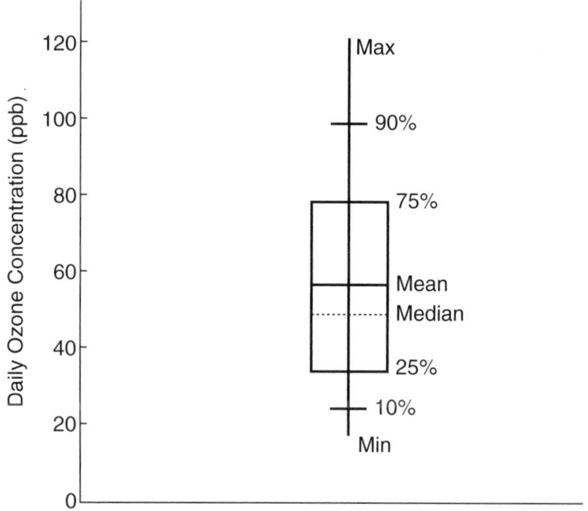

FIGURE 2-16 Box-and-whisker plot for maximum daily ozone concentration (ppb).

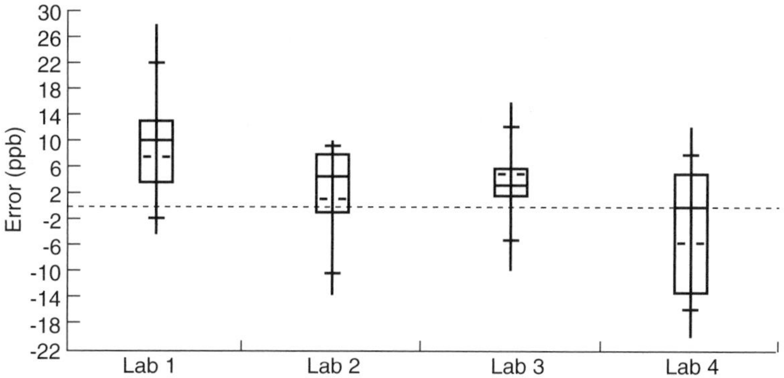

FIGURE 2-17 Display of multiple box-and-whisker plots.

chemical analyzed at four different laboratories. Because the samples were of different concentrations, the distributions are presented as the differences between the true concentration and the concentration reported by the laboratory (i.e., error). The difference between the line for zero error and the solid line for the mean in each plot represents the corresponding lab's bias. Lab 1 tends to overpredict by almost 10 ppb. While lab 2 shows a slight positive bias, it shows a skewed distribution of results with many values underestimating the true value. Laboratories 1, 2, 3, and 4 processed the same number of samples each; therefore, the widths of the boxes are the same, reflecting the equal sample sizes. However, it should be noted that greater accuracy of data descriptors can be expected for the descriptors based on larger sample sizes.

2.6. APPLICATIONS

2.6.1. Two Random Samples

The two random samples shown in Table 2-2 are used to illustrate the meaning of dispersion in data. Both samples have the same mean value of 10. The mean value for sample 1 can be computed as:

$$\text{mean of sample 1} = \frac{15(9) + 15(11)}{30} = 10$$

Similarly, the mean for sample 2 is:

$$\text{mean of sample 2} = \frac{3(5) + 2(6) + 4(7) + \cdots + 2(13) + 2(14) + 2(15)}{30} = 10$$

Although both samples have the same mean and the same sample size, they are different in their levels of scatter or dispersion. Figure 2-18 shows the histograms

TABLE 2-2 Two Random Samples

	Number of occurrences	
Sampled value	**Sample 1**	**Sample 2**
5	0	3
6	0	2
7	0	4
8	0	0
9	15	3
10	0	4
11	15	2
12	0	6
13	0	2
14	0	2
15	0	2

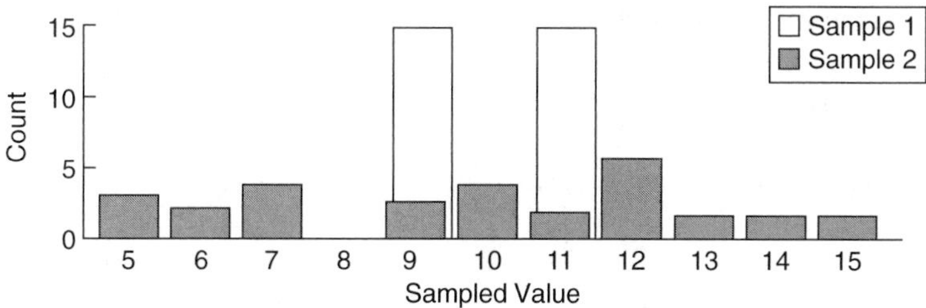

FIGURE 2-18 Histograms for two samples.

for the two samples, which show clearly the different levels of scatter. The variances, standard deviations, or coefficients of variation can be computed and used to measure the dispersion. The variance for sample 1 can be computed as:

$$\text{variance of sample 1} = \frac{15(9-10)^2 + 15(11-10)^2}{30-1} = 1.034$$

Similarly, the variance for sample 2 is:

$$\frac{3(5-10)^2 + 2(6-10)^2 + 4(7-10) + \ldots + 2(13-10)^2 + 2(14-10)^2 + 2(15-10)^2}{30-1} = 9.379$$

Therefore, the variance of sample 2 is about nine times the variance of sample 1. The standard deviations for samples 1 and 2 are 1.017 and 3.063, respectively, and the coefficient of variations are 0.10 and 0.31, respectively. The larger scatter in sample 2 could reflect natural variation in the value of the variable or a difficulty in making precise measurements. In some cases, variation is not desirable in a sample of data if it reflects a lack of precision. In other cases, sample variation is necessary to ensure that the sample is representative of the full range of the underlying population.

2.6.2. Stage and Discharge of a River

River stage is defined as the flow depth, and the discharge is defined as the volume rate of flow. For the Little Patuxent River near Guilford, MD, the stage in meters (m) and the discharge in cubic meters per second (cms) were obtained as shown in Table 2-3. The table shows the stage and discharge for the years 1933 to 1989. Descriptive statistics of stage and discharge are given in Table 2-4. The relative frequency histogram for the stage is shown in Figure 2-19. The average, which is indicated in Table 2-4, is 2.84 m. The median stage is 2.71 m. The standard deviation and coefficient of variation are 0.71 and 0.25 m, respectively. The maximum and minimum values (extreme values) are 5.60 and 1.69 m, respectively.

The relative frequency histogram for the discharge is shown in Figure 2-20. The average, which is indicated in Table 2-4, is 54.82 cms. The median discharge is 34.8 cms. The standard deviation and coefficient of variation are 53.78 and 0.98

TABLE 2-3 Stage and Discharge for Little Patuxent River, Guilford, MD

Year	Stage (m)	Discharge (cms)	Year	Stage (m)	Discharge (cms)
1933	3.81	119.2	1962	2.78	36.2
1934	2.86	41.9	1963	2.25	23.2
1935	2.35	25.9	1964	2.70	32.8
1936	2.74	37.4	1965	2.35	24.8
1937	3.14	56.6	1966	2.72	34.0
1938	3.08	51.5	1967	2.71	33.4
1939	2.41	27.4	1968	2.86	39.9
1940	3.51	77.6	1969	2.02	19.4
1941	1.87	15.9	1970	2.02	19.4
1942	3.25	61.7	1971	3.49	86.9
1943	2.79	39.6	1972	5.60	351.1
1944	3.19	58.3	1973	3.07	53.2
1945	3.72	107.9	1974	2.65	31.4
1946	2.52	30.6	1975	4.08	152.0
1947	2.44	23.1	1976	3.08	54.1
1948	3.02	43.9	1977	2.14	21.3
1949	2.22	19.9	1978	3.63	103.9
1950	1.95	15.9	1979	3.91	132.5
1951	3.19	55.8	1980	2.65	31.4
1952	4.04	150.1	1981	2.50	28.0
1953	3.11	56.6	1982	2.18	22.0
1954	1.91	18.3	1983	3.33	78.7
1955	3.69	107.3	1984	2.65	42.2
1956	2.55	28.6	1985	2.60	40.5
1957	1.90	17.3	1986	1.69	19.3
1958	2.75	34.8	1987	2.42	34.8
1959	2.14	21.4	1988	2.42	34.8
1960	2.54	28.3	1989	4.01	143.5
1961	2.49	27.4			

TABLE 2-4 Descriptive Statistics of Stage and Discharge for Little Patuxent River, Guilford, MD

Parameter	Stage (m)	Discharge (cms)
Average	2.84	54.82
Median	2.71	34.8
Mode	2.65	34.8
Standard deviation	0.71	53.78
Sample variance	0.505	2892.5
Coefficient of variation	0.25	0.98
Range	3.91	335.2
Minimum	1.69	15.9
Maximum	5.6	351.1
Count	57	57

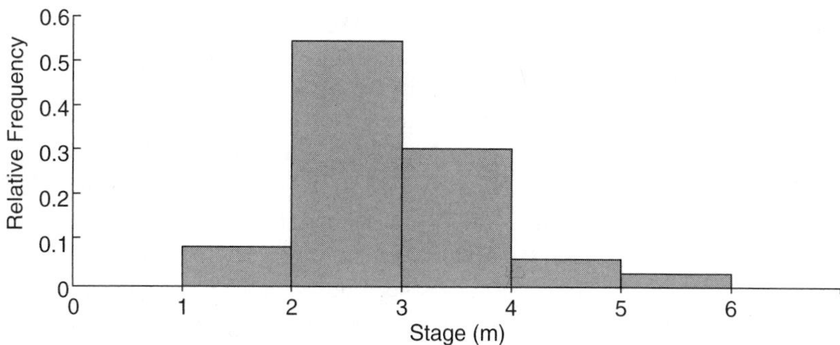

FIGURE 2-19 Relative frequency histogram of the stage of a river.

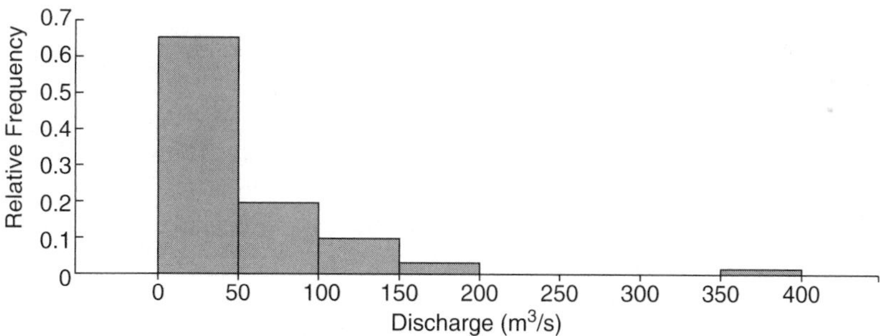

FIGURE 2-20 Relative frequency histogram of the discharge of a river.

cms, respectively. The maximum and minimum values (extreme values) are 351.10 and 15.90 cms, respectively.

By comparing the two histograms in Figures 2-19 and 2-20, it can be observed that the relative dispersion in discharge is larger than that in stage. The coefficient of variation of discharge is 0.98, which is four times larger than the coefficient of variation of stage (0.25). Also, it can be observed that the frequency histogram for stage is not bell shaped, whereas the histogram for discharge is highly skewed and appears as an exponential decay function. The closeness of the average stage (2.84 m) to the median stage (2.71 m) is an indication of a symmetric histogram. On the other hand, the average discharge (54.82 cms) is considerably different from the median discharge (34.8 cms), indicating a lack of symmetry in the discharge measurements.

2.7. ANALYSIS OF SIMULATED DATA

During large floods along major rivers, governmental agencies involved in flood preparedness make estimates of maximum flood stages at downstream locations using observed flood stages at upstream locations. For example, when locations on the Upper Mississippi River are at flood stage, people in Memphis, TN, are interested in the maximum flood stage expected in Memphis. A model might use the records of flood stages at Cincinnati on the Ohio River, St. Louis on the Mississippi River, and Kansas City on the Missouri River to simulate the flood profile that might occur in Memphis. The simulated maximum flood stage can allow the local residents to prepare for the flood. After the flood wave has passed through Memphis, those involved in the simulation would compare the simulated and actual flood waves. If the differences are significant, then the information can be used to adjust the simulation model so that the model can produce more accurate predictions of future floods. The point is that it is common practice both to use simulated data in decision making and to compare simulated and actual measured data. As another example, before Apollo astronauts landed on the moon, they went through many simulations in a flight simulator of the landing. These simulations influenced the moon landing program.

Before simulated data are used in decision making, they should be analyzed in much the same way as the measured data, with descriptive measures computed and graphical analyses made. Additionally, the simulated measures should be compared with the descriptive measures and graphs of the actual data. These analyses can be used as an indication of the reasonableness of the simulation.

The histogram of Figure 2-20 suggests that the data follow an exponential decay, which mathematically is:

$$f_X(x) = \frac{1}{b}e^{-x/b} \tag{2-15}$$

where b is a parameter. The exponential distribution is used frequently to represent engineering data. For data that follow this form, the sample mean can be used to estimate b. To simulate values, the cumulative function $F_X(x)$ for the density function of Equation 2-15 is as follows:

$$F_X(x) = 1 - e^{-x/b} \tag{2-16}$$

where values of $F_X(x)$ vary from 0 to 1 and represent the area under $f_X(x)$ from 0 to x.

Equation 2-16 can be used as a transformation graph to simulate values of the variable x. The cumulative function of the population shown in Figure 2-21 is the graphical representation of the transformation graph. Equation 2-16 can be algebraically rearranged as an expression for x given values of $F_X(x)$ and b as follows:

$$\begin{aligned} e^{-x/b} &= 1 - F_X(X) \\ -x/b &= \ln[1 - F_X(x)] \\ x &= -b\ln[1 - F_X(x)] \end{aligned} \tag{2-17}$$

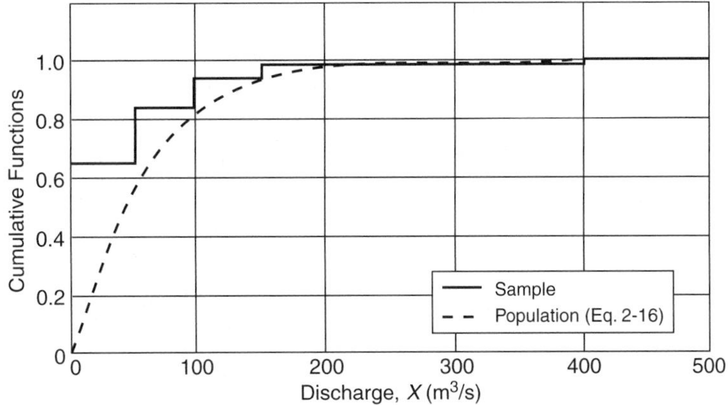

FIGURE 2-21 Cumulative functions for sample and assumed exponential decay model.

Because $F_X(x)$ has a uniform distribution (0 to 1), then $1 - F_X(x)$ also has a uniform distribution (0 to 1). Thus, Equation 2-17 can be rewritten with the definition $F'_X(x) = 1 - F_X(x)$ as follows:

$$x = -b\ln[F'_X(x)] \tag{2-18}$$

Example 2-17: Simulation of River Discharge Rates

The discharge data of Table 2-3, which have the descriptive statistics of Table 2-4, follow an exponential decay (see Figure 2-20). Thus, it would be appropriate to use the exponential function of Equation 2-16 in the simulation of discharge rates. The mean of the sample data of 54.82 cms can be used as an estimate of the parameter b. Thus, the transformation equation, based on Equation 2-18, is:

$$x_i = -54.82\ln(u_i) \tag{2-19}$$

where u_i is the ith uniform variate (i.e., a random number between 0 and 1) and x_i is the ith simulated discharge.

Values of the discharge x can be simulated using random numbers generated by the midsquare method and the function $F_X(x)$. To illustrate this procedure, a seed value of 8765 for random-number generation (r) is used to start the process. A sample of 57 values of x is simulated. The midsquare value is given in column 2 of Table 2-5, with the value transformed to a number between 0 and 1 by dividing it by 10,000 (column 3). The uniform variate is entered into Equation 2-19 to produce the discharge values of column 4 in Table 2-5. Some important descriptive statistics of the simulated discharges are given by:

TABLE 2-5 Simulation of Discharges

r^2	r	Random number (u)	Discharge, x (cms)	Rank
	8765			
76,825,225	8252	0.8252	10.5	47
68,095,504	955	0.0955	128.8	9
00,912,025	9120	0.9120	5.0	53
83,174,400	1744	0.1744	95.7	17
03,041,536	415	0.0415	174.4	5
00,172,225	1722	0.1722	96.4	16
02,965,284	9652	0.9652	1.9	56
93,161,104	1611	0.1611	100.1	15
02,595,321	5953	0.5953	28.4	37
35,438,209	4382	0.4382	45.2	29
19,201,924	2019	0.2019	87.7	18
04,076,361	763	0.0763	141.1	8
00,582,169	5821	0.5821	29.7	35
33,884,041	8840	0.8840	6.8	51
78,145,600	1456	0.1456	105.6	13
02,119,936	1199	0.1199	116.3	10
01,437,601	4376	0.4376	45.3	28
19,149,376	1493	0.1493	104.3	14
02,229,049	2290	0.2290	80.8	19
05,244,100	2441	0.2441	77.3	20
05,958,481	9584	0.9584	2.3	54
91,853,056	8530	0.8530	8.7	49
72,760,900	7609	0.7609	15.0	46
57,896,881	8968	0.8968	6.0	52
80,425,024	4250	0.4250	46.9	27
18,062,500	625	0.0625	152.0	6
00,390,625	3906	0.3906	51.5	26
15,256,836	2568	0.2568	74.5	22

Average	64.81 cms
Median	45.25 cms
Standard deviation	57.57 cms
Minimum	1.4 cms
Maximum	224.7 cms
Range	223.3 cms

These values can be compared with those of the measured data (see Table 2-4). Both the mean and standard deviation of the simulated values are slightly larger than the values of the measured data. Such differences are for small samples and expected and reflect sampling variation, which is discussed in Chapter 8.

Graphical analyses are also useful for examining simulated data and making comparisons with measured data. Figure 2-22 shows the relative frequency histograms for the simulated and measured data. The results suggest that the measured data are more skewed than the simulated data, as evident by the slightly higher

TABLE 2-5 Simulation of Discharges (cont.)

r^2	r	Random number (u)	Discharge, x (cms)	Rank
06,594,624	5946	0.5946	28.5	36
35,354,916	3549	0.3549	56.8	25
12,595,401	5954	0.5954	28.4	38
35,450,116	4501	0.4501	43.8	32
20,259,001	2590	0.2590	74.1	23
06,708,100	7081	0.7081	18.9	43
50,140,561	1405	0.1405	107.6	12
01,974,025	9740	0.9740	1.4	57
94,867,600	8676	0.8676	7.8	50
75,272,976	2729	0.2729	71.2	24
07,447,441	4474	0.4474	44.1	31
20,016,676	166	0.0166	224.7	1
00,027,556	275	0.0275	197.0	2
00,075,625	756	0.0756	141.6	7
00,571,536	5715	0.5715	30.7	34
32,661,225	6612	0.6612	22.7	41
43,718,544	7185	0.7185	18.1	44
51,624,225	6242	0.6242	25.8	39
38,962,564	9625	0.9625	2.1	55
92,640,625	6406	0.6406	24.4	40
41,036,836	368	0.0368	181.0	4
00,135,424	1354	0.1354	109.6	11
01,833,316	8333	0.8333	10.0	48
69,438,889	4388	0.4388	45.2	29
19,254,544	2545	0.2545	75.0	21
06,477,025	4770	0.4770	40.6	33
22,752,900	7529	0.7529	15.6	45
56,685,841	6858	0.6858	20.7	42
47,032,164	321	0.0321	188.5	3

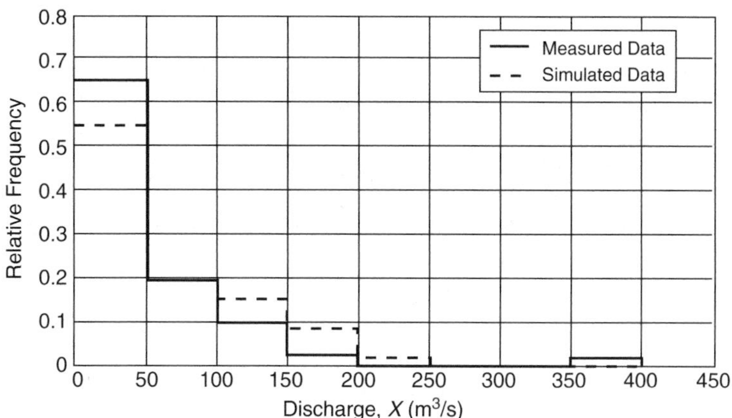

FIGURE 2-22 Comparison of relative frequency histograms for measured and simulated data.

proportion of small flows and the one event between 350 and 400 cms. However, these differences in the ordinates of the histograms are not large when one considers the relatively small sample size (i.e., $n = 57$). The differences would be less if a larger sample had been used.

2.8. PROBLEMS

2-1. For each of the measurement scales, identify five variables that are measured with the scale.

2-2. Using age as a variable of interest, identify a function that would be measured on each of the scales: nominal, ordinal, interval, and ratio.

2-3. Using the copper content of steel as a variable of interest, identify one function that would be measured for each of the four measurement scales.

2-4. For the eight methods of graphical analysis given in Section 2.3, develop a classification system for distinguishing among them. The classification system should center on basic, yet important, characteristics.

2-5. Using the data from Problem 2-12 and the following population totals, construct an area chart that shows the total and the breakdown of the total population in rural, suburban, and central city.

	Year							
	1900	1910	1920	1930	1940	1950	1960	1970
Total (10^6)	76	92	106	123	132	151	179	203

2-6. The following data are the solid waste (millions of tons) produced annually in the U.S.:

Municipal trash and garbage	150
Industrial	350
Mining	1700
Agriculture	2300

Construct a pie chart to present the data. Discuss the merits of presenting the data in a pie chart vs. the tabular summary given.

2-7. The following data give the age distribution of U.S. citizens as a function of age group. Select a method for graphing the results. Interpret the results. Then combine the values into three meaningful groups and provide a graphical analysis of the distribution. How do the two analyses differ in their information content and emphasis?

Age group	<5	5–9	10–14	15–19	20–24	25–29	30–34
Percentage	7.1	7.4	8.1	9.3	9.4	8.6	7.8

Age group	35–39	40–44	45–49	50–54	55–59	60–64	≥65
Percentage	6.2	5.2	4.9	5.2	5.1	4.5	11.2

2-8. A local highway department compiled the following percentages from accident records according to traffic-control method (flashing red light, two-way stop signs, or four-way stop signs) and accident severity (loss of life, major damage, minor damage):

Traffic control method	Loss of life	Major damage	Minor damage	Total
Flashing red light	23	41	36	100
Two-way stop signs	18	39	43	100
Four-way stop signs	12	21	67	100

a. Present the data using three pie charts, one for each of the traffic-control methods.

b. Present all the data in a bar chart in a way that emphasizes differences between accident severity.

c. Present all the data in a bar chart in a way that emphasizes differences between the traffic-control methods.

d. Present the data as a column chart.

e. Discuss the advantages and disadvantages of each of the above graphical methods with respect to these data.

2-9. Compare the graphical analyses of Figures 2-3, 2-5A, and 2-5B, and identify the circumstances under which each would be the most appropriate for making decisions.

2-10. Create a bar chart to display the following data, which provide the estimated remaining strippable resources and reserves of bituminous coal in the United States (billions of short tons):

Alaska	0.9
Rocky Mountains and North Great Plains	1.1
Interior and Gulf regions	32.0
Eastern region	27.0

2-11. Create column charts to present the following estimates of U.S. production of bituminous and lignite coal from surface and underground mines:

Year	Surface mines	Underground mines
1940	43	418
1950	123	393
1960	131	285
1970	264	339

Use one column chart to emphasize the temporal variation of total production. Use a second column chart to emphasize the proportion produced from surface mines. Use a third column chart to emphasize the total production from surface vs. underground mines.

2-12. The following percentages indicate the change in rural, suburban, and central city U.S. populations from 1900 to 1970. Present the data graphically to emphasize the decline in the proportion of the population living in rural areas. Present the data graphically to illustrate any association between the change in central city population and the increase in the proportion living in suburban areas.

Year	Rural	Suburban	Central city
1900	58	16	26
1910	55	17	28
1920	51	18	31
1930	46	20	34
1940	46	22	32
1950	45	24	31
1960	40	30	30
1970	34	38	28

2-13. Plot the data shown in the following table using a column chart:

	Number of constructed bridges by year		
Superstructure type	**1989**	**1990**	**1991**
Steel	5	10	12
Concrete	10	6	7
Prestressed concrete	4	6	5
Total	19	22	24

2-14. Using the bridge data in Problem 2-13, show the number of bridges as a function of year and by bridge type using a line chart, a pie chart, and three-dimensional surface chart.

2-15. The power model $y = ax^b$ is widely used in engineering. (a) Develop line graphs for $a = 1$ and $b = \{0.5, 1.0, 1.5\}$ for $0 \leq x \leq 2$. (b) Develop line graphs for $b = 0.5$ and $a = \{0.5, 1.0, 1.5\}$ for $0 \leq x \leq 2$. (c) For parts (a) and (b), plot the three curves on the same graph. Develop general observations from the two line graphs.

2-16. The following concrete strength data (in ksi) were collected using an ultrasonic nondestructive testing method at different locations of an existing structure: 3.5, 3.2, 3.1, 3.5, 3.6, 3.2, 3.4, 2.9, 4.1, 2.6, 3.3, 3.5, 3.9, 3.8, 3.7, 3.4, 3.6, 3.5, 3.5, 3.7, 3.6, 3.8, 3.2, 3.4, 4.2, 3.6, 3.1, 2.9, 2.5, 3.5, 3.4, 3.2, 3.7, 3.8, 3.4, 3.6, 3.5, 3.2, 3.6, and 3.8. Plot the histogram and relative frequency diagrams for concrete strength. Discuss the characteristics.

2-17. For the data of Problem 2-16, determine the central tendency measures: the average value, median, and mode.

2-18. For the data of Problem 2-16, determine the dispersion measures: the variance, standard deviation, and coefficient of variation.

2-19. Create a box-and-whisker plot of the data in Problem 2-42. Also create a relative frequency histogram of the data. Discuss and compare the information content of the two graphical analyses.

2-20. Piles are commonly used in foundations of civil work structures. Test piles were driven and used to measure pile strength at a selected site. The following strength data (in kips) were collected: 8829, 10236, 5101, 9144, 7790, 9327, 8470, 10570, 10186, 9305, 10746, 9069, 11436, 10044, 10281, 12311, 10639, 10215, 8723, and 8877. Plot the histogram and frequency diagrams for pile strength.

2-21. For the data of Problem 2-20, determine the central tendency measures: the average value, median, and mode.

2-22. For the data of Problem 2-20, determine the dispersion measures: the variance, standard deviation, and coefficient of variation.

2-23. An electronic board will be mass produced using a newly developed manufacturing process. The process was tested by producing 20 products, and the number of defects were counted on each board. The following defect numbers were observed for these boards: 0, 1, 3, 2, 1, 2, 2, 2, 3, 4, 1, 0, 0, 1, 0, 2, 0, 0, 2, and 0. Plot the histogram and frequency diagrams for the number of defects.

2-24. For the data of Problem 2-23, determine the central tendency measures: the average value, median, and mode.

2-25. For the data of Problem 2-23, determine the dispersion measures: the variance, standard deviation, and coefficient of variation.

2-26. Following are scores on a test in the form of ranges, with the corresponding number of students in parentheses: 55–60 (2); 60–65 (4); 65–70 (7); 70–75 (7); 75–80 (1); 80–85 (6); 85–90 (3); 90–95 (3); 95–100 (5). Plot three histograms: (a) cell width of 5; (b) cell width of 10 starting at 50; and (c) cell width of 10 starting at 55. What general observations can be made from a comparison among the three histograms?

2-27. Using the *rand* function of Chapter 1, generate 20 random numbers and 20 random pile strengths similar to the data provided in Problems 2-20 using linear transformation for pile strength in the range [5000, 12,000] kips. Use the data generated to construct a histogram and a frequency diagram and to determine the average value, standard deviation, and coefficient of variation.

2-28. Compare the results from Problems 2-20 and 2-27. Why are the histograms and a frequency diagram from the two problems different?

2-29. Generate 10 uniform random numbers in the range [0,1] using the *rand* function of Chapter 1 that correspond to one value per year for a period of 10 years for a bridge experiencing scour. Assume that the annual scour d_s (in feet) at a bridge is a random variable that can be simulated by:

$$d_s = 0.25 \exp(-4u_i)$$

where u_i is a random number. Using the generated numbers, compute the scour at the bridge for each of the 10 years. Compute the sample mean and standard deviation of the simulated scour for the 10 years.

2-30. Over the last 24 months, the number of accidents per month along one section of highway has been {2,0,1,0,0,4,3,4,2,1,1,2,0,5,1,1,1,0,2,1,3,1,3,0}. Use the *rand* function of Chapter 1 to generate 12 random numbers in the range [0,1].

Use the 12 values to simulate 12 months of accidents based on a linear transformation function between random numbers and number of accidents in the range 0 to 5. Assume that the 24-month sample is a good representation of the distribution of accidents in the future. Compare the means and standard deviations of the sample and the generated number of accidents. Construct a bar chart comparing the distribution of accident probabilities for the sample versus the generated values.

2-31. Figure 2-21 compares the relative frequency histograms for assumed and simulated data. The two do not agree exactly. Propose a means of deciding how large a sample of simulated data would be needed until the frequency histogram for the simulated data agrees with that for the measured data.

2-32. On a monthly basis, accidents were occurring at two intersections (A and B) at similar rates. Traffic control measures were taken at one of the two intersections (A), and then monthly accidents were recorded at both intersections for a 2-year period. Compute appropriate statistics and make graphical analyses to characterize the difference in accident rates (B – A) at the two sites. The accident counts are:

$$A = \{2,0,4,3,0,1,0,4,2,1,2,2,3,0,1,5,4,2,0,2,3,1,6,1\}$$

$$B = \{5,2,8,11,7,8,5,10,6,8,9,4,6,12,7,7,10,11,6,8,13,11,7,9\}$$

2-33. If a small sample of engineering measurements contains one extreme event, compare the use of the average and median values as measures of central tendancy. Illustrate your general point with the following six measurements of a ground water pollutant: {0.6, 1.1, 1.3, 1.9, 3.1, 157.9} mg/l.

2-34. Two sections of a class are given a quiz, which has a total of 10 1-point questions. The distributions of the grades are as follows:

Number correct	10	9	8	7	6	5	4	3	2	1	0
Section A	2	7	6	4	3	0	1	0	0	1	0
Section B	0	1	2	4	7	3	3	1	0	0	0

Compute the mean, median, and mode of the grades for each section. Construct bar graphs of the grades. Discuss the relationship among measures of central tendency and the visual interpretation of the graphs.

2-35. The calculation of the mean in Equation 2-5 uses the formula in Equation 2-4. Develop a formula for computing the mean for the calculations of Equation 2-6.

2-36. Using the data of Problem 2-32, compute the mean, standard deviation, and coefficient of variation for the accident rate at intersection B.

2-37. The winner's shares ($\$ \times 10^3$, not corrected for inflation) of the purse for the Kentucky Derby from 1920 to 1999, by decade, are:

1920 to 1929	30	38	47	54	53	53	50	51	55	54
1930 to 1939	51	49	52	49	28	39	38	52	47	46
1940 to 1949	60	61	64	61	65	65	96	92	83	92
1950 to 1959	93	98	96	90	102	108	123	108	116	120
1960 to 1969	115	121	120	109	114	112	121	120	123	113
1970 to 1979	128	145	140	155	274	210	165	215	187	229
1980 to 1989	251	317	418	426	537	407	609	619	611	574
1990 to 1999	581	656	725	736	629	707	870	700	739	775

Compute the mean, standard deviation, and coefficient of variation for each decade. Create bar charts for the means, standard deviations, and coefficient of variations as a function of decades.

2-38. Using the data of Problem 2-37, construct box-and-whisker plots for the 1920–59 and 1960–99 periods.

2-39. Using all of the data of Problem 2-37, construct histograms with cell widths of (a) 100 and (b) 50. What general observation can be made from a comparison of the two histograms?

2-40. Equation 2-7 provides the definition of the variance. Equation 2-8 is more commonly used to compute a sample variance than Equation 2-7. Derive Equation 2-8 from Equation 2-7.

2-41. If the length of an object (x) is recorded in units of feet and the measurements are transformed to units of meters (y), values of the mean, standard deviation, and variance will change by what ratio? What is the general rule for transforming a variable from one set of units to another set of units for computing its mean, standard deviation, and variance?

2-42. The following data are the maximum daily ozone concentrations for the months indicated. Graph the data in a way that will emphasize the monthly variation in the concentration. Also, graph the data in a way that emphasizes the annual variation. Does the effect of monthly variation make it more difficult to assess the importance of the annual variation? Explain.

Year	February	April	June	August	October	December
1980	61	72	77	83	64	55
1981	63	71	78	87	66	58
1982	64	72	78	86	66	59
1983	68	72	80	89	72	63
1984	74	73	85	94	68	66
1985	73	74	86	92	70	64
1986	67	73	85	90	67	64
1987	66	72	83	86	63	59
1988	62	68	81	87	65	57
1989	59	66	82	86	62	51
1990	56	65	80	86	60	50

2-43. Assume a professor gives, on the average, 20% A's, 35% B's, 30% C's, 10% D's, and 5% F's. Construct a transformation graph that transforms a value from 0 to 1 into a grade. Use the midsquare method (seed = 6348) and the constructed transformation graph to generate grades for a class of 32 students. Construct a pie chart of the simulated grades. Construct a bar chart that shows the proportion of each grade in the simulated class along with the average grades usually given by the professor. Discuss the differences between the simulated and long-term average grades.

2-44. Using a seed of 3597, generate a sample of 25 discharges using the midsquare method and Equation 2-19. Compute the mean and standard deviation and compare them with the values for the measured data (Table 2-4). Graph the frequency histogram of the simulated data and compare it with the histogram for the measured data (Figure 2-20).

2.9. SIMULATION PROJECTS

This section provides additional, or continuation, work items for the problem statements of the four simulation projects introduced in Section 1.5.

2.9.1. Structural Beam Study

For the project described in Section 1.5.1, perform the following graphical analyses:

a. Develop the relative frequency histogram of each variable.
b. Construct scatter plots of the computed stress vs. each of the three predictor variables. Discuss the information provided by each graph.
c. Compute the means, standard deviations, and coefficient of variations for all variables.

2.9.2. Stream Erosion Study

For both the actual and simulated data of the project described in Section 1.5.2, perform the following graphical analyses:

a. Develop the relative frequency histogram of each variable.
b. Construct scatter plots of Y vs. each of the three predictor variables (X). Discuss the information provided by each graph.
c. Compute the means, standard deviations, and coefficient of variations for all variables.

2.9.3. Traffic Estimation Study

For both the actual and simulated data of the project described in Section 1.5.3, perform the following graphical analyses:

a. Develop the relative frequency histogram of each variable.
b. Construct scatter plots of *Y* vs. each of the three predictor variables (*X*). Discuss the information provided by each graph.
c. Compute the means, standard deviations and coefficient of variations for all variables.

2.9.4. Water Evaporation Study

For both the actual and simulated data of the project described in Section 1.5.4, perform the following graphical analyses:

a. Develop the relative frequency histogram of each variable.
b. Construct scatter plots of *Y* vs. each of the three predictor variables (*X*). Discuss the information provided by each graph.
c. Compute the means, standard deviations, and coefficient of variations for all variables.

Chapter 3
Fundamentals of Probability

TABLE OF CONTENTS

3.1. INTRODUCTION

The analysis of an engineering system should begin with acquisition of knowledge about the system. This includes its definition, functions, and operational characteristics. The process of knowledge acquisition defines the available information about the system. Equally important is defining the lack of information or uncertainty about the system. This process is achieved by characterizing the gained information. These two components are vital for the analysis and decision-making processes.

The treatment of uncertainty requires the understanding of its nature and sources, and appropriate mathematical models can then be used to model it. Uncertainties in engineering systems can be attributed mainly to ambiguity and vagueness in defining the variables or parameters of the systems. The ambiguity component is due to (1) physical randomness, (2) statistical uncertainty due to the use of limited information to estimate the characteristics of these parameters, and (3) model uncertainties that are due to simplifying assumptions in analytical and prediction models, simplified methods, and idealized representations of real performances. The vagueness-related uncertainty is common in (1) the definition of certain parameters, such as structural performance (failure or survival), quality, deterioration, skill and experience of construction workers and engineers, environmental impact of projects, or conditions of existing structures; and (2) defining the interrelationships among the parameters of the problems, especially for complex systems. Probability theory can be used to deal with the ambiguity component of uncertainty.

The objective of this chapter is to introduce the basic concepts of probability theory, relevant computational techniques, and engineering and science applications.

3.2. SAMPLE SPACES, SETS, AND EVENTS

Sets constitute a fundamental concept in probabilistic analysis of engineering problems. To perform probabilistic analyses of these problems, the definition of the underlying sets is essential for the establishment of a proper model and obtaining realistic results. The goal of this section is to provide the necessary set foundation for probabilistic analysis.

3.2.1. Sets

Informally, a set can be defined as a collection of elements or components. Capital letters are usually used to denote sets (e.g., A, B, X, and Y), and lower-case letters are commonly used to denote their elements (e.g., a, b, x, and y).

Example 3-1: Sets

The following are examples of sets:

$$A = \{2, 4, 6, 8, 10\} \quad (3\text{-1a})$$

$$B = \{b: b > 0\} \quad (3\text{-1b})$$

where ":" means "such that."

$$C = \{\text{Maryland, Virginia, Washington}\} \quad (3\text{-1c})$$

$$D = \{\text{P, M, 2, 7, U, E}\} \quad (3\text{-1d})$$

$$F = \{1, 3, 5, 7, 11, \ldots\}\text{; the set of odd numbers} \quad (3\text{-1e})$$

In these example sets, each set consists of a collection of elements. In set A, 2 belongs to A, and 12 does not belong to A. Using mathematical notations, this can be expressed as $2 \in A$ and $12 \notin A$.

Sets can be classified as *finite* and *infinite* sets. For example, sets A, C, and D in Example 3-1 are finite sets, and sets B and F are infinite sets. The elements of a set can be either *discrete* or *continuous*. For example, the elements in sets A, C, D, and F are discrete, and the elements in set B are continuous. A set without any elements is called a *null* (or empty) set and is denoted as Ø.

If every element in a set A is also a member of set B, then A is called a *subset* of B, mathematically expressed as $A \subset B$. Mathematically expressed, if A is contained in or equal to B (i.e., $A \subseteq B$), then every a that belongs to A (i.e., $a \in A$) also belongs to B (i.e., $a \in B$). Every set is considered to be a subset of itself. The null set Ø is considered to be a subset of every set.

Example 3-2: Subsets

The following are examples of subsets of the sets defined in Example 3-1:

$$A_1 = \{2, 4\} \text{ is a subset of } A = \{2, 4, 6, 8, 10\} \quad (3\text{-2a})$$

$$B_1 = \{b: 7 < b \leq 200\} \text{ is a subset of } B = \{b: b > 0\} \quad (3\text{-2b})$$

$$F = \{1, 2, 3, 4, 5, 6, 7, \ldots\} \text{ is a subset of } F = \{1, 2, 3, 4, 5, 6, 7, \ldots\} \quad (3\text{-2c})$$

Membership (or characteristic) functions are used to describe sets. Let X be a universe, or a set of x values, and let A be a subset of X. Each element, x, is associated with a membership value to the subset A, $\mu_A(x)$. For any set A, the membership function is given by:

$$\mu_A(x) = \begin{cases} 1 & \text{if } x \in A \\ 0 & \text{if } x \notin A \end{cases} \quad (3\text{-3})$$

The meaning of this membership function is that there are only two possibilities for an element x: either x is a member of A, such that $\mu_A(x) = 1$, or x is not a member of A, such that $\mu_A(x) = 0$. In this case, the set A has sharp (clearly defined) boundaries.

3.2.2. Sample Spaces and Events

In engineering, the set of all possible outcomes of a system (or for an experiment) constitutes the sample space S. A sample space consists of points that correspond to all possible outcomes. Each outcome for the system should constitute a unique element in the sample space. A subset of the sample space is called an *event*. These definitions are the set basis of probabilistic analysis. An event without sample points is an empty set and is called the *impossible event* $\emptyset$. A set that contains all the sample points is called the *certain event* S. The certain event is equal to the sample space.

Example 3-3: Sample Spaces

The following are examples of sample spaces:

A = {number of cars waiting (queuing) for a left turn at a specified traffic light}
B = {number of units produced by an assembly line}
C = {the strength of concrete delivered at a construction site}
D = {the deformation of a structure under extreme load conditions}

Based on the sample spaces A and D, the following events A_1 and D_1 can be defined:

A_1 = {number of cars waiting (queuing) for a left turn at the specified traffic light between 3:30 p.m. and 6:30 p.m. on a work day}
D_1 = {failure of the structure}

3.2.3. Venn–Euler Diagrams

Events and sets can be represented using spaces that are bounded by closed shapes, such as circles. These shapes are called *Venn–Euler* (or simply Venn) diagrams. Belonging, nonbelonging, and overlaps between events and sets can be represented by these diagrams.

Example 3-4: Venn Diagrams

In the Venn diagram shown in Figure 3-1, two events (or sets) A and B that belong to a sample space S are represented. The event C is contained in B (i.e., $C \subset B$), and A is not equal to B (i.e., $A \neq B$). Also, the events A and B have an overlap in the sample space S.

3.2.4. Basic Operations

In this section, basic operations that can be used for sets and events are introduced. These operations are analogous to, for example, addition, subtraction, and multiplication in arithmetic calculations:

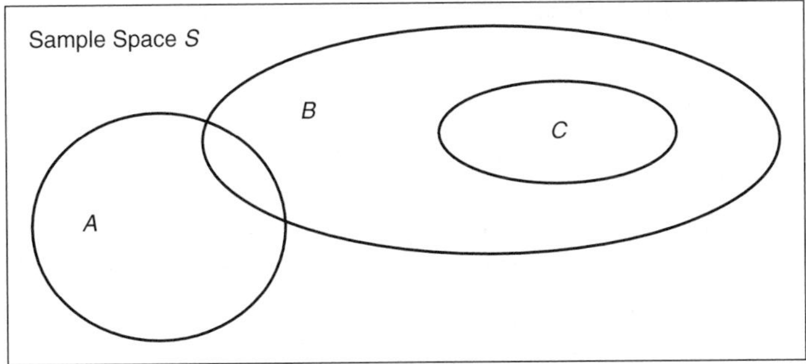

FIGURE 3-1 Venn diagram representation of events.

1. The *union* of events *A* and *B*, which is denoted as $A \cup B$, is the set of all elements that belong to *A* or *B* or both. Two or more events are called *collectively exhaustive* events if the union of these events results in the sample space.
2. The *intersection* of events *A* and *B*, which is denoted as $A \cap B$, is the set of all elements that belong to both *A* and *B*. Two events are termed *mutually exclusive* if the occurrence of one event precludes the occurrence of the other event. The term can also be extended to more than two events.
3. The *difference* of events *A* and *B*, $A - B$, is the set of all elements that belong to *A* but not to *B*.
4. The event that contains all of the elements that do not belong to an event *A* is called the *complement* of *A* and is denoted $\overline{A}$.

Example 3-5: Operations on Sets and Events

The following are example sets:

$$A = \{2, 4, 6, 8, 10\} \qquad (3\text{-}4a)$$

$$B = (1, 2, 3, 4, 5, 6, 7, 8, 9, 10\} \qquad (3\text{-}4b)$$

$$C = \{1, 3, 5, 7, 9, 11, \ldots\}\text{; the set of odd numbers} \qquad (3\text{-}4c)$$

$$F_1 = \{\text{failure of a structure due to an earthquake}\} \qquad (3\text{-}4d)$$

$$F_2 = \{\text{failure of a structure due to strong winds}\} \qquad (3\text{-}4e)$$

$$F_3 = \{\text{failure of a structure due to an extreme overload}\} \qquad (3\text{-}4f)$$

The following operations can be executed for the sets in Equations 3-4a to 3-4f:

$$A \cup B = B = \{1, 2, 3, 4, 5, 6, 7, 8, 9, 10\} \qquad (3\text{-}5a)$$

$$A \cap B = A = \{2, 4, 6, 8, 10\} \qquad (3\text{-}5b)$$

$$\overline{C} = \{2, 4, 6, 8, 10, 12, \ldots\}\text{; the set of even numbers} \qquad (3\text{-}5c)$$

$$F_1 \cup F_2 = \{\text{failure of the structure due to an earthquake or strong wind}\} \qquad (3\text{-}5d)$$

$$\overline{F}_3 = \{\text{nonfailure of the structure due to an extreme overload}\} \qquad (3\text{-}5e)$$

TABLE 3.1 Additional Operational Rules

Rule type	Operations
Identity laws	$A \cup \varnothing = A,\quad A \cap \varnothing = \varnothing,\quad A \cup S = S,\quad A \cap S = A$
Idempotent laws	$A \cup A = A,\quad A \cap A = A$
Complement laws	$A \cup \overline{A} = S,\quad A \cap \overline{A} = \varnothing,\quad \overline{\overline{A}} = A,\quad \overline{S} = \varnothing,\quad \overline{\varnothing} = S$
Commutative laws	$A \cup B = B \cup A,\quad A \cap B = B \cap A$
Associative laws	$(A \cup B) \cup C = A \cup (B \cup C),\quad (A \cap B) \cap C = A \cap (B \cap C)$
Distributive laws	$(A \cup B) \cap C = (A \cap C) \cup (B \cap C)$
	$(A \cap B) \cup C = (A \cup C) \cap (B \cup C)$
de Morgan's law	$\overline{(A \cup B)} = \overline{A} \cap \overline{B},\quad \overline{(E_1 \cup E_2 \cup \ldots \cup E_n)} = \overline{E}_1 \cap \overline{E}_2 \cap \ldots \cap \overline{E}_n$
	$\overline{(A \cap B)} = \overline{A} \cup \overline{B},\quad \overline{(E_1 \cap E_2 \cap \ldots \cap E_n)} = \overline{E}_1 \cup \overline{E}_2 \cup \ldots \cup \overline{E}_n$
Combinations of laws	$\overline{(A \cup (B \cap C))} = \overline{A} \cap \overline{(B \cap C)} = (\overline{A} \cap \overline{B}) \cup (\overline{A} \cap \overline{C})$

The events F_3 and $\overline{F}_3$ can be considered to be mutually exclusive. Table 3-1 shows additional rules based on the above fundamental rules. The validity of these rules can be checked using Venn diagrams.

3.2.5. Cartesian Product

The Cartesian-product set of A and B is the set of all ordered pairs (a,b) such that $a \in A$ and $b \in B$. In mathematical notation, this can be expressed as:

$$A \otimes B = \{(a,b)\text{: } a \in A \text{ and } b \in B\} \tag{3-6}$$

Therefore, the Cartesian product, $\otimes$, of the two sets results in a two-dimensional matrix of all the possible combinations [ordered pairs (a, b)] of the elements of the two sets.

Example 3-6: Production Lines

A precast-concrete factory produces components for floor or roof systems. The manufacturing of these components is performed by three independent production lines. The quality-control department for the factory determines whether the components are defective or nondefective for the lines. Therefore, the following two sets can be defined: (1) the set of production lines L = {line 1, line 2, line 3}, and (2) the set of component quality Q = {defective, nondefective}. Therefore, the Cartesian product of L and Q is given by the following:

	Line 1	Line 2	Line 3
Defective	(Line 1, defective)	(Line 2, defective)	(Line 3, defective)
Nondefective	(Line 1, nondefective)	(Line 2, nondefective)	(Line 3, nondefective)

3.3. MATHEMATICS OF PROBABILITY

The concept of probability has its origin in games of chance. In these games, probabilities are determined based on many repetitions of an experiment and counting the number of outcomes of an event of interest. Then, the probability of the outcome of interest was measured by dividing the number of event occurrences by the total number of repetitions. The objective of this section is to introduce the fundamentals of probability for engineering applications.

3.3.1. Definition of Probability

Upon receiving a graded test from the instructor, a student in a class is interested in three things: (1) the score received, (2) the distribution of the scores in the class or summary characteristics of the distribution, and (3) the grading policy. The scores that the students receive are meaningless by themselves, because they fail to tell the students individually both where they stand with respect to others and the opinion of the instructor about how well they understood the material. For example, if the student received a grade of 50 out of possible 100, he or she may or may not consider this to be good. If the class average was 45, the student (with a score of 50) knows that he or she did better than average. However, by itself, the class average does not indicate how well the student performed. The assessment of the performance depends, in part, on the dispersion of the grades. The student should feel better if the grades ranged from 35 to 53 than if the grades ranged from 10 to 90. In the first case, the score of 50 is closer to the highest score (53), thus the performance is probably better than a student with an average score. Although the closeness of the score with respect to the mean is important, the student is also interested in the grading policy, which indicates the instructor's assessment of how well he or she performed with respect to all the students that the instructor has taught in the past. For example, if the instructor believes that all the test scores were poor in comparison to students of previous years, then the instructor may set the grading policy such that there are no A's or B's; that is, even the student having the highest grade in the class (53) would get a C. Thus, besides the mean and dispersion, the grading policy, which is based on all past students, is important.

Statistics, such as the mean and dispersion, and policy criteria are used extensively in engineering decision making. If the value of 50 were a measurement of a water quality based on a certain pollutant, could we safely conclude that the average level of the pollutant in the river was not different from the level of 45 allowed by the state policy? Similarly, corrosion reduces the thickness of steel beams, but the loss of material due to corrosion is far from uniform over the cross section of a beam. If a 50-mil loss of thickness is measured at one point in a beam and the policy allows for a 45-mil loss, can we conclude that the steel beam is unsafe? Numerous other examples could be presented. The important point made through these examples is that decisions must often be based on sample measurements, and data must be analyzed systematically if the decisions are to be universally accepted.

The test score example is useful for introducing some important terminology. In the instructor's viewpoint, the collection of test scores represents *observations* on the *random variable* "test score." The test scores for the class represent a *sample* of observations; the test scores for all the students who have had the test (or similar test) in the past or will have the test in the future represent the *population*. Thus, the sample is a random collection or subset of the population. The class average and the range of scores are *sample characteristics*; that is, they are values that provide summary information about the sample. The grading policy is an assessment of the population, and characteristics derived from the population are *population characteristics*.

The student uses the available information (i.e., his or her score, the class distribution of scores, and the grading policy) to make inferences about the likelihood of getting a specific final grade at the end of the course. This represents an example of *probabilistic assessment* and *statistical inference*.

In the test score example, a student may be interested in determining the likelihood of his or her final grade. If the deadline for dropping courses is near, the likelihood of receiving a passing grade is of interest. *Probability* is a scale of measurement used to describe the likelihood of an event, where an event is defined as the occurrence of a specified value of the random variable "final grade." It is important to note that there are two random variables: the test score and the final grade. The student must use the value of one random variable to draw an inference about the other. The scale on which probability is measured extends from 0 to 1, inclusive, where a value of 1 indicates a certainty of occurrence of the event and a value of 0 indicates certainty of its nonoccurrence.

Quite often, probability is specified as a percentage; for example, when the weather bureau predicts a 30% chance of rain, experience indicates that under similar meteorological conditions it has rained three out of ten times. In this example, the probability was estimated empirically using the concept of relative frequency expressed as:

$$P(X = x_0) = \frac{n}{N} \tag{3-7}$$

in which n is the number of observations on the random variable X that results in an outcome of interest x_0, and N is the total number of observations of X.

In the case of the student who is trying to estimate the probability of getting a passing grade for the course, the student considers the grades that he or she received and how he or she has performed under similar circumstances in the past. If the student has received grades D and F on the two tests taken to date and has failed three out of four courses in which he or she also had a D and F on the first two tests, the likelihood is high (0.75) that the student will also fail this course. Thus, with a high probability of failure, the student would most likely elect to drop the course. This represents decision making under conditions of uncertainty.

Like students, engineers must make decisions under conditions of uncertainty. For example, engineers who have the responsibility of monitoring water quality in our nation's streams and bodies of water estimate pollution levels using samples collected from the water. The samples are then analyzed in a laboratory and the results are used to make a decision. Most sampling programs involve ten or fewer measurements. Uncertainty arises because of the highly variable nature of pollution; that is, the concentration of a pollutant may vary with time, the degree of turbulence in the water, and the frequency with which wastes are discharged into the water. These sources of variation must be accounted for when the engineer makes a decision about water quality.

Traffic engineers must also make decisions under conditions of uncertainty. For example, intersections are frequently the sites of accidents. The traffic engineer knows that accidents can be reduced by installing stop signs or traffic lights. However, there is a cost associated with installing such hardware. Also, traffic controls can cause delay and inconvenience to those who must travel through the intersections. Thus, the traffic engineer must consider numerous factors in making a decision, including the likelihood and severity of accidents at the intersection and the traffic load in each direction. The frequency and severity of accidents can be assessed using data from accidents that have occurred at that intersection in the past; however, these are data of the past, and there is no assurance that they will accurately reflect accident rates in the future. For example, reduced travel due to an increase in the cost of gasoline may reduce the number of accidents. Data on the traffic volumes originating from each street entering the intersection can be obtained using traffic counters, but these data may not completely characterize the traffic volumes that will take place in the future. For example, if the traffic volume data are collected during the summer, the opening of schools may alter the relative proportion of traffic volumes on each street entering the intersection. Such sources of variation introduce uncertainty into the decision-making process.

It is clear from these examples that the probability of an event can be defined as the relative frequency of its occurrence or the subjective probability of its occurrence. The type of definition depends on the underlying event. For example, in an experiment that can be repeated N times with n occurrences of the underlying event, the relative frequency of occurrence can be considered as the probability of occurrence. In this case, the probability of occurrence is n/N. However, there are many engineering problems that do not involve large numbers of repetitions, and still we are interested in estimating the probability of occurrence of some event. For example, during the service life of an engineering product, the product either fails or does not fail in performing a set of performance criteria. The events of failure and survival are mutually exclusive and collectively exhaustive of the sample space. The probability of failure (or survival) is considered as a subjective probability. An estimate of this probability can be achieved by modeling the underlying system, its uncertainties, and performances. The resulting subjective probability is expected to reflect the status of our knowledge about the system regarding the true likelihood of occurrence of the events of interest. In this section, the mathematics of probability is applicable to both definitions. However, it is important to keep in mind both definitions, so that results are not interpreted beyond the range of their validity.

3.3.2. Axioms of Probability

In general, an axiomatic approach can be used to define probability as a function from sets to real numbers. The domain is the set of all events within the sample space of the problem, and the range consists of the numbers on the real line. For an event A, the notation $P(A)$ means the probability of occurrence of the event A. The function $P(.)$ should satisfy the following properties:

$$0 \leq P(A) \leq 1, \text{ for every event } A \subseteq S \tag{3-8a}$$

$$P(S) = 1 \tag{3-8b}$$

If $A_1, A_2, \ldots, A_n$ are mutually exclusive events on S, then

$$P(A_1 \cup A_2 \cup \ldots \cup A_n) = P(A_1) + (A)_2 + \ldots + P(A_n) \tag{3-8c}$$

The first property states that the probability of any event is inclusively between 0 and 1; therefore, negative probabilities or probabilities larger than one are not allowed. The second property comes from the definition of the sample space. Because the sample space is the set of *all possible outcomes*, one or more of these outcomes must, therefore, occur, resulting in the occurrence of S. If the probability of the sample space does not equal 1, this means that the sample space was incorrectly defined. The third property results from the definition of mutually exclusive events. Computational rules can be developed based on these properties. Example rules are given in the following:

$$P(\varnothing) = 0 \tag{3-9}$$

$$P(A \cup B) = P(A) + P(B) - P(A \cap B) \tag{3-10a}$$

$$\begin{aligned} P(A \cup B \cup C) = {} & P(A) + P(B) + P(C) - P(A \cap B) - P(A \cap C) \\ & - \mathrm{P}(B \cap C) + \mathrm{P}(A \cap B \cap C) \end{aligned} \tag{3-10b}$$

$$P(\bar{A}) = 1 - P(A) \tag{3-11}$$

$$\text{If } A \subseteq B \text{, then } P(A) \leq P(B) \tag{3-12}$$

Example 3-7: Dice Probabilities

Rolling a fair pair of dice produces 36 possible outcomes for the sides of the dice that face up as shown in Figure 3-2A. The sample space contains 36 events. Event A is defined as the sum of the two sides that face up being seven. Six events have a sum of seven. The probability of the occurrence of A can be computed as 6 divided by 36, producing 1/6. The probability of having a sum of 11 as defined by event B is 2/36. Events A and B are mutually exclusive; therefore, the probability of their union is:

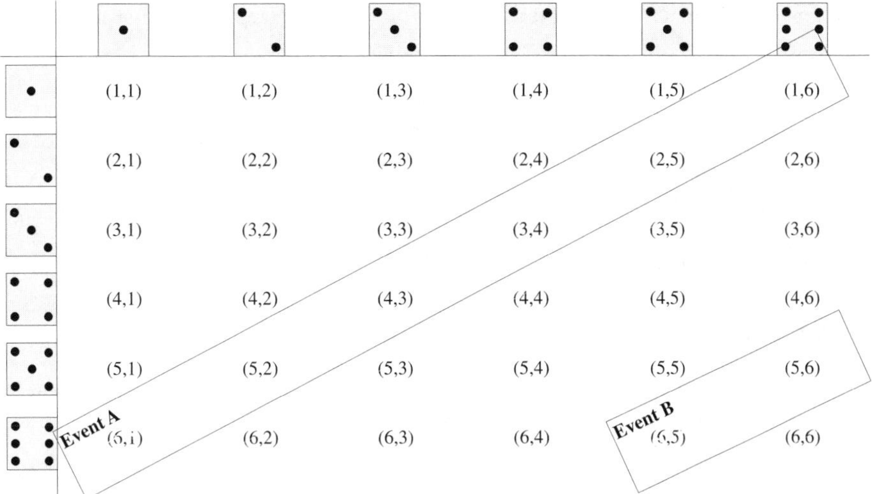

FIGURE 3-2A Sample space for rolling a pair of dice and related events.

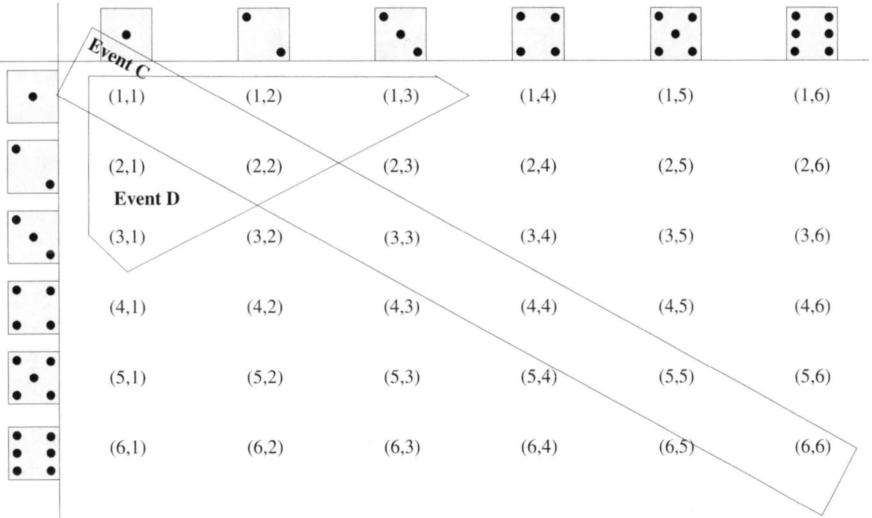

FIGURE 3-2B Sample space for rolling a pair of dice and related events.

$$P(A \cup B) = P(A) + P(B) = \frac{6}{36} + \frac{2}{36} = \frac{8}{36}$$

Figure 3-2B shows two additional events of C and D that are defined as follows:

C = The two sides facing up have the same value.

D = The sum of the sides facing up is less than five.

The probability of C is 1/6, and the probability of D is 1/6. The probability of the union of C and D can be computed using Equation 3-10a as follows:

$$P(C \cup D) = P(C) + P(D) - P(C \cap D) = \frac{6}{36} + \frac{6}{36} - \frac{2}{36} = \frac{10}{36}$$

The probability of the intersection of C and D was computed by counting the common elements in C and D and dividing by the sample space size of 36.

Example 3-8: Probabilities of Events

In the design of structures, the occurrence of damaging winds and earthquakes is of interest. The following events can be defined:

$$E = \{\text{the occurrence of an earthquake in a year}\}$$
$$W = \{\text{the occurrence of damaging winds in a year}\}$$

Assume that the events E and W have the probabilities $P(E)$ and $P(W)$ of 0.1 and 0.2, respectively. Also, the probability of occurrence of both E and W, $P(E \cap W)$, is 0.02. Therefore, the probability of occurrence of an earthquake or strong winds in a year can be determined as the union of E and W using Equation 3-10a as:

$$P(E \cup W) = 0.1 + 0.2 - 0.02 = 0.28 \tag{3-13}$$

The probability of nonoccurrence of an earthquake in a year according to Equation 3-11 is:

$$P(\bar{E}) = 1 - 0.1 = 0.9 \tag{3-14}$$

where $\overline{E}$ is the complement of E (i.e., nonoccurrence of an earthquake).

3.3.3. Counting

In experiments that result in finite sample spaces, the processes of identification, enumeration, and counting are essential for the purpose of determining the probabilities of some outcomes of interest. The identification process results in defining all possible outcomes and their likelihood of occurrence. The identification of equally likely outcomes is needed to determine any probabilities of interest. The order of occurrence of the outcomes can be important in certain engineering applications, requiring its consideration in the counting process.

The enumeration process can be performed in any systematic form that results in all possible outcomes. The multiplication principle can be used for this purpose. Let events $A_1, A_2, \ldots, A_n$ have $n_1, n_2, \ldots, n_n$ elements, respectively. Therefore, the total number of possible outcomes of selecting one element from each of $A_1, A_2, \ldots, A_n$ is the product $n_1, n_2, \ldots, n_n$, where the outcomes represent the ways to select the first element from A_1, the second element from A_2, …, and finally to select the nth element from A_n. In Example 3-7, the sample space size of rolling a pair of dice can be determined using the enumeration counting principle as 6(6) = 36 possible events, as shown in Figure 3-2A.

The permutation of r elements from a set of n elements is the number of arrangements that can be made by selecting r elements out of the n elements. The order of selection counts in determining these arrangements. The permutation $P_{r|n}$ of r out of n (where $r \leq n$) is:

$$P_{r|n} = \frac{n!}{(n-r)!} \tag{3-15}$$

where $n!$ is the factorial of $n = n(n-1)(n-2)\cdots(2)(1)$. It should be noted that $0! = 1$ by convention. Equation 3-15 results from the fact that there are n ways to select the first element, $(n-1)$ ways to select the second element, $(n-2)$ ways to select the third element, and so on to the last element (i.e., the rth element). An assumption is made herein that a selected element is neither replaced nor put back into the available pool for selecting the next element; that is, selections are made without replacement.

The combination of r elements from a set of n elements is the number of arrangements that can be made by selecting r elements out of the n elements without replacement. The order of selection in this case does not count in determining these arrangements. One arrangement differs from another, only if the contents of the arrangements are different. The combination $C_{r|n}$ of r out of n (where $r \leq n$) is:

$$C_{r|n} = \frac{P_{r|n}}{r!} \tag{3-16}$$

Therefore, the combination $C_{r|n}$ can be determined as:

$$C_{r|n} = \frac{n!}{(r!)(n-r)!} \tag{3-17}$$

It is very common to use the notation $\binom{n}{r}$ for the combination $C_{r|n}$. It can be shown that the following identity is valid:

$$\binom{n}{r} = \binom{n}{n-r} \tag{3-18}$$

Example 3-9: Counting Pump Failures

A contractor operates three concrete pumps. A pump is either operational (O) or not operational (N). The following table shows all the possible states of the pumps and their occurrence combinations:

Pump 1	Pump 2	Pump 3	Events
O	O	O	OOO
O	O	N	OON
O	N	O	ONO
O	N	N	ONN
N	O	O	NOO
N	O	N	NON
N	N	O	NNO
N	N	N	NNN

Because each pump can have two states (either O or N), the enumerations of all the possible states results in 2(2)(2) = 8 events.

Example 3-10: Election of Officers and Committee Members

A city council of 10 members has decided to elect three officers (chair, vice chair, and secretary) and a committee of three members. In the case of the officers, any one person cannot hold more than one position of chair, vice chair, or secretary. The numbers of possible cases of three officers without replacement from 10 council members can be computed using the permutation equation, as order counts in this case. The number of permutations according to Equation 3-15 is:

$$P_{3|10} = \frac{10!}{(10-3)!} = 720$$

The number of possible cases of three committee members without replacement from 10 council members can be computed using the combination equation, because order does not count in this case. The number of combinations according to Equation 3-16 is:

$$C_{3|10} = \frac{10!}{3!(10-3)!} = 120$$

The council has 720 unique cases for officers, and 120 unique cases for a committee.

Example 3-11: A Standard 52-Card Deck

A standard deck of cards consists of the four suits of spades, hearts, diamonds, and clubs. Each suit consists of 13 cards as follows: the ace, 2, 3, 4, 5, 6, 7, 8, 9, 10, jack, queen, and king. What is the probability of getting 5 spade cards by drawing 5 cards from this deck without replacement? Because the order of cards in a draw does not count in defining the event of interest, the probability can be computed by counting. The sample space has the following count:

$$C_{5|52} = \frac{52!}{5!(52-5)!} = 2{,}598{,}960$$

The event of 5 spade cards out of 13 possible spade cards has the following count of cases:

$$C_{5|13} = \frac{13!}{5!(13-5)!} = 1287$$

These two counts can be used to compute the probability of getting 5 spade cards by drawing 5 cards from this deck without replacement as follows:

$$\frac{C_{5|13}}{C_{5|52}} = \frac{1287}{2{,}598{,}960} \approx 0.0005$$

What is the probability of getting 2 kings and 3 queens in drawing 5 cards? This probability can be computed as follows:

$$\frac{(C_{3|4})(C_{2|4})}{C_{5|52}} = \frac{24}{2{,}598{,}960} \approx 0.00000923$$

Example 3-12: Counting for Bridge Failure

Consider a simple case of a bridge that is supported by three cables. The failure of interest is the failure of only two cables out of the three cables because it results in failure of the bridge. What is the number of combinations of $r = 2$ out of $n = 3$ that can result in bridge failure?

$$C_{r|n} = \frac{n!}{r!(n-r)!} = \frac{3!}{2!(3-2)!} = \frac{3(2)(1)}{(2)(1)(1)} = 3 \qquad \text{(3-19a)}$$

This number of combinations can be established by enumeration. The following events can be defined:

$$C_i = \text{failure of cable } i, \text{ where } i = 1, 2, \text{ and } 3 \tag{3-19b}$$

Therefore, the following events result in bridge failure:

$$C_1 \cap C_2 \tag{3-20a}$$

$$C_1 \cap C_3 \tag{3-20b}$$

$$C_2 \cap C_3 \tag{3-20c}$$

The number of events is three, which is the same as the result from Equation 3-19a. In Equation 3-21, the order of occurrence of cable failure is not a factor in determining the number of combinations. However, if we assume the order of failure is a factor, then the possible events become:

$$C_1 \cap C_2 \tag{3-21a}$$

$$C_1 \cap C_3 \tag{3-21b}$$

$$C_2 \cap C_3 \tag{3-21c}$$

$$C_2 \cap C_1 \tag{3-21d}$$

$$C_3 \cap C_1 \tag{3-21e}$$

$$C_3 \cap C_2 \tag{3-21f}$$

In this case, the number of combinations is six. The same result can be obtained by computing the number of permutations, where order counts, using Equation 3-15 as:

$$P_{r|n} = \frac{n!}{(n-r)!} = \frac{3!}{(3-2)!} = 6 \tag{3-22}$$

Now, if we assume that the bridge is supported by $n = 20$ cables, and the failure of $r = 8$ cables results in the failure of the bridge, what is the number of combinations of $r = 8$ out of $n = 20$ that can result in bridge failure? In this case, the number of combinations is:

$$C_{r|n} = \frac{n!}{r!(n-r)!} = \frac{20!}{8!12!} = \frac{(20)(19)(18)\ldots(13)}{(8)(7)(6)\ldots(1)} = 125{,}970 \tag{3-23}$$

For a real bridge, failure can result from the failure of at least $r = 8$ out of $n = 20$. The number of combinations in this case is:

$$\sum_{r=8}^{20} C_{r|n} = \frac{20!}{(8!)(12!)} + \frac{20!}{(9!)(11!)} + \frac{20!}{(10!)(10!)} + \cdots + \frac{20!}{(20!)(0!)} \tag{3-24}$$

3.3.4. Conditional Probability

The probabilities previously discussed are based on and relate to the sample space S. However, it is common in many engineering problems to have interest in the probabilities of occurrence of events that are conditioned on the occurrence of a subset of the sample space. This introduces the concept of conditional probability. For example, the probability of A given that B has occurred, denoted as $P(A|B)$, means the occurrence probability of a sample point that belongs to A given that we know it belongs to B. The conditional probability can be computed as follows:

$$P(A|B) = \frac{P(A \cap B)}{P(B)} \quad \text{if } P(B) \neq 0 \tag{3-25}$$

Clearly, the underlying sample space for the conditional probability is reduced to the conditional event B. The conditional probability satisfies all the properties of probabilities. The following properties can be developed for conditional probabilities:

1. The complement of an event:

$$P(\bar{A}|B) = 1 - P(A|B) \tag{3-26}$$

2. The multiplication rule for two events A and B:

$$P(A \cap B) = P(A|B)P(B) \quad \text{if } P(B) \neq 0 \tag{3-27a}$$

$$P(A \cap B) = P(B|A)P(A) \quad \text{if } P(A) \neq 0 \tag{3-27b}$$

3. The multiplication rule for three events A, B, and C:

$$P(A \cap B \cap C) = P(A|(B \cap C))P(B|C)P(C) = P((A \cap B)|C)P(C) \quad \text{if } P(C) \neq 0 \text{ and } P(B \cap C) \neq 0 \tag{3-28}$$

4. For mutually exclusive events A and B:

$$P(A|B) = 0 \tag{3-29a}$$

$$P(B|A) = 0 \tag{3-29b}$$

5. For statistically independent events A and B:

$$P(A|B) = P(A) \tag{3-30a}$$

$$P(B|A) = P(B) \tag{3-30b}$$

$$P(A \cap B) = P(A)P(B) \tag{3-30c}$$

A and $\bar{B}$ are independent events (3-30d)

$\bar{A}$ and B are independent events (3-30e)

$\bar{A}$ and $\bar{B}$ are independent events (3-30f)

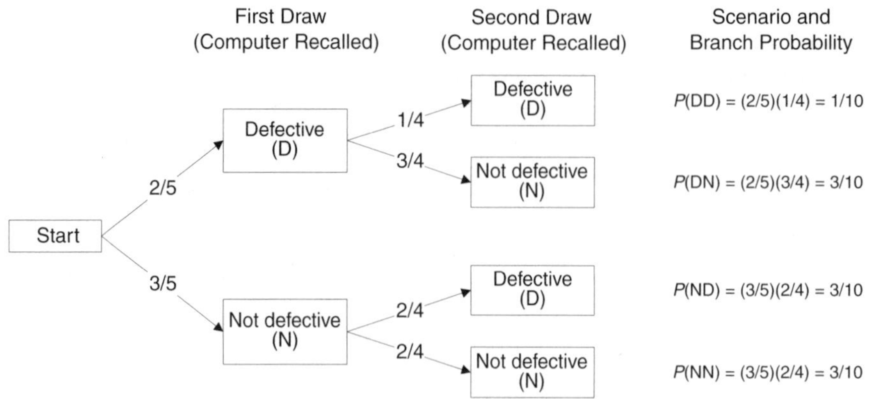

FIGURE 3-3 Probability tree for computer recall.

Example 3-13: Probability Trees To Model Conditional Events Illustrated Using Product Recall

After shipping five computers to users, a computer manufacturer realized that two out of the five computers were not configured properly (i.e., were defective), without knowing specifically which ones. The manufacturer allocated resources to recall only two computers in succession out of the five for examination. What is the probability that the second recalled computer is a defective computer?

Probability trees can be used to construct the sample space and compute the underlying probabilities based on conditional events. Figure 3-3 shows the probability tree for this example of computer recall. The tree shows four possible scenarios or branches with their corresponding occurrence probabilities. Now, the question can be answered by summing up the probabilities of all the branches that meet the question statement "second recalled computer is a defective computer," producing $(1/10) + (3/10) = 4/10$.

Example 3-14: Conditional Probabilities for Production Lines

Consider the three manufacturing lines of Example 3-6. Assume that 20, 30, and 50% of the components are manufactured by lines 1, 2, and 3, respectively. The quality assurance department of the producing factory determined that the probability of having defective products from lines 1, 2, and 3 are 0.1, 0.1, and 0.2, respectively. The following events can be defined in this case:

$$L_1 = \text{Component produced by line 1} \quad (3\text{-}31a)$$

$$L_2 = \text{Component produced by line 2} \quad (3\text{-}31b)$$

$$L_3 = \text{Component produced by line 3} \quad (3\text{-}31c)$$

$$D = \text{Defective component} \quad (3\text{-}31d)$$

Therefore, the following probabilities are given:

$$P(D|L_1) = 0.1 \tag{3-32a}$$

$$P(D|L_2) = 0.1 \tag{3-32b}$$

$$P(D|L_3) = 0.2 \tag{3-32c}$$

Because these events are not independent, the joint probabilities can be determined as follows:

$$P(D \cap L_1) = P(D|L_1)P(L_1) = 0.1(0.2) = 0.02 \tag{3-33a}$$

$$P(D \cap L_2) = P(D|L_2)P(L_2) = 0.1(0.3) = 0.03 \tag{3-33b}$$

$$P(D \cap L_3) = P(D|L_3)P(L_3) = 0.2(0.5) = 0.1 \tag{3-33c}$$

This example illustrates the importance of using conditional probabilities in determining the probability of the intersection of events.

3.3.5. Partitions, Total Probability, and Bayes' Theorem

A set of disjoint (i.e., mutually exclusive) events $A_1, A_2, \ldots, A_n$ form a partition of a sample space if $A_1 \cup A_2 \cup \ldots \cup A_n = S$. An example partition is shown in Figure 3-4.

If $A_1, A_2, \ldots, A_n$ represents a partition of a sample space S, and $E \subset S$ represents an arbitrary event as shown in Figure 3-5, the theorem of total probability states that:

$$P(E) = P(A_1)P(E|A_1) + P(A_2)P(E|A_2) + \cdots + P(A_n)P(E|A_n) \tag{3-34}$$

This theorem is very important in computing the probability of an event E, especially in practical cases where the probability cannot be computed directly, but the probabilities of the partitioning events and the conditional probabilities can be computed.

Bayes' theorem is based on the same conditions of partitioning and events as the theorem of total probability and is very useful in computing the reverse probability of the type $P(A_i|E)$, for $i = 1, 2, \ldots, n$. The reverse probability can be computed as follows:

$$P(A_i|E) = \frac{P(A_i)P(E|A_i)}{P(A_1)P(E|A_1) + P(A_2)P(E|A_2) + \cdots + P(A_n)P(E|A_n)} \tag{3-35}$$

The denominator of this equation is $P(E)$, which is based on the theorem of total probability.

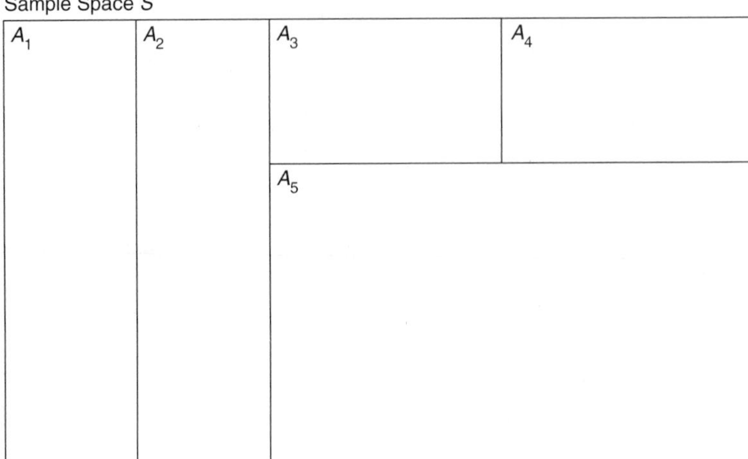

FIGURE 3-4 Partitioned sample space.

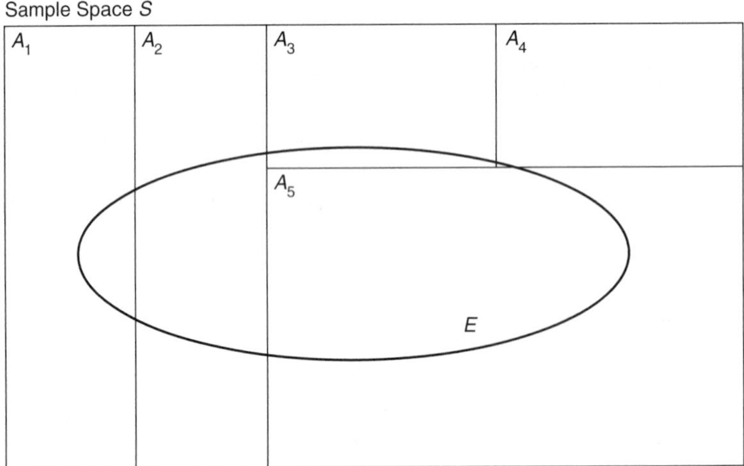

FIGURE 3-5 Theorem of total probability.

Example 3-15: Defective Products in Production Lines

The three manufacturing lines discussed in Example 3-14 are used to illustrate the total probability theorem. The theorem can be used to determine the probability of a defective component as follows:

$$\begin{aligned} P(D) &= P(D|L_1)P(L_1) + P(D|L_2)P(L_2) + P(D|L_3)P(L_3) \\ &= 0.1(0.2) + 0.1(0.3) + 0.2(0.5) = 0.02 + 0.03 + 0.1 \\ &= 0.15 \end{aligned} \tag{3-36}$$

This probability includes the three lines; therefore, on the average, 15% of the components produced by the factory are defective.

3.4. RANDOM VARIABLES AND THEIR PROBABILITY DISTRIBUTIONS

In some engineering applications, the outcomes (say, E_1, E_2, ..., E_n) that constitute a sample space S take numerical real values. In other applications, the outcomes can be nonnumeric or combinations of both. It is convenient and sometimes necessary to express all outcomes using numerical values on the real line R. The functions that establish such a transformation to the real line are called *random variables*. Formally, a random variable is defined as a function that assigns a real value to every possible outcome for an engineering system. This mapping can be one-to-one or one-to-many. Based on this definition, the properties of the underlying outcomes (e.g., intersection, union, and complement) are retained in the form of, for example, overlapping ranges of real values, combination of real ranges, and values outside these ranges.

Random variables are commonly classified into two types: *discrete* and *continuous* random variables. A discrete random variable may only take on distinct, usually integer, values; for example, the outcome of a roll of a die may take on only the integer values from 1 to 6 and is, therefore, a discrete random variable. The number of floods per year at a point on a river can take on only integer values, so it is also a discrete random variable. A continuous random variable takes values within a continuum of values. For example, the average of all scores on a test having a maximum possible score of 100 may take on any value, including nonintegers, between 0 and 100; thus, the class average would be a continuous random variable. A distinction is made between these two types of random variables because the computations of probabilities are different for the two types.

3.4.1. Probability of Discrete Random Variables

The probability of a discrete random variable is given by the *probability mass function*, which specifies the probability that the discrete random variable X equals some value x_i and is denoted by:

$$P_X(x_i) = P(X = x_i) \tag{3-37}$$

A capital X is used for the random variable, whereas an x_i is used for the ith value of the random variable. The probability mass function must satisfy the axioms of probability. Therefore, the probability of an event x_i must be less than or equal to one and greater than or equal to zero; that is,

$$0 \le P_X(x_i) \le 1 \tag{3-38}$$

This property is valid for all possible values of the random variable X. Additionally, the sum of all possible probabilities must be equal to one; that is,

$$\sum_{i=1}^{N} P_X(x_i) = 1 \tag{3-39}$$

in which N is the total number of possible outcomes; for the case of the roll of a die, N equals 6.

It is often useful to present the likelihood of an outcome using the *cumulative mass function*, $F_X(x_i)$, which is given by:

$$F_X(x_i) = P(X \le x_i) = \sum_{j=1}^{i} P_X(x_j) \tag{3-40}$$

The cumulative mass function is used to indicate the probability that the random variable X is less than or equal to x_i. It is inherent in the definition (Equation 3-40) that the cumulative probability is defined as zero for all the values less than the smallest x_i and one for all values greater than the largest value.

Example 3-16: Roll of a Die

To illustrate the concepts introduced for discrete random variables, assume that the random variable is the outcome of the roll of a die. The *sample space* consists of the collection of all possible values of the random variable (i.e., x = 1, 2, 3, 4, 5, and 6). The probability of each element of the sample space is given by:

$$P_X(x) = \frac{1}{6} \quad \text{for } x = 1, 2, 3, 4, 5, \text{ and } 6 \tag{3-41}$$

The probability mass function is shown in Figure 3-6A. The cumulative mass function can be evaluated using Equation 3-40. The results are shown in Figure 3-6B and are given by:

$$F_X(x) = P(X \le x) = \frac{x}{6} \quad \text{for } x = 1, 2, 3, 4, 5, \text{ and } 6 \tag{3-42}$$

Example 3-17: Pair of Dice

In this example, a random variable is defined as the sum of the dots showing when a pair of dice is rolled. What is the cumulative mass function? What is the probability of getting an even value? What is the probability that for a single roll of the two dice, the sum of the points is equal to 11? What is the probability of obtaining a value of 6 or less?

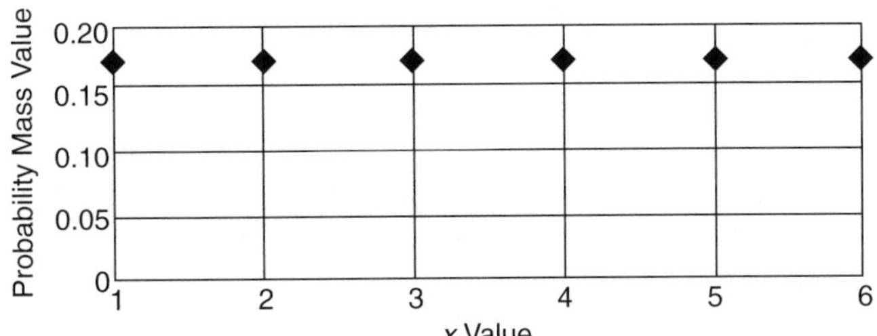

FIGURE 3-6A Probability mass function for the roll of a die.

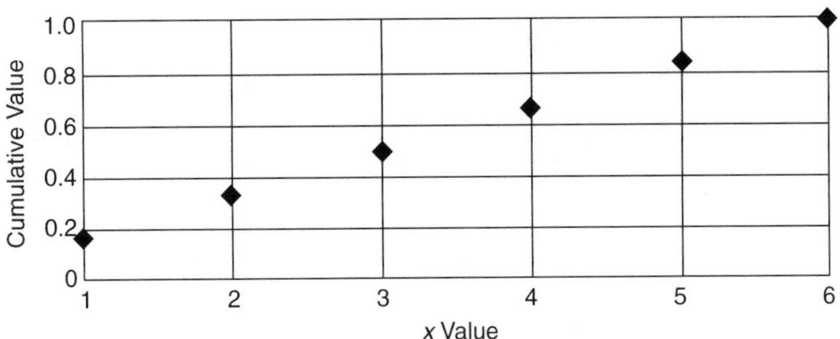

FIGURE 3-6B Cumulative mass function for the roll of a die.

The sample space consists of the integer values from 2 to 12. If one identifies all the 36 possible outcomes, the relative frequency concept of Equation 3-7 yields the following mass function:

$$P_X(x) = \left\{\frac{1}{36}, \frac{1}{18}, \frac{1}{12}, \frac{1}{9}, \frac{5}{36}, \frac{1}{6}, \frac{5}{36}, \frac{1}{9}, \frac{1}{12}, \frac{1}{18}, \frac{1}{36}\right\} \tag{3-43}$$

It is important to note that the sum of the probabilities equals 1; otherwise, $P_X(x)$ would not be a legitimate mass function. The cumulative mass function is:

$$F_X(x) = \left\{\frac{1}{36}, \frac{1}{12}, \frac{1}{6}, \frac{5}{18}, \frac{15}{36}, \frac{7}{12}, \frac{13}{18}, \frac{5}{6}, \frac{11}{12}, \frac{35}{36}, 1\right\} \tag{3-44}$$

The probability of the random variable X taking an even value is the sum of the probabilities for the even values of X. Thus,

$$P(x \text{ is even}) = \frac{1}{36} + \frac{1}{12} + \frac{5}{36} + \frac{5}{36} + \frac{1}{12} + \frac{1}{36} = \frac{1}{2} \tag{3-45}$$

Also,

$$P(x = 11) = \frac{1}{18} \tag{3-46}$$

and using the probability mass function, we have:

$$P(x \le 6) = \frac{1}{36} + \frac{1}{18} + \frac{1}{12} + \frac{1}{9} + \frac{5}{36} = \frac{15}{36} = 0.4167 \tag{3-47}$$

The same result can be obtained directly from the cumulative mass function.

Equations 3-43 and 3-44 and the values in Example 3-17 apply to the populations that underlie the random variables. They do not give the values obtained from a sample of data. For small samples, probabilities computed using sample results may poorly reflect the true probabilities. For example, if a die is rolled five times and produces the values 3, 2, 5, 1, 2, then the sample mass function would be as shown in Figure 3-7. The sample probability for $x = 2$ is over twice as high as the population probability (1/6). Because $x = 4$ and $x = 6$ did not occur, their sample probabilities of 0 are also poor reflections of the true probabilities. This illustrates the difficulty of using information based on small samples to draw inferences about population probabilities. Larger samples usually provide better estimates of population probabilities than those provided by small samples. Consider the following sample of rolling a single die 18 times: {3, 1, 4, 3, 6, 2, 4, 5, 5, 2, 6, 3, 1, 3, 5, 2, 6, 5}. The sample includes two 1's, three 2's, four 3's, two 4's, four 5's, and three 6's, giving sample probabilities of 0.111, 0.167, 0.222, 0.111, 0.222, and 0.167, respectively. Thus, the larger sample of 18 rolls does not contain the extreme probabilities that were observed in the sample of 5. Note in Figure 3-7 that the probabilities from the sample of 18 are generally closer to the true value of 0.167 than are the sample probabilities based on five values.

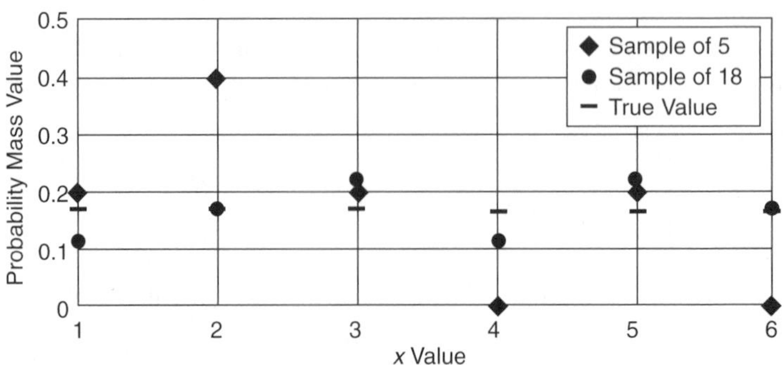

FIGURE 3-7 Probability mass function for two samples.

TABLE 3-2 Computation of Sample Probability Function for Hurricanes in Example 3-18

Number of hurricanes, X	Number of occurrences	Probability mass value	Cumulative mass value
0	3	0.2	0.2
1	3	0.2	0.4
2	3	0.2	0.6
3	2	0.13333	0.73333
4	2	0.13333	0.86667
5	1	0.06667	0.93333
6	1	0.06667	1.
7	0	0.	1.
8	0	0.	1.
⋮	0	0.	1.
	0	0.	1.
∞	0	0.	1.

Example 3-18: Hurricanes

The numbers of hurricanes in the U.S. from 1970/1971 through 1984/1985 are as follows:

1970–71	5	1975–76	3	1980–81	2
1971–72	4	1976–77	1	1981–82	1
1972–73	0	1977–78	3	1982–83	2
1973–74	0	1978–79	1	1983–84	4
1974–75	2	1979–80	6	1984–85	0

The sample probability mass function and the cumulative mass function can be tabulated as shown in Table 3-2. The probability mass values in the table were computed by dividing the number of occurrence by 15, which is the total number of hurricane seasons in the period 1970 to 1985. The cumulative mass value was computed using Equation 3-40. The results of the calculations are also shown in Figures 3-8A and 3-8B.

In spite of the small sample, the mass function shows a consistent trend, with the probability decreasing as the value of the random variable x; that is, the number of hurricanes per year increases. It appears that hurricanes do not have the same mass function as the roll of a die. Besides the decreasing trend, the mass function of Table 3-2 does not allow for values greater than six. Even though there was no year in which more than six hurricanes occurred in the 15-year period, the possibility that it could occur exists, so a mass function that allows for more than six occurrences of X should be selected for the population.

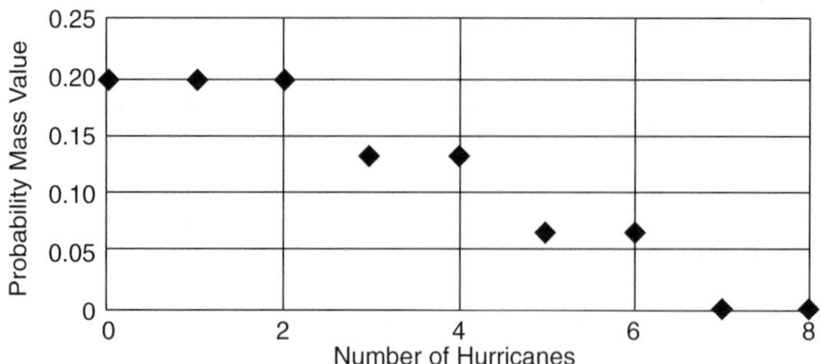

FIGURE 3-8A Sample probability mass function for hurricanes in the United States.

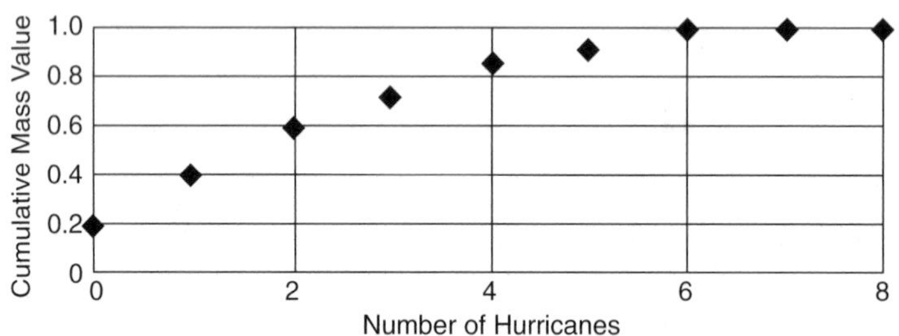

FIGURE 3-8B Sample cumulative mass function for hurricanes in the United States.

3.4.2. Probability for Continuous Random Variables

A *probability density function* (pdf) defines the probability of occurrence for a continuous random variable. Specifically, the probability that the random variable X lies within the interval form x_1 to x_2 is given by:

$$P(x_1 \le X \le x_2) = \int_{x_1}^{x_2} f_X(x)dx \tag{3-48}$$

in which $f_X(x)$ is the probability density function. If the interval is made infinitesimally small, x_1 approaches x_2 and $P(x_1 \le X \le x_2)$ approaches zero. This illustrates a property that distinguishes discrete random variables from continuous random variables. Specifically, the probability that a continuous random variable takes on a specific value equals zero; that is, probabilities for continuous random variables must be defined over an interval.

It is important to note that the integral of the pdf from $-\infty$ to $+\infty$ equals 1; that is,

$$(-\infty < X < +\infty) = \int_{-\infty}^{+\infty} f_X(x)dx = 1 \tag{3-49}$$

Also, because of Equation 3-49, the following holds:

$$P(X \geq x_0) = \int_{x_0}^{+\infty} f_X(x)dx = 1 - P(X < x_0) \tag{3-50}$$

The *cumulative distribution function* (cdf) of a continuous random variable is defined by:

$$F_X(x_0) = P(X \leq x_0) = \int_{-\infty}^{x_0} f_X(x)dx \tag{3-51a}$$

The cdf is a nondecreasing function in that $P(X \leq x_1) \leq P(X \leq x_2)$, where $x_1 \leq x_2$. The cdf equals 0 at $-\infty$ and 1 at $+\infty$. The relationship between $f_X(x)$ and $F_X(x)$ can also be expressed as:

$$f_X(x) = \frac{dF_X(x)}{dx} \tag{3-51b}$$

Example 3-19: Continuous Random Variables

The continuous random variable X has the following probability density function:

$$f_X(x) = \begin{cases} kx & \text{for } 0 \leq \text{x} \leq 2 \\ 0 & \text{otherwise} \end{cases} \tag{3-52}$$

where k is a constant. Find the value of k that is necessary for $f_X(x)$ to be a legitimate probability density function. Plot both the density and cumulative functions. What is the probability that X equals 1? What is the probability that X takes on a value less than 0.5? What is the probability that X is greater than 1.0 and less than 1.5?

For $f_X(x)$ to be a legitimate pdf, it must satisfy the following constraint:

$$P(-\infty < X < +\infty) = \int_{-\infty}^{+\infty} f_X(x)dx = 1 \tag{3-53}$$

Therefore,

$$1 = \int_0^2 kxdx = k\int_0^2 xdx = \left.\frac{kx^2}{2}\right|_0^2 \tag{3-54a}$$

and

$$k = 0.5 \tag{3-54b}$$

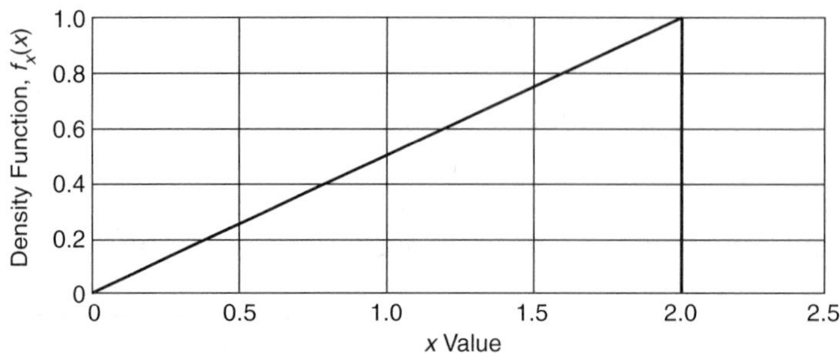

FIGURE 3-9A Probability density function for a continuous random variable.

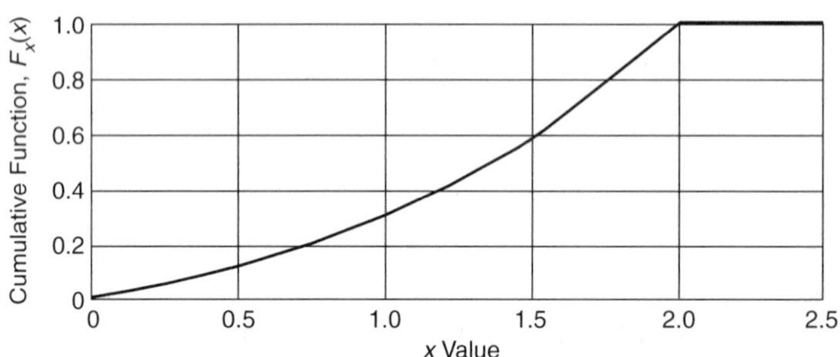

FIGURE 3-9B Cumulative distribution function for a continuous random variable.

The cumulative distribution function is given by:

$$F_X(x_0) = \int_0^{x_0} 0.5x\,dx = \left.\frac{x^2}{4}\right|_0^{x_0} = \frac{x_0^2}{4} \tag{3-55}$$

The probability density and cumulative functions appear graphically in Figures 3-9A and 3-9B, respectively. Because probabilities of continuous random variables are defined for regions rather than point values, $P(X = 1) = 0$. The $P(X < 0.5)$ can be determined from the cumulative function; specifically,

$$P(X < 0.5) = \left.\frac{x_0^2}{4}\right|_{x_0=0}^{x_0=0.5} = \frac{1}{16} \tag{3-56}$$

Similarly, the cumulative function can be used to find the probability for the following region:

$$\begin{aligned} P(1.0 < X < 1.5) &= P(X < 1.5) - P(X < 1.0) \\ &= \left.\frac{x_0^2}{4}\right|_{x_0=0}^{x_0=1.5} - \left.\frac{x_0^2}{4}\right|_{x_0=0}^{x_0=1.0} \\ &= \frac{9}{16} - \frac{1}{4} = \frac{5}{16} \end{aligned} \tag{3-57}$$

Example 3-20: Settlement of Footings

The settlement of a footing (*S*) depends on the average bearing pressure (*B*), a dimension of the footing (*D*), and the soil modulus of compressibility (*M*). The settlement is assumed to be a random variable that follows the following density function:

$$f_S(s) = \lambda exp(-\lambda s) \tag{3-58}$$

where λ is a constant depending on the three characteristics *B*, *D*, and *M*. Integration of the density function yields the following cumulative function:

$$F_S(s) = 1 - exp(-\lambda s) \tag{3-59}$$

If the value of λ is 0.5 for a specified set of conditions (*B*, *D*, and *M*) and the settlement *S* is in inches, then the probability that the settlement *S* is less than s_0 inches is:

$$F_S(s_0) = P(S < s_0) = 1 - exp(-0.5s_0) \tag{3-60}$$

For example, the probability of 1 in. or less of settlement is:

$$F_S(1.0) = P(S < 1.0) = 1 - exp[-0.5(1.0)] = 0.393 \tag{3-61}$$

If damage occurs to the structure where the settlement exceeds 4 in., then the probability of damage is:

$$P(S > 4.0) = 1 - F_S(4.0) = 1 - \{1 - exp[-0.5(4.0)]\} = 0.135 \tag{3-62}$$

Therefore, for 1000 footings at the specified conditions, damage can be expected with 135 footings.

3.5. MOMENTS

Whether summarizing a data set or attempting to find the population, one must characterize the sample. The moments are useful descriptors of data; for example, the mean, which is a moment, is an important characteristic of a set of test scores. A moment can be referenced to any point on the measurement axis; however, the origin (i.e., zero point) and the mean are the most common reference points.

Although most data analyses use only two moments, it is important for some probabilistic and statistical studies to examine three moments:

1. The *mean*, which is the first moment about the origin
2. The *variance*, which is the second moment about the mean
3. The *skewness*, which is the third moment about the mean

Two characteristics, the mean and variance, were briefly introduced in Chapter 2.

In this section, equations and computational procedures for these moments are introduced. These moments are analogous to the area moments used to compute quantities such as the centroidal distance, the first static moment, and the moment of inertia. For a continuous random variable X, the kth moment about the origin is given by:

$$M'_k = \int_{-\infty}^{+\infty} x^k f_X(x)dx \tag{3-63}$$

in which X is the random variable, and $f_X(x)$ is its density function. The corresponding equation for a discrete random variable is:

$$M'_k = \sum_{i=1}^{n} x_i^k P_X(x_i) \tag{3-64}$$

in which n is the number of elements in the underlying sample space of X, and $P_X(x)$ is the probability mass function. The first moment about the origin (i.e., $k = 1$ in Equations 3-63 and 3-64) is called the *mean* of X and is denoted μ.

For a continuous random variable, the kth moment about the mean (μ) is:

$$M_k = \int_{-\infty}^{+\infty} (x-\mu)^k f_X(x)dx \tag{3-65}$$

in which μ is the first moment about the origin (i.e., the mean). The corresponding equation for a discrete random variable is:

$$M_k = \sum_{i=1}^{n} (x_i-\mu)^k P_X(x_i) \tag{3-66}$$

The above moments are considered as a special case of mathematical expectation. The mathematical expectation of an arbitrary function $g(x)$, which is a function of the random variable X, is defined as:

$$E[g(x)] = \int_{-\infty}^{+\infty} g(x) f_X(x)dx \tag{3-67}$$

The corresponding equation for a discrete random variable is:

$$E[g(x)] = \sum_{i=1}^{n} g(x_i) P_X(x_i) \tag{3-68}$$

3.5.1. Mean

The mean value, which was introduced in Chapter 2, can be formally defined as the first moment measured about the origin; it is also the average of all observations on a random variable. It is important to note that the population mean is most often indicated as μ, while the sample mean is denoted by $\overline{X}$. For a continuous random variable, the mean (μ) is computed as:

$$\mu = \int_{-\infty}^{+\infty} x f_X(x)dx \tag{3-69}$$

For a discrete random variable, the mean is given by:

$$\mu = \sum_{i=1}^{n} x_i P_X(x_i) \tag{3-70}$$

For n observations, if all observations are given equal weights, such that $P_X(x_i) = 1/n$, then the mean for a discrete random variable (Equation 3-70) produces:

$$\overline{X} = \frac{1}{n}\sum_{i=1}^{n} x_i \tag{3-71}$$

which is the average of the observed values $x_1, x_2, x_3, \ldots, x_n$.

The mean corresponds to the average value that lies along the measurement axis; this is most easily represented by the density functions shown in Figure 3-10. If the random variable X is the magnitude of a flood, then $f_{X_1}(x_1)$ and $f_{X_2}(x_2)$ could represent the distributions of floods on two rivers. Obviously, the floods of the river characterized by $f_{X_2}(x_2)$ are usually larger than floods on the river characterized by $f_{X_1}(x_1)$. Similarly, if the density functions of Figure 3-10 represent the distribution of grades for two large sections of a class, it is evident that students in section 2 scored higher, on the average, than those in section 1.

Although the mean conveys certain information about a random variable, it does not completely characterize the random variable. The distributions shown in Figure 3-11 are obviously not identical, even though they have the same mean. Thus, the mean, by itself, cannot characterize a distribution. Other characteristics are also important.

Example 3-21: Mean Number of Hurricanes

Consider the data on hurricanes discussed in Example 3-18. The mean annual number of hurricanes in the United States can be computed using Equation 3-71 as follows, to produce a mean of 2.27 hurricanes per year:

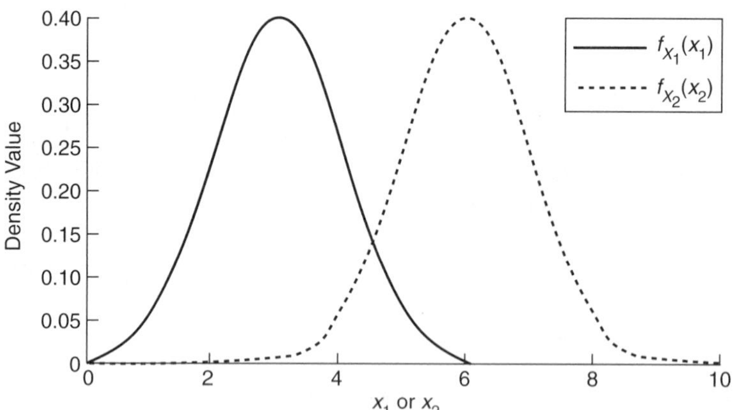

FIGURE 3-10 Mean values for continuous random variables.

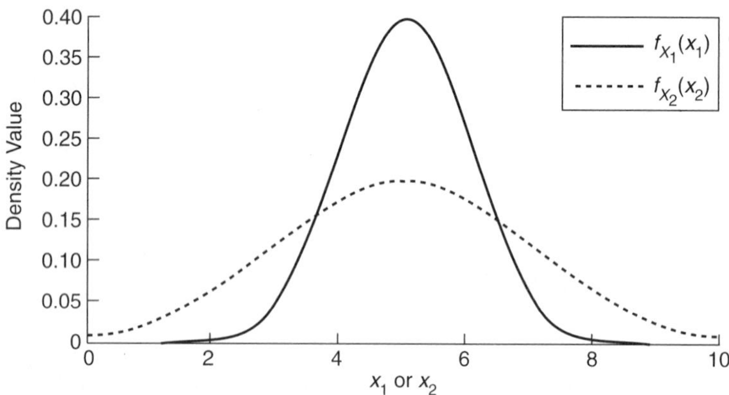

FIGURE 3-11 Random variables with different variances and same mean value.

$$\bar{X} = \frac{5+4+0+0+2+3+1+3+1+6+2+1+2+4+0}{15} = 2.27 \qquad (3\text{-}72)$$

Example 3-22: Mean Settlement of Footings

Example 3-20 used an exponential distribution to model the settlement of footings as provided by the density function of Equation 3-58. The mean value can be computed according to Equation 3-69 as follows:

$$\mu_S = \int_0^\infty s f_S(s) ds = \int_0^\infty s\lambda exp(-\lambda s) ds$$

Integration by parts can be used by assigning $u(s) = s$, and $v(s) = -\exp(-\lambda s)$. The integral can then be evaluated as:

$$\mu_S = \int_0^{\infty} u(s)v'(s)ds = (u(s)v(s))\Big|_0^{\infty} - \int_0^{\infty} u'(s)v(s)ds$$

where u' and v' are the derivatives of u and v, respectively. Substituting u and v and their derivatives gives the following integral:

$$\mu_S = (-s\,exp(-\lambda s))\big|_0^{\infty} - \int_0^{\infty}(1)\,exp(-\lambda s)ds = 0 - \frac{-1}{\lambda}exp(-\lambda s)\big|_0^{\infty} = \frac{1}{\lambda}$$

The mean is the reciprocal of the rate λ.

3.5.2. Variance

The variance is the second moment about the mean. The variance of the population is denoted by σ^2. The variance of the sample is denoted by S^2. The units of the variance are the square of the units of the random variable; for example, if the random variable is measured in pounds per square inch (psi), the variance has units of $(\text{psi})^2$. For a continuous random variable, the variance is computed as the second moment about the mean as follows:

$$\sigma^2 = \int_{-\infty}^{+\infty}(x-\mu)^2 f_X(x)dx \tag{3-73}$$

For a discrete variable, the variance is computed as:

$$\sigma^2 = \sum_{i=1}^{n}(x_i-\mu)^2 P_X(x_i) \tag{3-74}$$

When the n observations in a sample are given equal weight, such that $P_X(x_i) = 1/n$, the variance is given by:

$$S^2 = \frac{1}{n}\sum_{i=1}^{n}(x_i-\bar{X})^2 \tag{3-75}$$

The value of the variance given by Equation 3-75 is biased; an unbiased estimate of the variance is given by:

$$S^2 = \frac{1}{n-1}\sum_{i=1}^{n}(x_i - \bar{X})^2 \tag{3-76}$$

Equation 3-76, which is the equation used most frequently, provides an unbiased estimate because the average value of many sample estimates of σ^2 would approach the true value of σ^2; whereas, the average value of many sample estimates of σ^2 obtained from Equation 3-75 would not approach the true value of σ^2. The concept of unbiased estimators is discussed in Section 8.2.1. Computationally, the variance of a sample can be determined using the following alternative equation, which produces the same result as Equation 3-76:

$$S^2 = \frac{1}{n-1}\left[\sum_{i=1}^{n} x_i^2 - n\bar{X}^2\right] \tag{3-77}$$

If $\bar{X}$ is not computed, then Equation 3-77 becomes:

$$S^2 = \frac{1}{n-1}\left[\sum_{i=1}^{n} x_i^2 - \frac{1}{n}\left(\sum_{i=1}^{n} x_i\right)^2\right] \tag{3-78}$$

Equation 3-77 provides the same answer as Equation 3-78 when computations are made using an appropriate number of significant digits.

The variance is an important concept in probabilistic and statistical analyses because many solution methods require some measure of variance. Therefore, it is important to have a conceptual understanding of this moment. In general, it is an indicator of the closeness of the values in sample or a population to the mean. If all values in the sample equal the mean, the sample variance would equal zero. Figure 3-11 illustrates density functions with different variances.

Because of the importance of the concept of variance, it may be worthwhile to show its computation for two samples, each having a sample size (n) of 3; both sets have a mean of 10, with set 1: {9, 10, 11} and set 2: {5, 10, 15}. Tables 3-3A and 3-3B show the calculations for the two sets. Thus, the variance of set 2 is 25 times greater than that of set 1. To illustrate the computational formula (Equation 3-78), the variance of set 1 could be determined as:

$$\begin{aligned} S^2 &= \frac{1}{n-1}\left[\sum_{i=1}^{n} x_i^2 - \frac{1}{n}\left(\sum_{i=1}^{n} x_i\right)^2\right] \\ &= \frac{1}{3-1}\left[(9^2 + 10^2 + 11^2) - \frac{(9+10+11)^2}{3}\right] = 1 \end{aligned} \tag{3-79}$$

Although the variance is used in all aspects of probabilistic and statistical analyses, its use as a descriptor is somewhat limited due to its units; specifically, the units of the variance are not the same as those of either the random variable or the mean.

TABLE 3-3A Calculation of the Variance for Data Set 1

x	$\bar{X}$	$x - \bar{X}$	$(x - \bar{X})^2$	$\frac{1}{n-1}$	$\frac{1}{n-1}(x - \bar{X})^2$
9	10	−1	1	0.5	0.5
10	10	0	0	0.5	0
11	10	1	1	0.5	0.5
					Summation = 1

TABLE 3-3B Calculation of the Variance for Data Set 2

x	$\bar{X}$	$x - \bar{X}$	$(x - \bar{X})^2$	$\frac{1}{n-1}$	$\frac{1}{n-1}(x - \bar{X})^2$
5	10	−5	25	0.5	12.5
10	10	0	0	0.5	0
15	10	5	25	0.5	12.5
					Summation = 25

3.5.3. Standard Deviation and Coefficient of Variation

By definition, the standard deviation is the square root of the variance. It has the same units as the random variable and the mean; therefore, it is a better descriptor of the dispersion or spread of either a sample of data or a distribution function than the variance. The standard deviation of the population is denoted by σ, while the sample value is denoted by S.

The coefficient of variation (δ or COV) is a dimensionless quantity defined as:

$$\delta = \frac{\sigma}{\mu} \tag{3-80}$$

It is also used as an expression of the standard deviation in the form of a proportion of the mean. For example, consider μ and σ to be 100 and 10, respectively; therefore, $\delta = 0.1$ or 10%. In this case, the standard deviation is 10% of the mean.

Example 3-23: Variance of Soil Strength

A sample of five tests was taken to determine the unconfined compression strength (in tons/ft^2) of soil, with the following test results: 1.8, 3.5, 6.1, 3.6, and 4.3 tons/ft^2. Compute the variance, standard deviation, and coefficient of variation of soil strength.

The mean value of soil strength $\bar{X}$ is given by:

$$\bar{X} = \frac{1.8 + 3.5 + 6.1 + 3.6 + 4.3}{5} = 3.86 \text{ tons/ft}^2 \tag{3-81}$$

The variance of soil strength is computed using Equation 3-78 as follows:

$$S^2 = \frac{1.8^2 + 3.5^2 + 6.1^2 + 3.6^2 + 4.3^2 - \frac{(1.8 + 3.5 + 6.1 + 3.6 + 4.3)^2}{5}}{5 - 1} \tag{3-82}$$

The variance is 2.413 (tons/ft^2)^2. Therefore, the standard deviation is:

$$S = \sqrt{2.413} = 1.553 \text{ tons/ft}^2 \tag{3-83}$$

The coefficient of variation δ is:

$$\delta = \frac{S}{\bar{X}} = \frac{1.553}{3.86} = 0.402 \tag{3-84}$$

The large coefficient of variation (i.e., 40.2% of the mean value) suggests that either the mean value is not reliable and that additional measurements might be needed or the underlying random variable has a large dispersion. If with additional measurements the coefficient of variation remains large, relatively large factors of safety should be used for soil strength design.

Example 3-24: Settlement Variance

Example 3-20 used an exponential distribution to model the settlement of footings as provided by the density function of Equation 3-58. The variance can be computed according to Equation 3-73 using a mean value of $1/\lambda$ as follows:

$$\sigma^2 = \int_0^{\infty} \left(s - \frac{1}{\lambda}\right)^2 f_S(s)ds = \int_0^{\infty} \left(s - \frac{1}{\lambda}\right)^2 \lambda exp(-\lambda s)ds$$

Integration by parts can be used to determine the variance as follows:

$$\sigma^2 = \frac{1}{\lambda^2}$$

3.5.4. Skew

The skew is the third moment measured about the mean. Unfortunately, the notation for skew is not uniform from one user to another. The sample skew can be denoted by G, while λ can be used to indicate the skew of the population. Mathematically, it is given as follows for a continuous random variable:

$$\lambda = \int_{-\infty}^{+\infty} (x-\mu)^3 f_X(x)dx \tag{3-85}$$

For a discrete random variable, it can be computed by:

$$\lambda = \sum_{i=1}^{n} (x_i-\mu)^3 P_X(x_i) \tag{3-86}$$

It has units of the cube of the random variable; thus, if the random variable has units of pounds, the skew has units of $(\text{pounds})^3$.

The skew is a measure of the lack of symmetry. A symmetric distribution has a skew of zero, while a nonsymmetric distribution has a positive or negative skew depending on the direction of the skewness. If the more extreme tail of the distribution is to the right, the skew is positive; the skew is negative when the more extreme tail is to the left of the mean. Skewed probability distributions are illustrated in Figure 3-12.

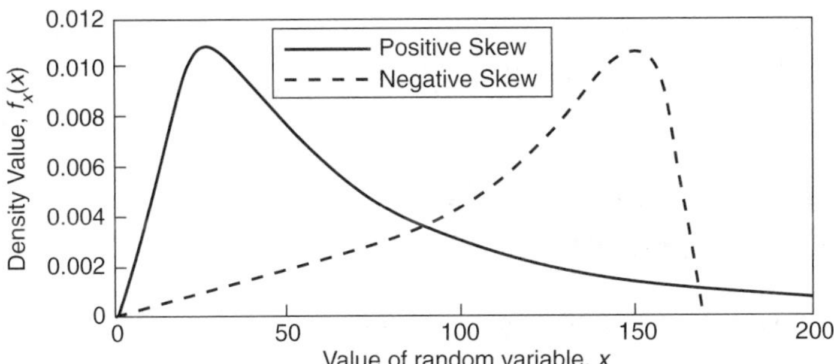

FIGURE 3-12 Skewed distributions.

Example 3-25: Moments of a Discrete Random Variable

The mean, variance, standard deviation, and coefficient of variation for the random variable defined in Example 3-17 are calculated in Table 3-4. Because the probabilities are for the population, the population mean can be determined using Equation 3-70 as:

$$\mu = \sum_{i=1}^{n} x_i P_X(x_i) = \frac{252}{36} = 7 \tag{3-87}$$

TABLE 3-4 Computations for the Moments for Example 3-25

	x	$P_X(x)$	$x\,P_X(x)$	$(x-\mu)$	$(x-\mu)^2$	$(x-\mu)^2\,P_X(x)$	$(x-\mu)^3\,P_X(x)$
	2	1/36	2/36	–5	25	25/36	–125/36
	3	1/18	6/36	–4	16	32/36	–64/36
	4	1/12	12/36	–3	9	27/36	–27/36
	5	1/9	20/36	–2	4	16/36	–8/36
	6	5/36	30/36	–1	1	5/36	–1/36
	7	1/6	42/36	0	0	0	0
	8	5/36	40/36	1	1	5/36	1/36
	9	1/9	36/36	2	4	16/36	8/36
	10	1/12	30/36	3	9	27/36	27/36
	11	1/18	22/36	4	16	32/36	64/36
	12	1/36	12/36	5	25	25/36	125/36
Summations	77	1.00	252/36	0		210/36	0

The population variance can be determined using Equation 3-74 as:

$$\sigma^2 = \sum_{i=1}^{n}(x_i-\mu)^2 P_X(x_i) = \frac{210}{36} = 5.833 \tag{3-88}$$

Therefore, the standard deviation is $\sqrt{5.833} = 2.415$, and the coefficient of variation is $\frac{2.415}{7} = 0.345$. The skew (λ) is zero, indicating a symmetric mass function.

Example 3-26: Moments of a Continuous Random Variable

The mean, variance, standard deviation, and coefficient of variation for the density function given in Example 3-19 are computed in this example. The mean can be computed using Equation 3-69 as follows:

$$\mu = \int_{-\infty}^{+\infty} x f_X(x)dx = \int_0^2 x(0.5x)dx = \frac{4}{3} \tag{3-89}$$

The variance can be computed using Equation 3-73:

$$\sigma^2 = \int_{-\infty}^{+\infty}(x-\mu)^2 f_X(x)dx = \int_0^2\left(x-\frac{4}{3}\right)^2(0.5x)dx = \frac{2}{9} \tag{3-90}$$

The standard deviation equals $\frac{\sqrt{2}}{3} = 0.4714$, and the coefficient of variation is $\frac{0.4714}{4/3} = 0.3771$.

3.6. APPLICATION: WATER SUPPLY AND QUALITY

The water for a residential unit is pumped from a well and passed through a neutralization unit (neutralizer) and then through a water filter. The annual probability of failure of the pump (which is electrically operated) is 0.1. The annual failure probabilities of the neutralizer and filter are 0.2 and 0.1, respectively. The probability of failure of both the neutralizer and filter at the same time is 0.02. The failure of the pump results in the loss of the water supply, whereas the failure of the neutralizer or filter results in reducing the water quality below acceptable limits without the loss of water supply. The failure events of the pump, neutralizer, and filter can be assumed to be independent. What is the probability of having water of acceptable quality at the residential unit?

The following events are defined:

$$\text{P} = \text{pump failure} \tag{3-91a}$$

$$\text{N} = \text{neutralizer failure} \tag{3-91b}$$

$$\text{F} = \text{filter failure} \tag{3-91c}$$

The probability of receiving water of acceptable quality is:

$$P(\bar{\text{P}} \cap \bar{\text{N}} \cap \bar{\text{F}}) = P[\bar{\text{P}} \cap \overline{(\text{N} \cup \text{F})}] \tag{3-92a}$$

$$\text{Probability of acceptable water} = (1 - 0.1)[(1 - (0.2 + 0.1 - 0.02)] = 0.648 \tag{3-92b}$$

If low-quality water is received at the residential unit, the probability that the low-level quality is due to the failure of the neutralizer is determined using Equation 3-25 as:

$$P[\text{N}|(\text{N} \cup \text{F})] = \frac{P(\text{N} \cap (\text{N} \cup \text{F}))}{P(\text{N} \cup \text{F})} = \frac{P(\text{N})}{P(\text{N} \cup \text{F})} \tag{3-93}$$

Therefore, the probability is:

$$P[\text{N}|(\text{N} \cup \text{F})] = \frac{0.2}{0.2 + 0.1 - 0.02} = 0.714 \tag{3-94}$$

This result indicates that the neutralizer is responsible for 71.4% of the failures in water quality.

If the neutralizer is replaced with a more reliable unit with $P(\text{N}) = 0.1$ and $P(\text{N} \cap F) = 0.01$, then the conditional probability of Equation 3-94 becomes:

$$P[\text{N}|(\text{N} \cup \text{F})] = \frac{0.1}{0.1 + 0.1 - 0.01} = 0.526 \tag{3-95}$$

3.7. SIMULATION AND PROBABILITY DISTRIBUTIONS

Simulation offers a unique opportunity to understand several important elements related to probability distributions and probabilistic analysis. First, it clearly shows the concept of sampling variation, which is random variations in statistics that characterize the difference between a sample value of a statistic and the corresponding population value. This is the underlying basis for probability distributions. Sampling variation was evident in the simulated discharge data of Example 2-17.

Second, simulation of random samples provides the opportunity to examine the effect of sample size on the accuracy of sample statistics. Small samples yield statistics that generally show greater deviation from the population value than those provided by large samples. Thus, the spread of a probability distribution of a random variable, on the average, decreases as the sample size increases.

Third, simulation can be used to show why it is important to assess the probability function of the random variable prior to the simulation. Different probability functions can lead to different decisions. Furthermore, random variables can be discrete or continuous, and the type of problem dictates the selection.

Fourth, measured data are often very limited, and making decisions with small sample sizes increases the risk of an incorrect decision. Using measured data for decision making is subject to the problems of (1) not having extreme values in the sample that enable the response of the system to be studied under extreme conditions and (2) outlying measurements that distort sample statistics. Simulation enables the effect of either of these factors on the probability distribution to be studied.

Example 3-27: Simulation of Distributions

During any time period, a minor earthquake (II), a major earthquake (III), or no earthquake (I) can occur. These events have probabilities of 30, 10, and 60%, respectively. If an earthquake occurs, damage can be negligible (N), little (L), or major (M). The probabilities of damage during a minor (II) earthquake are N, 35%; L, 40%; and M, 25%. The probabilities during a major (III) earthquake are N, 15%; L, 45%; and M, 40%. The following effects or probabilities can be studied by simulation: (a) the effect of sample size; (b) the effect of uncertainty in the distribution of damages; (c) the conditional probability that an event where major damage occurred was a minor (II) earthquake.

The simulated values are shown in Table 3-5. Twenty-four periods are simulated. For each period, two uniform (0 to 1) variates are generated, one (U_E) for the earthquake magnitude (I, II, or III) and one (U_D) for the level of damage (N, L, or M). The cumulative mass function for the earthquake magnitude is {0.6, 0.9, 1.0} for {I, II, III}, respectively. The value of U_E is used with this cumulative transformation function to decide the earthquake magnitude, which is indicated in column 3 of Table 3-5. The cumulative transformation function for damage for earthquake magnitudes I, II, and III are {1.0, 1.0, 1.0}, {0.35, 0.75, 1.0}, and {0.15, 0.60, 1.0}, respectively. The U_D values are used as input to the transformation function that

TABLE 3-5 Simulation Results for Example 3-27

U_E	U_D	E	D_1	D_2
0.41	0.48	I	N	N
0.03	0.97	I	N	N
0.91	0.24	III	L	N
0.98	0.99	III	M	M
0.74	0.20	II	N	N
0.94	0.67	III	M	L
0.58	0.18	I	N	N
0.31	0.47	I	N	N
0.45	0.62	I	N	N
0.31	0.49	I	N	N
0.68	0.37	II	L	L
0.93	0.67	III	M	L
0.77	0.56	II	L	L
0.37	0.07	I	N	N
0.72	0.08	II	N	N
0.55	0.22	I	N	N
0.67	0.24	II	N	N
0.01	0.86	I	N	N
0.51	0.40	I	N	N
0.58	0.78	I	N	N
0.46	0.94	I	N	N
0.97	0.21	III	L	N
0.27	0.47	I	N	N
0.15	0.32	I	N	N

corresponds to the earthquake magnitude indicated in column 3. The damage level (N, L, or M) is indicated in column 4 of Table 3-5.

Based on the 24-period record, the proportions of level I, II, and III magnitudes using the relative frequency concept of Equation 3-7 are 0.583, 0.209, and 0.208, which differ from the population values because of sampling variation.

To examine the effect of record length, the 24-period record is divided into two 12-period records. The first 12 periods give probabilities of 0.583, 0.167, and 0.250. The second 12 periods give probabilities of 0.833, 0.167, and 0.0. If enough 12-period simulations are made, the sampling distribution of the three proportions could be assessed. The effect of the sample size could then be examined by simulating for different sample sizes. It is shown here that the sample probabilities for record lengths of 12 periods show greater variation from the population than the probabilities for 24 periods.

To study the effect of uncertainty in the distribution of damages, assume that the cumulative transformation function for damage resulting from a major (III) earthquake is {0.25, 0.75, 1.0}. Then the damages for the 24-period record would be as shown in column 5 of Table 3-5 (i.e., D_2). The distribution of damages is, thus, {0.792, 0.267, 0.041}. This compares with the distribution of damages for the original transformation function of {0.708, 0.167, 0.125}. Thus, changing the distribution for just the major earthquakes produced a noticeable change in the damage probabilities.

For the original damage function, major damage occurred in three periods, all during major earthquakes. Thus, the sample conditional probability of major damage given minor and major earthquakes are 0 and 1, respectively. Given that major damage could occur in a minor earthquake, these conditional probabilities are inaccurate, with the inaccuracy being the result of the small sample of 24 periods.

3.8. PROBLEMS

3-1. A construction manager needs to procure building materials for the construction of a bridge. The following sources are identified:

Material type	Sources
Concrete	Sources A and B
Reinforcing steel	Sources C and D
Timber	Sources D and E
Structural steel	Sources C and F
Hardware	Sources F, G, and H

Define the sample space of all possible combinations of sources supplying the construction project, assuming that each material type can be procured from one source only, but a source may supply more than one material type at the same time.

3-2. A construction tower crane can operate up to a height H of 300 ft, a range (radius) R of 50 ft, and an angle ϕ of $\pm 90°$ in a horizontal plane. Sketch the sample space of operation of the crane. Sketch the following events:

Event	Definition
A	$30 < H < 80$, and $R < 30$
B	$H > 50$, and $0° < \phi < 50°$
C	$H < 40$, and $R > 60$
D	$H > 80$, and $-30° < \phi < 50°$

3-3. Construct Venn diagrams for each of the following:

a. Deck of playing cards
b. Roll of a die
c. Letter grades on a test, assuming equal probabilities for each grade
d. Letter grades on a test, assuming the following probabilities: A, 15%; B, 25%; C, 30%; D, 20%; F, 10%
e. Options at an intersection with the following probabilities: left turn, 20%; straight ahead, 40%; right turn, 25%; U-turn, 10%; and remaining stopped, 5%

3-4. For the data and events of Problem 3-2, sketch the following events: $A \cup B$, $A \cap B$, $C \cup D$, $C \cap D$, $A \cup C$, $A \cup (B \cap C)$, $\overline{A}$, and $\overline{A} \cap B$.

3-5. The traffic that makes a left turn at an intersection consists of two types of vehicles, types A and B. A type A vehicle is twice the length of type B. The left-turn lane can accommodate eight vehicles of type B, four of type A, or combinations of A and B. Define the sample space of all possible combinations of vehicles waiting for a left turn at the intersection. Also, define the following events: (1) at least one vehicle of type A waiting for a left turn, (2) two vehicles of type B waiting for a left turn, and (3) exactly one of type A and one of type B waiting for a left turn.

3-6. Construct a Venn diagram for a deck of playing cards (four suits, 13 cards per suit). Show the following events:

a. A = all diamonds and all aces.

b. B = all face cards.

c. C = the intersection of red cards and face cards.

d. D = the union of black cards and cards with values of 4 or smaller.

3-7. What is the difference between the population and a sample? A construction engineer obtains five specimens of steel reinforcing bars from a shipment to a construction site. What is the corresponding population? How could the engineer use the sample to characterize the population?

3-8. The scores on a test were distributed as shown below for a class of 60 students. Convert the frequency histogram to a probability distribution. What is the probability that a student had a score of at least 90? What is the probability that a student failed the test (F grade), assuming that a score less than 70 was considered an F?

	Range							
	50–59	60–69	70–74	75–79	80–84	85–89	90–94	95–100
Number	4	9	7	11	13	7	6	3

3-9. Two loads A and B are applied on a structural column and are mutually exclusive. The probability that the beam is safe against load A is 0.7 and against load B is 0.9. What is the probability that the structure is safe, assuming that the "safe" events for the two loads are independent?

3-10. A concrete beam may fail by shear or flexure. The failure probability in shear is equal to failure probability in flexure, and the probability of failure in shear when the beam is loaded beyond its flexural capacity (i.e., the beam already failed in flexure) is 80%. Find the failure probability of the beam in flexure, given that the probability of failure of the beam is 0.2.

3-11. Materials for a construction site are supplied by two sources A and B. Upon arrival of all materials from both sources, the construction is started after the arrival of workers. If the probability that materials would be supplied by A on time is 0.8 and by B is 0.75, and if the probability that the workers arrive on time for work is 0.95, find the probability that the job is started on time.

3-12. A shipping company facilitates for customers the transfer of any items from any location to another within the United States. The company transfers 20, 50, and 30% of the items by air, ground, and sea transportation, respectively. The breakdown of these items by shipping method is shown in the accompanying figure. Assume that 5% of the items shipped by air get damaged during

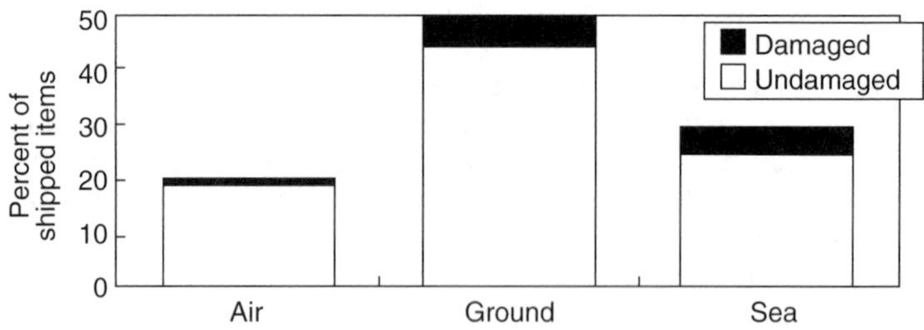

Shipping methods (Problem 3-12).

shipping, 10% of the items shipped by ground get damaged during shipping, and 15% of the items shipped by sea get damaged during shipping. Determine the following probabilities:

a. The probability of receiving a damaged item sent by air, ground, or sea
b. If an item was received damaged, the probability that it was sent by (i) air, (ii) by ground, and (iii) by sea

3-13. Find the following conditional probabilities:

a. On a single roll of a die, the probability of a 3 given that the value is an odd integer (draw the sample space for the conditional event)
b. The probability that the outcome of a single draw from a card deck is a red ace given that it is not a face card
c. The probability that the outcome of a single draw from a card deck is an ace given that five cards, including one ace, have already been drawn and discarded from the deck

3-14. A construction site receives fill material from three sources, with sources A, B, and C providing 15, 25, and 60% of the total, respectively. On the average, fill material from sources A, B, and C do not have the proper moisture characteristics 2, 6, and 3% of the time, respectively. If one dump truck is sampled at random and the fill material does not have the proper moisture, what is the probability that it came (a) from source A, (b) from source B, (c) from source C?

3-15. Given the following cumulative mass function, derive the mass function:

$$F_X(x) = \begin{cases} 0.2 & & x = 2 \\ 0.2 & & x = 3 \\ 0.2 & & x = 4 \\ 0.5 & \text{for} & x = 5 \\ 0.7 & & x = 6 \\ 0.8 & & x = 7 \\ 0.9 & & x = 8 \\ 1.0 & & x = 9 \end{cases}$$

Graph both the probability and cumulative mass functions.

3-16. Given the following probability mass function:

$$F_X(x) = \begin{cases} i & x = 1 \\ 2i & x = 2 \\ 3i & \text{for} \quad x = 3 \\ 0 & \text{otherwise} \end{cases}$$

determine the value of i that results in a legitimate mass function. Graph both the probability and cumulative mass functions.

3-17. For the probability mass function of Problem 3-15, determine the following probabilities: (a) $P(X = 4)$; (b) $P(X = 5)$; (c) $P(X \le 5)$; (d) $P(X < 5)$; (e) $P(4 < X \le 7)$; and (f) $P(X \ge 7)$.

3-18. The number of years of drought in each decade is:

1900 to 1909	2
1910 to 1919	0
1920 to 1929	3
1930 to 1939	5
1940 to 1949	4
1950 to 1959	1
1960 to 1969	1
1970 to 1979	2
1980 to 1989	2

Show the sample mass function and cumulative function for the number of drought years in a decade. What is the sample space?

3-19. Find the value k that is necessary to make the following a legitimate density function:

$$f_X(x) = \begin{cases} kx & \text{for } 0 \le x \le 1 \\ k & \text{for } 1 < x \le 2 \end{cases}$$

Graph both the density and the cumulative functions.

3-20. The duration of a construction activity was estimated to be in the range [2,6] days with a most likely duration of 4 days. A construction engineer used the following symmetric density function to model the duration:

$$f_X(x) = \begin{cases} bx & \text{for } 2 \le x \le 4 \\ a - bx & \text{for } 4 < x \le 6 \end{cases}$$

Determine the necessary constants a and b to have a legitimate density function. Graph both the density and cumulative functions.

3-21. For the probability mass function of Problem 3-15, determine the mean, variance, and standard deviation.

3-22. For the probability mass function of Problem 3-16, determine the mean, variance, standard deviation, coefficient of variation, and skewness.

3-23. For the probability density function of Problem 3-19, determine the mean, variance, and standard deviation.

3-24. For the probability density function of Problem 3-20, determine the mean, variance, standard deviation, coefficient of variation, and skewness.

3-25. Calculate the mean, variance, standard deviation, coefficient of variation, and skewness for the following data sample:

$$X = \{3, 5, 8, 6, 8\}$$

3-26. Ten steel specimens were tested for their yield strength. Determine the mean, variance, standard deviation, coefficient of variation, and skewness for the tested steel based on the following test results:

$$\{38, 36, 34, 37, 38, 39, 35, 38, 40, 36\} \text{ ksi}$$

3-27. The probability of a flood in any 1 year is 0.1. In a 10-year period, what is the probability of (a) no floods, (b) two or fewer floods, (c) from one to three floods?

3-28. Piles are used to support the foundation of a structure. The failure probability of a pile during proof testing is 0.1. Determine the following:

a. The probability of 3 failed piles out of 10 tested piles
b. The probability of no failures in 20 tested piles
c. The probability of 10 tested piles to obtain the first failure
d. The probability of three consecutive pile failures

3-29. A fair coin is tossed 10 times. What is the probability of getting (a) exactly five heads, (b) exactly two tails, (c) no heads, (d) two or fewer heads, (e) five or more tails, and (f) at least two but not more than six heads?

3-30. The probability that a flood of a specified magnitude occurs in any 1 year is 0.05. What is the probability that in the next 10 years (a) exactly two such floods will occur, (b) no more than one such flood will occur, (c) no such floods will occur, (d) at least four floods will occur, and (e) at least three but no more than six such floods will occur?

3-31. Use the midsquare method with a seed of 1937 to generate 30 uniform random numbers (0 to 1). Plot frequency histograms using cell widths of 0.1, 0.2, and 0.25. Comment on: (a) the degree to which the numbers follow a uniform distribution, and (b) the effect of cell size on the ability of the sample to appear to have a uniform distribution.

3-32. Use the *rand* function to generate 30 uniform random numbers (0 to 1). Plot frequency histograms using cell widths of 0.1, 0.2, and 0.25. Comment on: (a) the degree to which the numbers follow a uniform distribution, and (b) the effect of cell size on the ability of the sample to appear to have a uniform distribution.

3-33. Compare the results from Problems 3-31 and 3-32, and discuss any differences.

3-34. Construct a transformation curve that transforms uniform variates on a scale from 0 to 1 to uniform variates on a scale from 10 to 20. Use the 30 variates of Problem 3-31 and the transformation curve to obtain 30 uniform variates on a scale from 10 to 20. Discuss the results of the transformation.

3-35. Construct a transformation curve that transforms uniform variates on a scale from 0 to 1 to uniform variates on a scale from 10 to 20. Use the 30 variates of Problem 3-32 and the transformation curve to obtain 30 uniform variates on a scale from 10 to 20. Discuss the results of the transformation.

3.9. SIMULATION PROJECTS

This section provides additional work items for the problem statements of the four simulation projects introduced in Sections 1.5 and 2.9.

3.9.1. Structural Beam Study

Using the project information provided in Sections 1.5.1 and 2.9.1, reevaluate the failure probability similar to Table 1-3 using: (a) 100 simulation cycles and uniform probability distributions for all random variables as provided in Table 1-2, and (b) 100 simulation cycles and exponential probability distributions, using Equation 3-58 with λ equal to mid-intervals as provided in Table 1-2 for corresponding random variables.

3.9.2. Stream Erosion Study

Using the data of Table C-1 (Appendix C), the project information provided in Sections 1.5.2 and 2.9.2, and the relative frequency concept of Equation 3-7, compute: (a) the probability that the erosion Y is greater than 6 lb/sec/ft and the conditional probability that $Y > 6$ given that the water discharge (X_3) is greater than 0.42 cfs/ft; (b) the probability that the erosion Y is less than 3 lb/sec/ft and the conditional probability that $Y < 3$ given that the slope is greater than 1%; and (c) the probability that the mean particle diameter is greater than 3.25×10^{-3} ft on a channel with a slope less than 0.75%.

3.9.3. Traffic Estimation Study

Using the data of Table C-2 (Appendix C), the project information provided in Sections 1.5.3 and 2.9.3, and the relative frequency concept of Equation 3-7, compute: (a) the probability that the work trips per day Y are less than 1×10^3 and the conditional probability that $Y < 1 \times 10^3$ given that the population in the zone exceeds 15×10^3; (b) the probability that the work trips Y are greater than 1.25×10^3 and the conditional probability that $Y > 1.25 \times 10^3$ given that the zone includes less than 4×10^3 dwelling units; and (c) the conditional probability that there are fewer than 6×10^3 vehicles in zones with populations greater than 18×10^3.

3.9.4. Water Evaporation Study

Using the data of Table C-3, the project information provided in Sections 1.5.4 and 2.9.4, and the relative frequency concept of Equation 3-7, compute: (a) the probability that the daily evaporation Y exceeds 0.18 in./day and the conditional probability that Y exceeds 0.18 given that the temperature is less than 65°F; (b) the probability that the daily evaporation Y is less than 0.11 in./day and the conditional probability that $Y < 0.11$ given a day on which the relative humidity is less than 40%; and (c) the conditional probability that the temperature exceeds 60°F given that the relative humidity is less than 50%.

Chapter 4
Probability Distributions for Discrete Random Variables

TABLE OF CONTENTS

4.1. INTRODUCTION

Any function that results in probabilities satisfying the axioms of probability is qualified to be a probability distribution. As a result of previous experience with applications of probability distributions and an improved understanding of the underlying physical processes, several probability distributions are commonly used today by engineers. The objective of this chapter is to introduce these probability distributions that are used to model discrete random variables. They are commonly called *discrete probability distributions*. This chapter describes commonly used discrete probability distributions, and Chapter 5 describes commonly used continuous probability distributions.

A probability distribution function is expressed as a real-valued function of the random variable. The location, scale, and shape of the function are determined by its parameters. Distributions commonly have one or more parameters. These parameters take certain values that are specific for the problem and conditions being investigated. The parameters of a distribution can be expressed in terms of its moments (e.g., mean, variance, and skewness) but not necessarily in closed-form expressions.

In this chapter, the Bernoulli, binomial, geometric, and Poisson distributions are discussed. The first three distributions are based on Bernoulli trials (or sequences); the fourth one is not. An engineering experiment (or system) that consists of N trials is considered to result in a *Bernoulli process* (or sequence) if it satisfies the following conditions: (1) the N trials (or repetitions) are independent; (2) each trial has only two possible outcomes — say, survival (S) or failure (F); and (3) the probabilities of occurrence for the two outcomes remain constant from trial to trial. Also, the negative binomial, Pascal, and hypergeometric distributions are described as other discrete probability distributions at the end of the chapter.

4.2. BERNOULLI DISTRIBUTION

For convenience, the random variable X is defined as a mapping from the sample space $\{S,F\}$ for each trial of a Bernoulli sequence to the integer values $\{1,0\}$, with one-to-one mapping in the respective order and where, for example, S = success and F = failure. Therefore, the probability mass function is given by:

$$P_X(x) = \begin{cases} p & \text{for } x = 1 \\ 1-p & \text{for } x = 0 \\ 0 & \text{otherwise} \end{cases} \tag{4-1}$$

The probability mass function of the Bernoulli distribution is shown in Figure 4-1. The mean and variance for the Bernoulli distribution are, respectively, given by:

$$\mu_X = p \tag{4-2}$$

$$\sigma_X^2 = p(1-p) \tag{4-3}$$

Example 4-1: Quality Assurance

The quality assurance department in a structural-steel factory inspects every product coming off its production line. The product either fails or passes the inspection. Past experience indicates that the probability of failure (having a defective product) is 5%. Assuming that failures of the products are independent, determine the average percent of the products that will pass the inspection. What are the variance and coefficient of variation?

The average percent of the products that will pass the inspection, X, is:

$$E(X) = p = 1 - 0.05 = 0.95 = 95\% \tag{4-4}$$

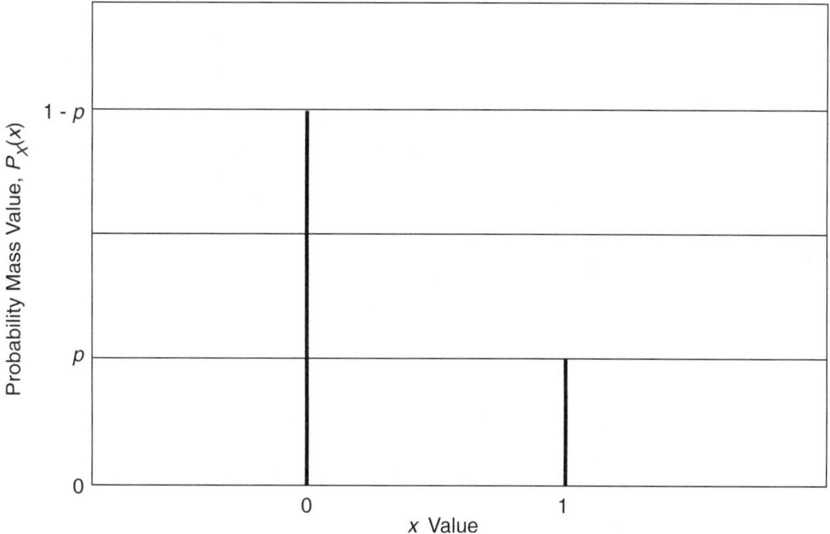

FIGURE 4-1 Probability mass function of the Bernoulli distribution.

The variance (Var) and coefficient of variation (COV) are:

$$\mathrm{Var}(X) = p(1-p) = 0.95(0.05) = 0.0475 \tag{4-5}$$

and

$$\mathrm{COV}(X) = \frac{\sqrt{\mathrm{Var}(X)}}{E(X)} = \frac{\sqrt{0.0475}}{0.95} = 0.229 \tag{4-6}$$

4.3. BINOMIAL DISTRIBUTION

The underlying random variable (X) for this distribution represents the number of successes in N Bernoulli trials. The probability mass function is given by:

$$P_X(x) = \begin{cases} \binom{N}{x} p^x (1-p)^{N-x} & \text{for } x = 0, 1, 2, \ldots, N \\ 0 & \text{otherwise} \end{cases} \tag{4-7}$$

where $\binom{N}{x}$ can be computed using Equation 3-17. The probability mass and cumulative functions of an example binomial distribution are shown in Figures 4-2A and 4-2B, respectively. The mean and variance for the binomial distribution, respectively, are given by:

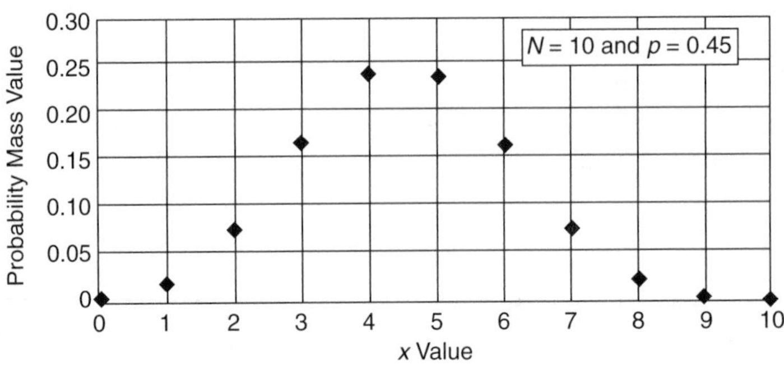

FIGURE 4-2A Probability mass function of the binomial distribution.

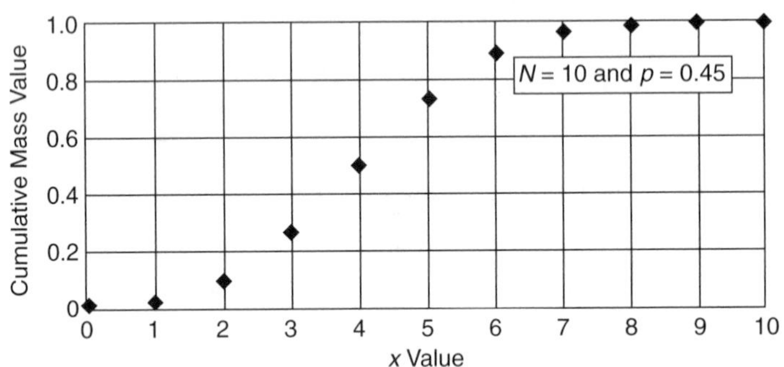

FIGURE 4-2B Cumulative mass function of the binomial distribution.

$$\mu_X = Np \tag{4-8}$$

$$\sigma_X^2 = Np(1-p) \tag{4-9}$$

A random variable can be represented by the binomial distribution, if the following three assumptions are met:

1. The distribution is based on N Bernoulli trials with only two possible outcomes.
2. The N trials are independent of each other.
3. The probabilities of the outcomes remain constant at p and $(1 - p)$ for each trial.

The flip of a coin would meet these assumptions, but the roll of a die would not because there are six possible outcomes.

Example 4-2: Quality Assurance

The production line in Example 4-1 manufactures (on the average) 1000 units of the product daily. What is the expected number of nondefective units? What is the coefficient of variation of the number of nondefective units?

Because there are only two possible outcomes (i.e., defective and nondefective units), a binomial distribution can be used to model this problem, with $N = 1000$ and $p = 0.95$. The characteristics of nondefective units are:

$$\text{Expected (mean) value } E(X) = Np = 1000(0.95) = 950 \text{ units/day} \tag{4-10a}$$

$$\text{Standard deviation, } \sigma = \sqrt{Np(1-p)} = \sqrt{1000(0.95)(0.05)} = 6.892 \tag{4-10b}$$

$$\text{Coefficient of variation, COV}(X) = \frac{\sigma}{E(X)} = 0.00725 \tag{4-10c}$$

The coefficient of variation in this case describes the dispersion of the daily nondefective units produced by a line. The probability of producing exactly 980 nondefective units in a day is:

$$P_X(980) = \binom{1000}{980}(0.95)^{980}(1-0.95)^{20} = 4.787 \times 10^{-7}$$

Such a small probability of having 980 nondefective units in a day resulted because of specifying *exactly* 980 units a day in computing the probability.

4.4. GEOMETRIC DISTRIBUTION

The underlying random variable for this distribution represents the number of Bernoulli trials that are required to achieve the first success. In this case, the number of trials needed to achieve the first success is neither fixed nor certain. The probability mass function is given by:

$$P_X(x) = \begin{cases} p(1-p)^{x-1} & \text{for } x = 0, 1, 2, 3, \ldots \\ 0 & \text{otherwise} \end{cases} \tag{4-11}$$

The probability mass function of an example geometric distribution is shown in Figure 4-3. The mean and variance for the geometric distribution, respectively, are given by:

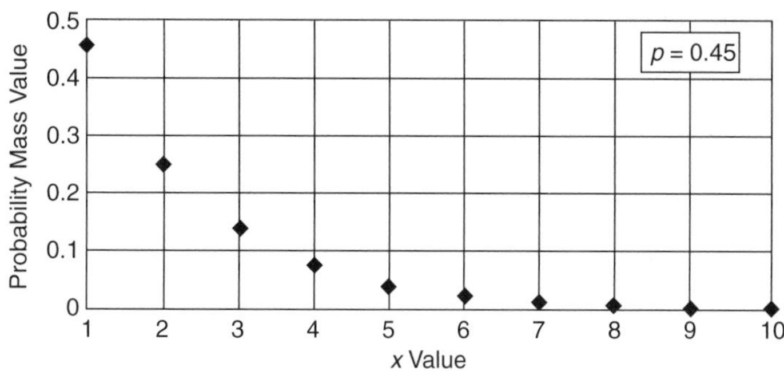

FIGURE 4-3 Probability mass function of the geometric distribution.

$$\mu_X = \frac{1}{p} \tag{4-12}$$

$$\sigma_X^2 = \frac{1-p}{p^2} \tag{4-13}$$

Example 4-3: Traffic Accidents

Based on previous accident records, assume that the annual probability of being in a fatal traffic accident is on the average 1.8×10^{-5} per 1000 miles of travel. What is the probability of being in a fatal accident for the first time at 10,000, 100,000, and 500,000 miles of travel? These miles of travel can be considered to correspond to (on the average) 10, 100, and 500 years.

The accident probability at 10,000 miles of travel is $1.8 \times 10^{-5} [(1 - 1.8 \times 10^{-5})^{10-1}] = 1.8 \times 10^{-5}$. At 100,000 miles, the probability is $1.8 \times 10^{-5} [(1 - 1.8 \times 10^{-5})^{100-1}] = 1.8 \times 10^{-5}$. At 500,000 miles, the probability is $1.8 \times 10^{-5} [(1 - 1.8 \times 10^{-5})^{500-1}] = 1.8 \times 10^{-5}$. It is interesting to note that the probability remains the same, because the geometric distribution assumes the accident to be in the last 1000 miles of travel.

4.5. POISSON DISTRIBUTION

The Poisson distribution is commonly used in engineering problems that deal with the occurrence of some random event in the continuous dimension of time or space. For example, the number of occurrences of a natural hazard, such as earthquakes, tornadoes, or hurricanes, in some time interval, such as 1 year, can be considered as a random variable with a Poisson distribution. In these examples, the number of occurrences in the time interval is the random variable. Therefore, the random variable is discrete, whereas its reference space (i.e., the time interval) is continuous. This distribution is considered to be the limiting case of the binomial distribution by dividing the reference space (time t) into nonoverlapping intervals of size Δt. The occurrence of the event (such as a natural hazard) in each interval is considered to constitute a Bernoulli sequence. The number of Bernoulli trials depends on the size of the interval Δt. By considering the limiting case where the size of the interval Δt approaches zero, the binomial distribution becomes the Poisson distribution.

The underlying random variable of this distribution is denoted by X_t, which represents the number of occurrences of an event of interest, and t is the time (or space) interval. The probability mass function for the Poisson distribution is:

$$P_{X_t}(x) = \begin{cases} \dfrac{(\lambda t)^x \exp(-\lambda t)}{x!} & \text{for } x = 0, 1, 2, 3, \ldots \\ 0 & \text{otherwise} \end{cases} \tag{4-14}$$

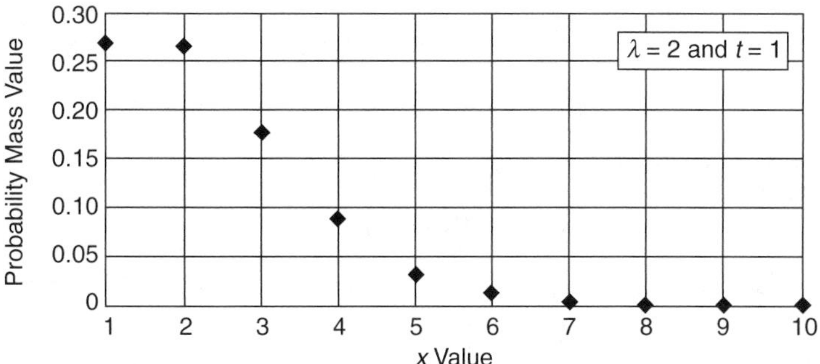

FIGURE 4-4 Probability mass function of the Poisson distribution.

The probability mass function of an example Poisson distribution is shown in Figure 4-4. The mean and variance for the Poisson distribution, respectively, are given by:

$$\mu_X = \lambda t \quad (4\text{-}15)$$

$$\sigma_X^2 = \lambda t \quad (4\text{-}16)$$

The parameter λ of the Poisson distribution represents the average rate of occurrence of the event of interest.

Example 4-4: Traffic Accidents

In Example 4-3, traffic accidents were used to illustrate the use of the geometric distribution. The traffic accident problem can be modeled more adequately using the Poisson distribution. In this case, the annual rate is assumed to equal the annual probability of fatal accidents per 1000 miles (i.e., $\lambda = 1.8 \times 10^{-5}$). The variable t is considered to be the travel distance in thousands of miles. Therefore, the probability of having one fatal accident in 10,000 miles of travel is:

$$P_{X_t}(x = 1) = \frac{e^{-\lambda t}}{x!}(\lambda t)^x \quad (4\text{-}17a)$$

Substituting produces:

$$P_{X_t}(x = 1) = \frac{\exp(-1.8 \times 10^{-5}(10))}{1!}(1.8 \times 10^{-5}(10))^1 = 1.7997 \times 10^{-4} \quad (4\text{-}17b)$$

It is important to note that this result is fundamentally different from the corresponding result in Example 4-3. This result gives the probability of one fatal accident in

10,000 miles that can happen at any 1000-mile interval. The result of Example 4-3 is the probability of having a fatal accident in the last 1000-mile interval of 10,000 miles of travel.

4.6. NEGATIVE BINOMIAL AND PASCAL PROBABILITY DISTRIBUTIONS

This section describes a select group of other discrete probability distributions that are related to the geometric distribution. The group consists of the negative binomial and the Pascal distribution.

The *negative binomial distribution* is considered a general case of the geometric distribution. Its underlying random variable is defined as the kth occurrence of an event of interest on the last trial in a sequence of X Bernoulli trials. The probability of this kth occurrence on the last trial is given by the probability mass function of the negative binomial distribution; that is,

$$P_X(x) = \begin{cases} \binom{x-1}{k-1} p^k (1-p)^{x-k} & \text{for } x = k, k+1, k+2, \ldots \\ 0 & \text{otherwise} \end{cases} \tag{4-18}$$

The mean and variance of this distribution, respectively, are given by:

$$\mu_X = \frac{k}{p} \tag{4-19a}$$

$$\sigma_X^2 = \frac{k(1-p)}{p^2} \tag{4-19b}$$

The negative binomial distribution is called the *Pascal distribution* if k takes on only integer values.

4.7. HYPERGEOMETRIC PROBABILITY DISTRIBUTION

The *hypergeometric distribution* deals with a finite population of size N, with a class of $D \le N$ elements of the population having a property of interest (e.g., defective units or nondefective units). A random sample is selected of size n without replacement; that is, a sampled element of the population is not replaced before randomly selecting the next element of the sample. The underlying random variable, X, for this distribution is defined as the number of elements in the sample that belong to the class of interest. The probability mass function is given by:

$$P_X(x) = \begin{cases} \dfrac{\binom{D}{x}\binom{N-D}{n-x}}{\binom{N}{n}} & \text{for } x = 0, 1, 2, \ldots, \min(n, D) \\ 0 & \text{otherwise} \end{cases} \tag{4-20}$$

The mean and variance of this distribution, respectively, are given by:

$$\mu_X = n\frac{D}{N} \tag{4-21a}$$

$$\sigma_X^2 = n\frac{D}{N}\frac{N-n}{N-1}\left(1-\frac{D}{N}\right) = \frac{nD(N-n)(N-D)}{N^2(N-1)} \tag{4-21b}$$

4.8. APPLICATIONS

4.8.1. Earthquakes and Structures

A structure is located in an earthquake zone, where ten damaging earthquakes of intensity VI have occurred in the last 100 years. For an earthquake of this intensity, the probability of failure of the structure is 0.01. The occurrence of earthquakes can be assumed to follow a Poisson distribution. Therefore, the time between consecutive earthquakes follows an exponential distribution. The rate of earthquake occurrence is:

$$\lambda = \frac{10}{100} = 0.1 \text{ earthquakes/year} \tag{4-22}$$

The probability of occurrence of X_t = two earthquakes in the next year is computed using the probability mass function of the Poisson distribution (Equation 4-14) as:

$$P_{X_t}(2) = \frac{[(0.1)(1)]^2}{2!}\exp[-(0.1)(1)] = 0.004524 \tag{4-23}$$

The probability of failure, P_f, of the structure during its design life of 50 years, due to earthquakes, can be computed using the theorem of total probability. Partitioning the sample space using the number of earthquakes (x) in 50 years results in the following failure probability for a given number of earthquakes (x):

$$P(\text{structural failure}|x \text{ earthquakes}) = [P(\text{structural failure}|\text{an earthquake})]^x \tag{4-24a}$$

Equation 4-24a is based on the assumption of independence between structural-failure events due to the x earthquakes. Using the theorem of total probability (Equation 3-34), the failure probability of the structure can be computed as:

$$P_f = \sum_{x=1}^{\infty} P(\text{earthquakes})[P(\text{structural failure}|\text{an earthquake})]^x \tag{4-24b}$$

In Equations 4-24a and 4-24b, a surviving structure after an earthquake is assumed to sustain no damage; therefore, based on the assumption that P(structural failure|an earthquake) = 0.01, Equation 4-24b can be rewritten as:

$$P_f = \sum_{x=1}^{\infty} \frac{[(0.1)(50)]^x}{x!} \exp[-(0.1)(50)](0.01)^x \tag{4-25}$$

Utilizing the following relationship for infinite series (see Appendix B):

$$\exp(x) = 1 + x + \frac{x^2}{2!} + \frac{x^3}{3!} + \cdots \tag{4-26}$$

P_f can be expressed as:

$$P_f = \exp[-0.1(50)]\{\exp[0.1(50)(0.01)] - 1\} = 0.0003455 \tag{4-27}$$

4.8.2. Floods and Coffer Dams

A specific coffer dam has a probability of 0.05 of being overtopped by a flood in any 1 week. We are interested in the probability that the first overflow will not occur until the xth week. Because there are only two possible outcomes, overtopped or not overtopped, and the probability of overtopping remains constant from week to week, the geometric probability distribution of Equation 4-11 can be used to approximate the probability that the first overtopping occurs in week $x = 1, 2, 3, \ldots, 7$ as:

$$P(X = 1) = (0.05)(0.95)^0 = 0.0500 \tag{4-28a}$$

$$P(X = 2) = (0.05)(0.95)^1 = 0.0475 \tag{4-28b}$$

$$P(X = 3) = (0.05)(0.95)^2 = 0.0451 \tag{4-28c}$$

$$P(X = 4) = (0.05)(0.95)^3 = 0.0429 \tag{4-28d}$$

$$P(X = 5) = (0.05)(0.95)^4 = 0.0407 \tag{4-28e}$$

$$P(X = 6) = (0.05)(0.95)^5 = 0.0387 \tag{4-28f}$$

$$P(X = 7) = (0.05)(0.95)^6 = 0.0368 \tag{4-28g}$$

In this case, the probability of the first occurrence remains fairly constant from week to week, for a relatively small number of weeks. On the average, the first overtopping occurs on the (1/0.05) week, or the 20th week (see Equation 4-12), with a standard deviation of first occurrence, σ_0, given by Equation 4-13 as:

$$\sigma_0 = \sqrt{\frac{1 - 0.05}{(0.05)^2}} = 19.5 \text{ weeks} \tag{4-29}$$

The probability of at least one overtopping in the next 10 weeks is computed using the binomial distribution of Equation 4-7 as:

$$P_X(\text{at least } 1) = 1 - P_X(0) \tag{4-30a}$$

$$= 1 - \frac{10!}{10!(0!)}(0.05)^0(0.95)^{10-0} = 0.40126 \tag{4-30b}$$

The probability that the dam will not be overtopped in a 2-year period (104 weeks) is:

$$P_X(0) = \binom{104}{0}(0.05)^0(0.95)^{104-0} = 0.0048 \tag{4-31}$$

The probability of one overtopping is:

$$P_X(1) = \binom{104}{1}(0.05)^1(0.95)^{104-1} = 0.0264 \tag{4-32}$$

The probability of more than one overtopping is:

$$P_X(x \text{ more than } 1) = 1 - 0.0048 - 0.0264 = 0.9688 \tag{4-33}$$

The probability of the dam not being overtopped can be increased by increasing the height of the dam. If the height of the dam is increased so that the probability of being overtopped in a 1-week period is decreased to 0.02, the probability of the dam not being overtopped in a 104-week period increases to:

$$P_X(0) = \binom{104}{0}(0.02)^0(0.98)^{104-0} = 0.1223 \tag{4-34}$$

Thus, the probability of the dam not being overtopped during the 104-week period increases 25 times when the probability of overtopping during 1 week is decreased from 0.05 to 0.02. Increasing the height of the dam requires additional resources but has the return that the probability of overtopping, which can result in damages, is reduced. Therefore, these probabilities can be used in a trade-off analysis, such as a benefit–cost study of alternative dam heights.

The binomial and geometric distributions were used in this example for the purpose of illustration. Because we are dealing with the random occurrence of an event in time, the Poisson distribution might be more appropriate to compute the above probabilities than the geometric distribution.

4.9. SIMULATION OF DISCRETE RANDOM VARIABLES

In Chapter 1, values of the flip of a coin and a roll of a die were simulated; both of these are discrete random variables. In Chapter 2, values for an exponential distribution were simulated. The discharge values of Example 2-17 are continuous variates. Both discrete and continuous variates can be simulated.

When using uniform variates to simulate discrete variates, the transformation graph uses a continuous scale for the uniform variate and a discrete scale on the axis for the discrete random variable. In such cases, the transformation graph can be portrayed as a series of spikes as shown in Chapter 1 or as a cumulative histogram as shown in Figure 2-21. They are used in the same way.

Example 4-5: Simulation of Pump Reliability

A mechanical engineering company produces hydraulic pumps. Over a long period of time, they have found that one in eight of the pumps is returned because of defects. If the company has orders to deliver six pumps per week for a period of 28 weeks, they would expect 1/8 of the 168 pumps, or 21, to be returned over the 28-week period. To assess the likelihood that multiple pumps will be returned in any week, the company decides to simulate a 28-week sequence.

Because a pump can only be defective or not defective and because defects in pumps are independent events, the sequence can be represented as a binomial process, with the probability of a defective pump being 1/8. Therefore, the probabilities for pumps being defective in a group of six are as follows:

Number of defective pumps	Binomial probability	Cumulative probability	
0	$\binom{6}{0}\left(\frac{1}{8}\right)^0\left(\frac{7}{8}\right)^6 = 0.448795$	0.448795	(4-35a)
1	$\binom{6}{1}\left(\frac{1}{8}\right)^1\left(\frac{7}{8}\right)^5 = 0.384682$	0.833477	(4-35b)
2	$\binom{6}{2}\left(\frac{1}{8}\right)^2\left(\frac{7}{8}\right)^4 = 0.137386$	0.970863	(4-35c)
3	$\binom{6}{3}\left(\frac{1}{8}\right)^3\left(\frac{7}{8}\right)^3 = 0.026169$	0.997032	(4-35d)
4	$\binom{6}{4}\left(\frac{1}{8}\right)^4\left(\frac{7}{8}\right)^2 = 0.002804$	0.999836	(4-35e)
5	$\binom{6}{5}\left(\frac{1}{8}\right)^5\left(\frac{7}{8}\right)^1 = 0.000160$	0.999996	(4-35f)
6	$\binom{6}{6}\left(\frac{1}{8}\right)^6\left(\frac{7}{8}\right)^0 = 0.000004$	1.000000	(4-35g)

TABLE 4-1 Simulation of Normal, Binomial, and Poisson Variates

(1) Week	(2) X^2	(3) X	(4) Random number	(5) Number of pumps returned (binomial)	(6) Normal random number	(7) Poisson variate
		1941 Seed				
1	03,767,481	7674	0.7674	1	0.7300	5
2	58,890,276	8902	0.8902	2	1.2274	7
3	79,245,604	2456	0.2456	0	–0.6884	3
4	06,031,936	0319	0.0319	0	–1.8538	1
5	00,101,761	1017	0.1017	0	–1.2718	2
6	01,034,289	342	0.0342	0	–1.8225	1
7	00,116,964	1169	0.1169	0	–1.1905	2
8	01,366,561	3665	0.3665	0	–0.3411	3
9	13,432,225	4322	0.4322	0	–0.1708	3
10	18,679,684	6796	0.6796	1	0.4667	5
11	46,185,616	1856	0.1856	0	–0.8946	2
12	03,444,736	4447	0.4447	0	–0.1390	4
13	19,775,809	7758	0.7758	1	0.7580	5
14	60,186,564	1865	0.1865	0	–0.8908	2
15	03,478,225	4782	0.4782	1	–0.0548	4
16	22,867,524	8675	0.8675	2	1.1148	6
17	75,255,625	2556	0.2556	0	–0.6569	3
18	06,533,136	5331	0.5331	1	0.0830	4
19	28,419,561	4195	0.4195	0	–0.2031	3
20	17,598,025	5980	0.5980	1	0.2482	4
21	35,760,400	7604	0.7604	1	0.7077	5
22	57,820,816	8208	0.8208	1	0.9185	6
23	67,371,264	3712	0.3712	0	–0.3287	3
24	01,778,944	7789	0.7789	1	0.6683	5
25	60,668,521	6685	0.6685	1	0.4358	5
26	44,689,225	6892	0.6892	1	0.4936	5
27	47,499,664	4996	0.4996	1	–0.0010	4
28	24,960,016	9600	0.9600	2	1.7511	8

The cumulative probability function can be used as the transformation function. Any random number less than 0.448795 would indicate zero defects. A value between 0.448795 and 0.833477 would indicate one defect.

To simulate a 28-week period, random numbers are generated using the midsquare method. The results are given in Table 4-1. A seed of 1941 was used. The sequence (column 5 of Table 4-1) indicates that only four pumps were returned in the first 10 weeks. However, in the last 10 weeks, a total of nine pumps were returned because of defects. In the 28-week period, 18 pumps were returned. Thus, the sample probability of a pump being returned was 18/[28(6)] = 0.1071, which is slightly less than the long-term average of 0.125. In 3 of the 28 weeks, more than one pump was returned. The simulation indicates that the probability of multiple pumps being returned in 1 week is 0.1071 (i.e., 3/28). Based on the long-term probability of 1/8, the expected rate can be computed from the cumulative function of Equations 4-35a to 4-35g. The probability of more than one pump being returned equals $1 - F_X(1)$, which would be $1 - 0.833477$, or 0.166523. If a simulation period longer than 28 weeks is used, the sample probability would be closer to the true probability. Again, the difference reflects sampling variation.

Example 4-6: Simulation of Traffic Flow I

To illustrate the generation of values that have a discrete distribution, uniform variates (continuous) are used to generate Poisson variates (discrete). The Poisson distribution was introduced in Section 4.5. Assume that the number of cars N arriving at a left-turn-only signal is Poisson distributed with a rate $\lambda = 4$. A traffic engineer is trying to design the length of the left-turn-only lane; cars wishing to turn left in excess of the design capacity would have to wait in the left lane, which represents a traffic hazard. For this simulation, a simplifying assumption that no cars remain from the previous cycle of the traffic signal is made. That is, all cars are able to turn left prior to the light turning red. The proportion of time that the left lane is occupied by cars wishing to use the left-turn-only lane is of interest to the traffic engineer.

The uniformly distributed random numbers for 28 cycles of the traffic signal were generated with the midsquare method, and the resulting values are given in column 4 of Table 4-1. The cumulative Poisson distribution for $\lambda = 4$ is given in Table 4-2. The uniform random numbers were entered into the cumulative function and the value of N found (see column 7 of Table 4-1). For example, for the first

TABLE 4-2 Cumulative Values of the Poisson Distribution with $\lambda = 4$ and $\lambda = 4.5$

(1)	(2)	(3)
N	$\sum_{x=0}^{N} p_{X_t}(x)$ with $\lambda = 4$	$\sum_{x=0}^{N} p_{X_t}(x)$ with $\lambda = 4.5$
0	0.0183	0.0111
1	0.0916	0.0611
2	0.2381	0.1736
3	0.4335	0.3423
4	0.6288	0.5321
5	0.7851	0.7029
6	0.8893	0.8311
7	0.9489	0.9134
8	0.9786	0.9597
9	0.9919	0.9829
10	0.9972	0.9933
11	0.9991	0.9976
12	0.9997	0.9992
13	0.9999	0.9997
14	1.0000	0.9999
15	1.0000	1.0000

TABLE 4-3 Sample Frequency Histogram and Simulated Frequency Distribution for Example 4-6

(1) N	(2) Sample frequency or count (f)	(3) fN	(4) Simulated proportion	(5) Cumulative proportion
0	0	0	0	0
1	2	2	0.071	0.071
2	4	8	0.143	0.214
3	6	18	0.214	0.428
4	5	20	0.179	0.607
5	7	35	0.250	0.857
6	2	12	0.071	0.928
7	1	7	0.036	0.964
8	1	8	0.036	1.000
Total	28	110	1.000	

cycle of the traffic light, the uniform variate is 0.7674. In Table 4-2, this corresponds to $N = 5$ because it is greater than 0.6288 and less than 0.7851. For the second cycle, a uniform variate of 0.8902 yields a Poisson variate of $N = 7$. The 28 values of N are given in column 7 of Table 4-1. A sample frequency histogram is tabulated in column 2 of Table 4-3. The sample cumulative distribution is tabulated in column 5 of Table 3-7. If the engineer set the length of the left-turn-only lane at five cars, then 14.3% of the time the lane capacity would be exceeded. For $\lambda = 4$, the true proportion is 0.2149 (1 – 0.7851 from Table 4-2). The difference between the sample and the population is the result of the small sample size (i.e., sampling variation).

Example 4-7: Simulation of Traffic Flow II

The above example does not illustrate the true value of simulation because the population was assumed to be Poisson with $\lambda = 4$. Thus, it was not really necessary to perform the simulation because the probability of N could be determined from the population. The example was also not overly realistic because an assumption was made that all cars in the queue were able to turn left before the light turned red. Anyone who has used a left-turn-only lane knows that this is not a realistic assumption. The value of simulation becomes more evident if random numbers are used for both the number of cars entering the left-turn-only lane and the number of cars that turn left during one cycle of the traffic light. Both of these are discrete random variables and two random variates are needed to simulate one cycle of the traffic light.

To illustrate the process, the problem of Example 4-6 can be modified by allowing a queue to form in the left-turn-only lane. The number of cars that turn during one cycle of the light is a discrete random variable, which is assumed to be of a Poisson distribution with $\lambda = 4.5$, with the cumulative distribution as given in column 3 of Table 4-2. A simulation run of 28 traffic-light cycles is given in Table 4-4. A separate

TABLE 4-4 Left-Turn-Only Simulation for Both Car Arrivals and Left Turns

(1)	(2)	(3)	(4)	(5)	(6)	(7)	(8)	(9)
4-digit value	Uniform variate	Car arrivals	Number in lane	4-digit value	Uniform variate	Number eligible for turn	Number turning	Number in lane
1941				5287				
				9523	0.9523			8
7674	0.7674	5	13	6875	0.6875	5	5	8
8902	0.8902	7	15	2656	0.2656	3	3	12
2456	0.2456	3	15	543	0.0543	1	1	14
319	0.0319	1	15	2948	0.2948	3	3	12
1017	0.1017	2	14	6907	0.6907	5	5	9
342	0.0342	1	10	7066	0.7066	6	6	4
1169	0.1169	2	6	9283	0.9283	8	6	0
3665	0.3665	3	3	1740	0.1740	3	3	0
4322	0.4322	3	3	276	0.0276	1	1	2
6796	0.6796	5	7	761	0.0761	2	2	5
1856	0.1856	2	7	5791	0.5791	5	5	2
4447	0.4447	4	6	5356	0.5356	5	5	1
7758	0.7758	5	6	6867	0.6867	5	5	1
1865	0.1865	2	3	1556	0.1556	2	2	1
4782	0.4782	4	5	4211	0.4211	4	4	1
8675	0.8675	6	7	7325	0.7325	6	6	1
2556	0.2556	3	4	6556	0.6556	5	4	0
5331	0.5331	4	4	9811	0.9811	9	4	0
4195	0.4195	3	3	2557	0.2557	3	3	0
5980	0.5980	4	4	5382	0.5382	5	4	0
7604	0.7604	5	5	9659	0.9659	9	5	0
8208	0.8208	6	6	2962	0.2962	3	3	3
3712	0.3712	3	6	7734	0.7734	6	6	0
7789	0.7789	5	5	8147	0.8147	6	5	0
6685	0.6685	5	5	3736	0.3736	4	4	1
6892	0.6892	5	6	9576	0.9576	8	6	0
4996	0.4996	4	4	6997	0.6997	5	4	0
9600	0.9600	8	8	9580	0.958	8	8	0

midsquare transformation is used for the two random variables. A seed of 1941 is used for cars entering the queue, and a seed of 5287 is used for simulating the number of cars that are able to make the left turn before the light turns red. A random variate is used to decide on the initial number of cars in the left-turn-only lane. The seed of 5287 leads to the first generated variate of 0.9523. This is used to compute the initial number of cars waiting to make a left turn. Referring to Table 4-2 (column 3) yields an initial queue length of eight cars. For the first traffic-light cycle, the random variate of 0.7674 indicates that five additional cars enter the queue (column 2 of Table 4-2). This is shown in column 3 of Table 4-4. The random variate of 0.6875 (column 6) indicates that time and traffic from the opposing direction would permit five cars to make a left turn. Thus, during the first cycle, five vehicles entered the queue and the five cars at the front of the queue made the left turn, which leaves eight cars in the queue. Column 9 of Table 4-4 gives the number in the left-turn-only lane at the end of the green-light part of the cycle. If we assume that those cars that make the left turn do so at the end of the cycle, then the number of cars in both the left-turn-only and left lanes would be the number at the start of the cycle plus the number of arrivals. This queue length is given in column 4 of Table 4-4. These values show if the design engineer only allowed for five cars in the left-turn-only lane, cars would usually be on hold in the left lane. Thus, a longer left-turn-only lane is warranted.

4.10. PROBLEMS

4-1. The probability of a flood in any 1 year is 0.05. In a 10-year period, what is the probability of (a) no floods, (b) two or fewer floods, (c) from one to three floods?

4-2. Piles are used to support the foundation of a structure. The failure probability of a pile during proof testing is 0.1. Determine the following:

a. The probability of three failed piles out of ten tested piles
b. The probability of no failures in 20 tested piles
c. The probability of ten tested piles to obtain the first failure
d. The probability of three consecutive pile failures

4-3. A fair coin is tossed ten times. What is the probability of getting (a) exactly six heads, (b) exactly three tails, (c) no heads, (d) three or fewer heads, (e) six or more tails, and (f) at least three but not more than eight heads?

4-4. The probability that a flood of a specified magnitude occurs in any 1 year is 0.1. What is the probability that in the next 10 years (a) exactly two such floods will occur, (b) no more than one such flood will occur, (c) no such floods will occur, (d) at least four floods will occur, and (e) at least three but no more than six such floods will occur?

4-5. Over the past 15 years, the numbers of hurricanes per year to cause damage in a certain city were 4, 2, 1, 3, 0, 2, 1, 3, 5, 2, 3, 1, 1, 2, and 0. What is the probability of not having hurricane damage in any 1 year? What are the average and variance of the number of hurricanes in 1 year?

4-6. The overload occurrence of an offshore facility due to extreme waves can be modeled by a Poisson process with a rate $\lambda = 0.5$. What is the probability of occurrence of two overloads in a year? What is the probability of no overloads in a year? What is the probability of one or two overloads in the design life of the structure of 20 years?

4-7. Derive Equations 4-15 and 4-16 using Equations 3-70 and 3-74.

4-8. The failure rate of an electronic device is $\lambda = 0.001$ per year of continuous operation. What is the probability of occurrence of two failures in a year? What is the probability of no failures in 5 years? What is the probability of one or two failures in the design life of the device of 10 years?

4-9. Extreme wave loads follows a Poisson distribution with an occurrence rate $\lambda = 0.01$ per year. What is the probability of occurrence of two extreme loads in a year? What is the probability of no extreme loads in 2 years? What is the probability of one or two load occurrences during a period of 25 years?

4-10. Use the negative binomial distribution to compute the probability of having a third success on the 10th trial for a success probability in a trial (p) of 0.1.

4-11. Use the negative binomial distribution to compute the probability of having a 10th success on the 10th trial for a success probability in a trial (p) of 0.1.

4-12. A lot of products contains 100 units with 10 defective units among these 100 units. Compute the probability of sampling at least 1 defective units in a sample of size 10 units. Is a sample size of 10 adequate? Discuss your results.

4-13. An automobile manufacturer is considering the recall of 100,000 cars to correct a safety device in them. The number of automobiles with defective safety devices is estimated to be about 10% of the 100,000 automobiles. Determine the expected number of defective items in a sample recall of 10 automobiles. What is the probability of not receiving any automobiles with defective devices in a sample of 10 automobiles? What is this probability if the sample size is increased from 10 to 1000 automobiles? Plot the variation of the probability as a function of sample size from a size of 10 to 1000 in increments of 10. (*Hint:* Use the hypergeometric distribution.)

4-14. A defense contractor developed a new welding process of high-strength metal for use in constructing submarines. The new process was used and the resulting welds were tested using a specialized nondestructive testing method. The testing produced a reliability level of 99.9% for the length of welds. The submarine construction is expected to require 10,000 ft of welding at various locations of the submarine hull. Determine the expected length of defective welds in the 10,000 ft of welds. Modeling each 1 ft of weld as a Bernoulli trial, what is the probability of having 1 ft or more of defective weld length based on nondestructive examination of sample welds of 500 ft? What is this probability if the sample size is varied from 10 to 10,000 ft as follows: 10, 20, 50, 100, 1000, 5000, and 10,000? Plot your results. Can you recommend a sample size?

4-15. A *k-out-of-n* system is defined as a system that functions if and only if at least k out of the n individual components in the system function. If individual components function independently of one another, each with a probability of 0.95, compute the probability that a 3-out-of-5 system functions.

4-16. During World War II, an economical procedure for testing for syphilitic (i.e., diseased) men among military inductees was initiated to test the blood of groups of n men by combining the blood samples of all n men in a group and performing one test on the combined sample. If no one in the group had the disease, the test would be negative, and only the one test was required. If at least one individual was diseased, the test on the combined sample would yield a positive result, in which case n individual tests would then be carried out. Assume that p is the probability that a randomly selected man has the disease. Compute the expected number of tests if $p = 0.1$ and $n = 3$. Evaluate and plot the expected number of tests as n is increased from 3 to 10 in increments of 1.

4-17. It was reported in 1993 that 1 in 200 carry the defective gene that causes inherited colon cancer. For a random sample of 1000 individuals, what is the distribution of the number of individuals who carry the gene? Compute the probability that 4 to 10 (inclusive) individuals out of the 1000 selected carry the gene and the probability that at least 6 carry the gene.

4-18. The arrival of aircraft to an airport can be modeled as a Poisson process with a rate of 10 arrivals per hour.

a. What is the probability that exactly 5 aircraft arrive in a 1-hour period?
b. What is the probability that at least 5 aircraft arrive in a 1-hour period?
c. What is the probability that at least 10 aircraft arrive in a 1-hour period?
d. The airport has a policy of requiring aircraft to wait in a holding position in the air if the number of arriving aircraft in any 10-minute period exceeds 2. What is the occurrence probability of an aircraft holding event?
e. Recompute and plot the occurrence probability of an aircraft holding event in part (d) as a function of the rate of the Poisson process for rates ranging from 10 to 50.

4-19. Construct a transformation curve that transforms uniform variates on a scale from 0 to 1 to uniform variates on a scale from 10 to 20 of integer values. Generate 30 variates using the transformation curve to obtain 30 uniform variates on a scale from 10 to 20 of integers. Discuss the results of the transformation.

4-20. Construct a transformation curve that transforms uniform variates on a scale from 0 to 1 to binomial values for $N = 10$ and $p = 0.1$. Generate 30 variates using the transformation curve to obtain 30 binomial variates. Discuss the results of the transformation.

4-21. Construct a transformation curve that transforms uniform variates on a scale from 0 to 1 to geometric values for $p = 0.1$. Generate 30 variates using the transformation curve to obtain 30 geometric variates. Discuss the results of the transformation.

4.11. SIMULATION PROJECTS

This section provides additional work items for the problem statements of the four simulation projects introduced in Sections 1.5, 2.9, and 3.9.

4.11.1. Structural Beam Study

Using the projects information provided in Sections 1.5.1, 2.9.1, and 3.9.1, assume that the occurrence of the load applied to the beam follows a Poisson distribution with a rate of four per year. Evaluate the failure probability of the beam in 2 years using the failure probability of the beam for a given load occurrence based on the 100 simulation cycles that utilized uniform probability distributions for all random variables of Table 1-2. The computations of the failure probability of the beam for a given load occurrence should be based on the requirements of Section 2.9.1. For this purpose, use Table 4-2 to generate 100 Poisson occurrences of loading and assume independent failure events in the 2-year period. Next, evaluate the failure probability of the beam in 3, 4, 5, ..., 20 years, and plot your results.

4.11.2. Stream Erosion Study

The project information of Sections 1.5.2, 2.9.2, and 3.9.2 provides background information. The discharge rate could reach extreme levels during flooding conditions. Generate 100 values for the number of floods in 10 years based on a Poisson process with a rate of 0.5 flood per year. Construct a table similar to Table 4-2 to generate 100 Poisson occurrences of floods.

4.11.3. Traffic Estimation Study

The project information of Sections 1.5.3, 2.9.3, and 3.9.3 provides background information. The number of vehicles in a zone is assumed to follow a Poisson distribution for the purpose of illustration. Generate 100 values for the number of vehicles in a zone in 5 days based on a Poisson process with a rate of 100 vehicles per day. Construct a table similar to Table 4-2 to generate 100 Poisson occurrences of the number of vehicles in a zone.

4.11.4. Water Evaporation Study

The project information of Sections 1.5.4, 2.9.4, and 3.9.4 provides background information. High temperature and wind conditions can greatly increase evaporation. Generate 100 values for the number of extreme temperature conditions in 10 years based on a Poisson process with a rate of 0.5 high-temperature occurrence per year. Construct a table similar to Table 4-2 to generate 100 Poisson occurrences.

Chapter 5
Probability Distributions for Continuous Random Variables

TABLE OF CONTENTS

5.1. INTRODUCTION

Any function that results in probabilities satisfying the axioms of probability is qualified to be a probability distribution to model continuous random variables. Experience with the various uses of probability distributions and improved understanding of the underlying physical processes resulted in the development of several probability distributions that are now put to practical use by engineers and scientists. The objective of this chapter is to introduce these probability distributions that are used to model continuous random variables. They are commonly called continuous probability distributions. This chapter describes these commonly used continuous probability distributions (Chapter 4 describes commonly used discrete probability distributions).

A probability distribution function is expressed as a real-valued function of the random variable. The location, scale, and shape of the function are determined by its parameters. Distributions commonly have one to three parameters. These parameters take certain values that are specific for the problem and conditions being investigated. The parameters of a distribution can be expressed in terms of its moments (e.g., mean, variance, and skewness), but not necessarily in closed-form expressions.

In this section, four continuous distributions are emphasized: uniform, normal, lognormal, and exponential. The uniform distribution is very important for performing random-number generation in simulation, as described in Chapter 7. The normal and lognormal distributions are important due to their common use and applications in engineering and science. These two distributions also have an important and unique relationship. The importance of the exponential distribution comes from its special relation to the Poisson distribution. The triangular, gamma, and beta distributions are also described in this chapter. In addition, sampling distributions that include the Student's t distribution, the chi-square distribution, and the F distribution are described due to their use in statistics. This chapter also includes extreme value distributions and some fundamental concepts of extreme value analysis.

5.2. UNIFORM DISTRIBUTION

The density function for the uniform distribution of a random variable X is given by:

$$f_X(x) = \begin{cases} \dfrac{1}{b-a} & \text{for } a \le x \le b \\ 0 & \text{otherwise} \end{cases} \tag{5-1}$$

where a and b are real values, called parameters, with $a < b$. The density function for the uniform distribution takes a constant value of $\frac{1}{b-a}$ in order to satisfy the probability axiom, which requires that the area under the density function be one. An example of a uniform density function is shown in Figure 5-1. The mean and variance for the uniform distribution, respectively, are given by:

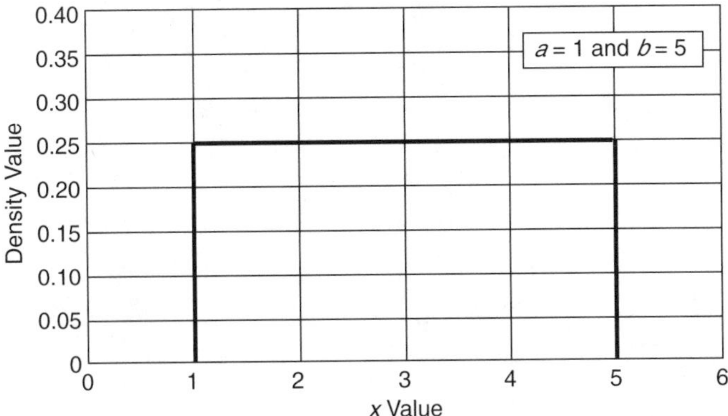

FIGURE 5-1 Probability density function of the uniform distribution.

$$\mu_X = \frac{a+b}{2} \tag{5-2}$$

$$\sigma_X^2 = \frac{(b-a)^2}{12} \tag{5-3}$$

Due to the simple geometry of the density function of the uniform distribution, it can be easily noticed that its mean value and variance, as given by Equations 5-2 and 5-3, respectively, correspond to the centroidal distance and centroidal moment of inertia, respectively, with respect to a vertical axis of the area under the density function. This property is valid for other distributions as well. The cumulative function for the uniform distribution is a line with a constant slope and is given by:

$$F_X(x) = \begin{cases} 0 & x \le a \\ \dfrac{x-a}{b-a} & a \le x \le b \\ 1 & x \ge b \end{cases} \tag{5-4}$$

Example 5-1: Subjective Assessment of Concrete Strength

Subjectively, as well as objectively, determined information is commonly used in engineering for the purpose of decision making. For example, a structural engineer might assess, based on his or her experience, the strength of concrete in an existing bridge to be in the range of 3 to 4 ksi. A uniform probability distribution can be used to estimate probabilities in this case, if the engineer believes that all the values in this range are equally likely. Therefore, the mean value of concrete strength is 3.5 ksi, and its variance is 1/12 $(\text{ksi})^2$, which gives a standard deviation of 0.2887 ksi. The probability that the strength of concrete X is larger than 3.6 ksi is:

$$P(X > 3.6) = 1 - F_X(3.6) = 1 - \frac{3.6-3}{4-3} = 0.4 \tag{5-5}$$

FIGURE 5-2 Carl F. Gauss (1777–1855) as shown on 1993 ten German mark.

5.3. NORMAL DISTRIBUTION

The normal distribution (also called the Gaussian distribution, after Carl F. Gauss [1777–1855]; see Figure 5-2) is used widely due to its simplicity and wide applicability. This distribution is the basis for many statistical methods. The normal density function for a random variable X is given by:

$$f_X(x) = \frac{1}{\sigma\sqrt{2\pi}}\exp\left(-\frac{1}{2}\left[\frac{x-\mu}{\sigma}\right]^2\right) \quad -\infty < x < \infty \tag{5-6}$$

It is common to use the notation $X \sim N(\mu,\sigma^2)$ to provide an abbreviated description of a normal distribution. The notation states that X is normally distributed with a mean value μ and variance σ^2. The normal distribution has several important properties, including the following:

1. $f_X(x)$ approaches 0 as x approaches either $-\infty$ or ∞.
2. $f_X(a + \mu) = f_X(-a + \mu)$ for any a (i.e., symmetric density function about the mean μ).
3. The maximum value of $f_X(x)$ (i.e., the mode of the distribution) occurs at $x = \mu$.
4. The inflection points of the density function occur at $x = \mu \pm \sigma$.
5. The density function has an overall bell shape.
6. The mean value μ and the variance σ^2 are also the parameters of the distribution.

In Figure 5-3A, the normal distribution is used to model the concrete strength of Example 5-1; however, now we are assuming that concrete strength has a normal distribution with a mean = 3.5 ksi and standard deviation = 0.2887 ksi. The density function of another normal distribution is shown in Figure 5-3B. The cumulative distribution function of the normal distribution is given by:

$$F_X(x) = \int_{-\infty}^{x} \frac{1}{\sigma\sqrt{2\pi}}\exp\left(-\frac{1}{2}\left[\frac{x-\mu}{\sigma}\right]^2\right)dx \tag{5-7}$$

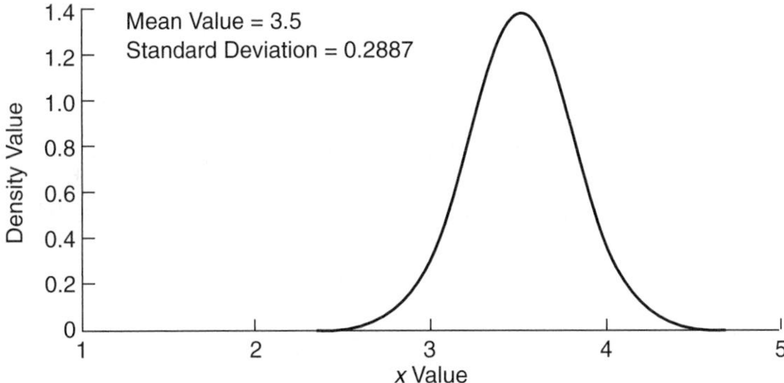

FIGURE 5-3A Probability density function of the normal distribution.

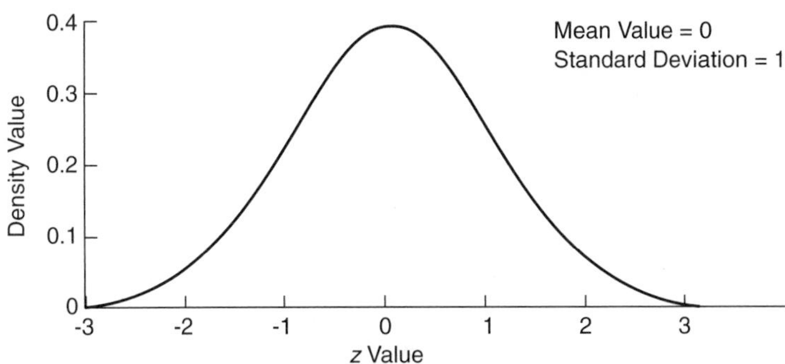

FIGURE 5-3B Probability density function of the standard normal distribution.

The evaluation of the integral of Equation 5-7 requires numerical methods for each pair (μ,σ^2). This difficulty can be reduced by performing a transformation that results in a standard normal distribution with a mean $\mu = 0$ and variance $\sigma^2 = 1$, denoted as $Z \sim N(0,1)$. Numerical integration can be used to determine the cumulative distribution function of the standard normal distribution. Therefore, by utilizing the transformation between the normal distribution $X \sim N(\mu,\sigma^2)$ and the standard normal distribution $Z \sim N(0,1)$, and the integration results for the standard normal, the cumulative distribution function for the normal distribution can be evaluated by using the following standard normal transformation:

$$Z = \frac{X - \mu}{\sigma} \tag{5-8}$$

The density function, which is shown in Figure 5-3B, and the cumulative distribution function of the standard normal, respectively, are determined as:

$$\phi(z) = \frac{1}{\sqrt{2\pi}} \exp\left(-\frac{1}{2}z^2\right) \qquad -\infty < x < \infty \tag{5-9}$$

$$\Phi(z) = \int_{-\infty}^{z} \frac{1}{\sqrt{2\pi}} \exp\left(-\frac{1}{2}z^2\right) dz \tag{5-10}$$

where $\phi(z)$ is a special notation for the probability density function of the standard normal, and $\Phi(z)$ is a special notation for the cumulative distribution function of the standard normal. The results of the integral $\Phi(z)$ are tabulated in Table A-1 of Appendix A. The tabulated values are for z $[0, 8.2]$. Negative z values can be obtained utilizing the symmetry property of the normal distribution; that is,

$$\Phi(-z) = 1 - \Phi(z) \tag{5-11}$$

The tabulated Φ values for $z > 0$ are in the range $[0.5, 1.0)$. The table can also be used for determining the inverse of the Φ function (i.e., Φ^{-1}). For values of specified $p = \Phi(z)$ that are less than 0.5, Table A-1 can be used with the help of the following relation to obtain the z values:

$$z = \Phi^{-1}(p) = -\Phi^{-1}(1 - |p|) \quad \text{for } p < 0.5 \tag{5-12}$$

Equation 5-12 is based on the symmetry property of the normal distribution. Spreadsheet program have standard functions for the standard normal cumulative distribution function and its inverse, called NORMSDIST(z) and NORMSINV(p), respectively.

Using the transformation of Equation 5-8, probabilities based on any normal distribution can be easily computed with the aid of the tabulated Φ values. For example, the following cumulative probability for any $X \sim N(\mu, \sigma)$ can be computed by changing the variable of integration according to Equation 5-8:

$$P(X \le x) = \int_{-\infty}^{x} \frac{1}{\sigma\sqrt{2\pi}} \exp\left[-\frac{1}{2}\left(\frac{x-\mu}{\sigma}\right)^2\right] dx \tag{5-13}$$

Then, the probability is given by:

$$P(X \le x) = \int_{-\infty}^{\frac{x-\mu}{\sigma}} \frac{1}{\sigma\sqrt{2\pi}} \exp\left(-\frac{z^2}{2}\right) \sigma dxz \tag{5-14}$$

Therefore,

$$P(X \le x) = \int_{-\infty}^{\frac{x-\mu}{\sigma}} \frac{1}{\sqrt{2\pi}} \exp\left(-\frac{z^2}{2}\right) dz = \Phi\left(\frac{x-\mu}{\sigma}\right) \tag{5-15}$$

It can also be shown that

$$
\begin{aligned}
P(a < X \leq b) &= F_X(b) - F_X(a) \\
&= \Phi\left(\frac{b-\mu}{\sigma}\right) - \Phi\left(\frac{a-\mu}{\sigma}\right)
\end{aligned} \tag{5-16}
$$

The normal distribution has several important and useful properties that include the following:

1. The addition of n normally distributed random variables, $X_1, X_2, \ldots, X_n$, is a normal distribution as follows:

$$Y = X_1 + X_2 + X_3 + \ldots + X_n \tag{5-17}$$

The mean of Y (i.e., μ_Y) is:

$$\mu_Y = \mu_{X_1} + \mu_{X_2} + \mu_{X_3} + \ldots + \mu_{X_n} \tag{5-18}$$

The variance of Y (i.e., σ_Y^2) is:

$$\sigma_Y^2 = \sigma_{X_1}^2 + \sigma_{X_2}^2 + \sigma_{X_3}^2 + \ldots + \sigma_{X_n}^2 \tag{5-19}$$

2. *Central limit theorem:* Informally stated, the addition of a number of individual random variables, without a dominating distribution type, approaches a normal distribution as the number of the random variables approaches infinity. The result is valid regardless of the underlying distribution types of the random variables. In practice, the number of the random variables needs to be sufficiently large.

Example 5-2: Subjective Assessment of Concrete Strength

Assume that the structural engineer of Example 5-1 decided to use a normal distribution to model the strength of concrete. The mean and variance are the same as before: 3.5 ksi and 1/12 $(\text{ksi})^2$, respectively (see Figure 5-3A); therefore, the probability that the concrete strength is larger than 3.6 ksi is:

$$
\begin{aligned}
P(X > 3.6 \text{ ksi}) &= 1 - P(X \leq 3.6) \\
&= 1 - \Phi\left[\frac{3.6 - 3.5}{\sqrt{1/12}}\right] = 1 - \Phi(0.3464)
\end{aligned} \tag{5-20}
$$

Using linear interpolation in Table A-1, $\Phi(0.3464)$ is given by:

$$\Phi(0.3464) = 0.633072 + (0.3464 - 0.34)\left(\frac{0.636831 - 0.633072}{0.35 - 0.34}\right) = 0.635478 \tag{5-21}$$

Therefore,

$$P(X > 3.6 \text{ ksi}) = 1 - 0.635478 = 0.364522 \tag{5-22}$$

The probability that the concrete strength is larger than 3.6 ksi was determined in Example 5-1 to be 0.4. The difference between the estimates of 0.4 and 0.36452 reflects the disparity that can result from selection of the probability distribution. Care should be exercised in selecting a probability distribution because the selection affects the resulting probabilities.

5.4. LOGNORMAL DISTRIBUTION

A random variable X is considered to have a lognormal distribution if $Y = \ln(X)$ has a normal probability distribution, where $\ln(x)$ is the natural logarithm to the base e. The density function of the lognormal distribution is given by:

$$f_X(x) = \frac{1}{x\sigma_Y\sqrt{2\pi}}\exp\left[-\frac{1}{2}\left(\frac{\ln x - \mu_Y}{\sigma_Y}\right)^2\right] \quad \text{for } 0 < x < \infty \tag{5-23}$$

where μ_y and σ_y are the parameters. It is common to use the notation $X \sim LN(\mu_Y, \sigma_Y^2)$ to provide an abbreviated description of a lognormal distribution. The notation states that X is lognormally distributed with the parameters μ_Y and σ_Y^2. The lognormal distribution has the following properties:

1. The values of the random variable X are positive (i.e., $x > 0$).
2. $f_X(x)$ is not a symmetric density function about the mean value μ_X.
3. The mean value μ_X and variance σ_X^2 are not equal to the parameters of the distribution (i.e., μ_Y and σ_Y^2). However, they are related to them as follows:

$$\sigma_Y^2 = \ln\left[1 + \left(\frac{\sigma_X}{\mu_X}\right)^2\right] \tag{5-24}$$

$$\mu_Y = \ln(\mu_x) - \frac{1}{2}\sigma_Y^2 \tag{5-25}$$

These two relations can be inverted as follows:

$$\mu_X = \exp\left(\mu_Y + \frac{1}{2}\sigma_Y^2\right) \tag{5-26}$$

$$\sigma_X^2 = \mu_X^2[\exp(\sigma_Y^2) - 1] \tag{5-27}$$

For a relatively small coefficient of variation δ_X — for example, $(\sigma_X/\mu_X) \leq 0.3$ — σ_Y is approximately equal to the coefficient of variation δ_X. An example density function of the lognormal distribution is shown in Figure 5-4.

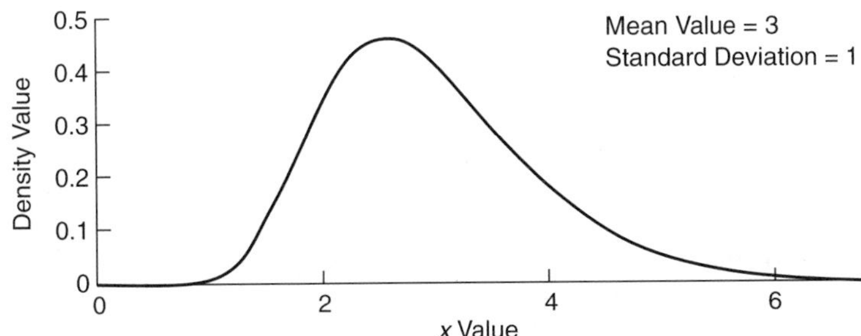

FIGURE 5-4 Probability density function of lognormal distribution.

The cumulative distribution function of the lognormal distribution can be determined based on its relationship to the normal distribution. Similar to the normal distribution, the difficulty of performing the integral of the cumulative distribution function can be reduced by performing a transformation that results in a standard normal distribution with a mean $\mu = 0$ and variance $\sigma^2 = 1$, denoted as $Z \sim N(0,1)$. Numerical integration can be used to determine the cumulative distribution function of the standard normal. Therefore, by utilizing the transformation relationship between the lognormal distribution $X \sim LN(\mu_Y, \sigma_Y^2)$ and the standard normal distribution $Z \sim N(0,1)$, and the integration results for the standard normal, the cumulative distribution function for the lognormal distribution can be evaluated using the following transformation:

$$Z = \frac{\ln X - \mu_Y}{\sigma_Y} \tag{5-28}$$

The density function and the cumulative distribution function of the standard normal, respectively, are given by Equations 5-9 and 5-10. They can, therefore, be used to compute probabilities for the lognormal distribution. The results of the integral $\Phi(z)$ are given in Table A-1 of Appendix A. The symmetry property of the standard normal should be used to obtain a complete range of values for the standard variate similar to the normal distribution.

Using the transformation of Equation 5-28, probabilities based on any lognormal distribution can be easily computed with the aid of the tabulated Φ values. For example, the cumulative probability $P(X \leq x)$ for any $X \sim LN(\mu_Y, \sigma_Y^2)$ can be computed by changing the variable of integration according to Equation 5-28. Therefore, the cumulative probability is given by:

$$P(X \leq x) = \int_{-\infty}^{\frac{\ln x - \mu_Y}{\sigma_Y}} \frac{1}{\sqrt{2\pi}} \exp\left(-\frac{z^2}{2}\right) dz = \Phi\left(\frac{\ln x - \mu_Y}{\sigma_Y}\right) \tag{5-29}$$

It can be also shown that

$$P(a < X \leq b) = F_X(b) - F_X(a)$$
$$= \Phi\left(\frac{\ln b - \mu_Y}{\sigma_Y}\right) - \Phi\left(\frac{\ln a - \mu_Y}{\sigma_Y}\right) \quad (5\text{-}30)$$

The lognormal distribution has several important properties that include the following:

1. The multiplication of n lognormally distributed random variables, X_1, X_2, …, X_n, is a lognormal distribution with the following statistical characteristics:

$$W = X_1 X_2 X_3 \cdots X_n \quad (5\text{-}31)$$

The first moment of W (i.e., μ_W) is:

$$\mu_W = \mu_{Y_1} + \mu_{Y_2} + \mu_{Y_3} + \cdots + \mu_{Y_n} \quad (5\text{-}32)$$

The second moment of W (i.e., σ_W^2) is:

$$\sigma_W^2 = \sigma_{Y_1}^2 + \sigma_{Y_2}^2 + \sigma_{Y_3}^2 + \cdots + \sigma_{Y_n}^2 \quad (5\text{-}33)$$

2. *Central limit theorem:* In this case, the multiplication of a number of individual random variables, without a dominating distribution type, approaches a lognormal distribution as the number of the random variables approaches infinity. The result is valid regardless of the underlying distribution types of the random variables. In practice, the number of the random variables needs to be sufficiently large.

Example 5-3: Subjective Assessment of Concrete Strength

Because the strength of concrete takes on only positive values, someone might argue against using the normal distribution to model concrete strength because it allows for negative values. The lognormal distribution has the property of being defined only for positive values for the underlying random variable, so let's assume that the structural engineer of Examples 5-1 and 5-2 decides to use a lognormal distribution to model concrete strength. The mean value and variance are the same as before: 3.5 ksi and 1/12 $(\text{ksi})^2$. Therefore, the parameters of the lognormal distribution can be determined using Equations 5-24 and 5-25 as follows:

$$\sigma_Y^2 = \ln\left(1 + \frac{1/12}{3.5^2}\right) = 0.00678 \quad (5\text{-}34a)$$

$$\mu_Y = \ln(3.5) - \frac{1}{2}(0.00678) = 1.25 \tag{5-34b}$$

The probability that the concrete strength is greater than 3.6 ksi can be determined as:

$$\begin{aligned} P(X > 3.6 \text{ ksi}) &= 1 - P(X \le 3.6) \\ &= 1 - \Phi\left[\frac{\ln(3.6) - 1.25}{\sqrt{0.00678}}\right] \\ &= 1 - \Phi(0.3833) = 0.3507 \end{aligned} \tag{5-35}$$

The answer in this case is slightly different from the corresponding value of Example 5-2. It should be noted that this positive property of the random variable of a lognormal distribution should not be used as the only basis for justifying its use. Statistical bases for selecting probability distribution can be used as discussed in Chapter 9.

5.5. EXPONENTIAL DISTRIBUTION

The importance of this distribution comes from its relationship to the Poisson distribution. For a given Poisson process, the time T between consecutive occurrences of events has an exponential distribution with the following density function:

$$f_T(t) = \begin{cases} \lambda\exp(-\lambda t) & \text{for } t \ge 0 \\ 0 & \text{otherwise} \end{cases} \tag{5-36}$$

The cumulative distribution function is given by:

$$F_T(t) = 1 - \exp(-\lambda t) \tag{5-37}$$

The density and cumulative functions of the exponential distribution with $\lambda = 1$ are shown in Figures 5-5A and 5-5B, respectively. The mean value and the variance, respectively, are given by:

$$\mu_T = \frac{1}{\lambda} \tag{5-38}$$

$$\sigma_T^2 = \frac{1}{\lambda^2} \tag{5-39}$$

Based on the means of the exponential and Poisson distributions, the *mean recurrence time* (or *return period*) is defined as $\frac{1}{\lambda}$.

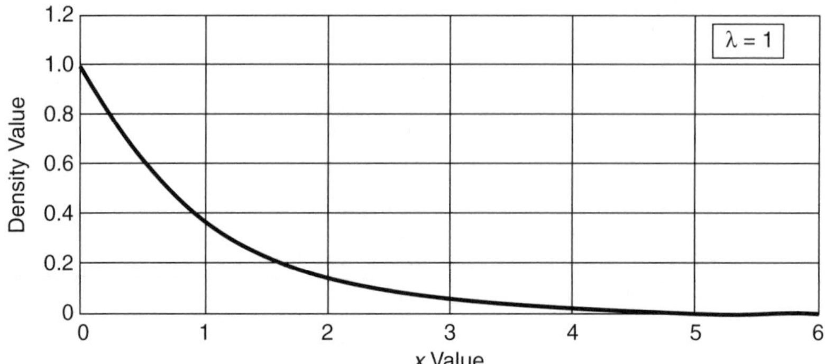

FIGURE 5-5A Probability density function of the exponential distribution.

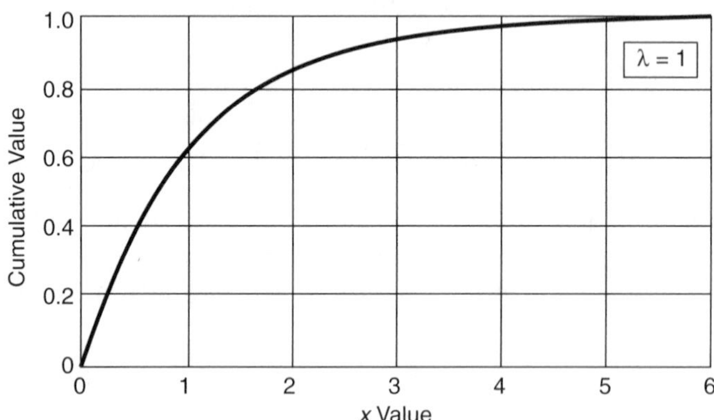

FIGURE 5-5B Cumulative distribution function of the exponential distribution.

Example 5-4: Earthquake Occurrence

The occurrence of earthquakes is commonly modeled by a Poisson process. Assume that the occurrence of a strong earthquake in a certain region was modeled by a Poisson process with an annual (λ) rate of 0.01. The probability of occurrence of exactly one earthquake in 1 year is given by:

$$P_{X_t}(x=1) = (\lambda t)^x \frac{e^{-\lambda t}}{x!} = [(0.01)(1)]^1 \frac{e^{-0.01(1)}}{1!} = 0.00990 \tag{5-40}$$

The probability of having one or more earthquakes in 1 year is:

$$P_{X_t}(x \geq 1) = 1 - P_{X_t}(x=0) = 1 - \frac{e^{-\lambda t}}{x!}(\lambda t)^x \tag{5-41a}$$

Substituting the values of λ and t produces:

$$P_{X_t}(x \geq 1) = 1 - \frac{e^{-0.01(1)}}{0!}[(0.01)(1)]^0 = 0.00995 \tag{5-41b}$$

The return period for such an earthquake is $\frac{1}{\lambda} = 100$ years. A return period of 100 years does not imply that it is only possible to have one earthquake in a 100-year period; in fact, there may be none or any number of earthquakes in any 100-year period. However, on the average, there will be one every 100 years. The probability of having an earthquake for the first time more than 100 years after the last occurrence is computed using the exponential distribution (Equation 5-37) as:

$$\begin{aligned} P(T > 100 \text{ years}) &= 1 - P(T \leq 100 \text{ years}) \\ &= 1 - [1 - e^{-0.01(100)}] = 0.3679 \end{aligned} \tag{5-42}$$

This result can be generalized to the following probability of an event, which follows a Poisson process, occurring (for the first time) at a time that exceeds *its return period*:

$$\begin{aligned} P\left(T > \frac{1}{\lambda}\right) &= 1 - P\left(T \leq \frac{1}{\lambda}\right) \\ &= 1 - [1 - e^{-\lambda(1/\lambda)}] = 0.3679 \end{aligned} \tag{5-43}$$

5.6. TRIANGULAR DISTRIBUTION

This distribution is used to qualitatively model an uncertain variable that can be bounded between two limits, such as the duration of a construction activity. For example, the duration of a construction activity can be described by the following density function:

$$f_X(x) = \begin{cases} \dfrac{2}{b-a}\left(\dfrac{x-a}{c-a}\right) & a \leq x \leq c \\ \\ \dfrac{2}{b-a}\left(\dfrac{b-x}{b-c}\right) & c \leq x \leq b \end{cases} \tag{5-44}$$

and 0 otherwise, where a, b, and c are lower limit, upper limit, and mode, respectively. The cumulative distribution function is given by:

$$F_X(x) = \begin{cases} 0 & x \leq a \\ \dfrac{(x-a)^2}{(c-a)(b-a)} & a \leq x \leq c \\ 1 - \dfrac{(b-x)^2}{(b-c)(b-a)} & c \leq x \leq b \\ 1 & x \geq b \end{cases} \tag{5-45}$$

The inverse of the cumulative distribution function can be computed as:

$$x = \begin{cases} a + \sqrt{(F_X)(c-a)(b-a)} & 0 \le F_X \le \dfrac{c-a}{b-a} \\ a - \sqrt{(1-F_X)(b-c)(b-a)} & \dfrac{c-a}{b-a} \le F_X \le 1 \end{cases} \tag{5-46}$$

The mean (μ) and variance (σ^2) for the distribution, respectively, are given by:

$$\mu_X = \frac{a+b+c}{3} \tag{5-47}$$

$$\sigma_X^2 = \frac{a^2+b^2+c^2-ab-ac-bc}{18} \tag{5-48}$$

The density and cumulative functions of the triangular distribution with $a = 1$, $b = 6$, and $c = 2$ are shown in Figures 5-6A and 5-6B, respectively.

5.7. GAMMA DISTRIBUTION

The density function of this probability distribution is given by:

$$f_X(x) = \frac{v(vx)^{k-1}\exp(-vx)}{\Gamma(k)} \qquad 0 \le x \tag{5-49}$$

where $k > 0$ and $v > 0$ are the parameters of the distribution. The function Γ is called the gamma function, as provided at the end of Appendix A, in Table A-8, and is given by:

$$\Gamma(k, x) = \int_0^x \exp(-y)y^{k-1}dy \tag{5-50a}$$

or

$$\Gamma(k) = \int_0^\infty \exp(-y)y^{k-1}dy \tag{5-50b}$$

The cumulative distribution function is given by:

$$F_X(x) = \int_0^x f_X(x)dx = \frac{\Gamma(k, vx)}{\Gamma(k)} \tag{5-51}$$

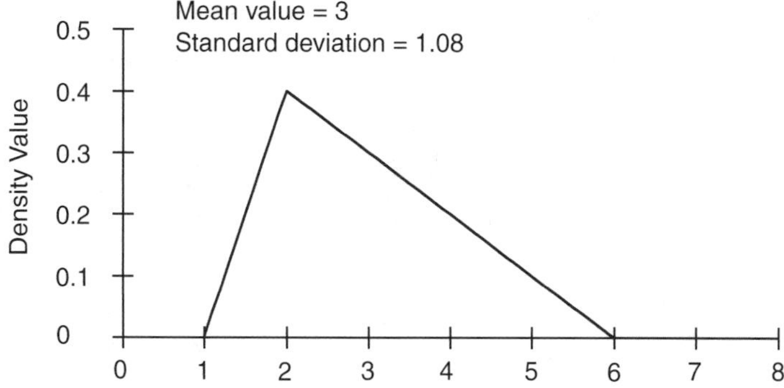

FIGURE 5-6A Probability density function of the triangular distribution.

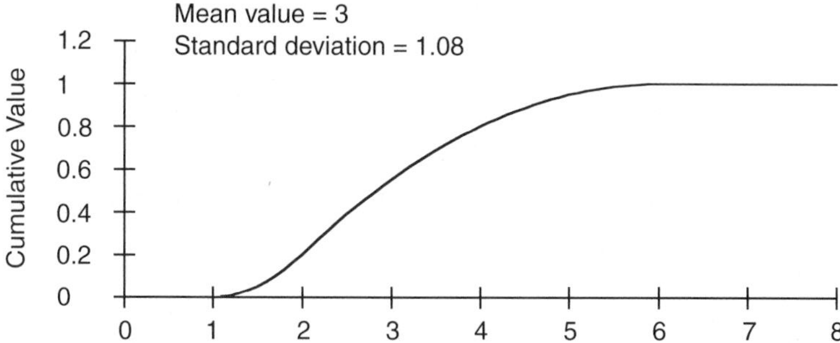

FIGURE 5-6B Cumulative distribution function of the triangular distribution.

The mean (μ) and variance (σ^2) for the distribution, respectively, are given by:

$$\mu_X = \frac{k}{v} \tag{5-52}$$

$$\sigma_X^2 = \frac{k}{v^2} \tag{5-53}$$

5.8. RAYLEIGH DISTRIBUTION

The density function of this probability distribution is given by:

$$f_X(x) = \frac{x}{\alpha^2} \exp\left[-\frac{1}{2}\left(\frac{x}{\alpha}\right)^2\right] \tag{5-54}$$

where α is the parameter of the distribution. The cumulative distribution function is given by:

$$F_X(x) = 1 - \exp\left(-\frac{x^2}{2\alpha^2}\right) \tag{5-55}$$

The mean (μ) and variance (σ^2) for the distribution, respectively, are given by:

$$\mu_X = \sqrt{\frac{\pi}{2}}\alpha \tag{5-56}$$

$$\sigma_X^2 = \left(2 - \frac{\pi}{2}\right)\alpha^2 \tag{5-57}$$

For a given mean, the parameter can be computed as:

$$\alpha = \sqrt{\frac{2}{\pi}}\mu_X \tag{5-58}$$

5.9. STATISTICAL PROBABILITY DISTRIBUTIONS

In statistical analysis, tables of values of the Student's t distribution, chi-square distribution, and F distribution are commonly used. These distributions and their basic properties are introduced in this section.

5.9.1. Student's *t* Distribution

The Student's t (or t) distribution (published under the pseudonym "Student" by its author) is a symmetric, bell-shaped distribution with the following density function:

$$f_T(t) = \frac{\Gamma[(k+1)/k]}{(\pi k)^{0.5}\Gamma(k/2)[1+(t^2/k)]^{0.5(k+1)}} \qquad -\infty < t < \infty \tag{5-59}$$

where k is a parameter of the distribution and is called the degrees of freedom. The function Γ(.) is called the gamma function and is given by:

$$\Gamma(n) = \int_0^{\infty} x^{n-1}\exp(-x)dx \qquad \text{for any value } n \tag{5-60}$$

This function is tabulated in Table A-8 and has the following useful properties:

$$\Gamma(n) = (n-1)\Gamma(n-1) \tag{5-61a}$$

$$\Gamma(n) = (n-1)! \qquad \text{for integer } n \tag{5-61b}$$

For $k > 2$, the mean and variance, respectively, are:

$$\mu_T = 0 \tag{5-62a}$$

$$\sigma_T^2 = \frac{k}{k-2} \tag{5-62b}$$

As k increases toward infinity, the variance of the distribution approaches unity, and the t distribution approaches the standard normal density function. Therefore, the t distribution has heavier tails (with more area) than the standard normal distribution. It is of interest in statistical analysis to determine the percentage points $t_{\alpha,k}$, which correspond to the following probability:

$$\alpha = P(T > t_{\alpha,k}) \tag{5-63a}$$

or

$$\alpha = \int_{t_{\alpha,k}}^{\infty} f_T(t)dt \tag{5-63b}$$

where α is the level of significance. These percentage points are tabulated (e.g., see Table A-2). Due to the symmetry of the t distribution, it is common to tabulate values for only the upper tail (i.e., nonnegative t values). For the lower tail, the following relationship can be used to determine the percentage points:

$$t_{(1-\alpha),k} = -t_{\alpha,k} \tag{5.64}$$

5.9.2. Chi-Square Distribution

The chi-square (χ^2) distribution is encountered frequently in statistical analysis, where we deal with the sum of squares of k random variables with standard normal distributions; that is,

$$\chi^2 = C = Z_1^2 + Z_2^2 + \ldots + Z_k^2 \tag{5-65}$$

where C is a random variable with a chi-square distribution, and Z_1 to Z_k are normally (standard normal) and independently distributed random variables. The probability density function of the chi-square distribution is:

$$f_C(c) = \frac{1}{2^{0.5k}\Gamma\left(\frac{k}{2}\right)} c^{0.5k-1} \exp\left(\frac{-c}{2}\right) \quad \text{for } c > 0 \tag{5-66}$$

where $\Gamma(.)$ is the gamma function as defined in Equation 5-61. The distribution is defined only for positive values and has the following mean and variance, respectively:

$$\mu_C = k \tag{5-67a}$$

$$\sigma_C^2 = 2k \tag{5-67b}$$

The parameter of the distribution, k, represents the degrees of freedom. This distribution is positively skewed with a shape that depends on the parameter k. It is of interest in statistical analysis to determine the percentage points $c_{\alpha,k}$ that correspond to the following probability:

$$\alpha = P(C > c_{\alpha,k}) \tag{5-68a}$$

$$\alpha = \int_{c_{\alpha,k}}^{\infty} f_C(c)dc \tag{5-68b}$$

where α is the level of significance. These percentage points are tabulated (e.g., see Table A-3).

5.9.3. *F* Distribution

The F distribution is used quite frequently in statistical analysis. It is a function of two shape parameters, $\nu_1 = k$ and $\nu_2 = u$, and has the following density function:

$$f_F(f) = \frac{\Gamma\left(\frac{u+k}{2}\right)\left(\frac{k}{u}\right)^{\frac{k}{2}}(f)^{\frac{k}{2}-1}}{\Gamma\left(\frac{k}{2}\right)\Gamma\left(\frac{u}{2}\right)\left[\frac{fk}{u}+1\right]^{\frac{u+k}{2}}} \quad \text{for } f > 0 \tag{5-69}$$

where $\Gamma(.)$ is the gamma function as given by Equation 5-61. For $u > 2$, the mean and variance of this distribution are:

$$\mu_F = \frac{u}{u-2} \tag{5-70a}$$

$$\sigma_F^2 = \frac{2u^2(u+k-2)}{k(u-2)^2(u-4)} \quad \text{for } u > 4 \tag{5-70b}$$

This distribution is positively skewed with a shape that depends on the parameters k and u. It is of interest in statistical analysis to determine the percentage points $f_{\alpha,k,u}$ that correspond to the following probability:

$$\alpha = P(F > f_{\alpha,k,u}) = \int_{f_{\alpha,k,u}}^{\infty} f_F(x)dx = \alpha \tag{5-71}$$

where α is the level of significance. These percentage points are tabulated (e.g., see Table A-4). The F distribution has a unique property that allows tabulating values for the upper tail only. For the lower tail, the following relationship can be used to determine the percentage points:

$$f_{(1-\alpha),u,k} = \frac{1}{f_{\alpha,k,u}} \tag{5-72}$$

It should be noted that the order of the parameters is not the same on both sides of Equation 5-72, allowing for the use of Table A-4 for $\alpha = 0.95$ and 0.99. Also, it means the parameters of this distribution cannot be transposed; that is,

$$f_{\alpha,u,k} \neq f_{\alpha,k,u} \tag{5-73}$$

5.10. EXTREME VALUE DISTRIBUTIONS

5.10.1. Introduction to Extreme Value Estimation

Extreme value distributions are a class of commonly used distributions in engineering and sciences. Three types of asymptotic extreme value distributions are used as described in subsequent sections. In this section, the fundamental concepts of extreme value estimation are provided.

Extreme values based on observational data are very important in system safety and life assessment. The prediction of future conditions, especially extreme conditions, are necessary in engineering planning and design. The prediction is performed based on an extrapolation from previously observed data. For a set of observations $(x_1, x_2, \ldots, x_k)$ from an identically distributed and independent set of random variables $(X_1, X_2, \ldots, X_k)$, the distribution of X_i is called the parent (or initial) distribution. It has the cumulative probability distribution function $F_X(x)$ and the density probability function $f_X(x)$. The maximum extreme value of the observed values is a random variable M_k, which can be represented as:

$$M_k = \text{Maximum}\ (X_1, X_2, \ldots, X_k) \tag{5-74}$$

The exact cumulative and density probability distribution functions of the maximum value, respectively, are given by:

$$F_{M_k}(m) = [F_X(m)]^k \tag{5-75}$$

$$f_{M_k}(m) = k[F_X(m)]^{k-1} f_X(m) \tag{5-76}$$

It can be shown that for relatively large values of k, the extreme distribution approaches an asymptotic form that is not dependent on the exact form of the parent distribution; rather, it depends on the tail characteristics of the parent distribution in the direction of the extreme. The central portion of the parent distribution has little influence on the asymptotic form of the extreme distribution. For parent probability distributions of exponential tails, the extreme distribution approaches an extreme value distribution of double exponential form as k becomes large. For example, a normal or lognormal probability distribution approaches a type I extreme value distribution as k becomes large. In this case, the difference between an exact distribution for M_k and the type I extreme value distribution is relatively small. The difference diminishes as k approaches infinity. Practically, the difference is negligible for k larger than approximately 25.

The above formulation can be used for the purpose of life prediction. The mathematical model for the extreme distribution needs to be a function of k in order to relate the outcome of the analysis of extreme statistics to time. Extreme value distributions, such as the type I largest extreme value distribution, can be used to model extreme load effects. Because the mathematical model is not sensitive to the type of the parent distribution, as long as it is within the same general class, the mathematical model used in this chapter is based on a parent distribution that follows the class of normal probability distributions.

For a normal parent probability distribution of the random variable X with a mean value μ and standard deviation σ, the cumulative distribution and density functions of the largest value M_k of k identically distributed and independent random variables $(X_1, X_2, ..., X_k)$, respectively, are given by:

$$F_{M_k}(m) = \exp\left\{-\exp\left[\left(-\frac{\alpha_k}{\sigma}\right)(m-\mu-\sigma u_k)\right]\right\} \tag{5-77}$$

$$f_{M_k}(m) = \left(\frac{\alpha_k}{\sigma}\right)\exp\left[\left(-\frac{\alpha_k}{\sigma}\right)(m-\mu-\sigma u_k)\right]\exp\left\{-\exp\left[\left(-\frac{\alpha_k}{\sigma}\right)(m-\mu-\sigma u_k)\right]\right\} \tag{5-78}$$

where

$$\alpha_k = [2\ln(k)]^{0.5} \tag{5-79a}$$

$$u_k = \alpha_k - \{\ln[\ln(k)] + \ln(4\pi)\}/(2\alpha_k) \tag{5-79b}$$

The mean value and standard deviation of M_k can be determined approximately using the central and dispersion characteristics of the type I extreme value distribution and are given, respectively, by the following:

$$\text{Mean value: } \mu_{M_k} = \sigma u_k + \mu + \frac{\gamma\sigma}{\alpha_k} \tag{5-80a}$$

$$\text{Standard deviation: } \sigma_{M_K} = \frac{\pi}{\sqrt{6}}\frac{\sigma}{\alpha_k} \tag{5-80b}$$

The constants π and γ have the values of 3.141593 and 0.577216, respectively.

The minimum extreme value of the observed values is a random variable M_1 which can be represented as:

$$M_1 = \text{Minimum}\,(X_1, X_2, \ldots, X_k) \tag{5-81}$$

The exact cumulative and density probability distribution functions of the minimum value, respectively, are given by:

$$F_{M_1}(m) = 1-[1-F_X(m)]^k \tag{5-82}$$

$$f_{M_1}(m) = k[1-F_X(m)]^{k-1} f_X(m) \tag{5-83}$$

Example 5-5. Maximum Value Based on Exponential Parent Distribution

Consider an exponential parent distribution as follows:

$$f_X(x) = \lambda\exp(-\lambda x) \tag{5-84}$$

The cumulative distribution function of the parent distribution is:

$$F_X(x) = 1-\exp(-\lambda x) \tag{5-85}$$

The kth largest value has the following distribution:

$$F_{M_k}(m) = (1-\exp(-\lambda m))^k \tag{5-86}$$

The density function of the kth largest value can be obtained by differentiation or using Equation 5-76 to obtain the following:

$$f_{M_k}(m) = \lambda k(1-\exp(-\lambda m))^{k-1}\exp(-\lambda m) \tag{5-87}$$

5.10.2. Type I Extreme Value (Gumbel) Distributions

Two forms of the type I extreme value distribution, also called the Gumbel distributions, are available: the largest and smallest extreme value. These two types are described in this section. The density function for the largest distribution of a random variable X_n is given by:

$$f_{X_n} = \alpha_n e^{-\alpha_n(x-u_n)}\exp[-e^{-\alpha_n(x-u_n)}] \tag{5-88}$$

The density function for the smallest distribution of a random variable X_1 is given by:

$$f_{X_1} = \alpha_1 e^{-\alpha_1(x-u_1)} \exp[-e^{-\alpha_1(x-u_1)}] \tag{5-89}$$

where u_n is the location parameter for X_n, α_n is the shape parameter of X_n, u_1 is the location parameter for X_1, and α_1 is the shape parameter of X_1. The cumulative function for the largest distribution is given by:

$$F_{X_n} = \exp[-e^{-\alpha_n(x-u_n)}] \tag{5-90}$$

The cumulative function for the smallest extreme is given by:

$$F_{X_1}(x) = 1 - \exp[-e^{\alpha_1(x-u_1)}] \tag{5-91}$$

For the largest extreme, the mean (μ) and variance (σ^2) for the distribution, respectively, are given by:

$$\mu_{X_n} = u_n + \frac{\gamma}{\alpha_n} \tag{5-92}$$

$$\sigma_{X_n}^2 = \frac{\pi^2}{6\alpha_n^2} \tag{5-93}$$

where $\pi = 3.14159$, and $\gamma = 0.577216$. For the smallest extreme, the mean (μ) and variance (σ^2) for the distribution, respectively, are given by:

$$\mu_{X_1} = \mu_1 - \frac{\gamma}{\alpha_1} \tag{5-94}$$

$$\sigma_{X_1}^2 = \frac{\pi^2}{6\alpha_1^2} \tag{5-95}$$

5.10.3. Type II Extreme Value (Frĕchet) Distributions

Two forms of the type II extreme value distribution, also called the Frĕchet distributions, are available: the largest and smallest extreme value. The two types are described in this section, although only the largest distribution has practical value. The density function for the largest extreme of a random variable X_n is given by:

$$f_{X_n}(x) = \frac{k}{v_n}\left(\frac{v_n}{x}\right)^{k+1} \exp\left[-\left(\frac{v_n}{x}\right)^k\right] \tag{5-96}$$

The density function for the smallest extreme of a random variable X_1 is given by:

$$f_{X_1}(x) = \frac{k}{v_1}\left(\frac{v_1}{x}\right)^{k+1} \exp\left[-\left(\frac{v_1}{x}\right)^k\right] \tag{5-97}$$

where v_n is the location parameter for X_n, v_1 is the location parameter for X_1, and k is the shape parameter of X_1 and X_n. The cumulative function for the largest distribution is given by:

$$F_{X_n}(x) = \exp\left[-\left(\frac{v_n}{x}\right)^k\right] \tag{5-98}$$

The cumulative function for the smallest extreme is given by:

$$F_{X_1}(x) = 1 - \exp\left[-\left(\frac{v_n}{x}\right)^k\right] \tag{5-99}$$

where $x \leq 0$, and $v_1 > 0$. For the largest extreme, the mean (μ) and variance (σ^2) for the distribution, respectively, are given by:

$$\mu_{X_n} = v_n \Gamma\left(1 - \frac{1}{k}\right) \tag{5-100}$$

$$\sigma_{X_n}^2 = v_n^2\left[\Gamma\left(1 - \frac{2}{k}\right) - \Gamma^2\left(1 - \frac{1}{k}\right)\right] \quad \text{for } k \geq 2 \tag{5-101}$$

where the $\Gamma(.)$ is the gamma function given by:

$$\Gamma(t) = \int_0^\infty r^{t-1} \exp(-r)\, dr \tag{5-102}$$

It should be noted that

$$\Gamma(t) = (t-1)\Gamma(t-1) \tag{5-103}$$

For an integer n, the gamma function becomes the factorial as follows:

$$\Gamma(n) = (n-1)! \tag{5-104}$$

Table A-8 contains tabulated values of the gamma function. The coefficient of variation (δ) based on Equations 5-100 and 5-101 is:

$$\delta_{X_n}^2 = \left[\frac{\Gamma\left(1-\frac{2}{k}\right)}{\Gamma^2\left(1-\frac{1}{k}\right)}\right] - 1 \tag{5-105}$$

For the smallest extreme, the mean (μ) and variance (σ^2) for the distribution, respectively, are given by:

$$\mu_{X_1} = v_1\Gamma\left(1-\frac{1}{k}\right) \tag{5-106}$$

$$\sigma_{X_1}^2 = v_1^2\left[\Gamma\left(1-\frac{2}{k}\right) - \Gamma^2\left(1-\frac{1}{k}\right)\right] \quad \text{for } k \geq 2 \tag{5-107}$$

The coefficient of variation (δ) is:

$$\delta_{X_1}^2 = \left[\frac{\Gamma\left(1-\frac{2}{k}\right)}{\Gamma^2\left(1-\frac{1}{k}\right)}\right] - 1 \tag{5-108}$$

5.10.4. Type III Extreme Value (Weibull) Distributions

Two forms of the type III extreme value distribution, commonly known as the Weibull, are available: the largest and smallest extreme values. These two types are described in this section. The smallest extreme value distribution is commonly used in engineering to model problems dealing with largest value (such as maximum loads) as a best-fit distribution to data. The density function for the largest extreme random variable X_n is given by:

$$f_{X_n}(x) = \frac{k}{\omega - n}\left(\frac{\omega - x}{\omega - u}\right)^{k-1}\exp\left[-\left(\frac{\omega - x}{\omega - u}\right)^k\right] \quad \text{for } x \leq \omega \tag{5-109}$$

The density function for the smallest extreme random variable X_1 is given by:

$$f_{X_1}(x) = \frac{k}{u - \omega}\left(\frac{x - \omega}{u - \omega}\right)^{k-1}\exp\left[-\left(\frac{x - \omega}{u - \omega}\right)^k\right] \quad \text{for } x \geq \omega \tag{5-110}$$

where $u > 0$ and $k > 0$, u is the scale parameter, k is the shape parameter, and ω is the upper or lower limit on x for the largest and smallest extremes, respectively. These distributions are commonly used as two-parameter distributions by setting $\omega = 0$. The cumulative distribution function for the largest extreme random variable X_n is given by:

$$F_{X_n}(x) = \exp\left[-\left(\frac{\omega - x}{\omega - u}\right)^k\right] \quad \text{for } x \leq \omega \text{ and } k > 0 \tag{5-111}$$

The cumulative distribution function for the smallest extreme random variable X_1 is given by:

$$F_{X_1}(x) = 1 - \exp\left[-\left(\frac{x-\omega}{u-\omega}\right)^k\right] \quad \text{for } x \geq \omega \tag{5-112}$$

For the largest extreme, the mean (μ) and variance (σ^2) for the distribution, respectively, are given by:

$$\mu_{X_n} = \omega - (\omega - u)\Gamma\left(1 + \frac{1}{k}\right) \tag{5-113}$$

$$\sigma_{X_n}^2 = (\omega - u)^2\left[\Gamma\left(1 + \frac{2}{k}\right) - \Gamma^2\left(1 + \frac{1}{k}\right)\right] \tag{5-114}$$

For the smallest extreme, the mean (μ) and variance (σ^2) for the distribution are given, respectively, by:

$$\mu_{X_1}(x) = \omega + (u - \omega)\left[\Gamma\left(1 + \frac{1}{k}\right)\right] \tag{5-115}$$

$$\sigma_{X_1}^2 = (u - \omega)^2\left[\Gamma\left(1 + \frac{2}{k}\right) - \Gamma^2\left(1 + \frac{1}{k}\right)\right] \tag{5-116}$$

5.11. APPLICATIONS

5.11.1. Earthquakes and Structures

For an earthquake of intensity VI, the probability of failure of the structure is taken to be 0.01. The occurrence of earthquakes can be assumed to follow a Poisson distribution with a rate of occurrence of:

$$\lambda = \frac{10}{100} = 0.1 \text{ earthquakes/year} \tag{5-117}$$

The probability of an earthquake occurring in the next year is computed using Equation 5-37 as:

$$P(T \leq 1) = 1 - \exp[-0.1(1)] = 0.09516 \tag{5-118}$$

where T is the time between consecutive earthquakes. The probability that an earthquake will *not* occur in the next 5 years is:

$$P(T < 5) = 1 - \{1 - \exp[-0.1(5)]\} = 0.6065 \tag{5-119}$$

5.11.2. Foundation of a Structure

The foundation of a column in a structure supports a concentrated axial force with a normally distributed magnitude, N(100,000 lb, 10,000^2 lb^2). The supporting foundation is 5×5 ft^2; therefore, its area is 25 ft^2. Because the foundation is concentrically loaded, the soil bearing pressure, P_b, can be assumed to a have a uniform distribution over the foundation area with a mean and variance computed, respectively, as:

$$\bar{P}_b = \frac{100{,}000}{25} = 4000 \text{ psf} \tag{5-120}$$

$$\text{Var}(P_b) = \frac{(10{,}000)^2}{(25)^2} = 160{,}000 \text{ (psf)}^2 \tag{5-121}$$

The standard deviation of the bearing pressure is $\sqrt{160,000} = 400$ psf, and its coefficient of variation is 400/4000 = 0.1. The bearing strength of the soil, P_s, was determined from test borings to be normally distributed: N(10,000 psf; 2500^2 psf^2). The probability of bearing failure, P_f, of the soil is:

$$P_f = P(P_b > P_s) = P(P_s - P_b < 0) \tag{5-122}$$

This probability can be computed by defining the following random variable:

$$M = P_s - P_b \tag{5-123}$$

where M is the safety margin. Therefore, the mean and variance of M, respectively, are:

$$\overline{M} = \overline{P_s} - \overline{P_b} = 10{,}000 - 4000 = 6000 \text{ psf} \tag{5-124}$$

$$\text{Var}(M) = \text{Var}(P_s) + \text{Var}(P_b) = (2500)^2 + 160{,}000 = 6{,}410{,}000 \text{ (psf)}^2 \tag{5-125}$$

According to Equations 5-17 and 5-19, M is normally distributed: N(6000 psf, 6,410,000 psf^2). Therefore, the probability of failure, P_f, is:

$$P_f = P(M < 0) = \Phi\left[\frac{0 - \overline{M}}{\sigma_M}\right] = \Phi\left[\frac{0 - 6000}{2532}\right] = 1 - \Phi(2.37) = 0.0089 \tag{5-126}$$

Table A-1 for the standard normal variate was used to determine the cumulative probability $\Phi(2.37)$.

5.11.3. Failure of Highway Drainage

A ditch alongside a roadway has a capacity of 4 m^3/min, and, based on its cross-sectional geometry and grade, the average water-flow velocity is 0.5 m/sec with a standard deviation of 0.25 m/sec. The maximum permissible velocity (so erosion is not a problem) is 1 m/sec. Historical rainfall records indicate that the ditch is expected to have a flow with a mean and standard deviation of 2 and 1 m^3/min, respectively. The water-flow volume rate and velocity can be assumed to be normally distributed. Therefore, the probability, P_e, that the water-flow velocity exceeds the erosion limit is:

$$P_e = P(V > 1) = 1 - \Phi\left[\frac{1 - 0.5}{0.25}\right] = 1 - \Phi(2) = 0.0228 \tag{5-127}$$

where V is the velocity of water flow. Table A-1 for the standard normal variate was used to determine the cumulative probability $\Phi(2)$. The probability, P_Q, of exceeding the volume-rate capacity of the ditch is:

$$P_Q = P(Q > 4) = 1 - \Phi\left[\frac{4 - 2}{1}\right] = 1 - \Phi(2) = 0.0228 \tag{5-128}$$

where Q is the volume rate of the water flow.

5.12. SIMULATION AND PROBABILITY DISTRIBUTIONS

The uniform distribution (see Section 5.2) is of special importance in simulation because many generators produce random numbers that are uniformly distributed. The uniform distribution was used as an input to the transformation equation of Equation 2-18. Random numbers from other distributions are often transformed from numbers generated with a uniform distribution. The midsquare method introduced in Chapter 1 produces uniformly distributed random numbers from 0 to 9999. Multiplying such numbers by 0.0001 will essentially produce numbers over the range from 0 to 1. Also, the *rand* function was introduced in Chapter 1 for generating random numbers in the interval [0,1] that should be uniformly distributed.

Because the midsquare method is not an especially reliable generator, generated numbers should be tested prior to their use. Consider the 28 random numbers given in column 4 of Table 4-1. Figure 5-7 shows histograms developed from these random numbers using different cell widths: 0.1, 0.2, and 0.25. The histograms suggest that the 28 generated numbers are approximately uniformly distributed. It should be noted that for cases with cell widths that have small numbers of events per cell, such as Figure 5-7A, the graph shows greater deviation from the population. For large cell widths, the frequencies are greater and the sample provides a better representation of the population. Considerable sampling variation can occur for small sample sizes.

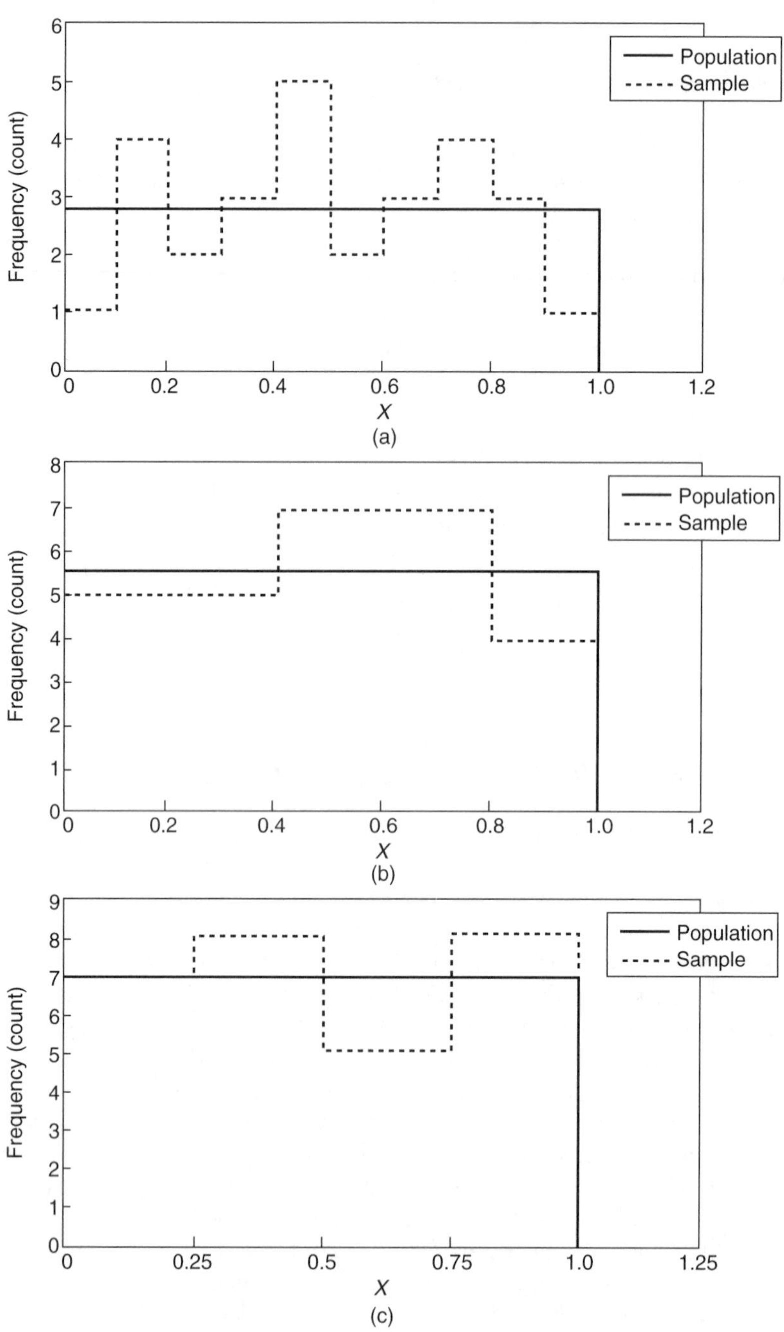

FIGURE 5-7 Histograms of uniformly generated random numbers for cell widths. (a) Cell width = 0.10; (b) cell width = 0.20; and (c) cell width = 0.25.

In Section 1.3.4, a transformation graph was shown for transforming uniform variates on the scale of 0 to 360 to a uniform variate on the scale from 0 to 1. The graph itself was a continuous line, with both axes being continuous scales. This is the character of a transformation graph for the transformation of one continuous variate to a second continuous variate.

Normally distributed random numbers are frequently used in simulation studies. Many statistical methods assume an underlying normal population. Therefore, normally distributed random numbers are widely used in statistically based simulations. Normally distributed random numbers can be generated by transforming uniformly distributed random numbers. In Chapter 7, a mathematical approach to making the transformation is provided. To provide a conceptual understanding of the transforming of one continuous variable to another, this approach is presented using the tabular representation of the normal distribution (Table A-1 in Appendix A).

Assuming that a random-number generator that provides uniformly distributed numbers from 0 to 1 is available, values of the standard normal deviate Z can be obtained in Table A-1 by using the uniform variate U as the probability and reading the value of Z that corresponds to the cumulative probability of U. Values of U less than 0.5 produce negative values of Z. Consider a U value of 0.4814. Entering Table A-1 with a probability of 0.4814 gives a Z value of –0.0467, assuming that linear interpolation is acceptable.

Example 5-6: Generation of Standard Normal Variates

The uniformly distributed numbers in column 4 of Table 4-1 were transformed to a set of 28 numbers that have a standard normal distribution (see column 6 of Table 4-1). A histogram of simulated values is shown in Figure 5-8. The values appear to closely match the standard normal distribution.

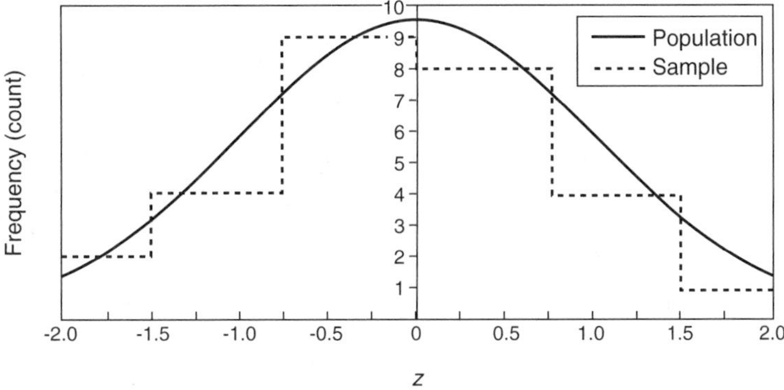

FIGURE 5-8 Comparison of random sample of standard normal variate and $N(0,1)$.

5.13. PROBLEMS

5-1. Assume that daily evaporation rates (E) have a uniform distribution with $a = 0$ and $b = 0.35$ in./day. Determine the following probabilities: (a) $P(E > 0.1)$; (b) $P(E < 0.22)$; (c) $P(E = 0.2)$; and (d) $P(0.05 < E < 0.15)$.

5-2. Graphically fit a uniform distribution to the following test grades data:

	Range							
	50–59	60–69	70–74	75–79	80–84	85–89	90–94	95–100
Number	4	9	7	11	13	7	6	3

5-3. The compressive strength of concrete specimen follows a normal distribution with a mean value (μ) of 2.8 ksi and coefficient of variation (COV) of 0.1. If the applied stress is 2.5 ksi, find the probability of failure.

5-4. The average annual precipitation for Washington, D.C., is 43 in., with a standard deviation of 6.5 in. If the amounts can be assumed to have a normal distribution, find the probability that the precipitation in any 1 year is (a) greater than 55 in., (b) less than 35 in., (c) either less than 32 in. or greater than 51 in., and (d) between 38 and 48 in. Find the precipitations that have (e) 5% chance of being exceeded, and (f) 1% chance of not being exceeded.

5-5. A concrete delivery truck shuttles between a concrete mixing plant and a construction site for the entire 10-hr working day. The one-way trip takes on the average 1 hr with a standard deviation of 0.2 hr. Assuming that the travel time can be modeled by a normal distribution, what is the probability of completing a two-way (complete) trip within 1.5 hr? Within 2.5 hr? Assuming that the time needed to unload the truck is also a normal variate with a mean of 0.5 hr and standard deviation of 0.2 hr, what are the mean and standard deviation of the time required for a complete two-way trip with unloading?

5-6. The Reynolds number for a pipe flow was found to follow lognormal distribution with a mean value and standard deviation of $\mu_{R_e} = 2650$ and $\sigma_{R_e} = 700$, respectively. The flow can be classified as laminar flow if $R_e < 2700$ and as turbulent flow if $R_e > 4000$. Find the probability that the flow is in a transitional state.

5-7. The random variable $Y = \ln(X)$ has a normal distribution with a mean of 5 and a standard deviation of 1. Determine the mean, variance, standard deviation, mode, median, and skewness coefficient of X.

5-8. Over the past 15 years, the numbers of hurricanes per year to cause damage in a certain city are 4, 2, 1, 3, 0, 2, 1, 3, 5, 2, 3, 1, 1, 2, and 0. What is the probability that the time between consecutive hurricanes exceeds 1 year? What is the return period of hurricanes for this city?

5-9. A random variable X with a mean of 2.5 follows an exponential distribution. Find the following probabilities: (a) $P(X > 2.5)$, (b) $P(X < 1)$, and (c) $P(0.5 < X < 1.5)$. Find the values of x_0 if (d) $P(X < x_0) = 0.1$, and (e) $P(X > x_0) = 0.05$.

5-10. Assume that the time required to pour a concrete floor for a structure (D) has a triangular distribution between 10 and 15 days with a mode of 13 days. Determine the following probabilities: (a) $P(D > 11)$, (b) $P(D < 13)$, (c) $P(D = 12)$, and (d) $P(10 < D < 15)$.

5-11. The 10-year extreme wave height (W) at a site of interest for oil drilling and extraction can be assumed to follow a Rayleigh distribution with a mean of 10 ft. Determine the following probabilities: (a) $P(W > 10)$, (b) $P(W < 10)$, (c) $P(W = 10)$, and (d) $P(5 < W < 15)$.

5-12. The 10-year extreme wind speed (W) at a site of interest for oil drilling and extraction can be assumed to follow a Rayleigh distribution with a mean of 100 mph. Determine the following probabilities: (a) $P(W > 100)$, (b) $P(W < 100)$, (c) $P(W = 100)$, and (d) $P(50 < W < 130)$.

5-13. Find the following probabilities for the t statistic: (a) $P(t > 2.45|k = 6)$, (b) $P(t < 2.72|k = 11)$, and (c) $P(t < -1.75|k = 16)$.

5-14. Find the values of t_α for a random variable that has the t distribution such that: (a) $P(t < t_\alpha|k = 15) = 0.05$; (b) $P(t > t_\alpha|k = 8) = 0.025$; and $P(t > t_\alpha|k = 24) = 0.90$.

5-15. Find the following probabilities for the chi-square (C) statistic: (a) $P(C > 5.024|k = 1)$, (b) $P(C < 0.831|k = 5)$, (c) $P(C > 2.555|k = 10)$, and (d) $P[(C < 6.251 \text{ or } C > 10.864)|k = 18]$.

5-16. Find the values of c_α for each of the following: (a) $P(C > c_\alpha|k = 4) = 0.020$; and (b) $P(C < c_\alpha|k = 7) = 0.01$.

5-17. Find the values of the F statistic (f_α) for each of the following: (a) $P[F > f_\alpha|(k = 8, u = 5)] = 0.05$; (b) $P[F > f_\alpha|(k = 15, u = 10)] = 0.01$.

5-18. The extreme wave height used in the design of an offshore facility for a design life of 30 years is a random variable with a mean of 25 ft and standard deviation of 5 ft. Determine the probability that the extreme wave height will exceed 30 ft in 30 years. Assume that the extreme wave height follows the largest extreme value distribution of type I.

5-19. Rework Problem 5-18 assuming that the extreme wave height follows the largest extreme value distribution of type II.

5-20. Rework Problem 5-18 assuming that the extreme wave height follows the smallest extreme value distribution of type III.

5-21. Use the *rand* function to generate 30 uniform random numbers. Transform the 30 generated values to standard normal variates (z). Compute the mean and standard deviation of the z values. Compare them with the population values of zero and one, respectively, and discuss the differences.

5.14. SIMULATION PROJECTS

This section provides additional work items for the problem statements of the four simulation projects introduced in Sections 1.5, 2.9, 3.9, and 4.11.

5.14.1. Structural Beam Study

Using the project information provided in Sections 1.5.1, 2.9.1, 3.9.1, and 4.11.1, evaluate the failure probability of the beam for a given load using the probabilistic characteristics in Table 5-1 and 100 simulation cycles (i.e., re-evaluate Table 1-3 using mixed normal and non-normal probability distributions of Table 5-1). Next, evaluate the failure probability of the beam using 1000, 2000, ..., 10,000 simulation cycles. Plot the trend diagrams. Plot the population and sample distributions for input and output random variables, and perform parametric analysis. Discuss your results.

TABLE 5-1 Probability Distribution for the Beam Failure Analysis Case Study

Random variable	Mean	Coefficient of variation	Distribution type
c	10	0.05	Normal
M	3000	0.30	Lognormal
I	1000	0.08	Normal
f_y	50	0.15	Lognormal

5.14.2. Stream Erosion Study

The project information of Sections 1.5.2, 2.9.2, 3.9.2, and 4.11.2 provides background information. Generate 100 uniform variates using the *rand* function. Transform the values to standard normal deviates and then to normally distributed values that have the same mean and standard deviation as the values of Y in Table C-1 in Appendix C. Compute the mean and standard deviation of the generated values, and compare your results with the table. Perform trend analysis on these values by increasing the number of simulation cycles from 10 to 100 in increments of 10, and from 100 to 1000 in increments of 100. Develop a histogram of the 100 generated values of Y, and compare the generated and sample values of Table C-1.

5.14.3. Traffic Estimation Study

The project information of Sections 1.5.3, 2.9.3, 3.9.3, and 4.11.3 provides background information. Generate 100 uniform variates using the *rand* function. Transform the values to standard normal deviates and then to normally distributed values that have the same mean and standard deviation as the values of Y in Table C-2 in Appendix C. Compute the mean and standard deviation of the generated values, and compare your results with the table. Perform trend analysis on these values by increasing the number of simulation cycles from 10 to 100 in increments of 10, and from 100 to 1000 in increments of 100. Develop a histogram of the 100 generated values of Y, and compare the generated and sample values of Table C-2.

5.14.4. Water Evaporation Study

The project information of Sections 1.5.4, 2.9.4, 3.9.4, and 4.11.4 provides background information. Generate 100 uniform variates using the *rand* function. Transform the values to standard normal deviates and then to normally distributed values that have the same mean and standard deviation as the values of Y in Table C-3 in Appendix C. Compute the mean and standard deviation of the generated values, and compare your results with the table. Perform trend analysis on these values by increasing the number of simulation cycles from 10 to 100 in increments of 10, and from 100 to 1000 in increments of 100. Develop a histogram of the 100 generated values of Y, and compare the generated and sample values of Table C-3.

Chapter 6
Multiple Random Variables

TABLE OF CONTENTS

6.1. INTRODUCTION

The discussion in Chapter 3 is limited to single random variables. However, in engineering and science, it is common to deal with two or more random variables simultaneously in solving problems. In general, multiple random variables are encountered in these problems in the following two forms:

1. Joint occurrences of multiple random variables that can be correlated or uncorrelated
2. Random variables that are known in terms of their functional relationship with other basic random variables

The objective of this chapter is to provide analytical tools to deal with these two forms of multiple random variables.

6.2. JOINT RANDOM VARIABLES AND THEIR PROBABILITY DISTRIBUTIONS

In some engineering applications, the outcomes (say, E_1, E_2, …, E_n) that constitute a sample space S are mapped to an n-dimensional (n-D) space of real numbers. The functions that establish such a transformation to the n-D space are called *multiple random variables* (or *random vectors*). This mapping can be one-to-one or one-to-many.

Multiple random variables are commonly classified into two types: discrete and continuous random vectors. A discrete random vector may only take on distinct, usually integer, values, whereas a continuous random vector takes on values within a continuum of values. A distinction is made between these two types of random vectors because the computations of probabilities depend on their type.

6.2.1. Probability for Discrete Random Vectors

The probability of a discrete multiple random variable or random vector $\boldsymbol{X} = (X_1, X_2, \ldots, X_n)$ is given by a *joint probability mass function*. A joint mass function specifies the probability that the discrete random variable X_1 is equal to some value x_1, X_2 is equal to some value x_2, …, X_n is equal to some value x_n and is denoted by:

$$P_{\boldsymbol{X}}(\boldsymbol{x}) = P(X_1 = x_1, X_2 = x_2, \ldots, X_n = x_n) \tag{6-1}$$

where $\boldsymbol{X}$ is a random vector that includes the random variables (X_1, X_2, …, X_n), and $\boldsymbol{x}$ is some specified values for the random vectors (x_1, x_2, …, x_n). The probability mass function must satisfy the axioms of probability. Therefore, the probability of an event ($X_1 = x_1$, $X_2 = x_2$, …, $X_n = x_n$) must be less than or equal to one, and it must be greater than or equal to zero; that is,

$$0 \leq P(X_1 = x_1, X_2 = x_2, \ldots, X_n = x_n) \leq 1 \tag{6-2}$$

This property is valid for all possible values of all of the random variables. Additionally, the sum of all possible probabilities must equal one.

It is often useful to present the likelihood of an outcome using the *cumulative mass function*, which is given by:

$$F_X(\boldsymbol{x}) = P(X_1 \le x_1, X_2 \le x_2, \ldots, X_n \le x_n) = \sum_{\text{all}(X_1 \le x_1, X_2 \le x_2, \ldots, X_n \le x_n)} P_X(x_1, x_2, x_3, \ldots, x_n) \quad (6\text{-}3)$$

The cumulative mass function is used to indicate the probability that the random variable X_1 is less than or equal to x_1, X_2 is less than or equal to x_2, …, and X_n is less than or equal to x_n. The cumulative distribution function must satisfy the following properties:

1. $F_X(\text{all } \boldsymbol{x} \to -\infty) = 0$.
2. $F_X(x_1, x_2, \ldots, x_i \to -\infty, \ldots, x_n) = 0$, for any $i = 1, 2, \ldots, n$.
3. $F_X(x_1, x_2, \ldots, x_i \to -\infty, \ldots, x_k \to -\infty, \ldots, x_n) = 0$, for any values of x_i, …, x_k.
4. $F_X(x_1, x_2, \ldots, x_i \to +\infty, \ldots, x_n) = F_{Xj}(x_j\text{: } j = 1, 2, \ldots, n \text{ and } j \ne i)$, called the *marginal distribution* of all the random variables except X_i.
5. $F_X(x_1, x_2, \ldots, x_i \to +\infty, \ldots, x_k \to +\infty, \ldots, x_n) = F_{Xj}(x_j\text{: } j = 1, 2, \ldots, n \text{ and } j \ne i \text{ to } k)$, called the *marginal distribution* of all the random variables except X_i to X_k.
6. $F_X(\text{all } \boldsymbol{x} \to +\infty) = 1$.
7. $F_X(\boldsymbol{x})$ is a nonnegative and nondecreasing function of $\boldsymbol{x}$.

The first, second, and third properties define the limiting behavior of $F_X(\boldsymbol{x})$; as one or more of the random variables approach $-\infty$, $F_X(\boldsymbol{x})$ approaches zero. The fourth and fifth properties define the possible marginal distributions as one or more of the random variables approaches $+\infty$. The sixth property is based on the probability axiom that states the probability of the sample space being one. The seventh property is based on the cumulative nature of $F_X(\boldsymbol{x})$.

The presentation of the materials in the remaining part of this section is limited to two random variables. The presented concepts can be generalized to n random variables. Based on the definition of conditional probabilities, the conditional probability mass function $P_{X_1|X_2}(x_1|x_2)$ for two random variables X_1 and X_2 is given by:

$$P_{X_1|X_2}(x_1|x_2) = \frac{P_{X_1X_2}(x_1, x_2)}{P_{X_2}(x_2)} \quad (6\text{-}4)$$

where $P_{X_1|X_2}(x_1|x_2)$ results in the probability of $X_1 = x_1$ given that $X_2 = x_2$; $P_{X_1X_2}(x_1,x_2)$ is the joint mass function of X_1 and X_2; and $P_{X_2}(x_2)$ is the marginal mass function for X_2 that is not equal to zero. In this case, the marginal distribution is given by:

$$P_{X_2}(x_2) = \sum_{\text{all } x_1} P_{X_1X_2}(x_1, x_2) \quad (6\text{-}5)$$

Similarly, the conditional probability mass function $P_{X_2|X_1}(x_2|x_1)$, for two random variables X_1 and X_2, is given by:

$$P_{X_2|X_1}(x_2|x_1) = \frac{P_{X_1X_2}(x_1, x_2)}{P_{X_1}(x_1)} \quad (6\text{-}6)$$

where the marginal mass function $P_{X_1}(x_1)$ is:

$$P_{X_1}(x_1) = \sum_{\text{all } x_2} P_{X_1X_2}(x_1, x_2) \tag{6-7}$$

The definitions provided by Equations 6-4 to 6-7 can be generalized for the n-D case. Based on the definition of conditional probabilities, it can be stated that, if X_1 and X_2 are statistically uncorrelated random variables, then

$$P_{X_1|X_2}(x_1|x_2) = P_{X_1}(x_1) \tag{6-8a}$$

and

$$P_{X_2|X_1}(x_2|x_1) = P_{X_2}(x_2) \tag{6-8b}$$

Therefore, using Equations 6-4 or 6-6, the following important relationship can be obtained for statistically uncorrelated random variables:

$$P_{X_1X_2}(x_1, x_2) = P_{X_1}(x_1)P_{X_2}(x_2) \tag{6-9}$$

Example 6-1: Reliability Engineering

A reliability engineer evaluates each unit produced at the end of a manufacturing line in a factory for mechanical and electrical defects. The evaluation is based on an accelerated test for a simulated life of the manufactured product. Based on a sample of 100 units, the following numbers of defects were determined:

(6-10a)

		X = number of mechanical defects				
		0	**1**	**2**	**3**	**4**
Y = number of electrical defects	**0**	70	4	2	1	0
	1	4	2	3	1	1
	2	2	1	1	1	2
	3	2	1	0	0	0
	4	1	1	0	0	0

In this table, two random variables are defined: X is the number of mechanical defects, and Y is the number of electrical defects. The table entries can be used to obtain likelihood measures of the occurrences of combinations of defects by dividing the entries by the total number of units (i.e., 100 units). The result can be considered to be the joint mass function of X and Y, $P_{XY}(x,y)$:

Y = number of electrical defects		X = number of mechanical defects				
		0	**1**	**2**	**3**	**4**
	0	0.70	0.04	0.02	0.01	0
	1	0.04	0.02	0.03	0.01	0.01
	2	0.02	0.01	0.01	0.01	0.02
	3	0.02	0.01	0	0	0
	4	0.01	0.01	0	0	0

(6-10b)

The marginal mass distributions for X and Y, $P_X(x)$ and $P_Y(y)$, can be determined by performing summations over all Y and X values, respectively, of $P_{XY}(x,y)$. The resulting marginal distributions are:

x	0	1	2	3	4
$P_X(x)$	0.79	0.09	0.06	0.03	0.03

(6-10c)

and

y	0	1	2	3	4
$P_Y(y)$	0.77	0.11	0.07	0.03	0.02

(6-10d)

The marginal distribution of a random variable provides a description of its probabilistic characteristics while the other random variable remains random. For example, $P_X(x)$ provides the probability mass function for mechanical defects without regard to the occurrences of the electrical defects (i.e., with random occurrence of electrical welds). The probability distribution of x mechanical defects given that there are y electrical defects, $P_{X|Y}(x|y)$, is given by:

For $y = 0$

$x\|y$	$0\|0$	$1\|0$	$2\|0$	$3\|0$	$4\|0$
$P_{X\|Y}(x\|y)$	70/77	4/77	2/77	1/77	0/77
or	0.909	0.052	0.026	0.013	0.000

(6-10e)

For $y = 1$

$x\|y$	$0\|1$	$1\|1$	$2\|1$	$3\|1$	$4\|1$
$P_{X\|Y}(x\|y)$	4/11	2/11	3/11	1/11	1/11
or	0.363	0.182	0.273	0.091	0.091

(6-10f)

For $y = 2$

$x\|y$	$0\|2$	$1\|2$	$2\|2$	$3\|2$	$4\|2$
$P_{X\|Y}(x\|y)$	2/7	1/7	1/7	1/7	2/7
or	0.285	0.143	0.143	0.143	0.286

(6-10g)

For $y = 3$

$x\|y$	$0\|3$	$1\|3$	$2\|3$	$3\|3$	$4\|3$
$P_{X\|Y}(x\|y)$	2/3	1/3	0/3	0/3	0/3
or	0.667	0.333	0.000	0.000	0.000

(6-10h)

For $y = 4$

$x\|y$	$0\|4$	$1\|4$	$2\|\|4$	$3\|4$	$4\|4$
$P_{X\|Y}(x\|y)$	1/2	1/2	0/2	0/2	0/2
or	0.5	0.5	0.000	0.000	0.000

(6-10i)

The conditional distributions enable the reliability engineer to investigate the probabilistic characteristics of a random variable as the other random variable is maintained constant at some specified value. For example, $P_{X|Y=4}(x|y = 4)$ provides the probability mass function for mechanical defects for the case that we have four electrical defects. Because $P_{XY}(x,y)$ is not equal to the product $P_X(x)\ P_Y(y)$, the random variables X and Y are statistically correlated.

6.2.2. Probability for Continuous Random Vectors

A *joint probability density function* (pdf) is used to define the likelihood of occurrence for a continuous random vector. Specifically, the probability that the random vector $\boldsymbol{X} = (X_1, X_2, \ldots, X_n)$ is within the interval from $\boldsymbol{x}^l = (x_1^l, x_2^l, x_3^l, \ldots, x_n^l)$ to $\boldsymbol{x}^u = (x_1^u, x_2^u, x_3^u, \ldots, x_n^u)$ is:

$$P(\boldsymbol{x}^l \le \boldsymbol{X} \le \boldsymbol{x}^u) = \int_{x_1^l}^{x_1^u}\int_{x_2^l}^{x_2^u}\ldots\int_{x_n^l}^{x_n^u} f_X(\boldsymbol{x})dx_1dx_2\ldots dx_n \tag{6-11}$$

in which $f_X(\boldsymbol{x})$ is the joint density function. It is important to note that the multiple integral of the joint pdf from $-\infty$ to $+\infty$ equals 1; that is,

$$P(-\infty < \boldsymbol{X} < +\infty) = \int_{-\infty}^{+\infty}\int_{-\infty}^{+\infty}\ldots\int_{-\infty}^{+\infty} f_X(\boldsymbol{x})dx_1dx_2\ldots dx_n = 1 \tag{6-12}$$

The *cumulative distribution function* (cdf) of a continuous random variable is defined by:

$$F_X(\boldsymbol{x}) = P(\boldsymbol{X} \le \boldsymbol{x}) = \int_{-\infty}^{x_1}\int_{-\infty}^{x_2}\cdots\int_{-\infty}^{x_n} f_X(\boldsymbol{x})dx_1dx_2\ldots dx_n \tag{6-13}$$

The cumulative distribution function must satisfy the following properties:

1. $F_X(all\ \boldsymbol{x} = -\infty) = 0$.
2. $F_X(x_1, x_2, \ldots, x_i \to -\infty, \ldots, x_n) = 0$, for any $i = 1, 2, \ldots, n$.
3. $F_X(x_1, x_2, \ldots, x_i \to -\infty, \ldots, x_k \to -\infty, \ldots, x_n) = 0$, for any values of $x_i, \ldots, x_k$.
4. $F_X(x_1, x_2, \ldots, x_i \to +\infty, \ldots, x_n) = F_{X_j}(x_j, j = 1, 2, \ldots, n$ and $j \ne i)$, called the *marginal distribution* of all the random variables except X_i.
5. $F_X(x_1, x_2, \ldots, x_i \to +\infty, \ldots, x_k \to +\infty, \ldots, x_n) = F_{X_j}(x_j, j = 1, 2, \ldots, n$ and $j \ne i$ to $k)$, called the *marginal distribution* of all the random variables except X_i to X_k.
6. $F_X(all\ \boldsymbol{x} \to +\infty) = 1$.
7. $F_X(\boldsymbol{x})$ is a nonnegative and nondecreasing function of $\boldsymbol{x}$.

These properties are similar to the properties of $F_X(\boldsymbol{x})$ for the discrete random variables as was discussed in Section 6.2.1. The joint density function can be obtained from a given joint cumulative distribution function by evaluating the partial derivative as follows:

$$f_X(\boldsymbol{x}) = \frac{\partial^n F_X(\boldsymbol{x})}{\partial \boldsymbol{X}} \tag{6-14a}$$

That is,

$$f_{X_1X_2\ldots X_n}(x_1, x_2, \ldots, x_n) = \frac{\partial^n F_{X_1X_2\ldots X_n}(x_1, x_2, \ldots, x_n)}{\partial X_1 \partial X_2 \ldots \partial X_n} \tag{6-14b}$$

The presentation of the materials in the remaining part of this section is limited to two random variables. The presented concepts can be generalized to n random variables. Based on the definition of conditional probabilities, the conditional probability density function $f_{X_1|X_2}(x_1|x_2)$ for two random variables X_1 and X_2 is given by:

$$f_{X_1|X_2}(x_1|x_2) = \frac{f_{X_1X_2}(x_1, x_2)}{f_{X_2}(x_2)} \tag{6-15}$$

where $f_{X_1X_2}(x_1,x_2)$ is the joint density function of X_1 and X_2, and $f_{X_2}(x_2)$ is the marginal density function for X_2 that is not equal to zero. In this case, the marginal distribution is given by:

$$f_{X_2}(x_2) = \int_{-\infty}^{+\infty} f_{X_1X_2}(x_1, x_2)dx_1 \tag{6-16}$$

Similarly, the conditional probability density function $f_{X_2|X_1}(x_2|x_1)$ for two random variables X_1 and X_2 is given by:

$$f_{X_2|X_1}(x_2|x_1) = \frac{f_{X_1X_2}(x_1, x_2)}{f_{X_1}(x_1)} \tag{6-17}$$

where the marginal density function $f_{X_1}(x_1)$ is:

$$f_{X_1}(x_1) = \int_{-\infty}^{+\infty} f_{X_1X_2}(x_1, x_2)dx_2 \tag{6-18}$$

The definitions as given by Equations 6-15 to 6-18 can be generalized for the n-D case, but they are not discussed herein. Based on the definition of conditional probabilities, it can be stated that if X_1 and X_2 are statistically uncorrelated random variables, then

$$f_{X_1|X_2}(x_1|x_2) = f_{X_1}(x_1) \tag{6-19}$$

and

$$f_{X_2|X_1}(x_2|x_1) = f_{X_2}(x_2) \tag{6-20}$$

Therefore, using Equations 6-15 or 6-17, the following important relationship can be obtained for uncorrelated random variables:

$$f_{X_1X_2}(x_1, x_2) = f_{X_1}(x_1)f_{X_2}(x_2) \tag{6-21}$$

Example 6-2: Times to Failure

An automated control system consists of two components with random times to failure X and Y. Previous experience with these components indicates that their times to failure are exponentially distributed and that the x time to failure is three times larger than the y time to failure. The following joint density function was suggested in this case:

$$f_{XY}(x, y) = \begin{cases} c\exp[-(x+3y)] & x \geq 0 \text{ and } y \geq 0 \\ 0 & \text{otherwise} \end{cases} \tag{6-22}$$

where c is a constant. The magnitude of c can be determined based on the following condition:

$$\text{The volume under } f_{XY}(x, y) = \int_0^\infty \int_0^\infty f_{XY}(x, y)dx\, dy = 1 \tag{6-23a}$$

$$= \int_0^\infty \int_0^\infty c\exp[-(x + 3y)]dx\, dy \tag{6-23b}$$

$$= \int_0^\infty \int_0^\infty c\exp(-x)\exp(-3y)dx\, dy \tag{6-23c}$$

$$= \int_0^\infty ([-c\exp(-3y)(-x)]|_{x=0}^{x=\infty})dy \tag{6-23d}$$

$$= \frac{-c\exp(-3y)}{3}\Bigg|_{y=0}^{y=\infty} = \frac{c}{3} = 1 \tag{6-23e}$$

Therefore, $c = 3$.

The marginal distributions of X and Y are:

$$f_X(x) = \int_0^\infty c\exp[-(x + 3y)]dy = \exp(-x) \tag{6-24}$$

$$f_Y(y) = \int_0^\infty c\exp[-(x + 3y)]dx = 3\exp(-3y) \tag{6-25}$$

The marginal distribution of a random variable provides a description of its probabilistic characteristics while the other random variable remains random. For example, $f_X(x)$ provides the probability density function for X without regard to the occurrence of Y (i.e., random occurrence of Y). Because $f_{XY}(x,y) = f_X(x)\, f_Y(y)$, the random variables X and Y are statistically uncorrelated.

The probability that X as a random variable is larger than Y as a random variable is:

$$P(X > Y) = \int_0^\infty \int_0^x f_{XY}(x, y)dy\, dx \tag{6-26a}$$

$$= \int_0^\infty \exp(-x)[-\exp(-3x) + 1]dx \tag{6-26b}$$

$$= \left(\frac{\exp(-4x)}{4} - \exp(-x)\right)\Bigg|_0^\infty = \frac{3}{4} \tag{6-26c}$$

This result is expected because the problem states that the X time to failure is three times larger than the Y time to failure.

6.2.3. Conditional Moments, Covariance, and Correlation Coefficient

In general, moments can be computed using the concept of mathematical expectation. For a continuous random vector $\boldsymbol{X}$, the kth moment about the origin is given by:

$$M'_k = \int_{-\infty}^{+\infty}\int_{-\infty}^{+\infty}\cdots\int_{-\infty}^{+\infty} x_1^k x_2^k \ldots x_n^k f_{X_1X_2\ldots X_n}(x_1, x_2, \ldots, x_n)dx_1dx_2\ldots dx_n \qquad (6\text{-}27)$$

in which $\{X_1, X_2, \ldots, X_n\}$ is the random vector and $f_{X_1X_2\ldots X_n}(x_1, x_2, \ldots, x_n)$ is its joint density function. The corresponding equation for a discrete random vector $\boldsymbol{X}$ is:

$$M'_k = \sum_{\text{all } x} x_1^k x_2^k \ldots x_n^k P_{X_1X_2\ldots X_n}(x_1, x_2, \ldots, x_n) \qquad (6\text{-}28)$$

in which $P_{X_1X_2\ldots X_n}(x_1, x_2, \ldots, x_n)$ is the joint probability mass function.

The above moments are commonly considered special cases of mathematical expectation. The mathematical expectation of an arbitrary function $g(\boldsymbol{X})$, a function of the random vector $\boldsymbol{X}$, is given by:

$$E[g(\boldsymbol{X})] = \int_{-\infty}^{+\infty}\int_{-\infty}^{+\infty}\cdots\int_{-\infty}^{+\infty} g(\boldsymbol{x}) f_{X_1X_2\ldots X_n}(x_1, x_2, \ldots, x_n)dx_1dx_2\ldots dx_n \qquad (6\text{-}29)$$

The corresponding equation for a discrete random vector $\boldsymbol{X}$ is:

$$E[g(\boldsymbol{X})] = \sum_{\text{all } x} g(\boldsymbol{x}) P_{X_1X_2\ldots X_n}(x_1, x_2, \ldots, x_n) \qquad (6\text{-}30)$$

For the two-dimensional case, X_1 and X_2, the conditional mean value for X_1 given that X_2 takes a value x_2, denoted $\mu_{X_1|x_2}$, is defined in terms of the conditional mass and density functions for the discrete and continuous random variables, respectively. The conditional mean for the continuous case is:

$$\mu_{X_1|x_2} = E(X_1|x_2) = \int_{-\infty}^{+\infty} x_1 f_{X_1|X_2}(x_1|x_2)dx_1 \qquad (6\text{-}31)$$

where $f_{X_1|X_2}(x_1|x_2)$ is the conditional density function of X_1 at a given (or specified) value of X_2 (see Equation 6-15). In this case, the conditional mean is the average value of the random variable X_1 given that the random variable X_2 takes the value x_2. For a discrete random variable, the conditional mean is given by:

$$\mu_{X_1|x_2} = E(X_1|x_2) = \sum_{\text{all } x_1} x_1 P_{X_1|X_2}(x_1|x_2) \qquad (6\text{-}32)$$

where $P_{X_1|X_2}(x_1|x_2)$ is the conditional mass function of X_1 at a given (or specified) value of X_2 (see Equation 6-4).

For statistically uncorrelated random variables X_1 and X_2, the conditional mean of a random variable is the same as its mean; that is,

$$\mu_{X_1|x_2} = E(X_1|x_2) = E(X_1) \tag{6-33}$$

and

$$\mu_{X_2|x_1} = E(X_2|x_1) = E(X_2) \tag{6-34}$$

Also, it can be shown that the expected value with respect to X_2 of the conditional mean $\mu_{X_1|X_2}$ is the mean of X_1; that is,

$$E_{X_2}(\mu_{X_1|X_2}) = E(X_1) \tag{6-35}$$

where E_{X_2} is the expected value with respect to X_2; that is, the variable of integration (or summation) for computing the expected value is x_2. In Equation 6-35, the quantity $\mu_{X_1|X_2}$ is treated as a random variable, because conditioning is performed on the random variable X_2 (not a specified value x_2).

As previously discussed, the variance is the second moment about the mean. For two random variables X_1 and X_2, the conditional variance $\sigma^2_{X_1|x_2}$ [or $\mathrm{Var}(X_1|x_2)$] is computed as follows:

$$\mathrm{Var}(X_1|x_2) = \int_{-\infty}^{+\infty} (x_1 - \mu_{X_1|x_2})^2 f_{X_1|X_2}(x_1|x_2)\,dx_1 \tag{6-36}$$

For a discrete variable, the conditional variance is computed by:

$$\mathrm{Var}(X_1|x_2) = \sum_{\text{all } x_1} (x_1 - \mu_{X_1|x_2})^2 P_{X_1|X_2}(x_1|x_2) \tag{6-37}$$

The variance of the random variable X_1 can also be computed using the conditional variance as follows:

$$\mathrm{Var}(X_1) = E_{X_2}[\mathrm{Var}(X_1|X_2)] + \mathrm{Var}_{X_2}[E(X_1|X_2)] \tag{6-38}$$

where E_{X_2} is the expected value with respect to X_2, and Var_{X_2} is the variance with respect to X_2; that is, the variable of integration (or summation) for computing the variance is x_2. In Equation 6-38, the quantity $\mathrm{Var}(X_1|X_2)$ is treated as a random variable, because the conditioning is performed on the random variable X_2 (not a value x_2).

The covariance (COV) of two random variables X_1 and X_2 is defined in terms of mathematical expectation as:

$$\mathrm{COV}(X_1, X_2) = E[(X_1 - \mu_{X_1})(X_2 - \mu_{X_2})] \tag{6-39}$$

It is common to use the notation $\sigma_{X_1X_2}$, σ_{12}, or COV(X_1,X_2) for the covariance of X_1 and X_2. The covariance for two random variables can also be determined using the following equation that results from Equation 6-39:

$$\mathrm{COV}(X_1, X_2) = E(X_1X_2) - \mu_{X_1}\mu_{X_2} \tag{6-40}$$

where the expected value of the product (X_1X_2) is given by:

$$E(X_1X_2) = \int_{-\infty}^{+\infty}\int_{-\infty}^{+\infty} x_1x_2 f_{X_1X_2}(x_1, x_2)dx_1dx_2 \tag{6-41}$$

Equation 6-40 can be derived from Equation 6-39 based on the definition of mathematical expectation and by separating terms of integration. If X_1 and X_2 are statistically uncorrelated, then

$$\mathrm{COV}(X_1, X_2) = 0 \tag{6-42a}$$

and

$$E(X_1X_2) = \mu_{X_1}\mu_{X_2} \tag{6-42b}$$

The correlation coefficient is defined as a normalized covariance with respect to the standard deviations of X_1 and X_2 and is given by:

$$\rho_{X_1X_2} = \frac{\mathrm{COV}(X_1, X_2)}{\sigma_{X_1}\sigma_{X_2}} \tag{6-43}$$

The correlation coefficient ranges inclusively between –1 and +1; that is,

$$-1 \le \rho_{X_1X_2} \le +1 \tag{6-44}$$

If the correlation coefficient is zero, then the two random variables are said to be uncorrelated. From the definition of correlation, in order for $\rho_{X_1X_2}$ to be zero, the COV(X_1,X_2) must be zero; therefore, X_1 and X_2 are statistically uncorrelated. However, the converse of this finding does not hold. The correlation coefficient can also be viewed as a measure of the degree of linear association between X_1 and X_2. The sign (– or +) indicates the slope for the linear association. It is important to note

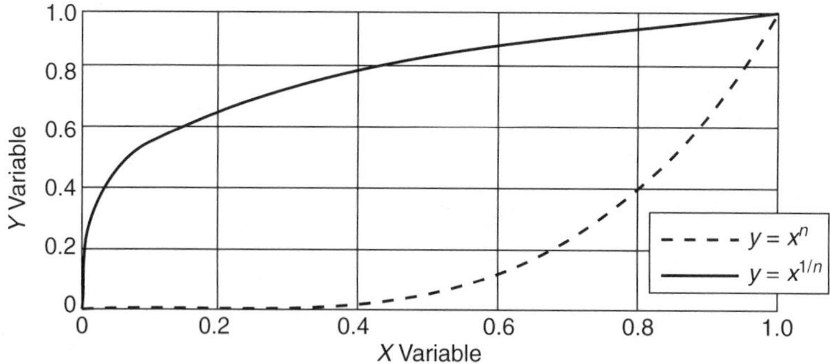

FIGURE 6-1 The two curves in Example 6-3.

that the correlation coefficient does not give any indications about the presence of a nonlinear relationship between X_1 and X_2 (or the lack of it).

Example 6-3: Covariance and Correlation

The sample space for two random variables, X and Y, is defined by the region between the two curves shown in Figure 6-1. Assume that any pair (x,y) in this region is equally likely to occur; that is, a uniform distribution is assumed over the region between the two curves. The two curves are given by the following equations:

$$y = x^{\frac{1}{n}} \quad \text{for } 0 \le x \le 1 \tag{6-45}$$

and

$$y = x^{n} \quad \text{for } 0 \le x \le 1 \tag{6-46}$$

Therefore, the joint density value for these random variables can be viewed on a third axis that is perpendicular to the plane of the curves in Figure 6-1. The joint density function in this case takes on a constant value over this region. The range of y for both curves is [0,1]. Therefore, the value of the density function can be determined based on the following condition:

$$\int_0^{\infty}\int_0^{\infty} f_{XY}(x, y)dx\, dy = 1 \tag{6-47a}$$

where $f_{XY}(x,y)$ is the constant (c). Therefore,

$$1 = \left[\int_0^1 x^{\frac{1}{n}}dx - \int_0^1 x^{n}dx\right]c \tag{6-47b}$$

Solving for c, the following expression can be obtained:

$$c = \frac{n+1}{n-1} \tag{6-48}$$

Therefore, the density function is given by:

$$f_{XY}(x, y)\begin{cases} \dfrac{n+1}{n-1} & \text{for } 1 \le x \le 1 \text{ and } x^n \le y \le x^{\frac{1}{n}} \\ 0 & \text{otherwise} \end{cases} \tag{6-49}$$

The marginal density function of X is given by:

$$f_X(x) = c\int_{x^n}^{x^{\frac{1}{n}}} dy = \frac{n+1}{n-1}\left(x^{\frac{1}{n}} - x^n\right) \quad \text{for } 0 \le x \le 1 \tag{6-50}$$

Similarly, the marginal density functions of Y is given by:

$$f_Y(y) = \frac{n+1}{n-1}\left(y^{\frac{1}{n}} - y^n\right) \qquad \text{for } 0 \le y \le 1 \tag{6-51}$$

Therefore, $f_Y(y)$ is similar to $f_X(x)$. Thus, the expected value of X is equal to the expected value of Y and is given by:

$$E(X) = E(Y) = \int_0^1 x f_X(x) dx \tag{6-52a}$$

$$= \frac{n+1}{n-1}\left[\frac{n}{1+2n} - \frac{1}{n+2}\right] \tag{6-52b}$$

Also, the second moments of X and Y are equal and given by:

$$E(X^2) = E(Y^2) = \int_0^1 x^2 f_X(x) dx \tag{6-53a}$$

$$= \frac{n+1}{n-1}\left[\frac{n}{1+3n} - \frac{1}{n+3}\right] \tag{6-53b}$$

Therefore, the variances of X and Y are:

$$\mathrm{Var}(X) = E(X^2) - [E(X)]^2 \tag{6-54a}$$

$$\mathrm{Var}(Y) = E(Y^2) - [E(Y)]^2 \tag{6-54b}$$

The expected value of the product XY is:

$$E(XY) = c\int_0^1 \int_{x^n}^{x^{\frac{1}{n}}} xy\, dy\, dx \tag{6-55a}$$

$$= \frac{1}{2}\frac{n+1}{n-1}\left[\frac{n}{2+2n} - \frac{1}{2n+2}\right] = \frac{1}{4} \tag{6-55b}$$

Therefore, the covariance of X and Y is:

$$\mathrm{COV}(X, Y) = E(XY) - E(X)E(Y) \tag{6-56}$$

For $n = 2$, these moments take the following values:

$$E(X) = E(Y) = \frac{9}{20} \tag{6-57}$$

$$E(X^2) = E(Y^2) = \frac{9}{35} \tag{6-58}$$

$$\mathrm{Var}(X) = \mathrm{Var}(Y) = 0.0546 \tag{6-59}$$

$$E(XY) = \frac{1}{4} \tag{6-60}$$

$$\mathrm{COV}(X, Y) = 0.0475 \tag{6-61}$$

Therefore, the correlation coefficient is:

$$\rho_{XY} = \frac{\mathrm{COV}(X, Y)}{\sqrt{\mathrm{Var}(X)\mathrm{Var}(Y)}} = \frac{0.0475}{0.0546} = 0.87 \tag{6-62}$$

For $n = 3$, the correlation coefficient ρ_{XY} is 0.71. Figure 6-2 shows selected probability descriptors, including ρ_{XY}, as functions of n. It is interesting to note that as the power order n approaches 1, the area between the two curves diminishes and the correlation coefficient approaches 1. Also, by increasing n, the area between the two curves increases, approaching a limiting case where it covers the entire area of Figure 6-2 and the correlation coefficient approaches zero.

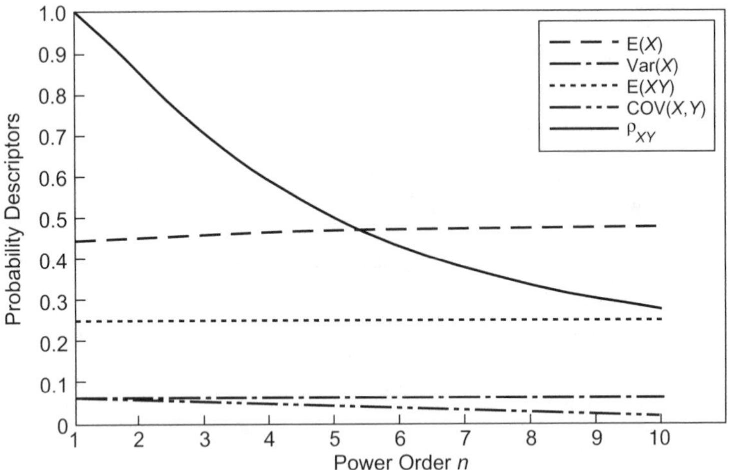

FIGURE 6-2 Probability descriptors as functions of n.

6.2.4. Common Joint Probability Distributions

A few joint probability distributions are used in engineering and science. They include the multinomial distribution, a discrete distribution, and the joint bivariate normal, a continuous distribution. These two distributions are described in this section.

The *multinomial distribution* is a generalization of the binomial distribution. In this case, the distribution is not based on a Bernoulli sequence (i.e., trials with two possible outcomes); it is based on trials with more than two possible outcomes (say, n outcomes). The probabilities of occurrence for these outcomes are assumed to be $p_1, p_2, \ldots, p_n$. The following condition is imposed on these probabilities:

$$p_1 + p_2 + \cdots + p_n = 1 \tag{6-63}$$

The trials are assumed to be independent. The random variables, in this case, $X_1, X_2, \ldots, X_n$, are the numbers of occurrences of the respective outcomes in N identical and independent trials. The joint probability mass function is:

$$P_{X_1X_2\ldots X_n}(x_1, x_2, \ldots, x_n) = \frac{N!}{x_1!x_2!\cdots x_n!}(p_1)^{x_1}(p_2)^{x_2}\ldots(p_n)^{x_n} \tag{6-64}$$

The following condition should be satisfied:

$$N = x_1 + x_2 + \cdots + x_n \tag{6-65}$$

The mean and variance of the respective random variables, $X_1, X_2, \ldots, X_n$, are:

$$\mu_i = NP_i \qquad \text{for } i = 1, 2, \ldots, n \tag{6-66a}$$

$$\sigma_i^2 = NP_i(1 - p_i) \qquad \text{for } i = 1, 2, \ldots, n \tag{6-66b}$$

The *joint bivariate normal distribution* is a generalization of the normal probability distribution. This distribution describes the joint probabilistic characteristics of random variables. For the case of two random variables, X_1 and X_2, the joint density function is given by:

$$f_{X_1X_2}(x_1, x_2) = \frac{1}{2\pi\sigma_1\sigma_2\sqrt{1-\rho^2}}\exp\left\{-\frac{1}{2(1-\rho^2)}\left[\left(\frac{x_1-\mu_1}{\sigma_1}\right)^2 - 2\rho\left(\frac{x_1-\mu_1}{\sigma_1}\right)\left(\frac{x_2-\mu_2}{\sigma_2}\right) + \left(\frac{x_2-\mu_2}{\sigma_2}\right)^2\right]\right\} \tag{6-67}$$

where $-\infty < x_1 < +\infty$, $-\infty < x_2 < +\infty$, ρ is the correlation coefficient for the two random variables, μ is the mean, and σ is the standard deviation. The marginal distributions for X_1 and X_2 are the same as the probability density function of the normal distribution (i.e., Equation 5-6) and are given, respectively, by:

$$f_{X_1}(x_1) = \frac{1}{\sigma_1\sqrt{2\pi}}\exp\left[-\frac{1}{2}\left(\frac{x_1-\mu_1}{\sigma_1}\right)^2\right] \qquad \text{for } -\infty < x_1 < \infty \tag{6-68}$$

$$f_{X_2}(x_2) = \frac{1}{\sigma_2\sqrt{2\pi}}\exp\left[-\frac{1}{2}\left(\frac{x_2-\mu_2}{\sigma_2}\right)^2\right] \qquad \text{for } -\infty < x_2 < \infty \tag{6-69}$$

The following moments can be determined for the two random variables:

$$E(X_1) = \mu_1 \tag{6-70a}$$

$$E(X_2) = \mu_2 \tag{6-70b}$$

$$\text{Var}(X_1) = \sigma_1^2 \tag{6-70c}$$

$$\text{Var}(X_2) = \sigma_2^2 \tag{6-70d}$$

It can be shown that the conditional random variables $X_1|x_2$ and $X_2|x_1$ have normal probability distributions with the following means, respectively:

$$E(X_1|x_2) = \mu_1 + \frac{\sigma_1}{\sigma_2}(x_2 - \mu_2)\rho \tag{6-71a}$$

$$E(X_2|x_1) = \mu_2 + \frac{\sigma_2}{\sigma_1}(x_1 - \mu_1)\rho \tag{6-71b}$$

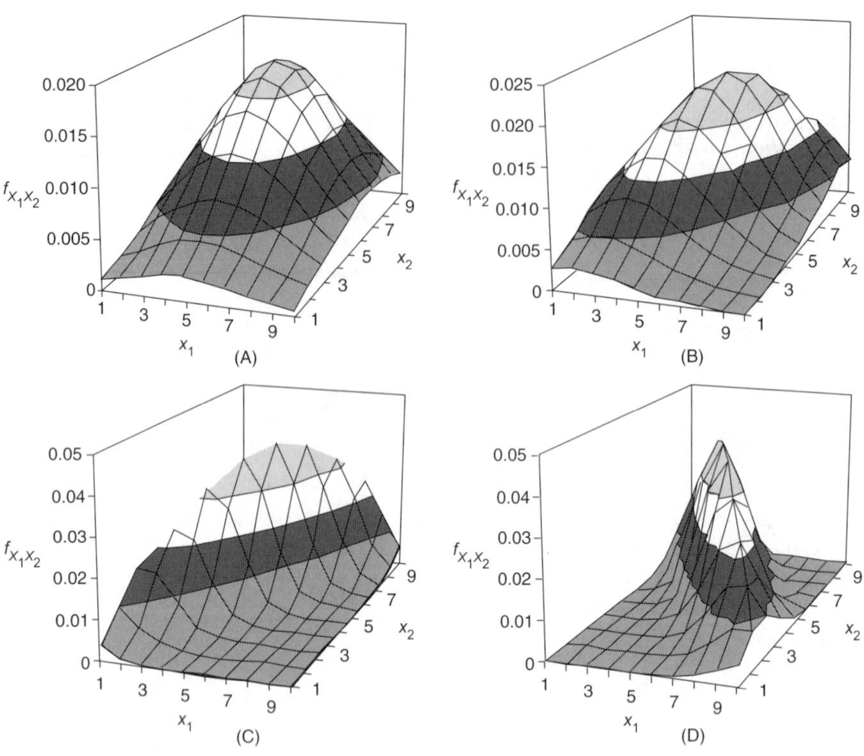

FIGURE 6-3 Joint normal density function. (A) ρ = 0; (B) ρ = 0.5; (C) ρ = 0.9; and (D) ρ = –0.9.

The respective variances of the conditional random variables are:

$$\mathrm{Var}(X_1|x_2) = \sigma_1^2(1-\rho^2) \tag{6-72a}$$

$$\mathrm{Var}(X_2|x_1) = \sigma_2^2(1-\rho^2) \tag{6-72b}$$

An example joint normal is shown in Figure 6-3 using mean values of X_1 and X_2 of 5 and 7, respectively. The corresponding standard deviations are 3 and 3, respectively. The correlation coefficient was varied as shown in the figure. As the correlation coefficient deviates from zero, the surface of the joint density function deviates from the bell shape to show the correlation between the two variables.

6.3. FUNCTIONS OF RANDOM VARIABLES

Many engineering problems deal with a dependent variable that is a function of one or more independent random variables. This section provides analytical tools for determining the probabilistic characteristics of the dependent random variable based on given probabilistic characteristics of independent random variables and a functional relationship between them. The discussion in this section is divided into the following cases: (1) probability distributions for dependent random variables, (2) mathematical expectations, and (3) approximate methods. Cases 1 and 2 result in complete distributions and moments, respectively, without any approximations, whereas the third case results in approximate moments.

6.3.1. Probability Distributions for Dependent Random Variables

A random variable X is defined as a mapping from a sample space of an engineering system or experiment to the real line of numbers. This mapping can be one-to-one mapping or many-to-one mapping, but not one-to-many mapping. If Y is defined to be a dependent variable in terms of a function $Y = g(X)$, then Y is also a random variable. Assuming that both X and Y are discrete random variables and for a given probability mass function of X, $P_X(x)$, the objective herein is to determine the probability mass function of Y, $P_Y(y)$. This objective can be achieved by determining the equivalent events of Y in terms of the events of X based on the given relationship between X and Y, $Y = g(X)$. For each value y_i, all of the values of x that result in y_i should be determined (say, $x_{i_1}, x_{i_2}, \ldots, x_{i_j}$). Therefore, the probability mass function of Y is given by:

$$P_Y(y_i) = \sum_{k=1}^{j} P_X(x_{i_k}) \tag{6-73}$$

If X is continuous but Y is discrete, the probability mass function for Y is given by:

$$P_Y(y_i) = \int_{R_e} f_X(x)dx \tag{6-74}$$

where R_e is the region of X that defines an event equivalent to the value $Y = y_i$.

If X is continuous with a given density function $f_X(x)$ and the function $g(X)$ is continuous, then $Y = g(X)$ is a continuous random variable with an unknown density function $f_Y(y)$. The density function of Y can be determined by performing the following four steps:

1. For any event defined by $Y \leq y$, an equivalent event in the space of X needs to be defined.
2. $F_Y(y) = P(Y < y)$ can then be calculated.
3. $f_Y(y)$ can be determined by differentiating $F_Y(y)$ with respect to y.
4. The range of validity of $f_Y(y)$ in the Y space should be determined.

Formally stated, if X is a continuous random variable, $Y = g(X)$ can be differentiated for all x, and $g(X)$ is either strictly (i.e., monotonically) increasing or strictly (i.e., monotonically) decreasing for all x, then $Y = g(X)$ is a continuous random variable with the following density function:

$$f_Y(y) = f_X[g^{-1}(y)]\left|\frac{\partial g^{-1}(y)}{\partial y}\right| \tag{6-75}$$

where $g^{-1}(y) = x$. Equation 6-75 can be derived by developing an expression for the cumulative distribution function of Y as:

$$F_Y(y) = \int_{x \le g^{-1}(y)} f_X(x)dx = \int_{-\infty}^{g^{-1}(y)} f_X(x)dx \tag{6-76}$$

This result can be expressed as:

$$F_Y(y) = \int_{-\infty}^{y} f_X(g^{-1})\frac{\partial g^{-1}(y)}{\partial y}dy \tag{6-77}$$

By taking the derivative $\frac{\partial F_Y(y)}{\partial y}$, Equation 6-75 results. The derivative $\frac{\partial g^{-1}(y)}{\partial y}$ is known as the Jacobian of the transformation (or inverse) and can be alternatively determined as follows:

$$\frac{\partial g^{-1}(y)}{\partial y} = \frac{1}{\frac{\partial g(x)}{\partial x}}\Bigg|_{\text{evaluated at } x = g^{-1}(y)} \tag{6-78}$$

Then the limits on Y should be determined based on the limits of X and $Y = g(X)$. If the inverse $g^{-1}(y)$ is not unique (say, $= x_1, x_2, \ldots, x_m$), then the density function of Y is determined as follows:

$$f_Y(y) = \sum_{i=1}^{m} f_X[g_i^{-1}(y)]\left|\frac{\partial g_i^{-1}(y)}{\partial y}\right| \tag{6-79}$$

where

$$g_i^{-1}(y) = x_i \tag{6-80}$$

The following cases are selected special functions of single and multiple random variables that are commonly used, where the resulting variable (Y) can have known distribution types for some cases:

1. For a multiple independent random variables $\boldsymbol{X} = (X_1, X_2, \ldots, X_n)$, where the function $g(\boldsymbol{X})$ is a linear combination as given by:

$$Y = g(\boldsymbol{X}) = a_0 + a_1X_1 + a_2X_2 + \cdots + a_nX_n \tag{6-81}$$

and $a_0, a_1, a_2, \ldots, a_n$ are real numbers, the mean value and variance of Y are:

$$E(Y) = g(\boldsymbol{X}) = a_0 + a_1E(X_1) + a_2E(X_2) + \cdots + a_nE(X_n) \tag{6-82}$$

and

$$\text{Var}(Y) = \sum_{i=1}^{n}\sum_{j=1}^{n} a_ia_j\text{COV}(X_i, X_j) \tag{6-83}$$

where $\text{COV}(X_i,X_j)$ is the covariance of X_i and X_j. It should be noted that $\text{COV}(X_i,X_i) = \text{Var}(X_i) = \sigma_{X_i}^2$. Equation 6-83 can be expressed in terms of the correlation coefficient as follows:

$$\text{Var}(Y) = \sum_{i=1}^{n}\sum_{j=1}^{n} a_ia_j\rho_{X_iX_j}\sigma_{X_i}\sigma_{X_j} \tag{6-84}$$

where $\rho_{X_iX_j}$ is the correlation coefficient of X_i and X_j. If the random variables of the vector $\boldsymbol{X}$ are statistically uncorrelated, then the variance of Y is:

$$\text{Var}(Y) = \sum_{i=1}^{n} a_i^2\text{Var}(X_i) \tag{6-85}$$

2. In Equations 6-82 to 6-85, if the random variables $X_1, X_2, X_3, \ldots, X_n$ have normal probability distributions, then Y has a normal probability distribution with a mean and variance as given by Equations 6-81 to 6-85. This special case was also described in Equations 5-17 to 5-19.
3. If X has a normal distribution, and $Y = g(X) = \exp(X)$, then Y has a lognormal distribution.
4. If $Y = X_1 X_2 X_3 \ldots X_n$, the arithmetic multiplication of $X_1, X_2, X_3, \ldots,$ and X_n with lognormal distributions, then Y has a lognormal distribution as given by Equations 5-31 to 5-33.
5. If $X_1, X_2, \ldots, X_n$ are independent random variables that have Poisson distributions with the parameters, $\lambda_1, \lambda_2, \ldots, \lambda_n$, respectively, then $Y = X_1 + X_2 + \ldots + X_n$ has a Poisson distribution with the parameter $\lambda = \lambda_1 + \lambda_2 + \ldots + \lambda_n$.

Example 6-4: Mean and Variance of a Linear Function

Consider the following function:

$$Z = 2X + 5Y + 10 \tag{6-86}$$

where X, Y, and Z are random variables. The means of X and Y are 3 and 5, respectively, and the standard deviations are 1 and 2, respectively. The random variables X and Y are assumed to be uncorrelated; therefore, the mean of Z, μ_Z, is computed using Equation 6-82 as:

$$\mu_Z = 2(3) + 5(5) + 10 = 41 \tag{6-87a}$$

The variance of Z, σ_Z^2, is computed using Equation 6-85 as:

$$\sigma_Z^2 = 2^2(1^2) + 5^2(2^2) = 104 \tag{6-87b}$$

The standard deviation of Z is $\sqrt{104} = 10.20$, and the coefficient of variation is $10.20/41 = 0.25$.

Example 6-5: Probability Density Function of a Nonlinear Function

For the following nonlinear function:

$$Y = aX^2 + b \tag{6-88}$$

where a and b are constants and X has an exponential distribution with a parameter λ, determine the probability density function of Y.

In order to use Equation 6-79 to determine $f_Y(y)$, we need to determine X as a function of Y and the derivative of X with respect to Y. The inverse of the function in Equation 6-88 is:

$$X = \pm\sqrt{\frac{Y-b}{a}} \tag{6-89}$$

The derivative of X with respect to Y is:

$$\frac{dX}{dY} = \mp\frac{1}{2\sqrt{a(Y-b)}} \tag{6-90}$$

Therefore, the density function of Y can be determined based on the density function of the exponential distribution (given by Equation 5-36) and by substituting in Equation 6-79:

$$f_Y(y) = \frac{1}{2\sqrt{a(y-b)}}\left[f_X\left(+\sqrt{\frac{y-b}{a}}\right) + f_X\left(-\sqrt{\frac{y-b}{a}}\right)\right] \tag{6-91a}$$

or

$$f_Y(y) = \frac{\lambda}{2\sqrt{a(y-b)}}\left[\exp\left(-\lambda\sqrt{\frac{y-b}{a}}\right) + \exp\left(+\lambda\sqrt{\frac{y-b}{a}}\right)\right] \tag{6-91b}$$

The resulting function $f_Y(y)$ is not an exponential density function.

6.3.2. Mathematical Expectation

In certain engineering and science applications, we might be interested in knowing only the moments (specifically, the mean and variance) of a random variable Y, based on known probabilistic characteristics of a random variable X and a function $Y = g(X)$. In such cases, mathematical expectation is very effective in achieving this objective.

The mathematical expectation of an arbitrary function $g(X)$ of a continuous random variable X is given by:

$$E[g(X)] = \int_{-\infty}^{+\infty} g(x) f_X(x) dx \tag{6-92}$$

The corresponding equation for a discrete random variable is:

$$E[g(X)] = \sum_{i=1}^{n} g(x_i) P_X(x_i) \tag{6-93}$$

Mathematical expectation can be used to determine the moments of Y for a given probabilistic characteristic of X and the function $Y = g(X)$. The mean (or expected value) can be determined by the direct use of Equations 6-92 and 6-93. The variance can be determined, respectively, for continuous and discrete X random variables as:

$$\mathrm{Var}(Y) = \mathrm{Var}[g(X)] = \int_{-\infty}^{+\infty} \{g(x) - E[g(X)]\}^2 f_X(x) dx \tag{6-94}$$

and

$$\mathrm{Var}(Y) = \mathrm{Var}[g(X)] = \sum_{i=1}^{n} \{g(x_i) - E[g(X)]\}^2 P_X(x_i) \tag{6-95}$$

If we consider a special case where the function, $g(X)$, is the following linear function:

$$Y = g(X) = aX + b \tag{6-96}$$

where a and b are real numbers, mathematical expectation can be used to determine, respectively, the mean and variance of Y as follows:

$$E(Y) = aE(X) + b \tag{6-97}$$

$$\text{Var}(Y) = a^2\text{Var}(X) \tag{6-98}$$

Equations 6-97 and 6-98 are valid regardless of the distribution type of X. Also, they can be used for both cases of discrete and continuous X. Based on Equation 6-96, it can be stated that mathematical expectation, $E(.)$, is a linear operator. However, the variance, Var(.), is not a linear operator.

Equations 6-97 and 6-98 can be generalized to the case of multiple random variables $\boldsymbol{X} = (X_1, X_2, \ldots, X_n)$, where the function $g(\boldsymbol{X})$ is given by:

$$Y = g(\boldsymbol{X}) = a_0 + a_1X_1 + a_2X_2 + \ldots + a_nX_n \tag{6-99}$$

where $a_0, a_1, a_2, \ldots, a_n$ are real numbers. The mean and variance of Y are:

$$E(Y) = a_0 + a_1E(X_1) + a_2E(X_2) + \ldots + a_nE(X_n) \tag{6-100}$$

and

$$\text{Var}(Y) = \sum_{i=1}^{n}\sum_{j=1}^{n} a_ia_j\text{COV}(X_i, X_j) \tag{6-101}$$

where $\text{COV}(X_i,X_j)$ is the covariance of X_i and X_j. It should be noted that $\text{COV}(X_i,X_i) = \text{Var}(X_i) = \sigma_{X_i}^2$. Equation 6-101 can be expressed in terms of the correlation coefficient as follows:

$$\text{Var}(Y) = \sum_{i=1}^{n}\sum_{j=1}^{n} a_ia_j\rho_{X_iX_j}\sigma_{X_i}\sigma_{X_j} \tag{6-102}$$

where $\rho_{X_iX_j}$ is the correlation coefficient of X_i and X_j. If the random variables of the vector $\boldsymbol{X}$ are statistically uncorrelated, then the variance of Y is expressed as:

$$\text{Var}(Y) = \sum_{n=1}^{n} a_i^2\text{Var}(X_i) \tag{6-103}$$

Again, these equations are valid regardless of the distribution types of $\boldsymbol{X}$, as well as for both cases of discrete and continuous $\boldsymbol{X}$.

Now, consider the case where the function $g(\boldsymbol{X}) = g(X_1, X_2, \ldots, X_n)$ is the arithmetic multiplication of all the random variables and is given by:

$$Y = X_1X_2\ldots X_n \tag{6-104}$$

For statistically independent random variables, X_1, X_2, X_3, ..., X_n, the mean and variance of Y are

$$E(Y) = E(X_1)E(X_2)\cdots E(X_n) \tag{6-105}$$

and

$$\text{Var}(Y) = E(X_1^2)E(X_2^2)\cdots E(X_n^2) - [E(X_1)E(X_2)\cdots E(X_n)]^2 \tag{6-106}$$

Again, these equations are valid regardless of the distribution types of $\boldsymbol{X}$, as well as for both cases of discrete and continuous $\boldsymbol{X}$.

Example 6-6: Cost of Precast Concrete

The total cost, C, to manufacture a concrete panel in a precast plant is:

$$C = 1.5X + 2Y \tag{6-107}$$

where X is the cost of materials, Y is the cost of labor, and the constants 1.5 and 2 are overhead cost multipliers for materials and labor, respectively. These multipliers include the costs of storing and handling the materials, management, supervision, and quality control. The costs X and Y are assumed to be uncorrelated random variables with means of \$100/panel and \$250/panel, respectively, and with standard deviations of \$10/panel and \$50/panel, respectively. The means of X and Y are denoted as μ_X and μ_Y, respectively, and the variances are denoted as σ_X^2 and σ_Y^2, respectively. Using mathematical expectation (Equations 6-100 and 6-101), the mean and variance of the total cost are:

$$\mu_C = 1.5\mu_X + 2\mu_Y = 1.5(100) + 2(250) = \$650\text{/panel} \tag{6-108}$$

$$\sigma_C^2 = 1.5^2\sigma_X^2 + 2^2\sigma_Y^2 = 1.5^2(10)^2 + 2^2(50)^2 = 10{,}225\ (\$\text{/panel})^2 \tag{6-109}$$

The standard deviation is $\sqrt{10{,}225} = \$101.12$/panel, and the coefficient of variation is 101.12/650 = 0.1556.

6.3.3. Approximate Methods

The closed-form solutions for the distribution types of dependent random variables, as well as mathematical expectation, provide solutions for the simple cases of functions of random variables. Also, they provide solutions for simple distribution types or a mixture of distribution types for the independent random variables. For cases that involve a more general function $g(\boldsymbol{X})$ or a mixture of distribution types, these methods are not suitable for obtaining solutions due to the analytical complexity of these methods. Also, in some engineering applications, precision might not be needed. In such cases, approximate methods based on Taylor-series expansion (see Appendix B), with or without numerical solutions of needed derivatives, can be used. The use of Taylor-series expansion, in this section, is divided into the following two headings: (1) single random variable X, and (2) multiple random variables (i.e., a random vector $\boldsymbol{X}$).

Single Random Variable X

The Taylor series expansion of a function $Y = g(X)$ about the mean of X, $E(X)$, is given by:

$$Y = g[E(X)] + [X - E(X)]\frac{dg(X)}{dX} + \frac{1}{2}[X - E(X)]^2\frac{d^2g(X)}{dX^2} + \ldots + \frac{1}{k!}[X - E(X)]^k\frac{d^kg(X)}{dX^k} + \ldots \tag{6-110}$$

in which the derivatives are evaluated at the mean of X. Truncating this series at the linear terms, the *first-order mean* and *variance* of Y can be obtained by applying the mathematical expectation and variance operators, respectively. The first-order (approximate) mean is:

$$E(Y) \approx g[E(X)] \tag{6-111}$$

The first-order (approximate) variance is:

$$\mathrm{Var}(Y) \approx \left(\frac{dg(X)}{dX}\right)^2 \mathrm{Var}(X) \tag{6-112}$$

Again the derivative in Equation 6-112 is evaluated at the mean of X.

Random Vector X

The Taylor series expansion of a function $Y = g(\boldsymbol{X})$ about the mean values of $\boldsymbol{X}$, i.e., $E(X_1)$, $E(X_2)$, …, $E(X_n)$, is given by:

$$Y = g[E(X_1), E(X_2), \ldots, E(X_n)] + \sum_{i=1}^{n}[X_i - E(X_i)]\frac{\partial g(\boldsymbol{X})}{\partial X_i} + \sum_{i=1}^{n}\sum_{j=1}^{n}\frac{1}{2}[X_i - E(X_i)][X_j - E(X_j)]\frac{\partial^2 g(\boldsymbol{X})}{\partial X_i \partial X_j} + \ldots \tag{6-113}$$

in which the derivatives are evaluated at the mean values of $\boldsymbol{X}$. Truncating this series at the linear terms, the *first-order mean* and *variance* of Y can be obtained by applying the mathematical expectation and variance operators, respectively. The first-order (approximate) mean is:

$$E(Y) \approx g[E(X_1), E(X_2), \ldots, E(X_n)] \tag{6-114}$$

The first-order (approximate) variance is:

$$\text{Var}(Y) \approx \sum_{i=1}^{n} \sum_{j=1}^{n} \frac{\partial g(\boldsymbol{X})}{\partial X_i} \frac{\partial g(\boldsymbol{X})}{\partial X_j} \text{COV}(X_i, X_j) \tag{6-115}$$

in which the derivatives are evaluated at the mean values of $\boldsymbol{X}$: $E(X_1), E(X_2), \ldots, E(X_n)$.

Example 6-7: Stress in a Beam

The stress, F, in a beam subjected to an external bending moment M is:

$$F = \frac{My}{I} \tag{6-116}$$

where y is the distance from the neutral axis of the cross section of the beam to the point where the stress is calculated, and I is the centroidal moment of inertia of the cross section. Assume that the M and I are random variables with means μ_M and μ_I, respectively, and variances σ_M^2 and σ_I^2, respectively. Also assume that y is not a random variable. The mean value of the stress, μ_F, based on first-order approximations (Equation 6-114) is:

$$\mu_F \approx \frac{\mu_M y}{\mu_I} \tag{6-117}$$

The first-order variance of the stress, σ_F^2, is (see Equation 6-115):

$$\sigma_F^2 \approx \left(\frac{y}{\mu_I}\right)^2 \sigma_M^2 + \left(\frac{\mu_M y}{\mu_I^2}\right)^2 \sigma_I^2 \tag{6-118}$$

Assume that M and I are random variables with means $\mu_M = 3000$ kip·in. and $\mu_I = 1000$ in.^{4}, respectively, and standard deviations $\sigma_M = 900$ kip·in. and $\sigma_I = 80$ in.^{4}, respectively. Also assume that c is constant (i.e., not a random variable) and is equal to 10 in. Based on a first-order approximation, the mean of F is:

$$E(F) \approx \frac{3000(10)}{1000} = 30 \text{ ksi}$$

Assuming the basic random variables to be noncorrelated, the variance of F is:

$$\sigma_F^2 \approx \left(\frac{\partial F}{\partial M}\bigg|_{\mu_M, \mu_I}\right)^2 \sigma_M^2 + \left(\frac{\partial F}{\partial I}\bigg|_{\mu_M, \mu_I}\right)^2 \sigma_I^2$$

$$= \left(\frac{1}{100}\right)^2 (900)^2 + \left(\frac{-3}{100}\right)^2 (80)^2 = 86.76\ (\text{ksi})^2$$

Assuming the basic random variables M and I to be correlated with a correlation coefficient of 0.87, and using the X_1 and X_2 to refer to M and I, respectively, the variance of F can be reevaluated as follows based on Equation 6-115:

$$\sigma_F^2 \approx \left(\frac{\partial F}{\partial M}\bigg|_{\mu_M, \mu_I}\right)^2 \sigma_M^2 + 2\frac{\partial F}{\partial I}\bigg|_{\mu_M, \mu_I} \frac{\partial F}{\partial M}\bigg|_{\mu_M, \mu_I} \text{COV}(I, M) + \left(\frac{\partial F}{\partial I}\bigg|_{\mu_M, \mu_I}\right)^2 \sigma_I^2$$

or

$$\sigma_F^2 \approx \left(\frac{1}{100}\right)^2 (900)^2 + 2\left(\frac{-3}{100}\right)\left(\frac{1}{100}\right)[0.87(900)(80)] + \left(\frac{-3}{100}\right)^2 (80)^2 = 49.18$$

The mean in this case is the same as the uncorrelated case. It should be noted that the variance in the correlated case is smaller than the variance of the uncorrelated case as a result of partial derivatives of opposite signs for the two basic random variables.

6.4. APPLICATIONS

6.4.1. Reactions Due to a Random Load

A random load P is applied at the end of a cantilever beam of length L as shown in Figure 6-4. The vertical force reaction, V, of the beam at the fixed support is:

$$V = P \tag{6-119}$$

The moment reaction M at the fixed support is:

$$M = PL \tag{6-120}$$

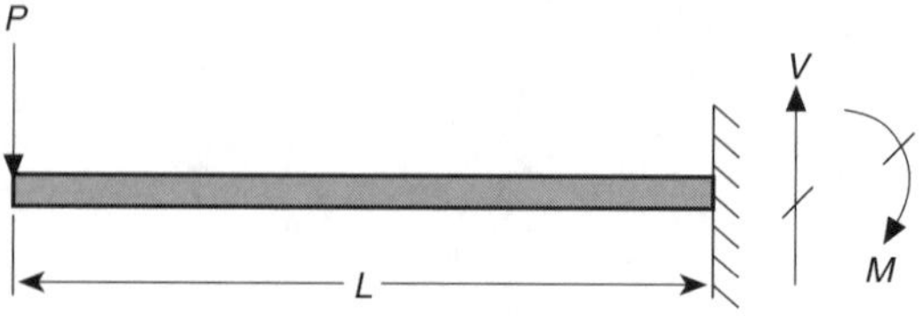

FIGURE 6-4 Cantilever beam with one load.

Assume that the load has a mean value μ_p and variance σ_P^2. The length L is considered to be nonrandom (i.e., a deterministic quantity). Therefore, the mean of the reaction moment based on Equation 6-82 is:

$$\mu_M = L\mu_P \tag{6-121}$$

The variance of the moment is computed using Equation 6-83 as follows:

$$\sigma_M^2 = L^2\sigma_P^2 \tag{6-122}$$

6.4.2. Buckling of Columns

The buckling load-capacity, P, of a column that has an effective length L and radius of gyration r is given by:

$$P = \frac{\pi^2 E}{\left(\frac{L}{r}\right)^2} = \frac{\pi^2 r^2 E}{L^2} \tag{6-123}$$

The column is made of a material with a modulus of elasticity, E, and is shown in Figure 6-5. Assume that r and E are deterministic quantities and that L follows a normal distribution, $N(\mu_L, \sigma_L^2)$. Therefore, the buckling load can be expressed as:

$$P = \frac{c}{L^2} \tag{6-124}$$

where the constant $c = \pi^2 r^2 E$. Thus, the derivative of L with respect to P can be computed as:

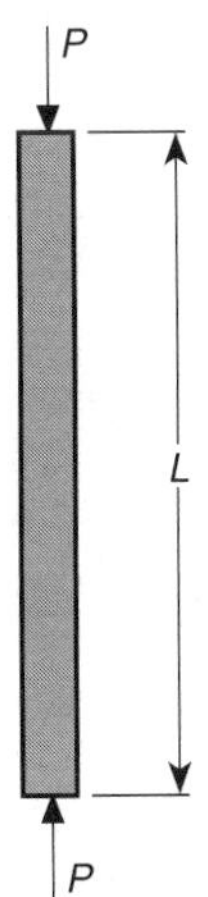

FIGURE 6-5 Buckling of a column.

$$L = \pm\sqrt{\frac{c}{P}} = \pm\sqrt{c}P^{-\frac{1}{2}} \tag{6-125}$$

$$\frac{dL}{dP} = \pm\frac{\sqrt{c}}{2P^{1.5}} \tag{6-126}$$

Therefore, the density function of P is given by:

$$f_P(p) = \left[f_L\left(\sqrt{\frac{c}{p}}\right) + f_L\left(-\sqrt{\frac{c}{p}}\right)\right]\frac{\sqrt{c}}{2p^{1.5}} \tag{6-127}$$

or

$$f_P(p) = \frac{1}{\sigma_L\sqrt{2\pi}}\frac{\sqrt{c}}{2p^{1.5}}\left[\exp\left(-\frac{1}{2}\left\{\frac{\sqrt{\frac{c}{p}}-\mu_L}{\sigma_L}\right\}^2\right) + \exp\left(-\frac{1}{2}\left\{\frac{-\sqrt{\frac{c}{p}}-\mu_L}{\sigma_L}\right\}^2\right)\right] \tag{6-128}$$

6.4.3. Reactions Due to Random Loads

Random concentrated loads P_1 and P_2 are applied at a cantilevered beam of length L_1 as shown in Figure 6-6. The loads are applied at distances L_1 and L_2 from the fixed support, respectively. The vertical force reaction V is:

$$V = P_1 + P_2 \tag{6-129}$$

The moment reaction M is:

$$M = P_1L_1 + P_2L_2 \tag{6-130}$$

The loads P_1 and P_2 have means of μ_{P_1} and μ_{P_2}, respectively, and variances of $\sigma^2_{P_1}$ and $\sigma^2_{P_2}$, respectively. The lengths are considered to be nonrandom (i.e., deterministic

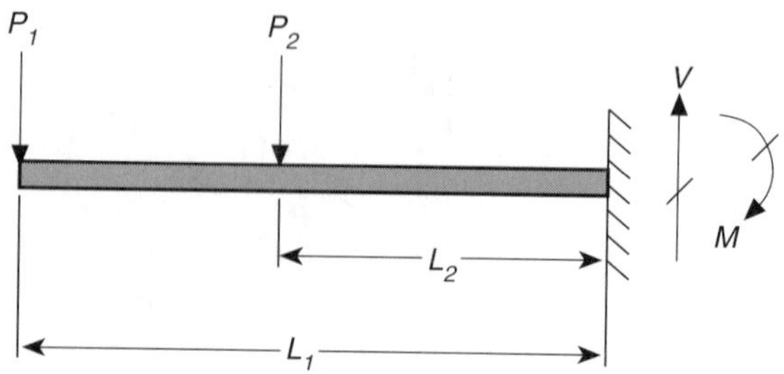

FIGURE 6-6 Cantilever beam with two loads.

quantities). Assume the loads to be statistically uncorrelated. Using mathematical expectation (see Equation 6-99), the means of the reactions are:

$$\mu_V = \mu_{P_1} + \mu_{P_2} \tag{6-131}$$

$$\mu_M = L_1\mu_{P_1} + L_2\mu_{P_2} \tag{6-132}$$

The variances of the reactions according to Equation 6-101 are:

$$\sigma_V^2 = \sigma_{P_1}^2 + \sigma_{P_2}^2 \tag{6-133}$$

$$\sigma_M^2 = L_1^2\sigma_{P_1}^2 + L_2^2\sigma_{P_2}^2 \tag{6-134}$$

Now, assume that the dimensions L_1 and L_2 are random variables with mean values μ_{L_1} and μ_{L_2}, respectively, and variances $\sigma_{L_1}^2$ and $\sigma_{L_2}^2$, respectively. In this case, the mean values of the reactions based on first-order approximations (Equation 6-114) are:

$$\mu_V \approx \mu_{P_1} + \mu_{P_2} \tag{6-135a}$$

$$\mu_M \approx \mu_{L_1}\mu_{P_1} + \mu_{L_2}\mu_{P_2} \tag{6-135b}$$

The first-order variances of the reactions according to Equation 6-115 are:

$$\sigma_V^2 \approx \sigma_{P_1}^2 + \sigma_{P_2}^2 \tag{6-136a}$$

$$\sigma_M^2 \approx \mu_{L_1}^2\sigma_{P_1}^2 + \mu_{P_1}^2\sigma_{L_1}^2 + \mu_{L_2}^2\sigma_{P_2}^2 + \mu_{P_2}^2\sigma_{L_2}^2 \tag{6-136b}$$

6.4.4. Open Channel Flow

In an open channel, water flows because of a difference in elevation between one section of the channel to another. The velocity V of the water flow in an open channel is:

$$V = \frac{1}{n}R^{\frac{2}{3}}S^{\frac{1}{2}} \tag{6-137}$$

where R is the hydraulic radius, S is the slope of the hydraulic gradient, and n is a channel roughness factor that depends on the lining of the channel. Equation 6-137 is called the *Manning formula*. The hydraulic radius, R, can be computed as:

$$R = \frac{A}{P} \tag{6-138}$$

where A is the cross-sectional area of flow and P is the wetted perimeter. For a channel near an existing highway, $n = 0.070$, $S = 4\%$, and R is a random variable with the following exponential distribution:

$$f_R(r) = \lambda \exp(-\lambda r) \quad \text{for } r \geq 0 \tag{6-139}$$

where $\lambda = 12.5$ sec/m. Determine the probability density function of the velocity V.

Equation 6-137 can be rewritten as:

$$V = cR^{\frac{2}{3}} \tag{6-140}$$

where c is a constant as follows:

$$c = \frac{S^{\frac{1}{2}}}{n} = \frac{\sqrt{0.04}}{0.070} = 2.8571 \tag{6-141}$$

Therefore, from Equation 6-140, r is given by:

$$R = \left(\frac{V}{c}\right)^{\frac{3}{2}} \tag{6-142}$$

and

$$\frac{dR}{dV} = \left(\frac{1}{c}\right)^{\frac{3}{2}} \frac{3}{2}\sqrt{V} \tag{6-143}$$

Then, using Equation 6-75,

$$f_V(v) = \frac{3}{2} f_R\left(\left(\frac{v}{c}\right)^{\frac{3}{2}}\right) c^{-\frac{3}{2}} \sqrt{v} \tag{6-144a}$$

$$= \frac{3}{2} \lambda \exp\left[-\lambda\left(\frac{v}{c}\right)^{\frac{3}{2}}\right] \sqrt{v} c^{-\frac{3}{2}} \quad \text{for } v \geq 0 \tag{6-144b}$$

The resulting distribution is not of the exponential type. However, the approximate mean of V, $\overline{V}$, can be computed using Equation 6-111. The result is:

$$\overline{V} \approx 2.8571 \overline{R}^{\frac{2}{3}} \tag{6-145a}$$

where $\bar{R}$ is the mean of R which is $1/\lambda$. Therefore, the mean value is:

$$\bar{V} = 2.8571\left(\frac{1}{\lambda}\right)^{\frac{2}{3}} = 0.530 \text{ m/sec} \quad (6\text{-}145b)$$

The approximate variance of V, Var(V), can be computed using Equation 6-112 as:

$$\text{Var}(V) \approx (2.8571)^2\left(\frac{2}{3}\right)^2\left(\bar{R}^{-\frac{1}{3}}\right)^2 \text{Var}(R) \quad (6\text{-}146a)$$

where Var(R) is the variance of R. Therefore, the variance is:

$$\text{Var}(V) \approx (2.8571)^2\left(\frac{2}{3}\right)^2\left(\frac{1}{\lambda}\right)^{\frac{-2}{3}}\frac{1}{\lambda^2} = 0.1250608 \text{ (m/sec)}^2 \quad (6\text{-}146b)$$

The standard deviation is $\sqrt{0.1250608} = 0.3536$ m/sec, and the coefficient of variation is 0.667.

6.4.5. Warehouse Construction

A warehouse is to be constructed from precast concrete elements that are produced by a nearby precast factory. The following construction tasks are identified for building the warehouse:

A: excavation of foundations
B: construction of foundations
C: construction of precast elements at factory
D: transportation of precast elements to construction site
E: assembly of elements at site
F: construction of roof
G: exterior and interior finishing

Figure 6-7 shows the logical network for performing these activities. The figure indicates that tasks *C* and *D* can be performed in parallel to tasks *A* and *B*; that is,

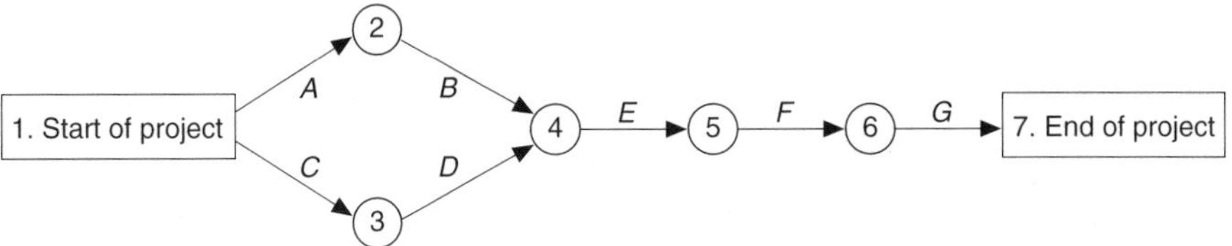

FIGURE 6-7 Construction network.

TABLE 6-1 Moments of Completion Times

Task	Name	Mean (days)	Standard deviation (days)
A	Foundation excavation	3	1
B	Foundation construction	2	0.5
C	Precast-elements construction	4	1
D	Transportation of elements	0.5	0.5
E	Assembly of elements	4	1
F	Roofing	2	1
G	Finishing	3	1

as excavation and construction of the footings are being performed at the site, the precast elements can be constructed at the factory and then transported to the construction site. Table 6-1 shows the means and standard deviations for the completion times of these tasks. Normal probability distributions and statistical independence are assumed for these times, although normally distributed random variables can take negative values, which is not appropriate for modeling durations. Also, the normal distribution can take any real value. For modeling durations of construction activities, distributions such as the triangular distribution might be more appropriate than the normal because it has lower and upper limits on possible values. The use of the normal distribution is for the purpose of illustration. The assumption of statistical independence of durations can be questionable due to common causes for delays such as weather or labor.

A construction engineer is interested in determining the completion time of the project (i.e., the duration of the project). Figure 6-7 shows the paths that are necessary to complete the project. Therefore, the completion time, T, is a dependent random variable that is given by:

$$T = \max\{A + B, C + D\} + E + F + G \tag{6-147}$$

where A, B, C, D, E, F, and G are random variables that model the times for completing the corresponding tasks. The mean duration of the project, $\overline{T}$, can be determined as:

$$\overline{T} = \max\{3 + 2, 4 + 0.5\} + 4 + 2 + 3 = 14 \text{ days} \tag{6-148}$$

The variance of the project duration, Var(T), can be computed approximately as:

$$\text{Var}(T) = \text{Var}(A) + \text{Var}(B) + \text{Var}(E) + \text{Var}(F) + \text{Var}(G) \tag{6-149a}$$

$$= (1)^2 + (0.5)^2 + (1)^2 + (1)^2 + (1)^2 = 4.25 \text{ (days)}^2 \tag{6-149b}$$

The computation of the variance was based on the path, from the construction network of Figure 6-7, that resulted in the mean duration of the project. The exact variance of T can be computed using simulation as discussed in Chapter 7. Assuming a normal probability distribution for T, the probability that the project finishes at the most 20 days after the start of the project can be computed as:

$$\text{Prob}(T < 20) = \Phi\left(\frac{20 - 14}{\sqrt{4.25}}\right) = \Phi(2.91) = 0.9986 \qquad (6\text{-}150)$$

6.5. MULTIVARIABLE SIMULATION

Simulation easily handles studies where multiple random variables are involved. Very often, it can provide information about the distributions of random variables that is beyond the ability of theory. Theoretical relationships are often based on restrictive assumptions, such as normal distributions, that may not be valid for a given problem. Simulation would enable the needed statistics to be computed for conditions where the restrictive assumptions are not applicable.

Simulation can provide information beyond that provided by theory. This is especially true in cases that involve multiple random variables. For example, if the distribution of a random variable that represents the output from a function with multiple random variable inputs is needed, simulation can enable the distribution to be empirically approximated.

6.5.1. Simulation of Expected Values

Equations 6-100 and 6-103 provide a method for estimating the mean and variance of uncorrelated multiple random variables. The information from Example 6-6 illustrates the computation of the mean and variance; however, the distribution of the total cost C might be of interest, and it will depend on the distribution of the two random variable inputs X and Y.

Example 6-8: Expectation with Normally Distributed Random Variables

To demonstrate that Equations 6-100 and 6-103 can be applied regardless of the distribution function of the two variables, two simulations were made, each with 50 estimates of the cost C. For the first simulation, the material (X) and labor (Y) costs of Equation 6-107 are assumed to be normally distributed, $X \sim N(\mu = 100, \sigma = 10)$, and $Y \sim N(\mu = 250, \sigma = 50)$. The simulations are given in Table 6-2. The means and standard deviations of the three variables based on the 50 trials are:

Variables	Mean	Standard deviation
X	99.07	10.87
Y	231.85	50.24
C	612.31	104.77

TABLE 6-2 Simulated Values of Material (*X*), Labor (*Y*), and Cost (*C*) Assuming Normal Populations

Random number for X	Standard normal variate for X	X	Random number for Y	Standard normal variate for Y	Y	C
0.8443	1.04	110.37	0.4661	–0.09	245.73	657.01
0.1980	–0.85	91.50	0.9932	2.49	374.30	885.85
0.1776	–0.94	90.60	0.1266	–1.16	191.77	519.44
0.1615	–1.01	89.89	0.8295	0.97	298.57	731.98
0.5239	0.06	100.60	0.6777	0.46	273.20	697.30
0.3515	–0.38	96.15	0.0094	–2.36	132.19	408.60
0.8930	1.25	112.51	0.2739	–0.61	219.64	608.04
0.6064	0.27	102.71	0.1731	–0.96	202.00	558.06
0.5811	0.20	102.05	0.9286	1.49	324.47	802.01
0.3413	–0.41	95.88	0.6909	0.50	274.99	693.79
0.0734	–1.47	85.25	0.2223	–0.77	211.46	550.79
0.6733	0.45	104.52	0.0389	–1.77	161.70	480.18
0.1736	–0.96	90.42	0.6915	0.50	275.07	685.77
0.5861	0.22	102.18	0.2933	–0.55	222.72	598.70
0.4850	–0.04	99.62	0.2706	–0.62	219.12	587.67
0.2585	–0.66	93.44	0.1157	–1.21	189.38	518.91
0.3677	–0.34	96.59	0.1653	–0.99	200.29	545.46
0.6824	0.48	104.77	0.6065	0.27	263.55	684.25
0.4980	–0.01	99.95	0.5294	0.07	253.70	657.33
0.6941	0.51	105.08	0.4002	–0.25	237.36	632.34
0.1540	–1.04	89.56	0.3219	–0.47	226.75	587.84
0.5457	0.12	101.15	0.3773	–0.31	234.26	620.24
0.3524	–0.38	96.18	0.7185	0.58	279.15	702.57
0.1393	–1.11	88.91	0.3694	–0.34	233.19	599.74
0.9629	1.79	117.91	0.4530	–0.12	244.08	665.03
0.0512	–1.64	83.64	0.2307	–0.74	212.79	551.03
0.9232	1.45	114.50	0.4632	–0.09	245.37	662.48
0.4701	–0.08	99.25	0.6520	0.39	269.71	688.30
0.6808	0.47	104.72	0.2168	–0.79	210.58	578.25
0.1739	–0.96	90.44	0.0920	–1.34	183.02	501.69
0.9865	2.23	122.31	0.0916	–1.34	182.87	549.21
0.6708	0.45	104.45	0.2004	–0.84	207.98	572.64
0.1648	–1.00	90.04	0.1855	–0.91	204.73	544.51
0.6297	0.33	103.34	0.1087	–1.24	187.84	530.68
0.9849	2.19	121.87	0.9313	1.51	325.45	833.71
0.7241	0.60	106.01	0.9626	1.79	339.35	837.71
0.2591	–0.65	93.46	0.8913	1.24	312.16	764.52
0.0157	–2.17	78.29	0.0723	–1.48	175.86	469.15
0.2928	–0.55	94.53	0.1062	–1.25	187.29	516.36
0.0017	–2.94	70.58	0.0542	–1.61	169.28	444.44
0.3103	–0.50	95.04	0.3740	–0.32	233.81	610.17
0.4511	–0.12	98.77	0.3404	–0.41	229.26	606.66
0.9829	2.13	121.33	0.8335	0.99	299.45	780.89
0.1553	–1.04	89.62	0.4301	–0.18	241.20	616.82
0.1830	–0.92	90.84	0.2275	–0.75	212.28	560.82
0.7176	0.58	105.80	0.2582	–0.66	217.15	593.00
0.0907	–1.35	86.51	0.2206	–0.78	211.19	552.14
0.5069	0.02	100.17	0.0563	–1.6	170.05	490.35
0.7756	0.76	107.64	0.1468	–1.08	196.22	553.90
0.9003	1.28	112.84	0.0806	–1.42	178.87	527.00

Using Equations 6-100 and 6-103, the computed moments are:

$$\bar{C} = 1.5(99.07) + 2.0(231.85) = \$612.30$$
$$S_c^2 = (1.5)^2(10.87)^2 + (2.0)^2(50.24)^2 = 10{,}362$$
$$S_c = \$101.80$$

The difference between the values of 101.8 and 104.77 is due to sampling variation; another set of 50 simulation trials will produce different answers such that on the average the true moments are attained.

Example 6-9: Expectation with Exponential Variables

For this second simulation, X and Y of Example 6-8 are assumed to be exponentially distributed, $X \sim \exp(\text{with } \lambda = 100)$ and $Y \sim \exp(\text{with } \lambda = 250)$. The exponential distribution assumes that the mean equals the standard deviation, which was not the case for the simulation based on the normal distribution. The simulations for the exponential distribution are given in Table 6-3. The means and standard deviations of the three variables based on the 50 trials are:

Variables	Mean	Standard deviation
X	111.98	115.31
Y	333.48	245.35
C	834.93	559.16

Using Equations 6-100 and 6-103, the computed moments are:

$$\bar{C} = 1.5(111.98) + 2.0(333.48) = \$834.93$$
$$S_c^2 = (1.5)^2(115.31)^2 + (2.0)^2(245.35)^2 = 270{,}703$$
$$S_c = \$520.30$$

The difference between the values of 520.3 and 559.16 is due to sampling variation; another set of 50 simulation trials will produce different answers such that on the average the true moments are attained.

While the relationships for the two moments, Equations 6-100 and 6-103, do not depend on the distributions of X and Y, the distribution of C depends on the distributions of X and Y. Histograms for the 50 simulated values of C from Tables 6-2 and 6-3 are shown in Figure 6-8. Because the means and standard deviations are different, the two histograms can only be assessed using their shapes, which are quite different. The distribution of C depends on the distributions of X and Y, the inputs. For the histogram of Figure 6-8A, which is based on normally distributed inputs, the histogram has a more centered peak than the histogram based on exponentially distributed inputs (Figure 6-8B). The latter histogram is relatively skewed, which is similar to the exponential distribution.

TABLE 6-3 Simulated Values of Material (X), Labor (Y), and Cost (C) Assuming Exponential Populations

Random number for X	X	Random number for Y	Y	C
0.8443	16.92	0.4661	190.84	407.06
0.1980	161.95	0.9932	1.71	246.33
0.1776	172.82	0.1266	516.68	1292.59
0.1615	182.33	0.8295	46.73	366.95
0.5239	64.65	0.6777	97.26	291.49
0.3515	104.55	0.0094	1166.76	2490.35
0.8930	11.32	0.2739	323.75	664.47
0.6064	50.02	0.1731	438.47	951.98
0.5811	54.28	0.9286	18.52	118.46
0.3413	107.50	0.6909	92.44	346.13
0.0734	261.18	0.2223	375.93	1143.64
0.6733	39.56	0.0389	811.69	1682.72
0.1736	175.10	0.6915	92.22	447.10
0.5861	53.43	0.2933	306.64	693.42
0.4850	72.36	0.2706	326.78	762.10
0.2585	135.29	0.1157	539.19	1281.31
0.3677	100.05	0.1653	450.00	1050.07
0.6824	38.21	0.6065	125.01	307.35
0.4980	69.72	0.5294	159.00	422.58
0.6941	36.51	0.4002	228.95	512.67
0.1540	187.08	0.3219	283.38	847.38
0.5457	60.57	0.3773	243.68	578.21
0.3524	104.30	0.7185	82.65	321.74
0.1393	197.11	0.3694	248.97	793.61
0.9629	3.78	0.4530	197.97	401.60
0.0512	297.20	0.2307	366.66	1179.12
0.9232	7.99	0.4632	192.40	396.78
0.4701	75.48	0.6520	106.93	327.08
0.6808	38.45	0.2168	382.20	822.06
0.1739	174.93	0.0920	596.49	1455.37
0.9865	1.36	0.0916	597.58	1197.20
0.6708	39.93	0.2004	401.86	863.61
0.1648	180.30	0.1855	421.18	1112.80
0.6297	46.25	0.1087	554.79	1178.96
0.9849	1.52	0.9313	17.79	37.87
0.7241	32.28	0.9626	9.53	67.48
0.2591	135.05	0.8913	28.77	260.12
0.0157	415.41	0.0723	656.73	1936.58
0.2928	122.83	0.1062	560.61	1305.46
0.0017	637.71	0.0542	728.77	2414.11
0.3103	117.02	0.3740	245.87	667.28
0.4511	79.61	0.3404	269.41	658.23
0.9829	1.72	0.8335	45.53	93.65
0.1553	186.24	0.4301	210.93	701.23
0.1830	169.83	0.2275	370.15	995.04
0.7176	33.18	0.2582	338.51	726.79
0.0907	240.02	0.2206	377.85	1115.73
0.5069	67.94	0.0563	719.27	1540.45
0.7756	25.41	0.1468	479.67	997.46
0.9003	10.50	0.0806	629.56	1274.88

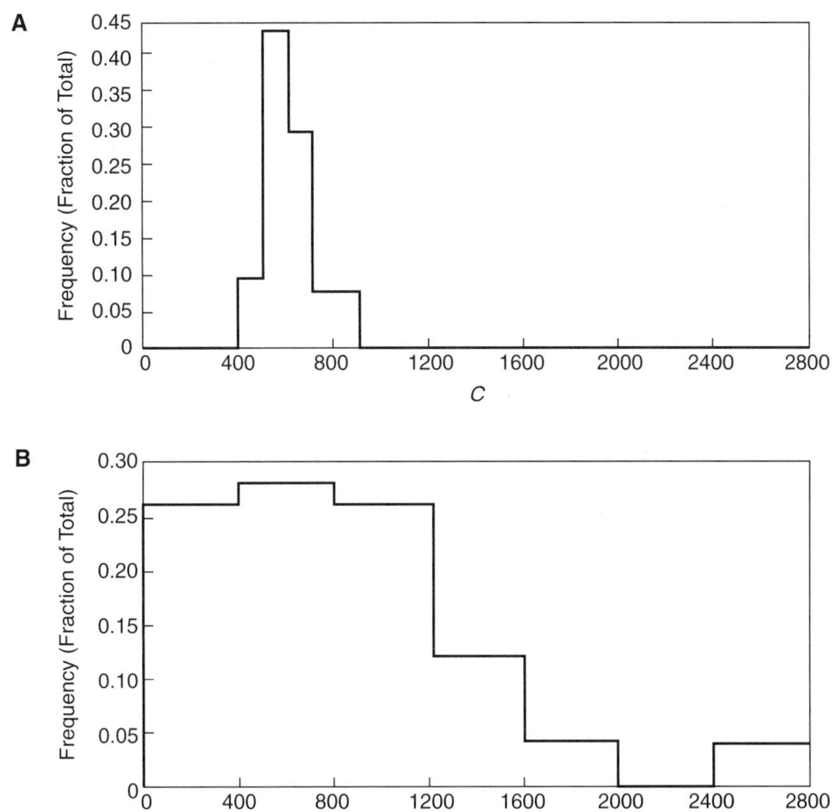

FIGURE 6-8 Distributions of total cost (C) for 50 simulation trials assuming cost components X and Y are (A) normally distributed, and (B) exponentially distributed.

6.5.2. Simulation and Correlation

Correlation was introduced in Section 6.2.3. It is an important topic in statistical analysis. Equation 6-43 provides the general concept of correlation based on probability theory. In practice, the numerical value depends on the calculation of the covariance. Several coefficients have been developed, with the Pearson product-moment correlation coefficient being the most widely used, as discussed in Chapter 12. For continuous random variables, the covariance is defined as the average of the products of the deviations of the two random variables (Equation 6-39); thus, the correlation coefficient for sample values of X_1 and X_2 becomes:

$$\rho_{X_1X_2} = \frac{\mathrm{COV}(X_1, X_2)}{\sigma_{X_1}\sigma_{X_2}} = \frac{1}{n}\sum_{i=1}^{n}\frac{(X_{1i}-\bar{X}_1)(X_{2i}-\bar{X}_2)}{S_{X_1}S_{X_2}} \tag{6-151}$$

or expressed as standardized variates:

$$\rho_{X_1X_2} = \frac{1}{n}\sum_{i=1}^{n}\left(\frac{X_{1i}-\bar{X}_1}{S_{X_1}}\right)\left(\frac{X_{2i}-\bar{X}_2}{S_{X_2}}\right) = \frac{1}{n}\sum_{i=1}^{n} z_{1i}z_{2i} \tag{6-152}$$

Of special interest in many statistical analyses is the distribution of $\rho_{X_1X_2}$. While a sample estimate of the value is easily computed with the above equation, knowledge of the distribution of the sample value is important when deciding how much weight to place on the sample value when making a decision about a causal relationship between X_1 and X_2. The distribution depends on the sample size and the true value $\rho_{X_1X_2}$ from which the sample values of X_1 and X_2 were drawn.

In order to simulate random samples that can be used to determine the distribution of the correlation coefficient, a model is needed. Because the distribution of sample values of the correlation coefficient depends on the population value, ρ, the model should be a function of ρ. The following relationship can be used to generate sample values of the random variable Y for given values of the second random variable X and a population value of the correlation coefficient ρ:

$$Y = \beta_0 + \beta_1 X + Z\sigma_Y\sqrt{1-\rho^2} \tag{6-153}$$

where β_0 and β_1 are the population values of a linear relationship between X and Y, σ_Y is the standard deviation of Y, and Z is the standard normal deviate. The slope coefficient β_1 is related to the population correlation by $\beta_1 = \rho\sigma_Y/\sigma_X$. Thus, the above equation becomes:

$$Y = \mu_Y + \rho\left(\frac{\sigma_Y}{\sigma_X}\right)(x-\mu_X) + Z\sigma_Y\sqrt{1-\rho^2} \tag{6-154}$$

The first two terms on the right-hand side of Equation 6-154 represent the correlation between Y and X, with μ_Y, the mean of Y, being a vertical location parameter, and σ_Y and σ_X being scaling parameters. The third term on the right-hand side of the above equation represents the extent to which Y and X are uncorrelated. If ρ equals 1 or –1, then the third term drops out and the relationship between Y and X is defined entirely by the deterministic linear relationship $Y = \beta_0 + \beta_1 X$. If ρ equals 0, then the random component and the vertical location parameter define the relationship between Y and X.

To generate random samples that are from a population with correlation ρ, values of σ_Y, σ_X, μ_Y, μ_X, and the sample size n must be set. Then values of X with set characteristics μ_X and σ_X can be generated randomly. Using random values of Z, where Z is normally distributed with mean of 0 and variance of 1, the above equation is used to generate values of Y. Other distributions can be used; however, the normal distribution is almost always assumed.

Example 6-10: Correlation Between Functions

Equations 6-45 and 6-46 provide two functions that depend on the value of a random variable X and parameter n. Figure 6-2 shows the relationship between the correlation coefficient ρ and parameter n. For $n = 2$, Equation 6-62 shows that the population correlation coefficient is 0.87. For $n = 3$, the population value equals 0.71. The development of Equation 6-62 applies to the population but not to sample estimates of the correlation coefficient. Furthermore, the theoretical development does not provide knowledge of the distribution of the sample correlation coefficient (R), which is a function of the size of the random sample (N).

Simulation can provide the distribution of sample estimates of the correlation between the two functions of Equations 6-45 and 6-46. Using uniform random numbers, samples of 5, 10, and 25 were generated for the case n = 3, and the correlation between the two functions was computed, with the following means:

N	**Mean correlation R**
5	0.891
10	0.832
25	0.803

As the sample size N increases, the mean would approach the population value of 0.71.

In addition to means, the distributions of the sample R were obtained for each N. Figure 6-9 shows the three simulated distributions with the values listed in Table 6-4. As expected, as the sample size increases, the spread of the distribution decreases, which reflects the increase in the accuracy of the statistic as N increases. The centers of the distributions move toward the population value of 0.71.

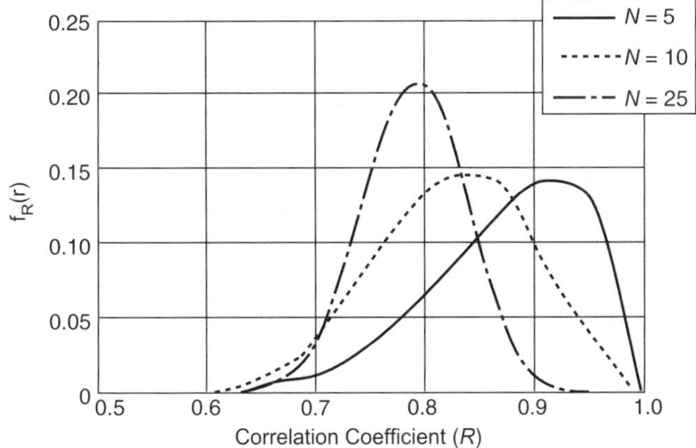

FIGURE 6-9 Sample distributions of the correlation coefficients of the functions of Equations 6-45 and 6-46 for sample sizes (N) of 5, 10, and 25.

TABLE 6-4 Sample Distributions of the Correlation of the Functions of Equations 6-45 and 6-46 for Sample Sizes (N) of 5, 10, and 25

Correlation coefficient	$N = 5$	$N = 10$	$N = 25$
0.575	0.0000	0.0002	—
0.600	0.0002	0.0007	—
0.625	0.0016	0.0036	0.0003
0.650	0.0022	0.0099	0.0028
0.675	0.0065	0.0179	0.0104
0.700	0.0106	0.0352	0.0269
0.725	0.0195	0.0591	0.0819
0.750	0.0325	0.0759	0.1468
0.775	0.0484	0.1016	0.1960
0.800	0.0672	0.1328	0.2068
0.825	0.0824	0.1376	0.1785
0.850	0.1022	0.1401	0.0989
0.875	0.1123	0.1199	0.0396
0.900	0.1374	0.0877	0.0097
0.925	0.1378	0.0529	0.0014
0.950	0.1371	0.0213	—
0.975	0.1021	0.0036	—
1.000	—	—	—

Example 6-11: Multivariable Simulation of Stage and Discharge

Example 6-10 evaluated the correlation between two functions of one random variable, X. In this example, values of two random variables were simulated and the correlation between the simulated data was assessed. The simulation assumes values of the population parameters of Equation 6-154 and the distribution of X, which along with the distribution of Z determine the distribution of Y. In this example, both X and Z are assumed to be normally distributed. The following statistics were assumed: $\mu_Y = 55$, $\rho = 0.921$, $\sigma_Y = 54$, $\sigma_X = 0.71$, $\mu_X = 2.84$. These are approximately the sample values for the data of Table 2-3.

Table 6-5 gives 50 simulated values of the stage and discharge. The values of stage X were simulated to have a normal distribution and Equation 6-154 was used to simulate the discharges Y. Figure 6-10 shows the distributions of the simulated values.

Simulated data should always be examined to ensure that the values are reasonable. Certainly, the distributions of Figure 6-10 are an important means of assessing the data. In this case, both appear to be normally distributed. It is also useful to compare the sample statistics of the simulated data with the assumed population. The following tabular summary compares the statistics:

Statistic	Assumed population	Simulated data
Stage: mean	2.84	2.780
Stage: standard deviation	0.71	0.619
Discharge: mean	55.	54.5
Discharge: standard deviation	54.	37.3

TABLE 6-5 Simulated Stages and Discharges

Observation number	Simulated stage	Random number	Simulated discharge
1	3.342	–0.257	84.8
2	2.927	0.546	72.6
3	2.477	–0.100	27.5
4	1.384	2.931	14.7
5	1.908	1.111	13.1
6	2.544	–0.735	18.8
7	2.546	1.711	70.4
8	3.182	0.005	79.1
9	3.564	0.150	108.9
10	3.305	–0.188	83.6
11	3.602	–0.466	98.6
12	2.237	–0.523	1.8
13	1.404	2.790	13.1
14	2.753	0.618	61.9
15	2.522	–0.307	26.3
16	2.529	0.174	36.8
17	2.734	0.052	48.7
18	2.135	0.705	20.5
19	3.219	–0.293	75.4
20	4.285	–0.223	151.5
21	3.132	–0.674	61.3
22	2.959	–1.042	41.4
23	3.415	–0.235	90.3
24	2.519	–0.553	20.9
25	3.459	–0.714	83.3
26	2.111	0.752	19.7
27	2.414	0.467	35.0
28	3.039	0.112	71.3
29	3.371	0.629	105.4
30	2.761	–1.198	24.2
31	3.008	0.056	67.9
32	3.417	–1.633	61.1
33	2.194	0.323	16.5
34	2.031	0.262	3.8
35	2.838	0.160	58.2
36	3.718	–0.585	104.2
37	3.112	–0.302	67.7
38	2.739	–0.739	32.4
39	2.639	–0.639	27.5
40	1.810	1.103	6.1
41	2.663	0.314	49.2
42	2.030	0.735	13.7
43	2.353	0.144	23.9
44	2.030	0.667	12.3
45	2.554	–0.434	25.8
46	3.758	–0.193	115.2
47	2.702	0.600	57.9
48	2.927	0.711	76.1
49	3.075	1.424	101.4
50	3.644	1.426	141.3

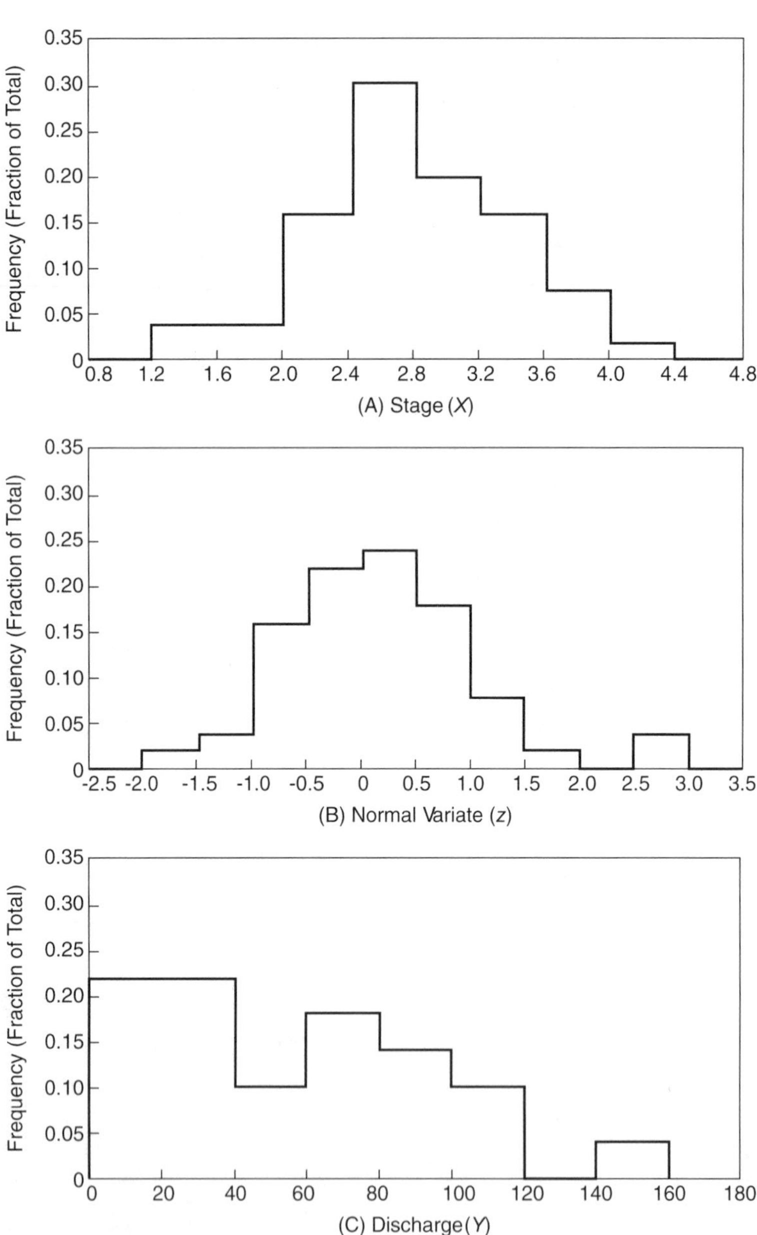

FIGURE 6-10 Distributions of (A) simulated stages; (B) random normal variate used to compute simulated discharges; (C) simulated discharges.

The standard deviation of the simulated discharges is smaller than that of the population, but the simulated statistics are within the expected sampling variation of the population values.

6.6. PROBLEMS

6-1. A shipping company records the conditions of cargoes at the delivery points as not damaged (ND), damaged (D), and partially damaged (PD). The following probability mass function was established for three shipping methods:

		X = condition of cargo		
		ND	**D**	**PD**
Y = shipping method	**Sea**	0.20	0.04	0.02
	Air	0.40	0.02	0.03
	Ground	0.27	0.01	0.01

Determine the marginal probability mass functions $P_X(x)$ and $P_Y(y)$. If a cargo is damaged, what is the probability that it was shipped by sea? By air? By ground?

6-2. An electronics company tests for the performance quality of its products by performing nondestructive thermal and magnetic-field tests. The products are classified as adequate (A), not adequate (NA), and partially adequate (PA). A sample of 100 produced the following results:

		X = thermal test		
		A	**NA**	**PA**
Y = magnetic field	**A**	80	2	0
	NA	4	2	2
	PA	4	2	4

Determine the joint probability mass function $P_{XY}(x,y)$ and the marginal probability mass functions $P_X(x)$ and $P_Y(y)$.

6-3. Are the random variables in Problem 6-1 statistically uncorrelated? Why?

6-4. Are the random variables in Problem 6-2 statistically uncorrelated? Why?

6-5. The following is a joint probability density function for two continuous random variables, X and Y:

$$f_{XY}(x, y) = cxy \quad \text{for } 0 < x < 1 \text{ and } 0 < y < 1$$

Determine the constant c such that $f_{XY}(x,y)$ is a legitimate joint density function. Evaluate the marginal density functions $f_X(x)$ and $f_Y(y)$. Evaluate the probability that $0 < x < 0.5$ and $0 < y < 0.25$.

6-6. The following is a joint probability density function for two continuous random variables, X and Y:

$$f_{XY}(x, y) = cx\sqrt{y} \quad \text{for } 0 < x < 1 \text{ and } 0 < y < 1$$

Determine the constant c such that $f_{XY}(x,y)$ is a legitimate joint density function. Evaluate the marginal density functions $f_X(x)$ and $f_Y(y)$. Evaluate the probability that $0 < x < 0.5$ and $0 < y < 0.25$.

6-7. Are the random variables in Problem 6-5 statistically uncorrelated? Why?

6-8. Are the random variables in Problem 6-6 statistically uncorrelated? Why?

6-9. Determine the conditional probability mass function $P_{X|Y}(X = \text{ND})$ for all y values for the random variables in Problem 6-1.

6-10. Determine the conditional probability mass function $P_{Y|X}(Y = \text{sea})$ for all x values for the random variables in Problem 6-1.

6-11. Determine the conditional probability mass function $P_{X|Y}(X = \text{PA})$ for all y values for the random variables in Problem 6-2.

6-12. Determine the conditional probability mass function $P_{Y|X}(Y = \text{NA})$ for all x values for the random variables in Problem 6-2.

6-13. Determine the conditional expected values $E(X|y)$ and $E(Y|x)$ for the random variables in Problem 6-5.

6-14. Determine the conditional expected values $E(X|y)$ and $E(Y|x)$ for the random variables in Problem 6-6.

6-15. Determine the correlation coefficient ρ_{XY} for the random variables in Problem 6-5.

6-16. Determine the correlation coefficient ρ_{XY} for the random variables in Problem 6-6.

6-17. The random variable X is normally distributed with mean and standard deviation $\mu_X = 0$ and $\sigma_X = 1$, respectively. A dependent random variable Y is defined as:

$$Y = X^2$$

Show that the probability density function of Y is a chi-square distribution (as given by Equation 5-66).

6-18. The random variable Y is normally distributed with mean and standard deviation μ_Y and σ_Y, respectively. The dependent random variable Y is defined as:

$$Y = \ln(X)$$

Show that the probability density function of X is a lognormal distribution (as given by Equation 5-23).

6-19. For a simply supported beam with a concentrated load P at midspan and uniformly distributed load of intensity w over the entire span L, the maximum shear force (V) and the maximum bending moment (M) are given by:

$$V = \frac{wL}{2} + \frac{P}{2}$$

$$M = \frac{wL^2}{8} + \frac{PL}{4}$$

Find the covariance of V and M, COV(V,M), using first-order approximations. Assume P and w to be random variables with known mean values and variances of μ_P, σ_P, μ_w, and σ_w, respectively.

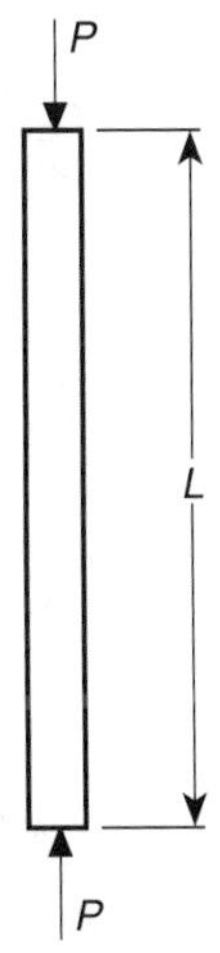

Deformation of a rod (Problem 6-20).

6-20. The change in the length of a rod shown in the figure above and due to an axial force P is given by:

$$\Delta L = \frac{PL}{AE}$$

where L is the length of rod, P is the applied axial force, A is the cross-sectional area of rod, and E is the modulus of elasticity. If P and E are normally distributed with μ_P and σ_P and μ_E and σ_E, respectively, determine the first-order mean and variance of ΔL for the following two cases: (1) P and E are uncorrelated, and (2) P and E are correlated with a correlation coefficient ρ.

6-21. The ultimate moment capacity, M, of an under-reinforced concrete rectangular section is given by:

$$M = A_s f_y \left(d - \frac{a}{2} \right)$$

where

$$a = \frac{A_s f_y}{0.85 b f'_c}$$

in which the following are random variables: A_s is the cross-sectional area of the reinforcing steel, f_y is the yield stress (strength) of the steel, d is the distance from the reinforcing steel to the top fibers of the beam, b is the width of the beam, and f'_c is the ultimate stress (strength) of the concrete. If the random variables are assumed to be statistically uncorrelated, determine the first-order mean and variance of the moment capacity in terms of the means and variances of the random variables.

6-22. For the reinforced concrete section described in Problem 6-21, determine the second-order mean and variance of the moment capacity by including the second-order term of Taylor series expansion (Equation 6-113) in the approximate expression for the mean and variance.

6-23. Using the midsquare method with a seed of 2941, generate 30 uniform variates. Use these to compute the material costs X assuming $X \sim N(\mu = 150, \sigma = 20)$. Using the midsquare method with a seed of 7676, generate 30 uniform variates and use them to compute the labor costs (Y) assuming $Y \sim N(\mu = 225, \sigma = 25)$. Generate the 30 total cost (C) values assuming that $C = 2.5X + 1.7Y$. Compute the mean and standard deviation of C and plot histograms of X, Y, and C. Discuss the results.

6-24. Using the midsquare method with a seed of 9507, generate a sample of ten values (X) that follow $X \sim N(\mu = 10, \sigma = 2)$. Assume $\sigma_Y = 3$, $\mu_Y = 20$, and $\rho = 0.5$. Generate ten values of Y using Equation 6-154 with values of Z generated using the midsquare method with a seed of 2782. Compute the correlation coefficient between the X and Y values. Compute the sample and population correlation coefficient.

6-25. Assume the river stage X is $N(\mu = 4, \sigma = 0.7)$. Assume $\mu_Y = 650$, $\sigma_Y = 60$, and $\rho = 0.82$. Using the midsquare method with a seed of 1073, generate 15 values of the stage and generate values of the discharge Y using Equation 6-154 with values of Z generated using the midsquare method and a seed of 6639. Compute the mean, standard deviation, and frequency histograms of the simulated data. Make comparisons with the assumed populations.

6-26. Assume the river stage X is $N(\mu = 4, \sigma = 0.7)$. Assume $\mu_Y = 650$, $\sigma_Y = 60$, and $\rho = 0.82$. Using the *rand* function, generate 15 values of the stage, and generate values of the discharge Y using Equation 6-154 with values of Z generated using the *rand* function. Compute the mean, standard deviation, and frequency histograms of the simulated data. Make comparisons with the assumed populations.

6.7. SIMULATION PROJECTS

This section provides additional work items for the problem statements of the four simulation projects introduced at the end of previous chapters.

6.7.1. Structural Beam Study

Using the project information provided in previous sections on this project, perform computations to produce results similar to the results in Section 6.5.2, as much as possible, to investigate the correlation among the output random variable and the predictor random variables.

6.7.2. Stream Erosion Study

Using the project information provided in previous sections on this project, perform computations to produce results similar to the results in Section 6.5.2, as much as possible, to investigate the correlation among the output random variable and the predictor random variables.

6.7.3. Traffic Estimation Study

Using the project information provided in previous sections on this project, perform computations to produce results similar to the results in Section 6.5.2, as much as possible, to investigate the correlation among the output random variable and the predictor random variables.

6.7.4. Water Evaporation Study

Using the project information provided in previous sections on this project, perform computations to produce results similar to the results in Section 6.5.2, as much as possible, to investigate the correlation among the output random variable and the predictor random variables.

Chapter 7
Simulation

TABLE OF CONTENTS

7.1. INTRODUCTION

7.1.1. Engineering Decision Making

Assume that you are a county transportation engineer and that you have found from studying traffic accident reports that traffic fatalities at a particular intersection in the county are especially high. The intersection is currently uncontrolled except for stop signs for the north–south street; there are no stop signs on the east–west street. As a possible measure for reducing traffic accidents, you propose installing a four-way traffic light; however, you want the light to operate in a pattern that will cause the least impedance on traffic. Therefore, you must decide the length of the red–green cycle in each direction and the proportion of time that it will be green in each direction. For example, will the light be green in the north–south direction less time, equal time, or more time than in the east–west direction? Also, will the entire cycle be 1 minute, 2 minutes? It is necessary for you, the transportation engineer, to establish the traffic-light operation policy that will reduce traffic accidents with minimal impedance of traffic.

One possible solution to your dilemma would be to install the traffic light, assume some light-operation sequence, and then observe its effectiveness over 2 weeks or a month. Then a second sequence could be used and traffic observed for another 2 weeks. This process could be repeated until the optimum light-operation sequence is found. Obviously, this is an unrealistic option, but an arbitrary setting of the light-operation sequence is also an undesirable option. A method of simulating the actual process in a very short period of time is needed.

Consider a second case that involves the task of a water-supply engineer who must establish a policy for public water use during periods of drought. Let's assume here that drought is defined as an extended period of time when the normal demand for water exceeds the available supply. It is the task of the engineer to set a policy that sets a limit on the daily use of water. Common limitations on water use include bans on car washing and lawn watering and limiting home use to some percentage of normal use. In order to ensure that the policy is enforceable, it must be applied consistently over a relatively long period of time; it cannot be changed daily. But what policy should the water-supply engineer recommend to the water commissioner? The engineer wants to minimize inconvenience to the public, but at the same time water must be available for consumption, fire safety, public health, and important commercial and industrial uses. The policy must take into consideration the engineer's expectation for rain over the next month or so, but this is an unknown, yet important, variable. The engineer would have some knowledge of the depth of rain that has occurred in the past, but past sequences of rain are unlikely to occur exactly in the future. How can the water supply engineer combine knowledge of existing water supplies, public demand for water, potential amounts of rainfall, and the effects of alternative policies for limiting water use? Just as in the traffic-light problem, the engineer needs a method of combining knowledge of each part of the process in a way that simulates the parts and their interaction so that estimates of the effects of alternative water policies can be assessed.

7.1.2. Sampling Variation

The topic of sampling variation has been introduced previously. The standard error of the sample mean was given as $S/(\sqrt{n})$. Confidence intervals are a statistical tool used to quantify the variation that can be expected in a statistic. The subject of the standard error of estimate is introduced in Chapter 12 as a measure of the expected variation of the residuals from a regression model. Numerous other examples of sampling variation have been introduced. A slightly different approach is considered here.

Consider the following experiment. A die is rolled three times, with the number of dots from the three rolls summed. We will denote this random variable as *S*. What is the sample space of *S*? Obviously, it cannot take on a value smaller than 3 or larger than 18. So the sample space consists of the integer values from 3 to 18. There is only one way to get *S* equal to 3 — namely, three occurrences of a 1: (1, 1, 1). There are three ways of getting *S* equal to 4: (1, 1, 2), (1, 2, 1), and (2, 1, 1). If we continued to identify all possible outcomes, we would see that there is a total of 216 possibilities. The number of possible outcomes can be converted to a probability by dividing the number of outcomes by 216; these are the population probabilities and are given in Table 7-1 as *p*. An experiment is to roll the die 150 times, with each group of three rolls used to compute a value of *S*; thus, the 150 rolls produce 50 values of the random variable. Each group of the 50 will be referred to as a simulation run; the number of values for each *S* and for each of 17 simulation runs (or cycles) are shown in Table 7-1. The point of this exercise is to show that each simulation run produces a different result. In fact, some runs are quite different; for example, runs 4 and 13 differ in that run 4 has 29 occurrences of *S* between 3 and 10, while run 13 has 21 occurrences. But each simulation run cannot be the same because run-to-run variation is expected when dealing with a random variable.

This variation occurs in engineering systems. We do not have the same flood patterns every year. Structures are not subjected to the same wind loads every year. Traffic patterns and accidents differ on a day-to-day basis. Construction delays are not systematic. Pollution loads in rivers vary continuously and the level of pollution at any time cannot be predicted with certainty. In problems where uncertainty is significant and important to the decisions that engineers must make, simulation can be an important modeling tool for the engineers.

7.1.3. Coin-Flipping Simulation

Problems such as the traffic-light and water-rationing examples can be studied on a computer without the need to conduct exhaustive real-world studies. The essence of the computer approach will be illustrated using an example. The reader is encouraged to perform a similar study of the problem while following the steps shown. The only requirement is a coin and pen and paper to record the results.

First, flip a coin 72 times and record the sequence of heads and tails (H and T, respectively). The following is an example result:

TABLE 7-1 Simulation of the Sampling Variation of a Random Variable

			Simulation runs																	
S	n	p	1	2	3	4	5	6	7	8	9	10	11	12	13	14	15	16	17	Sum	$\hat{p}$
3	1	0.005	0	0	1	0	0	0	1	0	0	1	1	0	0	0	0	0	0	4	0.005
4	3	0.014	0	2	1	1	0	0	1	1	1	1	0	0	0	0	0	0	1	9	0.010
5	6	0.028	2	2	1	1	1	0	0	1	2	2	0	0	0	0	0	2	3	17	0.020
6	10	0.046	0	1	0	4	3	0	2	3	1	1	2	1	2	1	1	3	2	27	0.032
7	15	0.069	3	7	2	4	6	7	5	2	4	3	1	3	4	3	3	2	4	63	0.074
8	21	0.097	6	5	6	6	9	6	5	7	7	6	4	2	3	9	4	6	1	92	0.108
9	25	0.116	10	4	8	9	2	6	4	10	6	4	6	8	8	5	11	6	7	114	0.134
10	27	0.125	3	3	9	4	5	4	5	4	4	10	8	9	4	8	3	7	5	95	0.112
11	27	0.125	8	4	7	7	8	10	5	6	6	5	10	6	4	7	10	5	9	117	0.137
12	25	0.116	5	6	5	2	7	3	7	5	9	8	6	3	5	3	2	5	5	86	0.101
13	21	0.097	4	7	3	5	2	7	6	6	5	2	2	6	6	8	7	5	5	86	0.101
14	15	0.069	3	4	4	5	2	2	3	1	2	3	4	5	5	3	4	4	3	57	0.067
15	10	0.046	2	1	1	1	1	2	3	1	3	1	1	3	3	2	2	3	2	32	0.038
16	6	0.028	2	2	1	0	3	2	1	1	0	2	2	2	3	0	1	2	3	27	0.032
17	3	0.014	2	2	1	1	1	1	1	2	0	0	3	0	3	1	2	0	0	20	0.024
18	1	0.005	0	0	0	0	0	0	1	0	0	1	0	2	0	0	0	0	0	4	0.005
Sum		1.000	50	50	50	50	50	50	50	50	50	50	50	50	50	50	50	50	50	850	1.000

Note: S = sample-space outcome; n = number of ways of getting S; p = probability of getting S; $\hat{p}$ = sample probability of getting S.

HTT HTH TTH THT TTH HTT HHT HTT TTH TTT HHT HHH

TTH HHT THT THH TTH HTH TTT HHH THH HHH HTH THT

Next, divide the sequence into 24 groups of 3 as shown above. This sequence is used to simulate the operation of a manufacturing plant and the selection of a piece of equipment based on a linear cost function. Assume that each group of three coin tosses represents the output from an assembly line. Also, assume that the machine randomly produces defective units, with a unit from machine A being defective for any of the following sequences: TTT, TTH, or HTT; all other combinations indicate a nondefective unit. For machine A, the sequence of defective (D) and nondefective (N) units from the above coin tosses can be, therefore, expressed as:

DNDNDDNDDDNNDNNNDNDNNNNN

The manufacturing engineer is considering the installation of a more reliable and more costly machine, which is called machine B. For machine B, the following sequences indicate a defective unit: TTT or TTH; thus, the sequence of defective and nondefective units from the above coin tosses is:

NNDNDNNNDDNNDNNNDNDNNNNN

Machine A produced ten defective units (41.7%), while machine B produced only seven defective units (29.2%). Let's assume that there is a \$5 profit for a nondefective unit and a \$2 loss for each defective unit. Machine A costs \$12,000, while machine B costs \$15,000. Based on a useful life of 1 year for the machines, during which 10,000 units would be produced, the expected profit for machine A is:

$$\text{Profit for A} = \$5\left(\frac{14}{24}\right)(10{,}000) - \$2\left(\frac{10}{24}\right)(10{,}000) - \$12{,}000 = \$8{,}833$$

The expected profit for machine B is:

$$\text{Profit for B} = \$5\left(\frac{17}{24}\right)(10{,}000) - \$2\left(\frac{7}{24}\right)(10{,}000) - \$15{,}000 = \$14{,}583$$

Therefore, based on the sample estimates obtained from the simulated sequences of defective and nondefective items, the profit expected from machine B exceeds the profit expected from machine A. Thus, an engineer would recommend the purchase of machine B. In practice, the engineer should also consider the standard deviation of the profit of each machine in the final selection.

This example demonstrates the point that it was not necessary to actually produce 10,000 units with each machine in order to estimate the profit for the two alternatives. The random nature of the production of defective units was represented by a simple experiment that had the same statistical characteristics as the real system.

7.1.4. Definitions

Before defining what is meant by engineering simulation, it is necessary to give a few other definitions. First, a *system* is defined herein as a set of processes or components that are interdependent with each other. This could be a natural system such as a watershed, a geographic system such as a road network, or a structural system such as a high-rise building. An accident on one road can lead to traffic congestion on a nearby road, so the roads are interdependent.

Second, a *model* is defined herein as a representation of a real system as was discussed in Chapter 1. The model could be either a physical model, such as those used in laboratories, or a mathematical model. Models can be developed from either theoretical laws or empirical analyses. The model includes components that reflect the processes and provides for interaction between components.

Third, an *experiment* is defined for our purposes here as the process of observing the system or the model. Where possible, it is usually preferable to observe the real system; however, the lack of control may make this option impossible or unrealistic.

Given these three definitions, we can now provide a preliminary definition for simulation. Specifically, simulation is the process of conducting experiments on a model when we cannot experiment directly on the system. The uncertainty or randomness inherent in model elements is incorporated into the model and the experiments are designed to account for this uncertainty. To illustrate this aspect, consider the coin-tossing example. The occurrence of defective items was considered to be random, so the model incorporated the randomness of coin tossing to represent the randomness of the real system. Both sequences of defective and nondefective items appear to be random. The \$5 profit and \$2 loss, as well as the cost of the machine, were not subject to uncertainty, so they were included in the model as constants.

The term *simulation run or cycle* is defined as an execution of the model through all operations for a length of simulated time. The 72 coin tosses shown above could represent one simulation run, with the 24 items being the length of the simulation run. Using the above model, a simulation run that consisted of 30,000 flips can be used to yield 10,000 units.

Some additional terms used in this chapter that should be defined here are as follows:

1. A *model parameter* is a value that is held constant over a simulation run but can be changed from run to run.
2. A *variable* is a model element whose value can vary during a simulation run.
3. *Input variables* are those model variables that require values to be input prior to the simulation run.
4. *Output variables* are model variables that reflect the end state of the system and can consist of single values or a vector of values.
5. *Initial conditions* are the values of model variables and parameters that establish the initial state of the model at the beginning of a simulation run.

7.1.5. Benefits of Simulation

Simulation is frequently encountered in our everyday lives. Flight simulators are widely used in the space and aircraft industries, and attractions at leading amusement parks simulate the experience of space travel. Even video games use simulation to mimic life-threatening activities.

Simulation is widely used in engineering decision making. It is a popular modeling tool because it enables a representation of the system to be manipulated when manipulating the real system is impossible or too costly. Simulation allows the time or space framework of the problem to be changed to a more convenient framework. Simulation enables the representation of the system to be changed in order to better understand the real system; of course, this requires the model to be a realistic representation of the system. Simulation enables an analyst to control any of the model parameters, variables, or initial conditions, something that is not possible with the real system. Simulation can be used to evaluate the response of the system for conditions that have not occurred in the past.

Simulation is a principal modeling tool because it allows a model to be altered to reflect conditions that have not occurred in the past but can be expected to occur in the future. Thus, the response of the real system to future extreme conditions can be evaluated, with the effects of possible prevention actions evaluated. The model can be used with simulation to assess the relative importance of the variables. The effects of uncertainty in design inputs can be evaluated with simulation.

While simulation is extremely useful, there are a few problems. First, it is quite possible to develop several different, but realistic, models of the system. The different models could lead to different decisions. Second, the data used to calibrate the model may be limited, so extrapolations beyond the range of the measured data may be especially inaccurate.

7.2. MONTE CARLO SIMULATION

Interest in simulation methods started in the early 1940s for the purpose of developing inexpensive techniques for testing engineering systems by imitating their real behavior. These methods are commonly called *Monte Carlo simulation techniques*, with an obvious connection to the European gambling enclave; however, Monte Carlo simulation goes way beyond gambling. The connection with uncertainty is the commonality. The principle behind the methods is to develop a computer-based analytical model that predicts the behavior of a system. When the model is evaluated using data measured from a system, it predicts the behavior of the system, usually for many simulation runs. Each evaluation (or simulation cycle) is based on a certain randomly selected set of conditions for the input parameters of the system. Certain analytical tools are used to assure the random selection of the input parameters according to their respective probability distributions for each evaluation. As a result, several predictions of the behavior are obtained. Then statistical methods are used to evaluate the moments and distribution type for the behavior of the system.

The analytical and computational steps that are needed for performing Monte Carlo simulation are (1) definition of the system using a model, (2) generation of random numbers, (3) generation of random variables, (4) evaluation of the model, (5) statistical analysis of the resulting behavior, and (6) study of simulation efficiency and convergence. The definition of the system should include its boundaries, input parameters, output (or behavior) measures, architecture, and models that relate the input parameters and architecture to the output parameters. The accuracy of the results of simulation are highly dependent on having an accurate definition for the system. All critical parameters should be included in the model. The definition of the input parameters should include their statistical or probabilistic characteristics, i.e., knowledge of their moments and distribution types. It is common to assume the architecture of the system in Monte Carlo simulation to be nonrandom. However, modeling uncertainty can be incorporated in the analysis in the form of bias factors and additional variability (e.g., coefficients of variation). The result of these generations is a set of certain values for the input parameters. These values should then be substituted in the model to obtain an output measure. By repeating the procedure N times (for N simulation cycles), N response measures are obtained. Statistical methods can now be used to obtain, for example, the mean value, variance, or distribution type of the response. The accuracy of the resulting measures for the behavior are expected to increase by increasing the number of simulation cycles. The convergence of the simulation methods can be investigated by studying their limiting behavior as N is increased. Also, the efficiency of simulation methods can be increased by using variance reduction techniques. These variance reduction techniques are not covered in this chapter, but they are described in Chapter 14 in regard to their use in reliability analyses.

7.3. RANDOM NUMBERS

Random numbers are real values; if they are normalized by dividing each value by the largest possible value, the resulting numbers are real values in the range $[0,1]$. Also, the normalized numbers have a uniform distribution on the range $[0,1]$. Therefore, it can be stated that a random number can be modeled as a random variable U that has a uniform distribution on the continuous range $[0,1]$. A set of random numbers should also satisfy the condition of no serial correlation for the purpose of simulation use.

The importance of uniform random numbers is that they can be transformed into real values that follow any distribution of interest. Therefore, they are the basis for most engineering simulation.

In the early years of simulation, mechanical random-number generators were used (e.g., drawing numbered balls, throwing dice, or dealing out cards). Many lotteries are still operated this way. Then, after several stages of development, arithmetic random-number generators were developed that are computer based and use some analytical models. In these generators, a random number is obtained based on a previous value (or values) and fixed mathematical equations. Therefore, a seed is needed to start the process of random-number generation, resulting in a stream of random values. The main advantages of arithmetic random-number generators over mechanical generators are their speed, the fact that they do not require memory for storage of numbers, and repeatability of results. The conditions of a uniform distribution and the absence of serial correlation should also be satisfied. Due to the nature of arithmetic generation of random numbers, a given seed should result in the same stream of random values all the time. This property of repeatability is important for debugging of the simulation algorithm and for comparative studies of design alternatives of a system. In this section, arithmetic random-number generators are briefly described.

7.3.1. Arithmetic Generators

There are many arithmetic random-number generators — for example, midsquare, linear congruential generator, mixed generators, multiplicative generators, general congruences, composite generators, and Tausworthe generators. The midsquare generator was used in Chapters 1 to 6 to illustrate the use of simulation. All of these generators are based on the same principle of starting with a seed and having fixed mathematical equations for obtaining the random value, then the resulting value is used in the same equations to obtain another random value. By repeating this recursive process N times, N random numbers in the range $[0, 1]$ are obtained. However, these methods differ in the mathematical equations that are used as the recursive model. In all recursive models, the period for the generator is of concern. The *period* is defined as the number of generated random values before the stream of values starts to repeat itself. It is always desirable to have random-number generators with large periods that are, for example, much larger than the number of simulation cycles needed in a simulation study of a system.

In almost all modern computers, a random-number generator is provided as a standard function. Modern generators are sometimes based on the direct use of bits and binary digits; however, the great majority of random-number generators are linear congruential generators. In these generators, a sequence of integers I_1, I_2, I_3, ... is defined by the following recursive equation:

$$I_i = (aI_{i-1} + b) - \text{Int}\left(\frac{aI_{i-1} + b}{c}\right)c \qquad \text{for } i = 1, 2, 3, \ldots \tag{7-1}$$

where $\text{Int}\left(\frac{aI_{i-1} + b}{c}\right)$ is the integer of the result of the division, a is the multiplier, b is the increment, and c is the modulus. The model constants a, b, and c are nonnegative integers. The starting value I_0 is called the *seed*, which should be provided by the user of the model. The value I_i is obtained by dividing $(aI_{i-1} + b)$ by c and letting I_i be the remainder of the division. The random number U_i is defined as:

$$U_i = \frac{I_i}{c} \tag{7-2}$$

The I_i value is normalized by dividing by c, as $0 \le I_i \le c$. The parameters of this recursive model should also satisfy the following conditions: $0 < c$, $a < c$, $b < c$, and $I_0 < c$. It is evident from this recursive model that the process is not random, as it can be repeated with the same results all the time. For this reason, this process is commonly called *pseudo* random-number generation. The period of this generator is less than or equal to c. For this reason and others, the value of c in real generators should be very large (e.g., $c \ge 10^9$). In addition to some computers and calculators that have random-number generators, tables of random numbers have also been published (for example, Table A-6). A special case of the linear congruential generators, where $b = 0$, is called the *multiplicative generator.* If $b > 0$, they are called *mixed generators.* The *rand* function, available in calculators and spreadsheet programs, can be used to generate random numbers. This function is described in Chapter 1.

Example 7-1: Random-Number Generators

For a recursive multiplicative generator that is defined by the following equation:

$$I_i = (2I_{i-1}) - \text{Int}\left(\frac{2I_{i-1}}{5}\right)5 \quad \text{for } i = 1, 2, 3, \ldots \tag{7-3}$$

and with a seed $I_0 = 2$, the random numbers shown in Table 7-2 are obtained. The period for this generator is 4. This is not a good generator because it has a small period. The period is less than or equal to c, which has a value of 5 in this case. In real generators, c should be very large. Also, the resulting random numbers from the generator are discrete: 0, $1/c$, $2/c$, …, $(c - 1)/c$ (in this example, 0, 0.2, 0.4, and 0.8). By selecting a large value for c, the number of discrete values becomes large, resulting in a better approximation of a continuous uniform distribution.

TABLE 7-2 Random Numbers Generated Using a Recursive Relation

i	I_i	U_i	i	I_i	U_i	i	I_i	U_i
0	2	seed	5	4	0.8	10	3	0.6
1	4	0.8	6	3	0.6	11	1	0.2
2	3	0.6	7	1	0.2	12	2	0.4
3	1	0.2	8	2	0.4	13	4	0.8
4	2	0.4	9	4	0.8	14	3	0.6

7.3.2. Testing of Generators

Before using a random-number generator, two important tests should be performed on the generator: (1) test of uniformity and (2) test of serial correlation. These tests can be performed either theoretically or empirically. A theoretical test is defined as an evaluation of the recursive model itself of a random-number generator. The theoretical tests include an assessment of the suitability of the parameters of the model without performing any generation of random numbers. An empirical test is a statistical evaluation of streams of random numbers resulting from a random-number generator. The empirical tests start by generating a stream of random numbers: N random values in the range $[0, 1]$. Then, statistical tests for distribution types (i.e., goodness-of-fit tests such as the chi-square test) are used to assess the uniformity of the random values. The objective in the uniformity test is to make sure that the resulting random numbers follow a uniform continuous probability distribution. This test is discussed in Chapter 9. The serial tests of the runs are used to assess the serial correlation of the resulting random vector, where each value in the stream is considered to come from a different, although identical, uniform distribution. The serial correlation tests are beyond the scope of this book.

7.4. GENERATION OF RANDOM VARIABLES

The process of generating a random variable according to its probability distribution can be viewed as a sampling procedure with a sample size N, where N = the number of simulation cycles. It is desirable, and in some cases necessary, for a random-number generator to possess the following properties: (1) exactness, (2) efficiency, (3) simplicity, (4) robustness, and (5) synchronism. The exactness property is necessary to assure that the resulting values for a generated random variable exactly follow the assumed probability distribution. The efficiency property is measured in terms of necessary computer memory and setup and execution times for generating a random variable. It is desirable to have generation methods that are efficient. It is also desirable to have a mathematically and computationally simple procedure for generating a random variable. The methods must also be robust; that is, they should be able to provide reasonable random values for wide ranges of values for the parameters of the distribution. In certain applications, it might be necessary to have synchronized random values of the input random variables. In such cases, the synchronism property must be satisfied. In developing or selecting a random-variable generator, some of these properties must be satisfied and others must be satisfied in different levels, using a trade-off approach. For example, increasing the simplicity might reduce exactness, whereas increasing exactness might increase complexity and reduce efficiency.

7.4.1. Inverse Transformation Method

This method is simple and direct. A random number u is first generated in the range $[0, 1]$; that is, $u \in U[0, 1]$, where $U[0, 1]$ is a continuous uniform probability distribution. Then a generated continuous random variable, X, is determined as follows:

$$x = F_X^{-1}(u) \tag{7-4}$$

where F_X^{-1} is the inverse of the cumulative distribution function of the random variable X. Because $F_X(x)$ is in the range $[0,1]$, a unique value for x is obtained all the time in each simulation cycle.

For a discrete random variable, a random number is generated, $u \in U[0,1]$. Then the value of the generated random variable, X, is determined as follows:

$$x_i \text{ such that } i \text{ is the smallest integer with } u \le F_X(x_i) \tag{7-5}$$

where x_i, $i = 1, 2, 3, \ldots, m$, are m discrete values of the random variable X with a cumulative mass distribution function $F_X(x)$.

Example 7-2: Inverse Transformation for a Discrete Distribution

In this case, we have a uniform random-number generator, $U(0,1)$, and we want to transform U values so that they have a specified discrete mass function. The discrete function of interest could be either some known function such as a binomial or Poisson function or a custom-made mass function. The transformation process is the same for both cases.

Consider the case where we have collected a sample of data on a discrete random variable X, and the data yield a sample estimate of the mass function as shown in Figure 7-1a. The cumulative function for values of X from 0 to 5 is $F_X(x) = \{0.10, 0.30, 0.60, 0.85, 0.95, 1.00\}$, for $x = 0, 1, 2, 3, 4$, and 5, respectively. This is shown in Figure 7-1b. For any value of X greater than 5, the cumulative function equals 1, and for values of X less than 0, the cumulative function equals 0.

To transform continuous uniform variates U to discrete variates having $f_X(x)$ as a mass function, the value of U is entered as the ordinate on the cumulative function of Figure 7-1b, with the point projected horizontally until it intersects the cumulative function. The point of intersection identifies the value of X.

Consider the following set of uniform variates: $U = \{0.47, 0.91, 0.08\}$. These are transformed to X variates as shown in Figure 7-1c. The generated values of X are $X = \{2, 4, 0\}$.

Example 7-3: Inverse Transformation for a Continuous Distribution

The process for generating continuous variates follows the same steps of the discrete case. We have a generator that gives values with a uniform distribution $U(0,1)$. A density function must be identified, usually from empirical analysis, but possibly from theoretical considerations. The cumulative distribution for the assumed density function is formed.

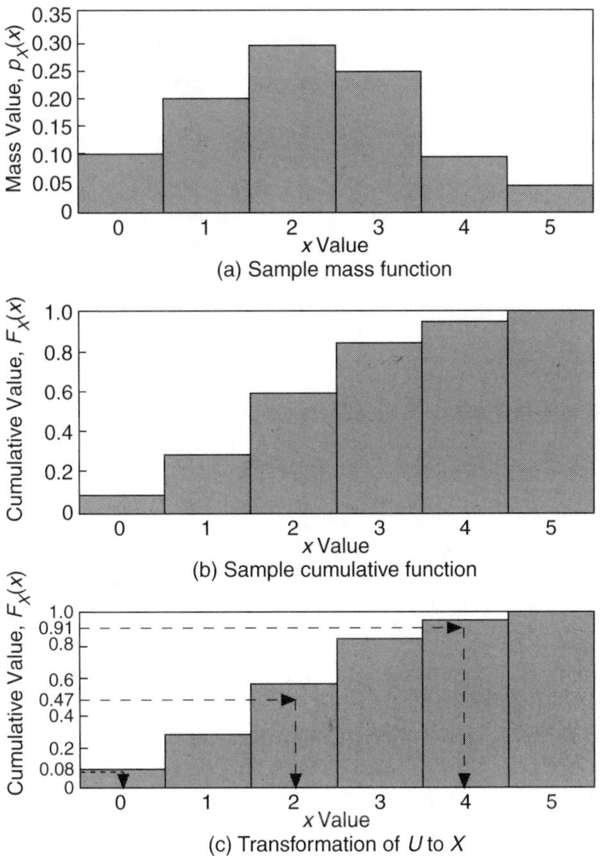

FIGURE 7-1 Inverse transformation for a discrete probability distribution.

Let's assume that the objective is to transform uniform variates of Figure 7-2a to variates that have a triangular density function as shown in Figure 7-2b. The following is the density function, which can be determined using the constraint $\int_{-\infty}^{\infty} f_X(x)dx = 1$ and the equation for the function as $f_X(x) = k(x - x_L)$:

$$f_X(x) = \begin{cases} \dfrac{2(x - x_L)}{(x_U - x_L)^2} & \text{for } x_L \le x \le x_U \\ 0 & \text{otherwise} \end{cases} \tag{7-6}$$

The cumulative function, $F_X(x)$, can be determined by integrating the density function of Equation 7-6 from x_L to any value x as follows:

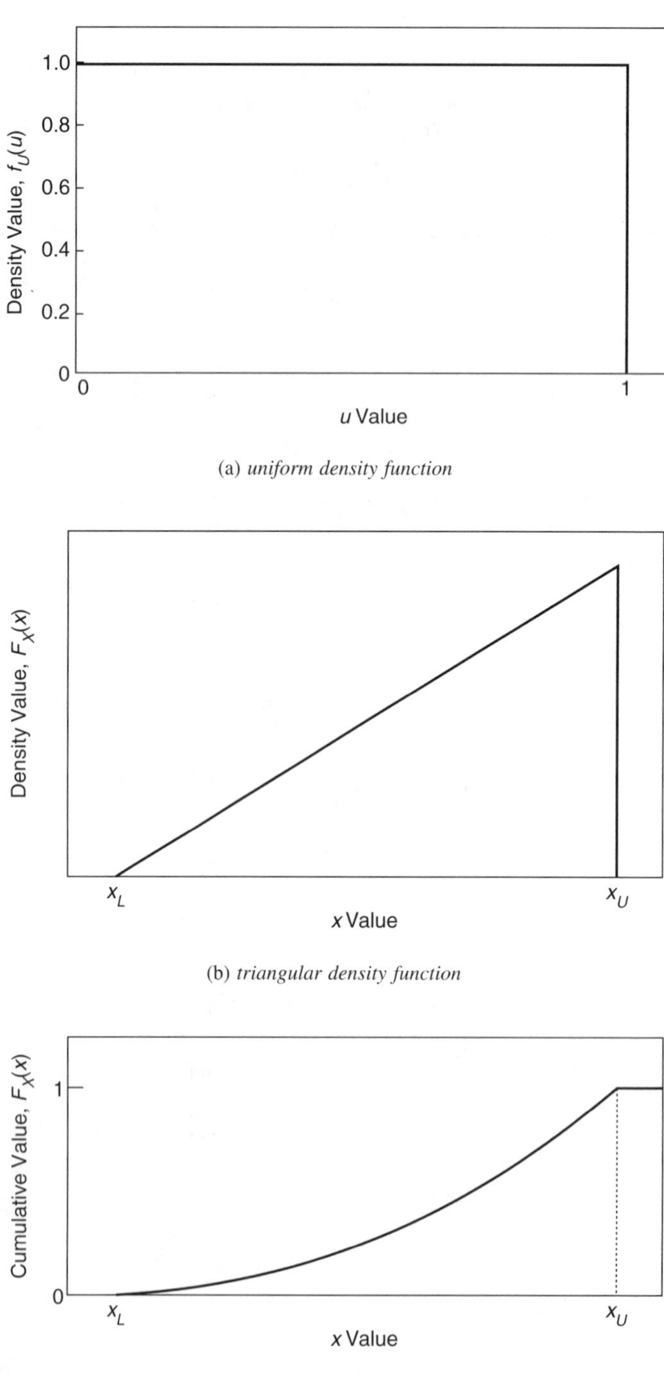

(a) *uniform density function*

(b) *triangular density function*

(c) *cumulative function*

FIGURE 7-2 Inverse transformation for a continuous probability distribution.

$$F_X(x) = \begin{cases} 0 & \text{for } x < x_L \\ \dfrac{(x - x_L)^2}{(x_U - x_L)^2} & \text{for } x_L \le x \le x_U \\ 1 & \text{for } x_U < x \end{cases} \tag{7-7}$$

The cumulative function is shown in Figure 7-2c.

To generate a value of X for a given uniform variate U, the cumulative function $F_X(x)$ is set equal to a generated random number u, and Equation 7-7 is solved for x. Solving the quadratic yields:

$$x = x_L + (x_U - x_L)\sqrt{u} \tag{7-8}$$

A graphical solution is also possible using the approach shown for the discrete variate.

For the case where $x_L = 1$, $x_U = 3$, and $U = \{0.47, 0.91, 0.08\}$, the values of X are computed from:

$$x = 1 + 2\sqrt{u} \tag{7-9}$$

which gives $X = \{2.371, 2.908, 1.566\}$.

Example 7-4: Transformation to a Normal Distribution

A common need is to transform uniform variates to normal variates. For transformation to a standard normal deviate, the standard normal table can be used. For a normal distribution with mean μ and standard deviation σ, the uniform variate is first transformed to the standard normal deviate z, which is then transformed to normal deviate x by:

$$x = \mu + z\sigma \tag{7-10}$$

A uniform deviate is transformed to a standard normal deviate by entering the standard normal table (Table A-1) using u as $\Phi(z)$ and then moving across to the left of the table to find z. That is, the value of u is used as the probability from which the deviate z is taken from the table.

Consider the case where the sample consists of the following uniform variates:

$$U = \{0.82, 0.43, 0.71, 0.18, 0.66\}$$

To find the corresponding z values, enter the standard normal table for a probability of 0.82, for example, and read the z value ($z = 0.915$). Continuing for each u_i value yields:

$$Z = \{0.915, -0.176, 0.556, -0.915, 0.413\}$$

7.4.2. Composition Method

This method can be used to generate a random variable X that has the following composite probability distribution $F_X(x)$:

$$F_X(x) = \sum_{i=1}^{m} w_i F_X^i(x) \tag{7-11}$$

where $F_X^i(x)$ is the ith cumulative distribution function (CDF) for the random variable X, and w_i is the ith weight factor. The weight factors must satisfy the following condition:

$$\sum_{i=1}^{m} w_i = 1 \tag{7-12}$$

The random variable X is generated in the following two stages:

1. A random number u_1 is generated in the range [0,1]. This random number is used to select the appropriate CDF, $F_X^i(x)$, $i = 1, 2, \ldots, m$, for random-variable generation. The selection can be based on treating the weight factors to constitute the following discrete probability function:

$$p_W(w) = \begin{cases} w_1 & \text{for } F_X^1(x) \\ w_2 & \text{for } F_X^2(x) \\ \vdots & \vdots \\ w_{m-1} & \text{for } F_X^{m-1}(x) \\ w_m & \text{for } F_X^m(x) \end{cases} \tag{7-13}$$

2. A second random number u_2 is then used to generate the random variable according to the selected distribution in step 1. The inverse transformation method or other methods can be used for this purpose.

7.4.3. Function-Based Generation

Some random variables can be expressed in terms of other random variables that can be easily generated. Such functions can be used for generation purposes. For example, the addition of m random variables $X_1, X_2, \ldots, X_m$ that have independent exponential distributions with a parameter λ is a gamma distribution with parameters (m,λ). Therefore, a random variable Y that has a gamma distribution with parameters (m,λ) can be expressed as:

$$Y = X_1 + X_2 + X_3 + \ldots + X_m \tag{7-14}$$

where m is an integer. This random variable can be generated by randomly generating m independent random variables using an exponential distribution with a parameter λ. The result is $x_1, x_2, \ldots, x_m$. Therefore, a generated value for Y is given by:

$$y = x_1 + x_2 + x_3 + \ldots + x_m \tag{7-15}$$

The function-based method can be used to generate random variables in a simple manner, but it can be inefficient. Direct generation of random variables should be used wherever possible.

7.4.4. Acceptance–Rejection Method

This method can be considered as an indirect generation of random variables, especially if it is compared with the inverse transformation, composition, or the function-based methods. For a random variable X that has a probability density function $f_X(x)$, the acceptance–rejection method for generating X can be summarized in the following steps:

1. A random variable Y is selected to facilitate the generation of X with the following density function $f_Y(y)$:

$$f_Y(y) = \frac{g(y)}{\int_{-\infty}^{\infty} g(y)dy} \tag{7-16}$$

where $g(y)$ is a generally simple arbitrary function such $g(x) \geq f_X(x)$ for any x, and

$$\int_{-\infty}^{\infty} g(y)dy < \infty \tag{7-17}$$

2. A random number u should be generated independent of Y. The random number $u \in U[0, 1]$.
3. A random value y should be generated based on the density function of Equation 7-16.
4. A generated value for the random variable X is taken as:

$$x = y \quad \text{if } u \leq \frac{f_X(y)}{g(y)} \tag{7-18}$$

Otherwise, if $u > \dfrac{f_X(y)}{g(y)}$, the value of y is rejected and a new pair of u and y values must be generated repeatedly until the condition of Equation 7-18 is satisfied and a value for x is obtained.

7.4.5. Generation Based on Special Properties

In this method, special properties or relations between distributions are used for the purpose of generating random variables. For example, if X has a standard normal distribution, $X \sim N(0, 1)$, and utilizing the property that $Y = X^2$ has a chi-square distribution with one degree of freedom, Y can be generated by: (1) generating X as a standard normal variate, and (2) using $y = x^2$. To generate a random variable Z that has a chi-square distribution with ν degrees of freedom, the following property can be used:

$$Z = Y_1 + Y_2 + Y_3 + \ldots + Y_\nu \qquad (7\text{-}19)$$

where Y_i, $i = 1, 2, \ldots, \nu$, have independent chi-square distribution with 1 degree of freedom each. Each Y_i can be generated as X^2, where X has a standard normal distribution. Then the results are added to obtain the distribution of Z. This method is not efficient. There are other more efficient methods for generating random variables of chi-square distributions with ν degrees of freedom that are not discussed here.

7.5. GENERATION OF SELECTED DISCRETE RANDOM VARIABLES

7.5.1. Bernoulli Distribution

The Bernoulli distribution can be generated using the inverse transformation method. Starting with a generated random number $u \in U[0, 1]$, a random variable X can be generated according to a Bernoulli distribution as follows:

$$x = \begin{cases} 1 & \text{if } u \le p \\ 0 & \text{otherwise} \end{cases} \qquad (7\text{-}20)$$

where p is a specified parameter for the Bernoulli distribution.

7.5.2. Binomial Distribution

A discrete random variable that has two possible outcomes can follow a binomial mass function. Four assumptions underlie a binomial process: (1) there are only two possible outcomes for each trial, (2) there are n trials in the experiment, (3) the n trials are independent, and (4) the probability of each outcome remains constant from trial to trial. The probability of exactly X occurrences in n trials is given by:

$$p_X(x) = \binom{n}{x} p^x (1-p)^{n-x} \qquad \text{for } x = 1, 2, \ldots, n \qquad (7\text{-}21)$$

where $\binom{n}{x}$ is the binomial coefficient and equals $\frac{n!}{x!(n-x)!}$, where $0! = 1$. In this section, the notation $b(X;n,p)$ is used for the binomial distribution.

A random variable X of a binomial distribution with the parameters n and p can be generated by generating n independent random variables, $X_1, X_2, \ldots, X_n$, of a Bernoulli distribution, each with a parameter p. Each X_i can be generated using the inverse transformation method for the Bernoulli distribution as discussed in Section 7.5.1. Then, a generated value of y is given by:

$$y = x_1 + x_2 + x_3 + \ldots + x_n \tag{7-22}$$

where x_i, $i = 1, 2, \ldots, n$, are the respective generated values of the random variables $X_1, X_2, \ldots, X_n$.

This method becomes inefficient for large values of n. In such cases, direct inverse transformation can be utilized.

Binomial experiments can be simulated using uniform variates. The procedure is as follows for N experiments, with each having n trials and an outcome A with an occurrence probability of p and an outcome B with an occurrence probability of $(1 - p)$:

1. For each trial in the experiment, generate a uniform variate, u_i.
2. If $u_i < p$, then assume outcome A has occurred; otherwise, assume outcome B has occurred.
3. Determine the number of A outcomes, X, that occurred in the n trials.
4. After completing the N experiments, with each having n trials, compute the relative frequency of X outcomes.

Example 7-5: Generation of Binomial Variates

This problem has three trials per experiment where the probability (p) of outcome A is 0.3. For example, if we have an unbalanced coin with a probability of 0.3 of getting a head, then we would be interested in the probabilities of getting 0, 1, 2, or 3 heads in three flips of the coin. From the binomial mass function of Equation 7-21, we can compute the following population probabilities:

$$b(0;3, 0.3) = \binom{3}{0} 0.3^0 0.7^3 = 0.343$$

$$b(1;3, 0.3) = \binom{3}{1} 0.3^1 0.7^2 = 0.441$$

$$b(2;3, 0.3) = \binom{3}{2} 0.3^2 0.7^1 = 0.189$$

$$b(3;3, 0.3) = \binom{3}{3} 0.3^3 0.7^0 = 0.027$$

The sum of these probabilities is 1.

TABLE 7-3 Generation of a Binomial Distribution

Experiment	u_i variates	x
1	0.78, 0.51, 0.43	0
2	0.85, 0.95, 0.22	1
3	0.88, 0.36, 0.20	1
4	0.79, 0.55, 0.71	0
5	0.11, 0.09, 0.70	2
6	0.11, 0.70, 0.98	1
7	0.79, 0.45, 0.86	0
8	0.62, 0.11, 0.77	1
9	0.45, 0.40, 0.13	1
10	0.37, 0.31, 0.12	1
11	0.54, 0.99, 0.95	0
12	0.79, 0.36, 0.37	0
13	0.73, 0.67, 0.60	0
14	0.56, 0.40, 0.71	0
15	0.24, 0.21, 0.23	3

To illustrate the simulation of binomial probabilities, let $N = 15$. For each of the 15 experiments, we generated three uniform variates and counted the number of values less than 0.3, which is then the sample estimate of X. One possible simulation is shown in Table 7-3.

From the 15 simulated estimates of X, there are seven 0s, six 1s, one 2, and one 3. This yields sample estimates of the probabilities of 7/15 = 0.467, 6/15 = 0.400, 1/15 = 0.067, and 1/15 = 0.066, respectively. As N becomes larger and larger, the sample estimates of the probabilities would more closely approximate the population probabilities.

Example 7-6: Life of a Project

Consider the case of a project that has a design life of 5 years, with the probability of failure in any 1 year equal to 0.1. A binomial process such as this case can be simulated using $n = 5$ and $p = 0.1$, with values of $X = 0, 1, 2, 3, 4$, and 5. To estimate the probability of no failures, we would be interested in the case of $X = 0$. Ten experiments were made, and the sample estimate of the probability of a project not failing were computed as shown in Table 7-4a. The sample ($\hat{p}$) and population (p) probabilities are shown in Table 7-4b. For a sample of 10, the sample estimates are in good agreement with the population values. The sample estimate of the probability of a project not failing in 5 years is 0.60.

7.5.3. Geometric Distribution

The geometric distribution can be generated using the inverse transformation method. An alternate method can be used by setting $i = 1$ and performing the following steps:

TABLE 7-4a Simulation Results for a Binomial Distribution

Project	u_i variates	x
1	0.05, 0.37, 0.23, 0.42, 0.05	2
2	0.60, 0.32, 0.79, 0.45, 0.20	0
3	0.67, 0.86, 0.66, 0.55, 0.23	0
4	0.41, 0.03, 0.90, 0.98, 0.74	1
5	0.94, 0.58, 0.31, 0.45, 0.31	0
6	0.69, 0.93, 0.77, 0.37, 0.72	0
7	0.55, 0.69, 0.01, 0.51, 0.58	1
8	0.33, 0.58, 0.72, 0.22, 0.97	0
9	0.09, 0.29, 0.81, 0.44, 0.68	1
10	0.29, 0.54, 0.75, 0.36, 0.29	0

TABLE 7-4b Sample ($\hat{p}$) and Population (p) Probabilities

X	Sample probability ($\hat{p}$)	Population probability (p)
0	0.60	0.5905
1	0.30	0.3280
2	0.10	0.0729
3	0.00	0.0081
4	0.00	0.0005
5	0.00	0.0000
Column summations	1.00	1.0000

1. Generate a random number $u \in U[0,1]$.
2. If $u \leq p$, where p is a specified parameter for the geometric distribution, return $x = i$; otherwise, replace i by $i + 1$ and go back to step 1.

A third method of random generation for the geometric distribution starts with generating a random number $u \in U[0,1]$, and then x is calculated as:

$$x = \frac{\ln(u)}{\ln(1-p)} \tag{7-23}$$

The result from Equation 7-23 is a real value; therefore, the integer of this real value should be used as the generated value and should be assigned to x.

7.5.4. Poisson Distribution

Discrete random variables that can take on integer values may be represented by a Poisson mass function, which has the following form:

$$p_{X_t}(x) = \frac{e^{-\lambda t}(\lambda t)^x}{x!} \qquad \text{for } x = 0, 1, 2, \ldots \tag{7-24}$$

where X_t is the number of occurrences in a reference period t, λ is a parameter of the function, and 0! is defined as 1. The cumulative function is:

$$F_{X_t}(x) = \sum_{i=0}^{x} \frac{e^{-\lambda t}(\lambda t)^i}{i!} \quad \text{for } x = 0, 1, 2, \ldots \tag{7-25}$$

The mean and variance of the function both equal λt, so the sample mean can be used as an estimator of λt.

The Poisson distribution can be generated using the inverse transformation method based on Equation 7-25 or using its relationship to the exponential distribution.

According to the first method, uniform random variates can be used to generate random variates that have a Poisson distribution for any λ. The procedure for any given λ is as follows:

1. Calculate the cumulative function $F_{X_t}(x_i)$ for $x_i = 0, 1, 2, \ldots$.
2. Generate a unit variate u_i in the range $[0, 1]$.
3. If $F_{X_t}(x_{i-1}) \le u_i < F_{X_t}(x_i)$, then the Poisson variate is x_i.

The procedure in the second case is based on the special relation between the Poisson distribution and the exponential distribution. In order to generate a random variable X that has a Poisson distribution with a parameter λ, the following must be set:

$$a = \exp(-\lambda t), b = 1, \text{ and } i = 0 \tag{7-26}$$

In the $(i + 1)$ random generation cycle, generate a random number $u \in U[0, 1]$ and assign a value to X as follows:

$$x = \begin{cases} i & \text{if } bu < a \\ \text{increment } i \text{ by 1 and repeat the process} & \text{otherwise} \end{cases} \tag{7-27}$$

Example 7-7: Floods in Northern Idaho

The number of floods per year (n) on a river in northern Idaho was obtained from streamflow records over a 38-year period. The sample analysis is shown in Table 7-5. A total of 120 floods occurred in the 38 years, which yields a mean annual flood count of 3.158; thus, the sample estimate of the Poisson parameter λ is 3.158, with $t = 1$ year. The assumed population mass function is:

$$p_{X_1}(0) = \frac{e^{-3.158}(3.158)^x}{x!} \quad \text{for } x = 0, 1, 2, \ldots \tag{7-28}$$

The population probabilities are shown in Table 7-6.

TABLE 7-5 Frequency of Flows above 28,000 cfs on Clearwater River at Kamiah, ID

Year	n	Year	n	Year	n	Year	n	Year	n
1911	4	1919	2	1927	5	1935	3	1943	6
1912	3	1920	2	1928	2	1936	5	1944	1
1913	4	1921	2	1929	3	1937	2	1945	2
1914	3	1922	3	1930	1	1938	4	1946	7
1915	1	1923	3	1931	2	1939	2	1947	3
1916	5	1924	2	1932	4	1940	2	1948	6
1917	5	1925	3	1933	4	1941	1		
1918	3	1926	3	1934	4	1942	3		

TABLE 7-6 Poisson Approximation of Flood Counts

	Sample		Population	
X	N_X	N_X/N	Mass function p_{X_t}	Cumulative function $F_{X_t}(x)$
0	0	0	0.043	0.043
1	4	0.105	0.134	0.177
2	10	0.263	0.212	0.389
3	11	0.290	0.223	0.612
4	6	0.158	0.176	0.788
5	4	0.105	0.111	0.899
6	2	0.053	0.059	0.958
7	1	0.026	0.026	0.984
≥8	0	0	0.016	1.000

Example 7-8: Hurricanes Hitting the U.S. Coasts

Records indicate that the annual number of hurricanes hitting the U.S. coast has an average of 4. The discrete variable can be represented by a Poisson mass function with $\lambda = 4$. Thus, the probability of the coast not being hit by a hurricane in any year is:

$$p_{X_1}(0) = \frac{e^{-4}(4)^0}{0!} = 0.0183 \tag{7-29}$$

where 0! is defined as 1.

Example 7-9: Annual Number of Floods

In Example 7-7, the mean annual number of floods on a particular river was found to be 3.158. Assume that it is necessary to simulate the number of floods in a 10-year period. The cumulative function is given in Table 7-6. For a given vector of uniform variates u_i, the generated number of floods per year is shown in Table 7-7. Based on this sample of ten simulated values, the mean number of floods is 26/10 = 2.6. As the sample size increases, the sample mean would be expected to be closer to the assumed population mean of 3.158.

TABLE 7-7 Synthetic Series of the Frequency of Floods

Year	u_i	x_i	$F_{X_t}(x_{i-1})$	$F_{X_t}(x_i)$
1	0.13	1	0.043	0.177
2	0.04	0	0.000	0.043
3	0.66	4	0.612	0.788
4	0.80	5	0.788	0.899
5	0.77	4	0.612	0.788
6	0.52	3	0.389	0.612
7	0.30	2	0.177	0.389
8	0.14	1	0.043	0.177
9	0.37	2	0.177	0.389
10	0.65	4	0.612	0.788
Column summation		26		

7.6. GENERATION OF SELECTED CONTINUOUS RANDOM VARIABLES

7.6.1. Uniform Distribution

Variates with a uniform distribution that has parameters other than 0 and 1 can be generated using the inverse transformation method. Starting with a generated random number $u \in U[0,1]$, a random variable X can be generated according to a uniform distribution as follows:

$$x = a + (b - a)u \tag{7-30}$$

where a and b are specified parameters for the uniform distribution that correspond to the lower and upper limits of the random variable, respectively.

7.6.2. Normal Distribution

Many problems require random normal numbers that have mean μ and standard deviation σ, which is denoted as $N(\mu,\sigma)$. Several methods can be used to generate standardized normal deviates, $N(0,1)$. Three methods are presented in this section: the inverse transformation method, the empirical method, and the polar method. Once the standard normal deviate is generated, a normal deviate x, which is $N(\mu,\sigma)$, can be computed with Equation 7-10.

The inverse transformation method sets a uniform deviate, $U(0,1)$, equal to the cumulative standard normal probability. The normal deviate is the value of z that corresponds to the cumulative probability:

$$z_i = \Phi^{-1}(u_i) \tag{7-31}$$

Approximate values can be obtained by linear interpolation of a table of standard normal deviates (Table A-1). Use the uniform variate u_i as the value of Φ and interpolate the value of z_i.

A single standard normal deviate z_i can be obtained empirically by:

$$z_i = \left(\sum_{i=1}^{12} u_i\right) - 6 \tag{7-32}$$

The disadvantage of using Equation 7-32 is that 12 uniform deviates are required to generate a single standard normal deviate.

A random variable X can be generated according to a normal distribution $N(\mu, \sigma^2)$ by generating the first variate Z with a standard normal distribution, and then using:

$$x = z\sigma + \mu$$

The normal distribution can be generated by several methods. A direct, although approximate, method is to develop regression equations for segments of the CDF distribution of the standard normal. Then, inverse transformation can be used to generate a variate Z according to a standard normal distribution using the following relation:

$$z = \Phi^{-1}(u)$$

The polar method can also be used to generate standard normal variates. The polar method starts by generating two independent random numbers u_1 and u_2 from the interval $[0,1]$. Then, the following variables y_1, y_2, and w are defined as:

$$y_1 = 2u_1 - 1 \tag{7-33}$$

$$y_2 = 2u_2 - 1 \tag{7-34}$$

$$w = y_1^2 + y_2^2 \tag{7-35}$$

If $w > 1$, then generate new random numbers u_1 and u_2 and repeat the evaluation of y_1, y_2, and w. If $w \le 1$, then evaluate a pair of independent standard normal variates as follows:

$$z_1 = \sqrt{\frac{-2\ln(w)}{w}}\, y_1 \tag{7-36}$$

$$z_2 = \sqrt{\frac{-2\ln(w)}{w}}\, y_2 \tag{7-37}$$

Example 7-10: Piston Ring Tolerance

A machine that prepares piston rings is reported to machine rings to be normally distributed with a mean diameter of 10 cm and a standard deviation of 0.3 mm. A piston ring is considered defective if its diameter is more than 0.6 mm larger or smaller than the design diameter. For a set of six rings, compute the probability that all rings in a set of six will meet the tolerance. The probability can be approximated using simulation.

The theoretical probability that one piston ring is machined within tolerance is:

$$P(99.4 < X < 100.6 \text{ mm}) = P\left(z < \frac{100.6 - 100}{0.3}\right) - P\left(z < \frac{99.4 - 100}{0.3}\right)$$
$$= P(z < 2) - P(z < -2) = 0.9544$$

Given that a machined ring is either within tolerance or not within tolerance, the probability that all six rings in a set of six have been machined to be within tolerance can be computed with the binomial distribution:

$$P(0 \text{ defective}) = b(0, 6, 0.9544) = 0.7558$$

This indicates that approximately 25% of the sets of six rings will have at least one ring that is not within the tolerance of 0.6 mm.

Ten sets of six rings were simulated assuming a normal population with a mean of 10 cm and a standard deviation of 0.3 mm. Table 7-8 shows the 10 sets of six standard normal deviates, which are used to compute the deviation of the ring diameter from 10 cm. Of the 60 deviations, only two exceed the tolerance 0.6 mm. These two are in sets 3 and 8. Thus, 8 out of 10 sets, or 80%, include rings that are all within tolerance. For a sample of 10, this is as close as possible to the theoretical probability of 75.6%. Of course, if 10^6 sets of six were simulated, the probability would be expected to be closer to the theoretical value.

TABLE 7-8 Generation of Deviations of Piston Ring Diameter from the Design Diameter for Example 7-10

Set	Standard normal deviates	Deviation of simulated ring diameters from 100 mm
1	1.80, 0.46, 1.10, –1.07, –0.34, 0.94	0.540, 0.138, 0.330, –0.321, –0.102, 0.282
2	–0.18, –0.75, –0.13, 1.23, 1.48, –0.15	–0.054, –0.225, –0.039, 0.369, 0.444, –0.045
3	–0.86, –1.46, 0.53, 0.24, –0.21, 2.02	–0.258, –0.438, 0.159, 0.072, –0.063, 0.606
4	–0.58, 0.34, –0.82, 0.14, -0.33, 0.43	–0.174, 0.102, –0.246, 0.042, –0.099, 0.129
5	0.83, 0.80, –0.46, 0.56, 0.64, –0.73	0.249, 0.240, –0.138, 0.168, 0.192, –0.219
6	–1.34, 0.95, –1.46, –1.21, 1.02, –0.15	–0.402, 0.285, –0.438, –0.363, 0.306, –0.045
7	1.48, 1.26, –1.43, 0.49, –0.52, –0.65	0.444, 0.378, –0.429, 0.147, –0.156, –0.195
8	0.53, –1.05, 0.88, 0.65, 2.96, –0.97	0.159, –0.315, 0.264, 0.195, 0.888, –0.291
9	–0.65, 0.12, 0.06, –0.08, –1.64, –0.75	–0.195, 0.036, 0.018, –0.024, –0.492, –0.225
10	–0.51, –1.51, –0.71, –1.58, 0.91, 1.75	–0.153, –0.453, –0.213, –0.474, 0.273, 0.525

7.6.3. Lognormal Distribution

The generation of lognormal variates can be based on generating standard normal variates and the special relationship between the lognormal and the standard normal distributions. In order to generate a variate X with given mean value and variance according to a lognormal distribution, the parameters of the lognormal distribution, μ_Y and σ_Y, should first be determined from the given mean and variance. The relations between the parameters of a distribution and its mean and variance were previously discussed in Chapter 3. Then, a normal variate Y with the parameters μ_Y and σ_Y is generated, resulting in a normal value y. The generated lognormal value is taken as:

$$x = \exp(y) \tag{7-38}$$

The generation of normal variates was discussed in Section 7.6.2.

To illustrate the generation of lognormally distributed variates, consider the case where x values with μ_x equal to 250 and σ_x equal to 35 are needed. Using Equations 5-24 and 5-25, the statistics in the log space are:

$$\sigma_y^2 = \ln[1 + (35/250)^2] = 0.01941$$

$$\mu_y = \ln(250) - 0.5(0.01941)^{0.5} = 5.4518$$

Then values of y are computed by:

$$y_i = \mu_y + z_i\sigma_y = 5.4518 + 0.1393z_i$$

For example, if a generated standard normal deviate is 0.74, then the computed value of y is:

$$y = 5.4518 + 0.1393(0.74) = 5.5549$$

Therefore, x is computed with Equation 7-38:

$$x = \exp(5.5549) = 258.5$$

The values of y are normally distributed, but the values of x are lognormally distributed.

7.6.4. Exponential Distribution

The exponential distribution can be generated using the inverse transformation method. Starting with a generated random number $u \in U[0, 1]$, a random variable X can be generated according to a exponential distribution as follows:

$$x = \frac{-\ln(u)}{\lambda} \tag{7-39}$$

where λ is a specified parameter for the exponential distribution.

Example 7-11: Simulation of Exponentially Distributed Times to Failure

The time (years) to failure of the motor in a blender in exponentially distributed with $\lambda = 6.5$. A restaurant purchases six blenders. What is the probability that no more than one blender will need to have its motor replaced in 10 years? Simulate 10 sets of six blenders and compare the theoretical and simulated probabilities.

The probability of one blenders not failing in 10 years because of motor failure is:

$$P(NF > 10) = \frac{1}{6.5}\int_{10}^{\infty} e^{-t/6.5}\,dt = 0.2147$$

where T is time to failure. The binomial distribution can be used to compute the probability that two or more of the blenders will still be working after 10 years, where x is the number of blenders in which the motor has not failed:

$$\begin{aligned} P(X \ge 2) &= \sum_{x=2}^{6} b(x;6, 0.2147) = 1 - \sum_{x=0}^{1} b(x;6, 0.2147) \\ &= 1 - 0.2345 - 0.3847 = 0.3808 \end{aligned}$$

To simulate the time to motor failure of 10 sets of six blenders, 60 uniform variates, $U(0,1)$, are generated and used with Equation 7-39 to generate the time to motor failure. The number of non-failures in each group is taken as the number of failure times that were greater than 10. The results are given in Table 7-9. For only one set of six blenders did two or more motors survive 10 years. Thus, the probability is 0.1 vs. the theoretical value of 0.38. The difference between the simulated and theoretical probabilities is the result of sampling variation. To achieve a failure time of 10 years or more requires a uniform variate of 0.214 or smaller. Of the 60 uniform variates of Table 7-9, only 10 were less than 0.214, rather than the expected number of 13.

7.7. APPLICATIONS

7.7.1. Simulation of a Queuing System

A classical problem in engineering is establishing policies for problems involving queues. The following problem illustrates the use of simulation in the evaluation of policies on the operation of a checkout station in a retail store. The manager of the

TABLE 7-9 Failure Times of Blender Motors for Example 7-11

Set	Uniform variates, u_i	Failure times (years)	F
1	0.63, 0.16, 0.54, 0.76, 0.30, 0.94	3.0, 11.9, 4.0, 1.8, 7.8, 0.4	5
2	0.24, 0.09, 0.87, 0.60, 0.98, 0.40	9.3, 15.7, 0.9, 3.3, 0.1, 6.0	5
3	0.62, 0.05, 0.85, 0.61, 0.88, 0.91	3.1, 19.5, 1.1, 3.2, 0.8, 0.6	5
4	0.85, 0.34, 0.73, 0.89, 0.20, 0.87	1.1, 7.0, 2.0, 0.8, 10.5, 0.9	5
5	0.76, 0.41, 0.23, 0.75, 0.32, 0.94	1.8, 5.8, 9.6, 1.9, 7.4, 0.4	6
6	0.75, 0.12, 0.07, 0.56, 0.07, 0.59	1.9, 13.8, 17.3, 3.8, 17.3, 3.4	3
7	0.30, 0.47, 0.77, 0.01, 0.65, 0.38	7.8, 4.9, 1.7, 29.9, 2.8, 6.3	5
8	0.80, 0.97, 0.30, 0.53, 0.23, 0.01	1.5, 0.2, 7.8, 4.1, 9.6, 29.9	5
9	0.17, 0.99, 0.55, 0.41, 0.80, 0.85	11.5, 0.1, 3.9, 5.8, 1.5, 1.1	5
10	0.23, 0.82, 0.93, 0.78, 0.81, 0.40	9.6, 1.3, 0.5, 1.6, 1.4, 6.0	6

store is interested in the operation of the checkout station because she is aware that customers who spend excessive time waiting to check out will be less likely to return to the store. Although the total time that a customer spends in the checkout process is of interest, the manager is also interested in both the time spent waiting to be served and the time being served; the total time is the sum of the two. The manager also recognizes that customers are discouraged by long lines; therefore, the number of customers waiting to be served is also a criterion of interest. The manager would be interested in examining alternatives that would reduce both the total time and the length of the queue. For example, the manager could install a second checkout counter, which would require increases in costs associated with installation, operation, and maintenance of the added counter. As an alternative, the manager might decide to install a new checkout system that is more efficient; this system would reduce the time required for checkout without the costs associated with the installation of a new counter and the added labor. Of course, there would be a cost associated with the purchase of the new equipment.

Because the manager is most concerned with peak demand hours, she needs to collect data on the current operation of the checkout counter. Each evening for 1 week, the manager measured the time between the arrival of a customer and the time required for a customer to be serviced. The manager finds that the service time can be approximated by a uniform probability function with a mean of 2 min; based on her measurements, the manager believes that the service time can be approximated with probabilities of 0.25 for service times of 1.25, 1.75, 2.25, and 2.75 min; these are the center points of the computational interval of 0.5 min. The time between arrivals is shown in Table 7-10 for 0.5-min intervals. During the time of peak demand all interarrival times were 5 min or less, with an average of 2.2 min. Because the interarrival times did not follow a known probability function, the empirical function was used for simulations. The discrete cumulative function for the interarrival times is shown in Figure 7-3. The discrete function is used because the data were collected on a 0.5-min interval; the center point of each interval is used to represent the interval. The derivation of the cumulative function for the interarrival times is given in Table 7-10.

TABLE 7-10 Computation of Cumulative Probability Function of Interarrival Times

Interarrival time (min)	Number of arrivals	Probability function	Cumulative function
0.0–0.5	20	0.031	0.031
0.5–1.0	50	0.077	0.108
1.0–1.5	90	0.138	0.246
1.5–2.0	140	0.215	0.461
2.0–2.5	120	0.185	0.646
2.5–3.0	100	0.154	0.800
3.0–3.5	70	0.108	0.908
3.5–4.0	30	0.046	0.954
4.0–4.5	20	0.031	0.985
4.5–5.0	10	0.015	1.000
Column summations	650	1.000	

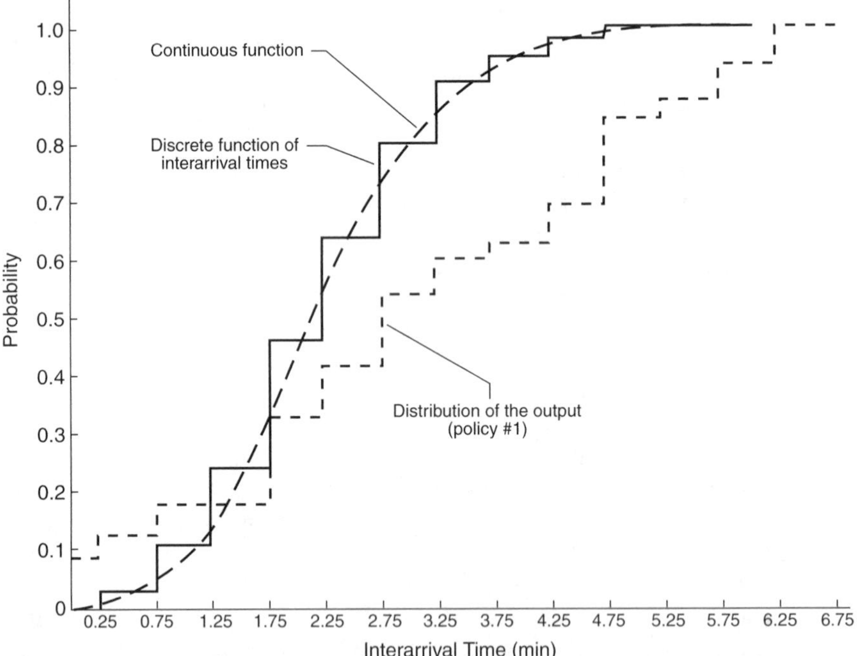

FIGURE 7-3 Comparison of the distributions of the interarrival times and output from the system.

To demonstrate the simulation technique, a 1-hr period was simulated. The interarrival times were simulated by generating a sequence of random numbers from a uniform distribution having location and scale parameters of 0 and 1, respectively. Each of the numbers in the sequence was entered into Figure 7-3, and the interarrival times determined. The interarrival times are given in Table 7-11, with values ranging between 0.25 and 4.75 on a 0.5-min interval. The second customer arrived at the checkout counter 1.25 min after the first customer. The third customer arrived at a time of 3 min, which is 1.75 min after the second customer. The time of arrival of each customer is also given in Table 7-11.

A second sequence of random numbers having a uniform probability function ($a = 0$, $b = 1$) is generated. Because these values are used for simulating the time required for service, there are four service times (1.25, 1.75, 2.25, and 2.75) with an equal probability of occurrence. The random numbers were used to generate the service times, which are shown in Table 7-11, for the existing situation; this is called policy 1. The time of departure equals the sum of the service time and either the time that the previous customer leaves the checkout counter or the time that the person arrives at the counter when no one is being served. The waiting time for a customer, which is also given in Table 7-11, equals the difference between the time of arrival and the time that the customer starts service. The total time equals the sum of the waiting time and servicing time. The number of customers in the queue is determined by comparing the time of departure of a person and the time of arrival of customers who follow the person being served. For example, the 11th customer departs at time 21.75. At that time, customers 12, 13, 14, and 15 had already arrived at the counter (at times 17.75, 19.50, 20.75, and 21.00), so there are four customers waiting in the queue to be served when customer 11 departs.

The manager is interested in determining the effect of a new means of checking out a customer. The equipment is expected to decrease the service time. The simulation based on this service time distribution is referred to as policy 2. The probabilities for service times of 1.25, 1.75, 2.25, and 2.75 min are 0.35, 0.30, 0.20, and 0.15, respectively. The same sequence of arrivals was used to simulate the effect of service policy 2. The service times and times of departure were computed and are given in Table 7-11. The total time, waiting time, and number in the queue were computed in the same way as for policy 1. The distribution of waiting times is given in Table 7-12. It is quite evident that the waiting time is considerably less with policy 2 than with policy 1. The mean waiting times for policies 1 and 2 are 3.27 and 1.08 min, respectively. The mean service times are 1.78 and 1.59 min, respectively. It is evident from both the mean values and the distributions of service times shown in Table 7-13 that the sample values for the simulated period of 1 hr are less than what would be expected. As the length of the period of simulation is increased, one can expect the sample values to be closer approximations to the true values.

7.7.2. Warehouse Construction

A warehouse is to be constructed from precast concrete elements that are produced by a nearby precast factory. The following construction tasks are identified for building the warehouse:

TABLE 7-11 Simulations of Interarrival and Servicing Times for Alternative Policies

					Policy 1					Policy 2				
Customer	Random number	Interarrival time (min)	Time of arrival (min)	Random number	Servicing time (min)	Time of departure (min)	Waiting time (min)	Total time (min)	Number waiting in queue	Servicing time (min)	Time of departure (min)	Waiting time (min)	Total time (min)	Number waiting in queue
1	0.19	1.25	0	0.57	2.25	2.25	0.00	2.25	1	1.75	1.75	0.00	1.75	1
2	0.37	1.75	1.25	0.84	2.75	5.00	1.00	3.75	2	2.25	4.00	0.50	2.75	2
3	0.09	0.75	3.00	0.54	2.25	7.25	2.00	4.25	2	1.75	5.75	1.00	2.75	2
4	0.30	1.75	3.75	0.25	1.25	8.50	3.50	4.75	2	1.25	7.00	2.00	3.25	1
5	0.76	2.75	5.50	0.95	2.75	11.25	3.00	5.75	2	2.75	9.75	1.50	4.25	1
6	0.53	2.25	8.25	0.14	1.25	12.50	3.00	4.25	2	1.25	11.00	1.50	2.75	1
7	0.09	0.75	10.50	0.21	1.25	13.75	2.00	3.25	2	1.25	12.25	0.50	1.75	1
8	0.26	1.75	11.25	0.51	2.25	16.00	2.50	4.75	2	1.75	14.00	1.00	2.75	1
9	0.43	1.75	13.00	0.16	1.25	17.25	3.00	4.25	2	1.25	15.25	1.00	2.25	1
10	0.13	1.25	14.75	0.39	1.75	19.00	2.50	4.25	2	1.75	17.00	0.50	2.25	1
11	0.25	1.75	16.00	0.85	2.75	21.75	3.00	5.75	4	2.25	19.25	1.00	3.25	1
12	0.42	1.75	17.75	0.82	2.75	24.50	4.00	6.75	4	2.25	21.50	1.50	3.75	3
13	0.31	1.75	19.50	0.16	1.25	25.75	5.00	6.25	4	1.25	22.75	2.00	3.25	2
14	0.15	1.25	20.75	0.14	1.25	27.00	5.00	6.25	3	1.25	24.00	2.00	3.25	2
15	0.01	0.25	21.00	0.80	2.75	29.75	6.00	8.75	4	2.25	26.25	3.00	5.25	2
16	0.54	2.25	23.25	0.24	1.25	31.00	6.50	7.75	3	1.25	27.50	3.00	4.25	2
17	0.28	1.75	25.00	0.12	1.25	32.25	6.00	7.25	3	1.25	28.75	2.50	3.75	2

18	0.50	2.25	27.25	0.87	2.75	35.00	5.00	7.75	4	2.75	31.50	1.50	4.25	2
19	0.12	1.25	28.50	0.06	1.25	36.25	6.50	7.75	3	1.25	32.75	3.00	4.25	1
20	0.67	2.75	31.25	0.33	1.75	38.00	5.00	6.75	3	1.25	34.00	1.50	2.75	1
21	0.22	1.25	32.50	0.44	1.75	39.75	5.50	7.25	3	1.75	35.75	1.50	3.25	1
22	0.54	2.25	34.75	0.49	1.75	41.50	5.00	6.75	3	1.75	37.50	1.00	2.75	1
23	0.58	2.25	37.00	0.22	1.25	42.75	4.50	5.75	2	1.25	38.75	0.50	1.75	1
24	0.18	1.25	38.25	0.13	1.25	44.00	4.50	5.75	2	1.25	40.00	0.50	1.75	0
25	0.53	2.25	40.50	0.03	1.25	45.25	3.50	4.75	1	1.25	41.75	0.00	1.25	0
26	0.75	2.75	43.25	0.79	2.75	48.00	2.00	4.75	2	2.25	45.50	0.00	2.25	0
27	0.76	2.75	46.00	0.01	1.25	49.25	2.00	3.25	2	1.25	47.25	0.00	1.25	1
28	0.05	0.75	46.75	0.27	1.75	51.00	2.50	4.25	1	1.25	48.50	0.50	1.75	0
29	0.57	2.25	49.00	0.18	1.25	52.25	2.00	3.25	1	1.25	50.25	0.00	1.25	0
30	0.69	2.75	51.75	0.28	1.75	54.00	0.50	2.25	0	1.25	53.00	0.00	1.25	0
31	0.61	2.25	54.00	0.14	1.25	55.25	0.00	1.25	0	1.25	55.25	0.00	1.25	0
32	0.83	3.25	57.25	0.06	1.25	58.50	0.00	1.25	1	1.25	58.50	0.00	1.25	1
33	0.00	0.25	57.50	0.65	2.25	60.75	1.00	3.25	0	1.75	60.25	1.00	2.75	0
34	0.81	3.25	60.75	0.34	1.75	62.50	0.00	1.75	—	1.25	62.00	0.00	1.25	—

A: excavation of foundations
B: construction of foundations
C: construction of precast elements at factory
D: transportation of precast elements to construction site
E: assembly of elements at site
F: construction of roof
G: exterior and interior finishing

TABLE 7-12 Computation of Cumulative Probability Function of Waiting Times

	Policy 1			Policy 2		
Waiting time (min)	Frequency	Probability function	Cumulative function	Frequency	Probability function	Cumulative function
0.0	3	0.091	0.091	8	0.242	0.242
0.5	1	0.030	0.121	6	0.182	0.424
1.0	2	0.061	0.182	6	0.182	0.606
1.5	0	0.000	0.182	6	0.182	0.788
2.0	5	0.152	0.333	3	0.091	0.879
2.5	3	0.091	0.424	1	0.030	0.909
3.0	4	0.121	0.545	3	0.091	1.000
3.5	2	0.061	0.606	—	—	—
4.0	1	0.030	0.636	—	—	—
4.5	2	0.061	0.697	—	—	—
5.0	5	0.152	0.848	—	—	—
5.5	1	0.030	0.878	—	—	—
6.0	2	0.061	0.939	—	—	—
6.5	2	0.061	1.000	—	—	—
Column summations	33	1.000	—	33	1.000	—

TABLE 7-13 Distributions of Total Time in Systems and Service Time for Two Policies

Total time (min)	Frequency for policy 1	Frequency for policy 2	Service time (min)	Frequency for policy 1	Frequency for policy 2
1.25	2	7	1.25	16	20
1.75	1	5	1.75	7	7
2.25	2	3	2.25	4	5
2.75	0	7	2.75	7	2
3.25	4	5	—	—	—
3.75	1	2	—	—	—
4.25	5	4	—	—	—
4.75	4	0	—	—	—
5.25	0	1	—	—	—
5.75	4	—	—	—	—
6.25	2	—	—	—	—
6.75	3	—	—	—	—
7.25	2	—	—	—	—
7.75	3	—	—	—	—
8.25	0	—	—	—	—
8.75	1	—	—	—	—
Column summations	34	34	—	34	34

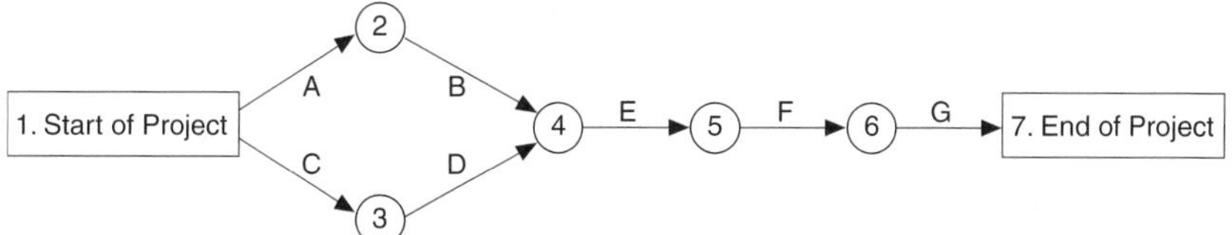

FIGURE 7-4 Construction network.

Figure 7-4 shows the logical network for performing these activities. The figure indicates that tasks C and D can be performed in parallel to tasks A and B; that is, as excavation and construction of the footings are being performed at the site, the precast elements can be constructed at the factory and then transported to the construction site. Table 7-14 shows the means and standard deviations for the completion times of these tasks. Normal probability distributions and statistical noncorrelation are assumed for these times. This example was used in Section 6.4.5 to illustrate the estimation of the mean and variance of the completion time of the project using approximate methods. In this section, simulation is used to compute these moments.

The project completion time, T, is a dependent random variable that is given by:

$$T = \{A + B, C + D\} + E + F + G \tag{7-40}$$

where A, B, C, D, E, F, and G are random variables that model the times for completing the corresponding tasks. The random numbers used for generating the completion times for tasks A, B, C, D, E, F, and G are shown in Table 7-15. In this example, 20 simulation cycles were used. The random numbers were used to generate the random times as shown in Table 7-16 utilizing the inverse transformation method. Then, Equation 7-40 was used to compute the completion time for the project for each simulation cycle as shown in Table 7-16. The mean value and variance of the completion time of the project were computed using values in the last column of Table 7-16. The resulting statistics are:

$$\text{Mean value} = 14.59 \text{ days} \tag{7-41a}$$

$$\text{Variance} = 5.55 \text{ (days)}^2 \tag{7-41b}$$

TABLE 7-14 Moments of Completion Times

Task	Name	Mean (days)	Standard deviation (days)
A	Foundation excavation	3.0	1.0
B	Foundation construction	2.0	0.5
C	Precast-elements construction	4.0	1.0
D	Transportation of elements	0.5	0.5
E	Assembly of elements	4.0	1.0
F	Roofing	2.0	1.0
G	Finishing	3.0	1.0

The approximate mean duration of the project, $\overline{T}$, was computed in Section 6.4.5 as:

$$\overline{T} = \max\{3 + 2, 4 + 0.5\} + 4 + 2 + 3 = 14 \text{ days} \tag{7-42}$$

The approximate variance of the project duration, Var(T), was computed as:

$$\text{Var}(T) = \text{Var}(A) + \text{Var}(B) + \text{Var}(E) + \text{Var}(F) + \text{Var}(G) \tag{7-43a}$$

$$= (1)^2 + (0.5)^2 + (1)^2 + (1)^2 + (1)^2 = 4.25 \text{ (days)}^2 \tag{7-43b}$$

These values differ from the computed values using simulation. The simulation results are also approximate because of the small number of cycles used to compute the results. By increasing the number of simulation cycles, the accuracy of the results is expected to increase. The assumed probability distributions for the task durations resulted in negative durations that are not realistic, as shown in Table 7-16. Other distributions for finite durations might be appropriate in this case. Also, the simulation results in the case of a small number of cycles are dependent on the random numbers used. For example, by using a different set of random numbers as those used in Table 7-17, the results shown in Table 7-18 are obtained. The statistics in this case are:

$$\text{Mean value} = 13.98 \text{ days} \tag{7-44a}$$

$$\text{Variance} = 3.67 \text{ (days)}^2 \tag{7-44b}$$

TABLE 7-15 Random Numbers Used for Completion Time of Tasks

Task A	Task B	Task C	Task D	Task E	Task F	Task G
0.642707	0.758002	0.547225	0.510713	0.924981	0.444910	0.671304
0.240297	0.092418	0.847150	0.071252	0.981120	0.793358	0.780596
0.169051	0.446979	0.990008	0.079644	0.391058	0.793205	0.276989
0.457609	0.521270	0.606333	0.006137	0.479270	0.121284	0.343670
0.386325	0.395759	0.956544	0.432595	0.723067	0.448813	0.008538
0.313708	0.061922	0.343042	0.230356	0.538481	0.636290	0.211676
0.137571	0.078837	0.471558	0.383158	0.203166	0.500447	0.101354
0.296782	0.610994	0.785467	0.282056	0.282186	0.560465	0.539651
0.908314	0.124274	0.709123	0.508328	0.496352	0.886927	0.720611
0.763968	0.327695	0.506164	0.246872	0.743617	0.275227	0.218178
0.139498	0.935402	0.789508	0.966422	0.440431	0.682035	0.476614
0.220256	0.040641	0.347426	0.282962	0.178687	0.092735	0.964860
0.344963	0.100168	0.963482	0.569873	0.933351	0.646640	0.858627
0.095613	0.791418	0.726318	0.376506	0.872995	0.895403	0.962331
0.225540	0.262949	0.632760	0.550859	0.198235	0.077169	0.086730
0.239485	0.985236	0.212528	0.445724	0.662470	0.325610	0.025242
0.191603	0.108613	0.897544	0.990706	0.933851	0.557361	0.050711
0.946010	0.241317	0.187334	0.015071	0.228146	0.832563	0.816427
0.973859	0.343243	0.197940	0.177672	0.125638	0.099943	0.747989
0.484109	0.214928	0.020997	0.424466	0.893968	0.866459	0.706856

TABLE 7-16 Generated Random Values for Completion Time of Tasks

Task A	Task B	Task C	Task D	Task E	Task F	Task G	Project completion time
3.3652600	2.3498042	4.1183849	0.5133909	5.4396627	1.8617675	3.4430860	16.459580
2.2949296	1.3368915	5.0242746	−0.2334048	6.0778971	2.8179440	3.7739874	17.460698
2.0421427	1.9334970	6.3270846	−0.2038625	3.7238635	2.8174073	2.4085497	15.073043
2.8937885	2.0266006	4.2693511	−0.7523029	3.9481516	0.8313093	2.5979742	12.297824
2.7115217	1.8680395	5.7122992	0.4152843	4.5916169	1.8716254	0.6144956	13.205321
2.5150499	1.2304214	3.5962670	0.1312882	4.0963777	2.3481148	2.1995787	12.389543
1.9086613	1.2934043	3.9288268	0.3516179	3.1698067	2.0011170	1.7259547	11.177323
2.4667156	2.1407416	4.7905834	0.2118119	3.4240093	2.1518299	3.0993185	13.677553
4.3306524	1.4230146	4.5504403	0.5104081	3.9908831	3.2104754	3.5842925	16.539318
3.7188585	1.7770724	4.0154067	0.1579639	4.6542217	1.4032765	2.2218555	13.775285
1.9173874	2.7587914	4.8045186	1.4155181	3.8504386	2.4729750	2.9415001	15.484950
2.2288943	1.1281585	3.6081638	0.2131524	3.0797160	0.6756985	4.8105053	12.387236
2.6014871	1.3596149	5.7930090	0.5878512	5.5015181	2.3758197	4.0742012	18.332399
1.6928452	2.4055804	4.6013610	0.3428837	5.1407426	3.2559441	4.7787973	18.119729
2.2466222	1.6830249	4.3387303	0.5637742	3.1522166	0.5753632	1.6386032	10.268687
2.2923139	3.0884019	3.2025201	0.4319117	4.4187744	1.5483608	1.0437219	12.391573
2.1281350	1.3829576	5.2678487	1.6770614	5.5053989	2.1439730	1.3616434	15.955925
4.6076817	1.6491037	3.1123607	−0.5843253	3.2552784	2.9642836	3.9017196	16.378067
4.9412389	1.7984060	3.1511551	0.0379101	2.8526594	0.7179441	3.6678706	13.978119
2.9602643	1.6053855	1.9659854	0.4049415	5.2480626	3.1098640	3.5438339	16.467410

TABLE 7-17 Random Numbers Used for Completion Time of Tasks

Task A	Task B	Task C	Task D	Task E	Task F	Task G
0.606883	0.093464	0.125703	0.736216	0.585157	0.033755	0.719628
0.277315	0.682777	0.759930	0.485396	0.288004	0.697372	0.101427
0.725585	0.326737	0.091488	0.718726	0.819744	0.912300	0.910932
0.179915	0.471119	0.072710	0.293896	0.559946	0.441863	0.749723
0.152424	0.240208	0.294833	0.769227	0.786163	0.121520	0.663357
0.168486	0.035771	0.513560	0.880006	0.748794	0.115441	0.953369
0.915682	0.243600	0.610186	0.848375	0.102922	0.009326	0.801494
0.124135	0.682049	0.610019	0.203327	0.081627	0.866440	0.514767
0.342101	0.739733	0.131999	0.569512	0.388688	0.518582	0.204704
0.985961	0.613146	0.914132	0.898415	0.543517	0.091718	0.970920
0.336867	0.616759	0.402409	0.268781	0.913337	0.098700	0.545388
0.583809	0.471045	0.343964	0.278476	0.128413	0.359243	0.341192
0.798033	0.053788	0.467997	0.405734	0.923671	0.587813	0.126547
0.688703	0.028898	0.021365	0.039026	0.483284	0.546590	0.267746
0.959589	0.749079	0.914929	0.729020	0.917082	0.870119	0.652013
0.331024	0.626462	0.697033	0.771629	0.382801	0.702866	0.060994
0.201754	0.233297	0.417021	0.770881	0.034672	0.724181	0.395496
0.633503	0.380850	0.538246	0.326588	0.633842	0.176778	0.346776
0.840578	0.895108	0.071531	0.714916	0.400981	0.243865	0.211002
0.531249	0.463470	0.952944	0.073020	0.345216	0.578557	0.214954

TABLE 7-18 Generated Random Values for Completion Time of Tasks

Task A	Task B	Task C	Task D	Task E	Task F	Task G	Project completion time
3.2707799	1.3400400	2.8529732	0.8156949	4.2147164	0.1713280	3.5813727	12.578237
2.4095244	2.2375281	4.7058017	0.4817430	3.4411537	2.5164544	1.7263661	12.871519
3.5991597	1.7757440	2.6681462	0.7893469	4.9142851	3.3552784	4.3467318	17.991199
2.0844111	1.9638631	2.5438200	0.2291748	4.1505155	1.8540637	3.6733168	13.726170
1.9739182	1.6473216	3.4610715	0.8680251	4.7929741	0.8324813	3.4212035	13.375756
2.0399005	1.0987845	4.0339016	1.0875602	4.6703954	0.8017872	4.6788003	15.272445
4.3768346	1.6527585	4.2793768	1.0147408	2.7347579	–0.3528514	3.8468091	12.258309
1.8453495	2.2365066	4.2789404	0.0851884	2.6055421	3.1097732	3.0369229	13.116367
2.5937050	2.3210996	2.8829455	0.5873924	3.7176888	2.0464701	2.1752441	12.854208
5.1966454	2.1435484	5.3668740	1.1363712	4.1090447	0.6695408	4.8949117	17.013691
2.5794094	2.1482732	3.7532931	0.1919199	5.3618234	0.7108243	3.1137570	13.914087
3.2112634	1.9637700	3.5987720	0.2064942	2.8660028	1.6399625	2.5912269	12.272226
3.8344438	1.1952375	3.9198922	0.3809357	5.4304724	2.2215315	1.8570525	14.538738
3.4917621	1.0511701	1.9732268	–0.3812494	3.9582012	2.1167830	2.3806975	12.998614
4.7463211	2.3356456	5.3719798	0.8047524	5.3859459	3.1270236	3.3903168	18.985253
2.5633476	2.1610288	4.5154829	0.8719888	3.7023022	2.5322679	1.4531975	13.075239
2.1647959	1.6361127	3.7908605	0.8707520	2.1834205	2.5949514	2.7353984	12.175383
3.3407015	1.8485931	4.0957866	0.2755382	4.3416040	1.0723773	2.6064022	13.209678
3.9968036	2.6271607	2.5352335	0.7837149	3.7496031	1.3063624	2.1972483	13.877178
3.0782144	1.9542634	5.6744610	–0.2269675	3.6021733	2.1978306	2.2108603	13.458358

7.8. PROBLEMS

7-1. Using the linear congruential generator given by Equation 7-1, generate a stream of ten random numbers based on the following model parameters: $a = 7$, $b = 5$, $c = 26$, and $I_0 = 3$.

7-2. Using the linear congruential generator given by Equation 7-1, generate a stream of ten random numbers based on the following model parameters: $a = 7$, $b = 5$, $c = 345$, and $I_0 = 3$.

7-3. Using the linear congruential generator given by Equation 7-1, generate a stream of ten random numbers based on the following model parameters: $a = 77$, $b = 345$, $c = 26$, and $I_0 = 3$.

7-4. Perform a parametric analysis of the linear congruential generator given by Equation 7-1 by varying the values of a, b, c, and I_0 to investigate their importance. Based on your results, discuss the importance of these parameters.

7-5. If a random-number generator is available to generate uniformly distributed, 0 to 1 random numbers, (a) how could it be used to generate rolls of a fair die that has the six discrete values? and (b) how could it be used to generate rolls of a loaded die that has a probability of 0.25 for a 1, 0.35 for a 4, and equal probabilities for the remaining four numbers?

7-6. A random variable x can take on values of 1, 2, 3, 4, and 5. A random variable y can take on values of 2, 4, and 6. If a random sequence of x is {2, 4, 3, 1, 5, 4, 3, 4} and the corresponding sequence of y is {4, 2, 6, 6, 2, 2, 6, 2}, what is the transformation rule?

7-7. The roll of a die is used to simulate the toss of a coin. If the roll produces an even value, a head is assumed. If the roll produces an odd value, a tail is assumed. Indicate the outcomes of coin flips if the random sequence of rolls is {2, 5, 4, 2, 1, 6, 3, 3, 6}.

7-8. The density function of a continuous random variable x is:

$$f_X(x) = \begin{cases} 3x^2/26 & \text{for } 1 \le x \le 3 \\ 0 & \text{otherwise} \end{cases}$$

Use the function to generate values of x for the following sequence of uniform (0 to 1) variates:

$$u_i = \{0.63, 0.21, 0.90, 0.56, 0.35, 0.48, 0.06\}$$

Compare the population mean and the mean of the generated values of x.

7-9. A discrete random variable X has a Bernoulli distribution with parameter $p = 0.3$. Using the following uniform variates, generate Bernoulli variates using Equation 7-20. Determine the sample proportion and compare it to the population value.

$$u_i = \{0.79, 0.34, 0.27, 0.55, 0.06, 0.13, 0.70, 0.18, 0.16, 0.50, 0.19, 0.63\}$$

7-10. On any day, assume that the probability of rainfall is 0.3. Assume that the occurrence of rainfall on any day can be simulated by a Bernoulli variate, with 0 indicating no rain on that day and 1 indicating rainfall occurring. Using the following uniform variates and the inverse transformation method, generate 2 weeks of the rainfall indicator. Compare the sample and population probabilities.

$$u_i = \{0.71, 0.53, 0.29, 0.42, 0.02, 0.20, 0.87, 0.43, 0.26, 0.92, 0.34, 0.53, 0.15, 0.76\}$$

7-11. A random variable that can take on values of x and y has $P(x) = 0.2$ and $P(y) = 1 - P(x) = 0.8$. Assume that an experiment was conducted with three trials per experiment. Use the following 30 uniform variates to generate 10 binomial variates. Compare the sample and population probabilities.

$$u_i = \{0.47, 0.06, 0.02, 0.77, 0.44, 0.73, 0.40, 0.01, 0.53, 0.00, 0.83, 0.71, 0.74, 0.34, 0.04, 0.62, 0.82, 0.90, 0.97, 0.29, 0.61, 0.32, 0.01, 0.18, 0.22, 0.95, 0.67, 0.64, 0.68, 0.67\}$$

7-12. Generate 10 binomial variates with $p = 0.75$ and $n = 2$ using the following uniform variates. Compare the population and sample probabilities.

$$u_i = \{0.17, 0.35, 0.92, 0.24, 0.20, 0.02, 0.78, 0.34, 0.62, 0.29, 0.43, 0.35, 0.51, 0.67, 0.82, 0.47, 0.90, 0.01, 0.13, 0.68\}$$

7-13. Generate 10 binomial variates with $p = 0.65$ and $n = 3$ using the first three columns of uniform variates in Table 7-4a. Compare the population and sample probabilities.

7-14. A discrete random variable (N) has a geometric distribution with parameter $p = 0.5$. Using the first ten values of u_i in Problem 7-17, generate values of N using the inverse transformation method. Compare these to values generated with Equation 7-23.

7-15. The following is a sequence of independent trials of successes (S) and failures (F). Identify the underlying geometric distribution. Compare the sample values of the random variable X, which is the number of trials to the first success, and the theoretical values for the population.

SFFSFSSFFFSFFSSFSFFFFSSFF

7-16. In a certain manufacturing process, 1 in every 12 items has a defect. (a) Compute the probability that exactly three items will be inspected before an inspector finds a defective items. (b) Compute the probability that no more than four items will be inspected before the first defective item is found. (c) Use Equation 7-23 and the following uniform variates to generate values of x:

$$u_i = \{0.76, 0.33, 0.47, 0.60, 0.08, 0.39, 0.84, 0.27\}$$

7-17. A random variable is Poisson distributed with $\lambda = 1.5$. Using the following random variates from a uniform distribution, generate Poisson variates. Compute the sample mean and standard deviation of the Poisson variates and compare them to the population values. Comment on the differences. Use:

$$u_i = \{0.43, 0.53, 0.65, 0.66, 0.48, 0.87, 0.67, 0.66, 0.85, 0.81, 0.44, 0.37, 0.54, 0.71, 0.27, 0.15, 0.34, 0.89, 0.52, 0.64\}$$

7-18. Generate 20 Poisson variates with parameter $\lambda t = 1.5$. Use the inverse transformation method and the 20 uniform variates of Problem 7-12. Compare the population and sample probabilities.

7-19. On the average, three radioactive particles pass through a counter per millisecond. Use the following uniform variates to generate a time sequence of the number of particles in 15 seconds. Compare the mean of the generated sequence and the expected value. Use:

$$u_i = \{0.16, 0.37, 0.82, 0.51, 0.70, 0.23, 0.44, 0.92, 0.78, 0.32, 0.03, 0.54, 0.63, 0.22, 0.15\}$$

7-20. Barges arrive at a certain lock on the Mississippi River at an average rate of two per hour. Using the following uniform variates, simulate the number of barges assuming that one day has 12 hours of lock operation. Use:

$$u_i = \{0.84, 0.47, 0.23, 0.59, 0.10, 0.66, 0.35, 0.14, 0.72, 0.51, 0.88, 0.12\}$$

Compare the population mean with the mean of the generated hourly arrivals.

7-21. Transform the following uniform variates from a $U(0,1)$ population to variates of a second uniform distribution X, which is $U(4,10)$. Use:

$$u_i = \{0.62, 0.31, 0.85, 0.76, 0.09, 0.43\}$$

7-22. A probability density function $f_X(x)$ for the random variable consists of two sections, each a constant as follows:

$$f_X(x) = \begin{cases} 0.20 & \text{for } 2 \le x \le 5 \\ 0.08 & \text{for } 5 < x \le 10 \\ 0 & \text{otherwise} \end{cases}$$

Transform the following uniform variates $U(0,1)$ to values of x:

$$u_i = \{0.03, 0.79, 0.52, 0.41, 0.86, 0.22, 0.90\}$$

7-23. Use the inverse transformation method to transform the following uniform variates $U(0,1)$ to normal variates $N(3,2^2)$ with mean of 3 and standard deviation of 2:

$$u_i = \{0.68, 0.04, 0.76, 0.37\}$$

7-24. Use the inverse transformation method to transform the following uniform variates $U(2,4)$ to normal variates $N(5,2)$ with mean of 5 and standard deviation of 2:

$$u_i = \{2.76, 3.47, 2.06, 3.84\}$$

7-25. What uniform variate u_i for $U(0,1)$ was used to generate a normal variate of 7.81 from $N(6,3^2)$ with mean of 6 and standard deviation of 3?

7-26. Assume that the mean weekly dissolved oxygen concentration can be represented by a normal distribution $N(6.5,1.5^2)$ with mean of 6.5 and standard deviation of 1.5. Simulate 6 weekly mean values using the following uniform variates:

$$u_i = \{0.55, 0.38, 0.82, 0.27, 0.64\}$$

7-27. Assume the random variable X has a lognormal distribution and $y = \log(x)$ with mean $\overline{Y}$ of 1.32 and standard deviation S of 0.17. Generate values of x for the following uniform variates:

$$u_i = \{0.13, 0.71, 0.44, 0.60, 0.27\}$$

7-28. Assume that floods are lognormally distributed with a mean of X of 254 m^3/sec and a standard deviation of X of 38 m^3/sec. Use the following uniform variates to generate values of x that are lognormally distributed:

$$u_i = \{0.39, 0.61, 0.50, 0.13, 0.82\}$$

7-29. The mean (μ_Y) of $Y = \ln(X)$ and standard deviation (σ_Y) of $Y = \ln(X)$ are 2.76 and 0.57, respectively. Generate values of x that have a lognormal distribution using the following uniform variates:

$$u_i = \{0.03, 0.82, 0.37, 0.54\}$$

7-30. Using inverse transformation, generate values of exponential variates for the parameter $\lambda = 1.4$ with the following uniform variates:

$$u_i = \{0.34, 0.83, 0.02\}$$

7-31. Assume the following values of x_i are a sample from an exponential distribution: $x_i = \{0.27, 0.62, 1.12, 1.35\}$. Generate exponential variates that are from the same population using the following uniform variates:

$$u_i = \{0.47, 0.92, 0.23, 0.52, 0.17\}$$

7-32. The change in the length of a rod due to axial force P is given by:

$$\Delta L = \frac{PL}{AE}$$

where L is the length of the rod, P is the applied axial force, A is the cross-sectional area of rod, and E is the modulus of elasticity. Using 20 simulation cycles, determine the mean and variance of ΔL assuming noncorrelated random variables with the following probabilistic characteristics:

Random variable	Mean value	Coefficient of variation	Distribution type
P	100 kips	0.35	Lognormal
L	20 in.	0.05	Normal
E	30,000 ksi	0.10	Lognormal
A	1 in.^{2}	0.05	Normal

7-33. For the rod in Problem 7-32, study the effect of increasing the number of simulation cycles on the estimated mean and variance of the deformation. Use the following numbers of simulation cycles: 20, 100, 500, 1000, 2000, and 10,000. Provide your results in the form of plots of estimated statistics as a function of the number of simulation cycles.

7-34. For the rod in Problems 7-32 and 7-33, plot two frequency histograms for the deformation of the rod based on 20 and 100 simulation cycles. Suggest a distribution type for the deformation.

7-35. For the rod in Problems 7-32 and 7-33, plot a frequency histogram for the deformation of the rod based on 1000 simulation cycles. Suggest a distribution type for the deformation.

7-36. Compare the results of Problems 7-34 and 7-35 to the rod in Problems 7-32 and 7-33. Provide a discussion.

7-37. The ultimate moment capacity, M, of an under-reinforced concrete rectangular section is given by:

$$M = A_s f_y \left(d - \frac{a}{2} \right)$$

where

$$a = \frac{A_s f_y}{0.85 b f'_c}$$

in which the following are random variables: A_s is the cross-sectional area of the reinforcing steel, f_y is the yield stress (strength) of the steel, d is the distance from the reinforcing steel to the top fibers of the beam, b is the width of the beam, and f'_c is the ultimate stress (strength) of the concrete. If the random variables are assumed to be statistically noncorrelated, determine the mean and variance of the moment capacity using the following information:

Random variable	Mean value	Coefficient of variation	Distribution type
A_s	0.25 in.^{2}	0.10	Lognormal
f_y	40,000 psi	0.10	Normal
d	20 in.	0.05	Lognormal
b	12 in.	0.05	Normal
f'_c	4000 psi	0.20	Lognormal

Use 100 simulation cycles. Is this a sufficient number of cycles? Why? Provide a discussion.

7.9. SIMULATION PROJECTS

This section provides additional work items for the problem statements of the four simulation projects introduced at the end of previous chapters.

7.9.1. Structural Beam Study

Using the project information provided in previous sections on this project, perform parametric analysis on results produced in Chapter 5 for this project. Investigate the effects of individually changing mean values and standard deviations of input random variables. Investigate the effects of individually changing distribution types of input random variables.

7.9.2. Stream Erosion Study

Using the mean and standard deviation of each of the variables in Table C-1, generate a sample of 62 values each for the random variables that have a normal distribution. Compute the correlation coefficients between each pair of variables: (x_1,x_2), (x_1,x_3), (x_2,x_3), (x_1,y), (x_2,y), and (x_3,y), for both the actual and generated data. Discuss any differences in both the moments and the correlations between the generated data and sample data.

7.9.3. Traffic Estimation Study

Using the mean and standard deviation of each of the variables in Table C-2, generate a sample of 45 values each for the random variables that have a normal distribution. Compute the correlation coefficients between each pair of variables: (x_1,x_2), (x_1,x_3), (x_2,x_3), (x_1,y), (x_2,y), and (x_3,y), for both the actual and generated data. Discuss any differences in both the moments and the correlations between the generated data and sample data.

7.9.4. Water Evaporation Study

Using the mean and standard deviation of each of the variables in Table C-3, generate a sample of 71 values each for the random variables that have a normal distribution. Compute the correlation coefficients between each pair of variables: (x_1,x_2), (x_1,x_3), (x_2,x_3), (x_1,y), (x_2,y), and (x_3,y), for both the actual and generated data. Discuss any differences in both the moments and the correlations between the generated data and sample data.

Chapter 8
Fundamentals of Statistical Analysis

TABLE OF CONTENTS

8.1. INTRODUCTION

Values of random variables obtained from sample measurements are commonly used in making important engineering decisions. For example, samples of river water are collected to estimate the average level of a pollutant in the entire river at that location. Samples of stopping distances are used to develop a relationship between the speed of a car at the time the brakes are applied and the distance traveled before the car comes to a complete halt. The average of sample measurements of the compressive strength of concrete collected during the pouring of a large concrete slab, such as the deck of a parking garage, is used to help decide whether or not the deck has the strength specified in the design specifications. It is important to recognize the random variables involved in these cases. In each case, the individual measurements or samples are values of a random variable, and the computed mean is also the value of a random variable. For example, the transportation engineer measures the stopping distance; each measurement is a sample value of the random variable. If ten measurements are made for a car stopping from a speed of 50 mph, then the sample consists of ten values of the random variable. Thus, there are two random variables in this example: the stopping distance and the estimated mean of the stopping distance. This is also true for the water-quality-pollutant and compressive-strength examples.

The estimated mean for a random variable is considered by itself to be a random variable, because different samples about the random variable can produce different estimated mean values, thus the randomness in the estimated mean. When a sample of n measurements of a random variable is collected, the n values are not necessarily identical. The sample is characterized by variation. For example, let's assume that five independent estimates of the compressive strength of the concrete in a parking garage deck are obtained from samples of the concrete obtained when the concrete was poured. For illustration purposes, let's assume that the five compressive strength measurements are 3250, 3610, 3460, 3380, and 3510 psi. This produces a mean of 3442 psi and a standard deviation of 135.9 psi. Assume that another sample of five measurements of concrete strength was obtained from the same concrete pour; however, the values were 3650, 3360, 3328, 3420, and 3260 psi. In this case, the estimated mean and standard deviation are 3404 and 149.3 psi, respectively; therefore, the individual measurements and the mean are values of two different random variables: X and $\overline{X}$.

It would greatly simplify decision making if the sample measurements were identical (i.e., there is no sampling variation so the standard deviation was zero). Unfortunately, that is never the case, so decisions must be made in spite of the uncertainty. For example, let's assume in the parking garage example that the building code requires a mean compressive strength of 3500 psi. Because the mean of 3442 psi based on the first sample is less than the required 3500 psi, should we conclude that the garage deck does not meet the design specifications? Unfortunately, decision making is not that simple. If a third sample of five measurements had been randomly collected from other locations on the garage deck, the following values are just as likely to have been obtained: 3720, 3440, 3590, 3270, and 3610 psi. This sample of five produces a mean of 3526 psi and a standard deviation of 174.4 psi. In this case, the mean exceeds the design standard of 3500 psi. Because the sample mean is greater than the specified value of 3500 psi, can we conclude that the concrete is of adequate strength? Again, we cannot conclude with certainty that the strength is adequate any more than we could conclude from the first sample that the strength was inadequate. The fact that different samples lead to different means is an indication that we cannot conclude that the design specification is not met just because the sample mean is less than the design standard. We need to have more assurance.

The need then is for a systematic decision process that takes into account the variation that can be expected from one sample to another. The decision process must also be able to reflect the risk of making an incorrect decision. Decisions can be made using, for example, hypothesis testing as described in Chapter 9.

8.1.1. Samples and Populations

The data that are collected represent sample information, but this information is not complete by itself, and predictions are not made directly from the sample. The intermediate step between sampling and prediction is identification of the underlying population. The sample is used to identify the population and then the population is used to make predictions or decisions. This sample-to-population-to-prediction sequence is true for the univariate methods of this chapter or for the bivariate and multivariate methods that follow.

A known function or model is most often used to represent the population. The normal and lognormal distributions are commonly used to model the population for a univariate problem. For bivariate and multivariate prediction, linear ($\hat{Y} = a + bX$) and power ($\hat{Y} = aX^b$) models are commonly assumed functions for representing the population, where $\hat{Y}$ is the predicted value of dependent variable Y, X is the independent random variable, and a and b are model parameters. When using a probability function to represent the population, it is necessary to estimate the parameters. For example, for the normal distribution, we must estimate the location and scale parameters, which are the mean and standard deviation, respectively. For the exponential distribution, the rate (λ) is a distribution parameter that needs to be estimated and is not equal to the mean or standard deviation. When using the linear or power models as the population, it is necessary to estimate the coefficients a and b. In both the univariate and multivariate cases, they are called *sample estimators* of the population parameters.

8.2. ESTIMATION OF PARAMETERS

In developing models for populations, models can be classified as univariate, bivariate, or multivariate, with parameters that provide the needed complete definition of a model. Models can have one, two, or more parameters. For example, the normal distribution as a univariate model has two parameters, the exponential distribution has one parameter, and the bivariate power model ($\hat{Y} = aX^b$) has two parameters. Samples are used to develop a model that can adequately represent the population and to estimate the parameters of the population model. The parameters can be estimated in the form of point estimates (single values) or interval estimates (ranges of values) using the samples. The equations or methods used to estimate the parameters are called *estimators*. In this section, estimators and their properties are introduced. The statistical uncertainty associated with the estimators is considered in the next sections for statistical decision making using hypothesis testing (see Sections 9.1 to 9.3). Interval estimation is discussed in Section 11.1.

8.2.1. Properties of Estimators

Several properties of estimators are of interest to engineers. The concepts that are widely used, and sometimes misunderstood, include accuracy, bias, precision, consistency, efficiency, and sufficiency. Because of the lack of uniform terminology, definitions that represent the most common usage are given here.

8.2.1.1. Bias

An estimate of a parameter θ made from the sample statistic is said to be an unbiased estimate if the expected value of the sample quantity $\hat{\theta}$ is θ; that is,

$$E(\hat{\theta}) = \theta \tag{8-1}$$

where $E(\cdot)$ denotes the mathematical expectation. The bias is defined as $[E(\hat{\theta}) - \theta]$. Therefore, bias deprives a statistic result of representativeness by systematically distorting it. It is not the same as a random error that may distort at any one occasion but balances out on the average. It is important to note that bias is a systematic error.

Figure 8-1 shows the results of four experiments, with each experiment being repeated six times. It is known that the true (i.e., population) mean of each experiment is 15. The sample means for the four experiments are 15, 24, 7, and 14. Thus, experiment A is unbiased because its expected value equals the true value. Experiments B, C, and D show varying degrees of bias. Whereas experiment B has a positive bias of 9, the biases of experiments C and D are negative; that is, experiment B tends to overestimate θ and experiments C and D underestimate θ. A question that is discussed in Sections 9.2 and 9.3 is whether or not the biases are significant.

8.2.1.2. Precision

In addition to the systematic variation, variation in an observation may be due to random error. Random error is measured by fluctuations in the value of a variable that occur when the same experiment is performed more than once. It may be the result of (1) errors in judgment, (2) uncontrollable variation in environmental conditions, (3) differences due to deficiencies in defining the quantity being measured, or (4) intrinsically random process.

Precision can be defined as the ability of an estimator to provide repeated estimates that are very close together. Therefore, precision may be expressed in terms of the variance of an estimator, with a large variance signifying lack of precision and a small variance signifying high precision. Obviously, absolute precision implies a zero variance, a level that practically cannot be achieved.

Using the variance of the observations, as measured about the sample mean, the precision of each experiment is shown in Figure 8-1. Experiments A and B show considerably more precision (i.e., they have lower variances among the observations) than do experiments C and D. Experiment C has the largest variation (i.e., it is the least precise experiment). In comparing experiments A and B, they have the same level of precision but experiment A is unbiased, whereas B is highly biased.

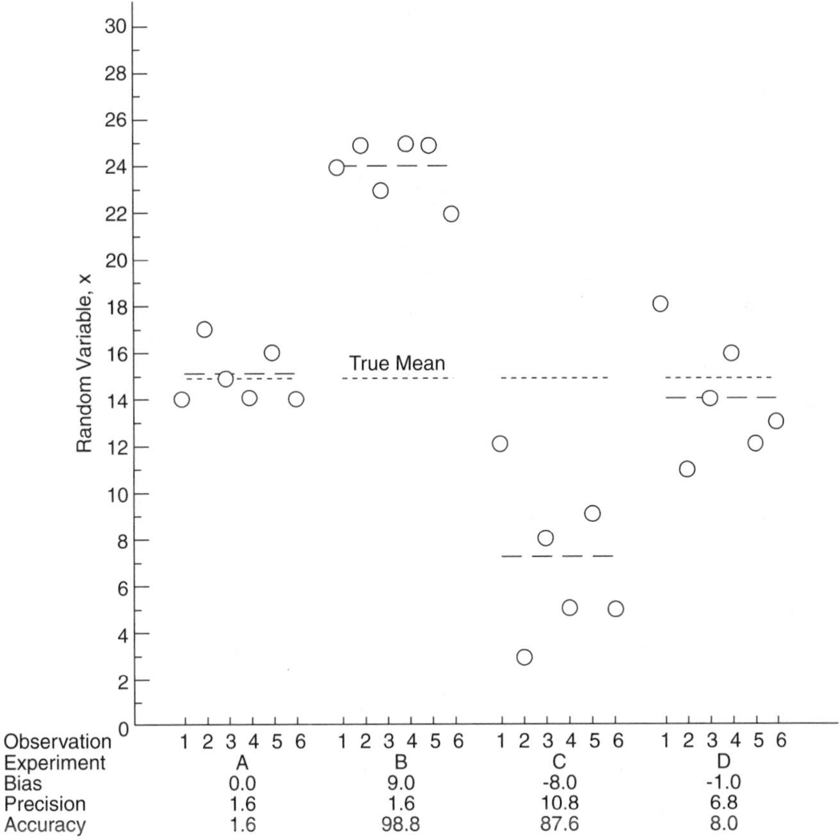

FIGURE 8-1 Bias, precision, and accuracy of experimental observations.

8.2.1.3. *Accuracy*

Accuracy can be defined as the closeness or nearness of the measurements to the true or actual value of the quantity being measured. The concept of accuracy can be considered to encompass both the precision and bias.

Definitions of accuracy often imply that the term should only be used in a dichotomous manner; that is, a measurement or a statistic is either accurate or inaccurate. In practice, the term *accuracy* is used for comparisons, and quantitative estimates of accuracy are used to describe a measurement or statistic; that is, accuracy is a characteristic that should be measured on a continuous scale, rather than as a discrete variable.

The variance of the observations in Figure 8-1, as measured about the true value of the statistic (i.e., 15), is used as a measure of the accuracy. The values of the four experiments are shown in Figure 8-1. Experiment B is the least accurate; in addition to its high bias, it is imprecise. Experiment D is reasonably accurate, in spite of its noticeable lack of precision. Experiment A is the most accurate.

8.2.1.4. Comparison of Bias, Precision, and Accuracy

It is interesting to distinguish among accuracy, bias (systematic error), and precision (random error). The measurements using experiment A are all close to the true value, which indicates no significant bias and high precision; thus, experiment A is highly accurate. There is little variation among the six observations in experiment B, which indicates that it provides precise measurements; however, because the values are consistently higher than the true value, the experiment is biased. The significant bias contributes to its poor accuracy. Experiment C is both biased and imprecise because the systematic and random variations are both significant; thus, it is also highly inaccurate. Experiment D is imprecise but relatively unbiased. In summary, inaccuracy can result from either a bias or lack of precision. Although accuracy can be of greatest overall concern, one should examine the bias and precision components to identify the most likely sources of the inaccuracy.

8.2.1.5. Mean Square Error

One objective of data analysis is to explain variation in a set of measurements. Accuracy reflects both systematic and random errors, and both sources of variation must be assessed when attempting to select one estimator from among many; specifically, a biased estimator may be preferable to an unbiased estimator if the precision of the biased estimator is significantly better than that of the unbiased estimator. The variance of an estimator may be viewed as a measure of its precision. The mean square error (MSE) can be defined as the expected value of the square of the deviation of the estimate from the true value of the quantity being estimated. It is equal to the variance of the estimate plus the square of the bias. Thus, the MSE is a measure of accuracy.

8.2.1.6. Consistency, Sufficiency, and Efficiency

In addition to unbiasness, there are other properties that a statistic should have in order to be a good estimator. The estimator should be both consistent and sufficient, and the method of estimation should be relatively efficient.

An estimator is consistent if the probability (P) that $\hat{\theta}$ will deviate from θ more than any fixed amount $\varepsilon > 0$, no matter how small, approaches zero as the sample size (n) becomes larger and larger. Mathematically, consistency is given by:

$$P(|\hat{\theta} - \theta| \leq \varepsilon) \to 1 \qquad \text{as } n \to \infty \tag{8-2}$$

in which n is the sample size. In contrast with biasedness, consistency is a "sample size" property of an estimator. It can be shown that unbiased estimators are consistent estimators. However, a consistent estimator is not necessarily unbiased; for example, the sample variance is consistent but must be corrected for its biasedness by multiplying it by the factor $n/(n - 1)$.

An estimate $\hat{\theta}$ is said to be sufficient if it exhausts all possible information on θ in a sample of any size. Sufficiency implies that no other estimator computed from the same sample can provide additional information about θ. Sufficiency is the most desirable property of an estimator, and such estimators are said to be optimum.

Efficiency is an important criterion for evaluating the quality of an estimator. Because it is desirable to have an estimate that is close to the true value and the variance is a measure of closeness, the efficiency of an estimator is inversely proportional to its variance. A consistent estimate, $\hat{\theta}_1$, is said to be more efficient than another estimate, $\hat{\theta}_2$, if $\mathrm{Var}(\hat{\theta}_1) < \mathrm{Var}(\hat{\theta}_2)$, where $\mathrm{Var}(\cdot)$ is the variance. An estimate is said to be efficient if it has the smallest variance of all available estimators.

8.2.2. Estimation of Moments

In engineering, it is common to characterize a sample of interest by certain estimated moments. These measures, which take numerical values, can be easily communicated to others and quantify the main characteristics of the sample. The concept of moments was introduced in Chapter 3. The first moment is a commonly used quantity that provides a measure of central tendency, or mean. The second central moment or variance is a common measure of dispersion. In this section, methods for estimating these two moments are defined and discussed. A formal discussion of these measures is provided in Chapter 3.

The mean or average value of n observations, if all observations are given equal weights, is given by:

$$\bar{X} = \frac{1}{n}\sum_{i=1}^{n} x_i \tag{8-3}$$

where x_i is a sample point, and $i = 1, 2, \ldots, n$. Although this moment conveys certain information about the underlying sample, it does not completely characterize the underlying variable. Two variables can have the same mean but different histograms. Other characteristics are also important and necessary. The estimator of the mean as given by Equation 8-3 was discussed in Chapter 2.

The dispersion measure (variance) describes the level of scatter in the data about the central tendency location. For n observations in a sample that are given equal weight, the variance (S^2) is given by:

$$S^2 = \frac{1}{n-1}\sum_{i=1}^{n}(x_i - \bar{X})^2 \tag{8-4}$$

The units of the variance are the square of the units of the parameter or variable x; for example, if the variable is measured in pounds per square inch (psi), the variance has units of $(\mathrm{psi})^2$. Computationally, the variance of a sample can be determined using the following alternative equation:

$$S^2 = \frac{1}{n-1}\left[\sum_{i=1}^{n} x_i^2 - \frac{1}{n}\left(\sum_{i=1}^{n} x_i\right)^2\right] \tag{8-5}$$

Equation 8-5 provides the same answer as Equation 8-4 when computations are made using an appropriate number of significant digits. Equations 8-4 and 8-5 provide an estimate of the variance, which is an average of the squared difference between the x values and their average value. In these equations, $(n - 1)$ is used to compute the average, instead of n, in order to obtain an unbiased estimate of the variance. The concept of variance was introduced in Chapter 2.

Two commonly used measures are based on the variance: the standard deviation and coefficient of variation. By definition, the standard deviation (S) is the square root of the variance as follows:

$$S^2 = \sqrt{\frac{1}{n-1}\left[\sum_{i=1}^{n} x_i^2 - \frac{1}{n}\left(\sum_{i=1}^{n} x_i\right)^2\right]} \tag{8-6}$$

The coefficient of variation (COV or δ) is a normalized quantity based on the standard deviation and mean value. Therefore, it is dimensionless. The COV is defined as:

$$\text{COV} = \frac{S}{\bar{X}} \tag{8-7}$$

The standard deviation and coefficient of variation were also introduced in Chapter 2.

8.2.3. Method-of-Moments Estimation

The method of moments is one method of estimating population parameters using the moments of samples. The method of moments is frequently used to provide estimates of the parameters of a distribution, primarily because of its computational simplicity. Equations that relate the moments of a sample to the parameters of a distribution can be derived. Estimates obtained by the method of moments are always consistent, but they may not be efficient in a statistical sense.

Because of its structural simplicity, the uniform distribution is used to illustrate the method of moments. The uniform distribution is a function of two parameters, α and β. The location parameter α defines the lower limit of the distribution, while the scale parameter β indicates the spatial extent of the density function which is given by:

$$f_X(x) = \begin{cases} \dfrac{1}{b-a} & a \le x \le b \\ 0 & \text{otherwise} \end{cases} \tag{8-8}$$

where a and b are the estimators of the parameters α and β, respectively. The sample mean can be related to the definition of the mean of the population (Chapter 3) as follows:

$$\bar{X} = \int_{-\infty}^{\infty} x f_X(x)dx = \int_a^b x\left(\frac{1}{b-a}\right)dx = \frac{1}{b-a}\int_a^b x\,dx$$
$$= \frac{1}{b-a}\left(\frac{x^2}{2}\right)\Bigg|_a^b = \frac{b^2-a^2}{2(b-a)} = \frac{(b-a)(b+a)}{2(b-a)} = \frac{b+a}{2} \tag{8-9}$$

The sample variance can be related to the variance of the population. The following relationship simplifies the computation:

$$\mu = \mu_2' - (\mu_1)^2 \tag{8-10}$$

where μ_1 is the mean and μ_2' is the second moment about the origin. Thus, Equation 8-10 becomes:

$$S^2 = \int_a^b x^2\left(\frac{1}{b-a}\right)dx - \left(\frac{b+a}{2}\right)^2 = \frac{1}{b-a}\left(\frac{x^3}{3}\right)\Bigg|_a^b - \left(\frac{b+a}{2}\right)$$
$$= \frac{b^3-a^3}{3(b-a)} - \left(\frac{b+a}{2}\right)^2 = \frac{(b-a)^2}{12} \tag{8-11}$$

Equations 8-9 and 8-11 provide the means for evaluating the parameters of the uniform density function, which in turn provides two equations with two unknowns, *b* and *a*, as follows:

$$\bar{X} = \frac{b+a}{2} \tag{8-12}$$

$$S^2 = \frac{(b-a)^2}{12} \tag{8-13}$$

Solving Equations 8-12 and 8-13 for *a* and *b* gives:

$$a = \bar{X} - S\sqrt{3} \tag{8-14}$$

and

$$b = \bar{X} + S\sqrt{3} \tag{8-15}$$

Table 8-1 provides a summary of the relationships between the parameters of commonly used distributions and the mean and variance. These relationships can be developed using an approach similar to the method used for the uniform distribution and using the concepts in Chapters 3 and 5.

TABLE 8-1 Relationships for the Method of Moments: (a) Discrete Distributions, (b) Continuous Distributions

Distribution type	Probability mass or density function	Parameters	Relationships
	(a) Discrete Distributions		
Bernoulli	$P_X(x) = \begin{cases} p & x = 1 \\ 1-p & x = 0 \\ 0 & \text{otherwise} \end{cases}$	p	$\bar{X} = p$ $S^2 = p(1-p)$
Binomial	$P_X(x) = \begin{cases} p(1-p)^{x-1} & x = 0, 1, 2, 3, \ldots \\ 0 & \text{otherwise} \end{cases}$	p	$\bar{X} = Np$ $S^2 = Np(1-p)$
Geometric	$P_X(x) = \begin{cases} p(1-p)^{x-1} & x = 0, 1, 2, 3, \ldots \\ 0 & \text{otherwise} \end{cases}$	p	$\bar{X} = 1/p$ $S^2 = (1-p)/p^2$
Poisson	$P_{X_t}(x) = \begin{cases} \dfrac{(\lambda t)^x \exp(-\lambda t)}{x!} & x = 0, 1, 2, 3, \ldots \\ 0 & \text{otherwise} \end{cases}$	λ	$\bar{X} = \lambda t$ $S^2 = \lambda t$

(b) Continuous Distributions

Uniform	$f_X(x) = \begin{cases} \dfrac{1}{b-a} & \text{for } a \le x \le b \\ 0 & \text{otherwise} \end{cases}$	a, b	$\overline{X} = (a+b)/2$ $S^2 = \dfrac{1}{12}(b-a)^2$
Normal	$f_X(x) = \dfrac{1}{\sigma\sqrt{2\pi}} \exp\left[-\dfrac{1}{2}\left(\dfrac{x-\mu}{\sigma}\right)^2\right] \quad \text{for } -\infty < x < \infty$	μ, σ	$\overline{X} = \mu$ $S^2 = \sigma^2$
Lognormal	$f_X(x) = \dfrac{1}{x\sigma_Y\sqrt{2\pi}} \exp\left[-\dfrac{1}{2}\left(\dfrac{\ln x-\mu_Y}{\sigma_Y}\right)^2\right] \quad \text{for } 0 < x < \infty$	μ_Y, σ_Y	$\overline{X} = \exp(\mu_Y + 0.5\sigma_Y^2)$ $S^2 = \mu_Y^2 [\exp(\sigma_Y^2) - 1]$
Exponential	$f_X(x) = \begin{cases} \lambda\exp(-\lambda t) & \text{for } x \ge 0 \\ 0 & \text{otherwise} \end{cases}$	λ	$\overline{X} = 1/\lambda$ $S^2 = 1/\lambda^2$

Example 8-1: Method of Moments for the Uniform Distribution

To illustrate the method of moments, consider the sample histogram of Figure 8-2. This sample was introduced as sample number 2 in Section 2.6.1 (in Table 2-2 and Figure 2-18). It is shown that the sample mean and standard deviation are 10 and 3.0625, respectively. Using Equations 8-14 and 8-15, the parameters are estimated to be $a = 4.7$ and $b = 15.3$. The density function $f_X(x)$ is thus given by:

$$f_X(x)\begin{cases} 0.0943 & 4.7 \le x \le 15.3 \\ 0 & \text{otherwise} \end{cases} \tag{8-16}$$

The population density function of Equation 8-16 is shown in Figure 8-2. Although the population density function differs from the sample probability histogram, the population should be used to make probability statements about the random variables X. For example, the probability that X is between 7 and 10 is $0.0943(10 - 7) = 0.283$.

Example 8-2: Method of Moments for the Normal Distribution

For the normal distribution, an estimator for the location parameter (μ) can be found by equating the population and the sample means:

$$\bar{X} = \int_{-\infty}^{\infty} x\, f_X(x)dx = \int_{-\infty}^{\infty} x\left[\frac{1}{\sigma\sqrt{2\pi}}\exp\left(-\frac{1}{2}\left(\frac{x-\mu}{\sigma}\right)^2\right)\right]dx \tag{8-17}$$

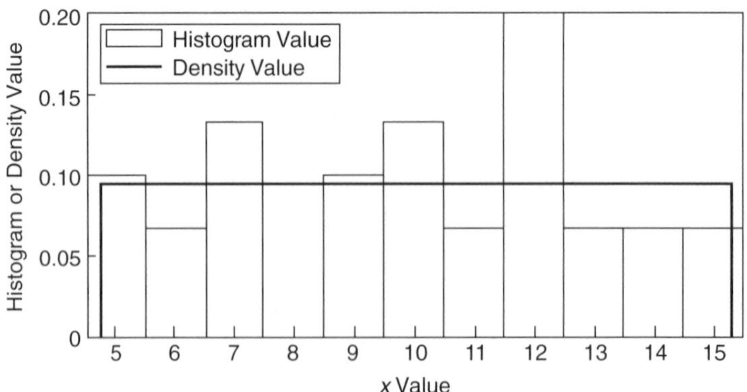

FIGURE 8-2 Uniform probability function fit using the method of moments.

Thus, the location parameter (μ) equals the mean ($\overline{X}$). To derive an estimator for the scale parameter (σ) the sample variance can be equated to the definition of the variance:

$$S^2 = \int_{-\infty}^{\infty} (x-\mu)^2 f_X(x)dx = \int_{-\infty}^{\infty} (x-\mu)^2 \left[\frac{1}{\sigma\sqrt{2\pi}} \exp\left(-\frac{1}{2}\left(\frac{x-\mu}{\sigma}\right)^2\right)\right]dx \tag{8-18}$$
$$= \sigma^2$$

Thus, the scale parameter (σ) equals the standard deviation of the sample (S).

8.2.4. Maximum-Likelihood Estimation

The most common statistical method of parameter estimation is the method of maximum likelihood. This method is based on the principle of calculating values of parameters that maximize the probability of obtaining the particular sample.

The likelihood of the sample is the total probability of drawing each item of the sample. The total probability is the product of all the individual item probabilities. This product is differentiated with respect to the parameters, and the resulting derivatives are set to zero to achieve the maximum.

Maximum-likelihood solutions for model parameters are statistically efficient solutions, meaning that parameter values have minimum variance. This definition of a best method, however, is theoretical. Maximum-likelihood solutions do not always produce solvable equations for the parameters. The following examples illustrate easy to moderately difficult solutions. For some distributions, including notably the normal distribution, the method of moments and maximum-likelihood estimation produce identical solutions for the parameters.

Example 8-3: Maximum-Likelihood Estimation for the Exponential Distribution

For this example we will find the maximum-likelihood estimate of the parameter λ in the density function λexp(−λx). Consider a sample of n items: $x_1, x_2, x_3, \ldots, x_n$. By definition, the likelihood function, L, is:

$$L = \prod_{i=1}^{n} \lambda \exp(-\lambda x_i) \tag{8-19}$$

The product form of the function in Equation 8-19 is difficult to differentiate. We make use of the fact that the logarithm of a variate must have its maximum at the same place as the maximum of the variate. Taking logarithms of Equation 8-19 gives:

$$\ln(L) = n\ln(\lambda) - \lambda\sum_{i=1}^{n} x_i \tag{8-20}$$

The differential of ln(L) with respect to λ, set to zero, produces the value of the parameter that maximizes the likelihood function. The derivative is given by:

$$\frac{d\ln(L)}{d\lambda} = \frac{n}{\lambda} - \sum_{i=1}^{n} x_i = 0 \tag{8-21}$$

Equation 8-21 yields the following:

$$\frac{1}{\lambda} = \frac{1}{n}\sum_{i=1}^{n} x_i = \bar{X} \tag{8-22}$$

Thus, the maximum-likelihood value of $1/\lambda$ is the mean of the sample of x's.

Example 8-4: Maximum-Likelihood Estimation of an Arbitrary Density Function

Consider the problem of finding the maximum-likelihood value of parameter A in the density function:

$$f_X(x) = cx\exp(-Ax) \qquad \text{for } x \geq 0 \tag{8-23}$$

where c is a constant. To use this equation as a probability density function, we must first find c from the condition that the total probability equals 1 as follows:

$$c\int_0^{\infty} x\exp(-Ax)dx = 1 \tag{8-24}$$

Solution of this equation gives $c = A^2$. Thus, the likelihood function is:

$$L = \prod_{i=1}^{n} A^2 x_i \exp(-Ax_i) \tag{8-25}$$

The logarithm of this function is:

$$\ln(L) = 2n\ln(A) + \sum_{i=1}^{n} \ln(x_i) - A\sum_{i=1}^{n} x_i \tag{8-26}$$

and

$$\frac{d\ln(L)}{dA} = \frac{2n}{a} - \sum_{i=1}^{n} x_i = 0 \tag{8-27}$$

We find that the maximum-likelihood value of $1/A$ is one half the mean of the sample.

Example 8-5: Maximum-Likelihood Estimation and the Gamma Distribution

The problem here is to find the maximum-likelihood expressions for the parameters of the gamma distribution. The density function of the gamma distribution can be written as:

$$f_X(x) = \frac{b^{a+1}}{a!}\exp(-bx)x^a \tag{8-28}$$

where $a!$ is the factorial of a and is related to the gamma integral by $a! = \Gamma(a + 1)$ and $\Gamma(a + 1) = a\Gamma(a)$. Solving for the likelihood expressions:

$$L = \prod_{i=1}^{n} \frac{b^{a+1}}{a!}\exp(-bx_i)x_i^a \tag{8-29}$$

and

$$\ln(L) = n[(a+1)\ln(b) - ln(a!)] - b\sum_{i=1}^{n} x_i + a\sum_{i=1}^{n} \ln(x_i) \tag{8-30}$$

Therefore,

$$\frac{\partial\ln(L)}{\partial a} = n\ln(b) - \frac{n\partial\ln(a!)}{\partial a} + \sum_{i=1}^{n} \ln(x_i) \tag{8-31}$$

and

$$\frac{\partial\ln(L)}{\partial b} = \frac{n(a+1)}{b} - \sum_{i=1}^{n} x_i \tag{8-32}$$

Setting the two partial differentials of Equations 8-31 and 8-32 to zero for their maxima and solving simultaneously for a and b give:

$$b = \frac{n(a+1)}{\sum_{i=1}^{n} x_i} \tag{8-33a}$$

and

$$\ln(a+1) - \frac{\partial \ln(a!)}{\partial a} = \ln\left(\frac{\sum_{i=1}^{n} x_i}{n}\right) - \frac{\sum_{i=1}^{n} \ln(x_i)}{n} \tag{8-33b}$$

Equations 8-33a and 8-33b illustrate the potential complexity of the maximum-likelihood solutions. Equation 8-33b can be solved by a trial-and-error process for a. With a evaluated, Equation 8-33a can be solved for b. The expression $\partial \ln(a!)/\partial a$ in Equation 8-33b is called the *gamma derivative function* of a and is designated $\Psi(a)$. Tables of values of the gamma function can be found in mathematical handbooks as shown in Table A-8.

A sample of 36 synthetic numbers is shown as a histogram in Figure 8-3. Values for the sample-summation terms in Equation 8-33b are:

$$\ln\left(\frac{\sum_{i=1}^{n} x_i}{n}\right) = \ln\left(\frac{116}{36}\right) = 1.17 \tag{8-34a}$$

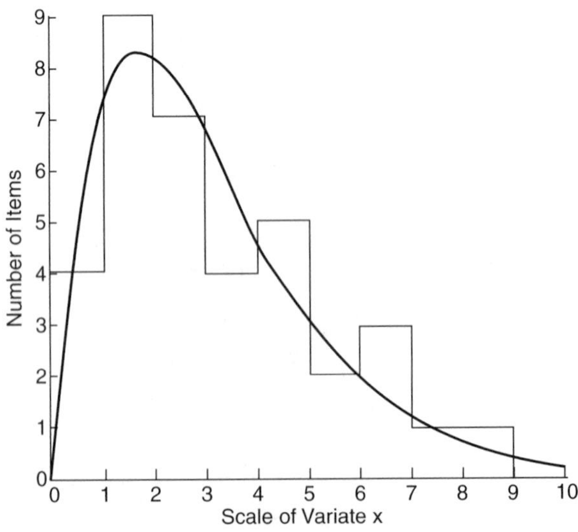

FIGURE 8-3 Gamma function fitted by maximum likelihood.

and

$$\frac{\sum_{i=1}^{n} \ln(x_i)}{n} = \frac{33.002}{36} = 0.9167 \qquad (8\text{-}34b)$$

Substituting in Equation 8-33b and rearranging give:

$$\ln(a+1) - \left(\frac{\partial \ln(a!)}{\partial a} + 0.2533\right) = 0 \qquad (8\text{-}35)$$

We must now search for values of a in Equation 8-35. One systematic method is to try values of a at the midpoints of intervals that are known to contain a root of the equation. This method is illustrated in Table 8-2. In this table, A denotes the first term on the left side of Equation 8-35, B is the second expression in brackets in this equation. Trial values of a of 0, 1, and 2 were assumed. The quantity $(A - B)$ changed from positive to negative between 1 and 2; therefore, a zero point of the equation falls in this interval. The successive midinterval values of 1.5, 1.25, and 1.125 were assumed, yielding the approximate values of a and b as shown. The function, $\psi(a)$, is $\partial \ln(a!)/\partial a$, which can be related to the gamma function. The values for the gamma function were obtained from Table A-8. The resulting value of a of 1.12 was then substituted in Equation 8-33a to obtain a value for b of 0.658.

The gamma distribution (Equation 8-28), using the estimated values of a and b, is superimposed on the histogram in Figure 8-3. Probability estimates for the occurrence of various values of X can now be made using the assumed population and not the sample histogram.

8.2.5. Other Estimation Methods

The estimation methods discussed in the previous sections are commonly used in engineering to obtain point estimates of parameters in univariate analysis. Other methods of point estimation are available. They include, for example, Bayes estimators and least-squares estimators.

TABLE 8-2 Maximum-Likelihood Solution for Gamma Parameter *a*

Trial value for a	$\psi(a)$	B	A	$A - B$
0	−0.5772	−0.3239	0	0.3239
1.0	0.4228	0.6761	0.6931	0.0170
2.0	0.9228	1.1761	1.0986	−0.0775
1.5	0.7032	0.9565	0.9163	−0.0402
1.25	0.5725	0.8258	0.8109	−0.0149
1.125	0.5034	0.7567	0.7538	−0.0029

The Bayes estimators treat the unknown parameter as a random variable with a known prior probability distribution. The sample provides additional information that is used to modify the prior distribution to obtain a posterior probability distribution for the parameter. The modification of the distribution from prior to posterior is performed using Bayes' theorem as discussed in Chapter 3. The Bayes estimator is, therefore, updated as additional information is obtained. This concept can be used for both univariate and multivariate analyses. The process of Bayesian updating of prior quantities constitutes the basis of Bayesian techniques discussed in Chapter 16.

The least-squares estimators are commonly used in bivariate and multivariate analysis. The parameter estimation is performed such that the mean-squared error is minimized. The MSE was defined as the expected value of the square of the deviation of the estimate $\hat{\theta}$ from the parameter being estimated, θ; that is,

$$\text{MSE} = E[(\hat{\theta} - \theta)^2] \tag{8-36}$$

This method constitutes the basis of least-squares regression in bivariate and multivariate analyses. Regression analysis is discussed in Chapter 12.

8.3. SAMPLING DISTRIBUTIONS

In Section 8.1, two samples of five measurements of the compressive strength of concrete were given. The means differed (3442 vs. 3526) and the standard deviations differed (135.9 and 174.4). If we took many samples of five and computed the mean and standard deviation for each sample, we could form two histograms, one for the sample means and one for the sample standard deviations. What would the histograms look like? The histogram for the sample means would approximate the sampling distribution of the estimated mean $\bar{X}$, where $\bar{X}$ is treated as a random variable. The histogram of the sample standard deviations would approximate the sampling distribution of the estimated standard deviation S, where S is treated as a random variable.

8.3.1. Sampling Distribution of the Mean

The sampling distribution of the mean depends on whether or not the population variance σ^2 is known. If it is known, then the mean of a random sample of size n from a population with mean μ and variance σ^2 has a normal distribution with mean μ and variance σ^2/n. The statistic Z has a standard normal distribution (i.e., mean = 0 and variance = 1) as follows:

$$Z = \frac{\bar{X} - \mu}{\sigma/\sqrt{n}} \tag{8-37}$$

If the population variance is not known, then the distribution of the mean depends on the distribution of the random variable. For a random variable with a normal distribution with mean μ, the distribution of the mean has a mean μ and a standard deviation $S/\sqrt{n}$. The statistic t has a t distribution with $(n - 1)$ degrees of freedom:

$$t = \frac{\bar{X} - \mu}{S/\sqrt{n}} \tag{8-38}$$

To illustrate the distribution of the sample mean, consider the 40 samples of five measurements from the standard normal population, $N(0, 1)$, that are given in Table 8-3a. A histogram of the 200 sample values of the random variable and the density function for the underlying population are given in Figure 8-4a. The 200 values give a reasonably good approximation of the population $N(0, 1)$. For each of the 40 samples, the mean was computed as shown in Table 8-3b. A sample histogram is shown in Figure 8-4b, with the density function $N(0, 1/5)$ also shown. This density function of the sample mean has a variance of σ^2/n, where n is the size of each sample ($n = 5$). Thus, the standard error of the mean is $\sigma/\sqrt{n}$ or $1/\sqrt{5}$. The two population distributions (X and $\bar{X}$) are given in Figure 8-4c for the purpose of comparison. It is important to note that, while both have identical means of 0, the sampling distribution of the mean has a much smaller spread. This should be expected because there is much less variation in means than in the values that are used to compute the means.

If two independent samples of sizes n_1 and n_2 are drawn from populations with means μ_1 and μ_2 and variances σ_1^2 and σ_2^2, respectively, then the difference of the sample means, $\bar{X}_1 - \bar{X}_2$, has a sampling distribution that is approximately normal with a mean $\mu_1 - \mu_2$ and a variance $(\sigma_1^2/n_1 + \sigma_2^2/n_2)$. Thus, the statistic Z has a standard normal distribution:

$$Z = \frac{(\bar{X}_1 - \bar{X}_2) - (\mu_1 - \mu_2)}{\left(\frac{\sigma_1^2}{n_1} + \frac{\sigma_2^2}{n_2}\right)^{0.5}} \tag{8-39}$$

If the population means and variances are equal, then the Z statistic of Equation 8-39 is

$$Z = \frac{X_1 - X_2}{\sigma\left(\frac{1}{n_2} + \frac{1}{n_2}\right)^{0.5}} \tag{8-40}$$

Equations 8-37 to 8-40 can be used to test hypotheses about the means and to form confidence intervals.

TABLE 8-3 Sampling Distributions for the Mean and Variance for Samples Taken from $N(0,1)$ Population

(a) Forty Samples from $N(0,1)$, Each with a Sample Size of 5

0.048	1.040	−0.111	−0.120	1.396	−0.393	−0.220	0.422	0.233	0.197
−0.521	−0.563	−0.116	−0.512	−0.518	−2.194	2.261	0.461	−1.533	−1.836
−1.407	−0.213	0.948	−0.073	−1.474	−0.236	−0.649	1.555	1.285	−0.747
1.822	0.898	−0.691	0.972	−0.011	0.517	0.808	2.651	−0.650	0.592
1.346	−0.137	0.952	1.467	−0.352	0.309	0.578	−1.881	−0.488	−0.329
0.420	−1.085	−1.578	−0.125	1.337	0.169	0.551	−0.745	−0.588	1.810
−1.760	−1.868	0.677	0.545	1.465	0.572	−0.770	0.655	−0.574	1.262
−0.959	0.061	−1.260	−0.573	−0.646	−0.697	−0.026	−1.115	3.591	−0.519
0.561	−0.534	−1.730	−1.172	−0.261	−0.049	0.173	0.027	1.138	0.524
−0.717	0.254	0.421	−1.891	2.592	−1.443	−0.061	−2.520	−0.497	0.909
−2.097	−0.180	−1.298	−0.647	0.159	0.769	−0.735	−0.343	0.966	0.595
0.443	−0.191	0.705	0.420	−0.486	−1.038	−0.396	1.406	0.327	1.198
0.481	0.161	−0.044	−0.864	−0.587	−0.037	−1.304	−1.544	0.946	−0.344
−2.219	−0.123	−0.260	0.680	0.224	−1.217	0.052	0.174	0.692	−1.068
1.723	−0.215	−0.158	0.369	1.073	−2.442	−0.472	2.060	−3.246	−1.020
−0.937	1.253	0.321	−0.541	−0.648	0.265	1.487	−0.554	1.890	0.499
−0.568	−0.146	0.285	1.337	−0.840	0.361	−0.468	0.746	0.470	0.171
−1.717	−1.293	−0.556	−0.545	1.344	0.320	−0.087	0.418	1.076	1.669
−0.151	−0.266	0.920	−2.370	0.484	−1.915	−0.268	0.718	2.075	−0.975
2.278	−1.819	0.245	−0.163	0.980	−1.629	−0.094	−0.573	1.548	−0.896

(b) Sample Means

0.258	0.205	0.196	0.347	−0.246	−0.399	0.556	0.642	−0.231	−0.425
−0.491	−0.634	−0.694	−0.643	0.897	−0.290	−0.027	−0.740	0.641	0.797
−0.334	−0.110	−0.211	−0.008	0.077	−0.793	−0.571	0.351	−0.063	−0.128
−0.219	−0.454	0.243	−0.456	0.264	−0.520	0.114	0.151	1.412	0.094

(c) Sample Variances

1.764	0.514	0.529	0.694	1.272	1.147	1.257	2.829	1.113	0.882
0.955	0.752	1.325	0.880	1.777	0.627	0.231	1.462	3.305	0.766
3.042	0.024	0.514	0.484	0.445	1.493	0.248	2.033	3.233	1.002
2.280	1.402	0.276	1.745	0.945	1.318	0.613	0.442	0.421	1.192

(d) Sample Standard Deviations

1.328	0.717	0.727	0.833	1.128	1.071	1.121	1.682	1.055	0.939
0.977	0.867	1.151	0.938	1.333	0.792	0.481	1.209	1.818	0.875
1.744	0.155	0.717	0.696	0.667	1.222	0.498	1.426	1.798	1.001
1.510	1.184	0.525	1.321	0.972	1.148	0.783	0.665	0.649	1.092

8.3.2. Sampling Distribution of the Variance

The estimated variance of a sample is a random variable, and so it has a distribution. The distribution depends on the characteristics of the underlying population from which the sample is derived. If the population is normal, then it can be shown that for the unbiased estimate of the variance S^2, the quantity $(n - 1)S^2/\sigma^2$ is a random

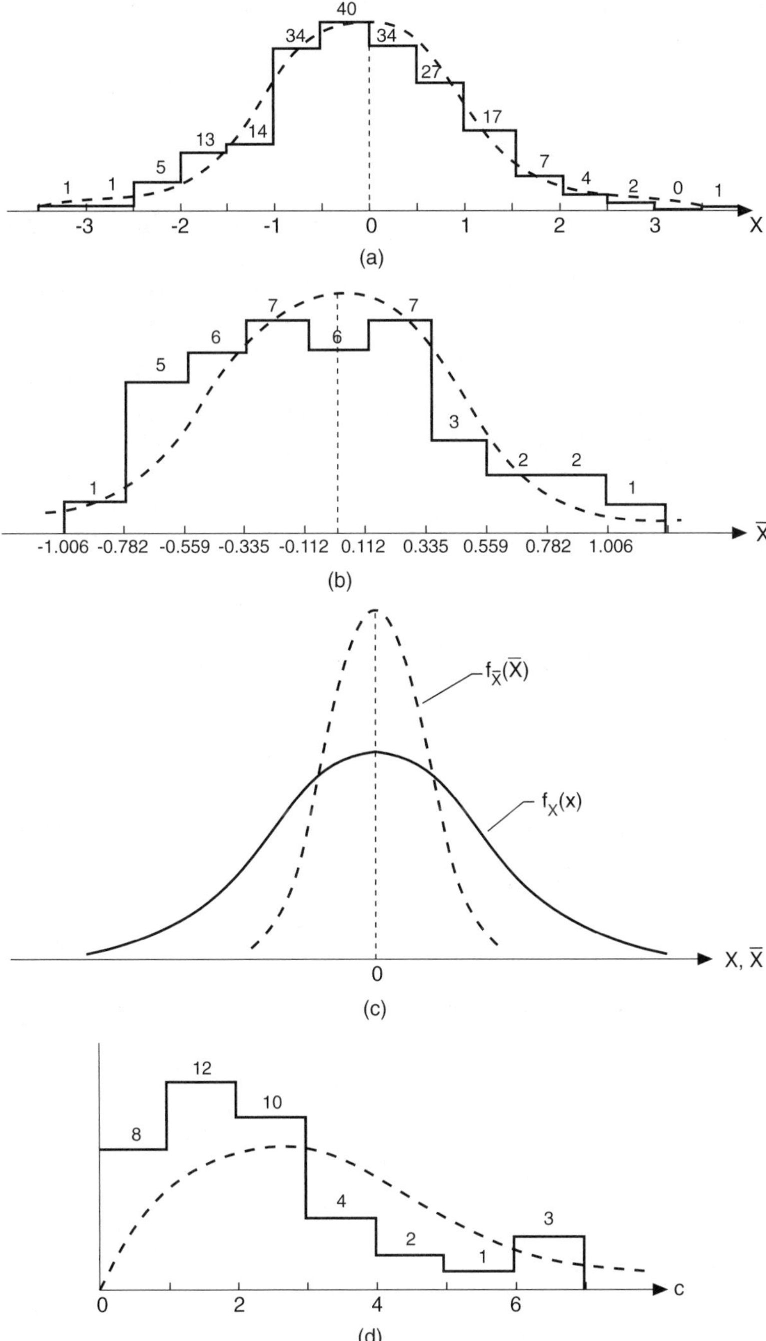

FIGURE 8-4 Sample distributions of (a) random variable from $N(0,1)$ population; (b) sample means; (c) the population of the random variable and the mean; and (d) variances.

variable distributed as chi-square (χ^2, called C in Chapter 5) with $(n - 1)$ degrees of freedom. Thus, inferences about the variance of a single normally distributed population are made with:

$$\chi^2 = \frac{(n-1)S^2}{\sigma^2} \tag{8-41}$$

The chi-square statistic of Equation 8-41 can be used to test hypotheses about the variance of a single random variable and to form confidence intervals.

The 40 samples of 5 were used to compute the sample variances as shown in Table 8-3c, from which the standard deviations were also computed, as shown in Table 8-3d. Because each sample was based on a size of 5, the number of degrees of freedom is four. The histogram of sample variances is shown in Figure 8-4d, as is the underlying chi-square distribution. The sample does not match the population as well for the variance as for the mean because the sampling variation of the variance is relatively greater than that of the mean. It would take considerably more than 40 samples to provide a reasonable approximation of the sampling distribution of the variance. However, it is evident from Figure 8-4d that the variance has a chi-square distribution.

8.3.3. Sampling Distributions for Other Parameters

Any estimated quantity using a sample can be treated as a random variable, and so it has a distribution. The distribution depends on the characteristics of the underlying population from which the sample is derived. For example, in Chapter 12, correlation analysis and regression analysis are introduced. The estimated correlation coefficient and the estimated parameters (or coefficients) in the regression models are treated as random variables. Therefore, they are random variables and have probability distributions. Their distributions are discussed after their introduction in Chapter 12.

8.4. APPLICATIONS

8.4.1 Toll Booth Rates of Service

During rush hours on a toll road, backups are common. The Highway Administration is considering installing electronic toll lanes to improve service. An administrator visits a facility at another toll road and measures the number of cars passing through electronic (E), exact-change (C), and person-operated (P) booths. Ten 1-minute readings are taken for each service line with the following results:

$$C = \{6, 8, 6, 7, 9, 9, 8, 6, 5, 6\}$$
$$E = \{12, 10, 13, 13, 11, 12, 10, 13, 14, 12\}$$
$$P = \{3, 5, 2, 4, 4, 6, 3, 5, 3, 5\}$$

These provide mean rates of $\overline{\text{C}} = 7$, $\overline{\text{E}} = 12$, and $\overline{\text{P}} = 4$.

Letting X be the number of cars passing through a toll booth per minute, the probability of X cars passing can be assumed to follow a Poisson distribution. Both the method of moments and maximum-likelihood analyses yield the relationship for estimating the Poisson parameter as $\lambda = \overline{X}$, where the mass function is:

$$P_X(x) = \frac{\lambda^x e^{-\lambda}}{x!} \tag{8.42}$$

Based on the sample data, the probabilities of getting more than 5 cars per minute through each type of toll facility are:

$$\text{C:} \quad 1 - \sum_{x=0}^{5} \frac{7^x e^{-7}}{x!} = 1 - e^{-7}\left[\sum_{x=0}^{5} (7^x/x!)\right]$$

$$= 1 - e^{-7}(1 + 7 + 24.5 + 57.17 + 100.0 + 140.1)$$

$$= 1 - 0.3007 = 0.6993$$

$$\text{E:} \quad 1 - \sum_{x=0}^{5} \frac{12^x e^{-12}}{x!} = 1 - e^{-12}\left[\sum_{x=0}^{5} (12^x/x!)\right]$$

$$= 1 - e^{-12}(1 + 12 + 72 + 288 + 864 + 2073.6)$$

$$= 1 - 0.02034 = 0.9797$$

$$\text{P:} \quad 1 - \sum_{x=0}^{5} \frac{4^x e^{-4}}{x!} = 1 - e^{-4}\left[\sum_{x=0}^{5} (4^x/x!)\right]$$

$$= 1 - e^{-4}\left(1 + 4 + 8 + \frac{32}{3} + \frac{32}{3} + 8.533\right)$$

$$= 1 - 0.7851 = 0.2149$$

8.4.2 Historic Information with Flood Records

At many locations, historic information such as newspaper accounts of flooding is available to supplement a systematic record of floods obtained with a flow recording gauge. Maximum-likelihood estimation can be used to derive expressions for the moments of a normal distribution for a period of H years that includes N years of systematic measurement of flood peaks, H historic floods, and M years in which flow measurements were not made.

The objective is to derive expressions for estimating the population parameters of the normal distribution given a systematic record of N events and Z historic events in a period of H years; the N systematic events occur during the H years. The location μ and scale σ, parameters of the normal distribution, must be estimated. The sample information consists of the N values of the systematic record, denoted as x_i, and the

Z historic events, denoted as y_i. It is assumed that the events within the H years that are not part of the systematic record or are not historic events are less than some threshold, which is denoted as x_0; these unmeasured values are denoted as w_i.

Using this notation and assumption, the following likelihood function can be formulated:

$$L(\mu, \sigma | x_1, x_2, \ldots, x_N, y_1, \ldots, y_Z) = \prod_{i=1}^{N} f_X(x_i|\mu, \sigma) \cdot \prod_{i=1}^{Z} f_X[(y_i|y > x_0)|\mu, \sigma] \cdot \prod_{i=1}^{M} f_X[(w_i|w < x_0)|\mu, \sigma] \tag{8-43}$$

in which M is the number of unmeasured values in the record length H (i.e., $M = H - N - Z$). Equation 8-43 gives the likelihood of population parameters μ and σ given that sampling has yielded the N systematic values x_i and the Z historic values y_i. The conditional probabilities in Equation 8-43 can be restructured by redefining the sample spaces; for simplicity, the likelihood function of Equation 8-43 will be denoted as L:

$$L = \prod_{i=1}^{N} f_X(x_i|\mu, \sigma) \cdot \prod_{i=1}^{Z} \frac{f_X(y_i|\mu, \sigma)}{\int_{x_0}^{\infty} f_X(x > x_0|\mu, \sigma)dx} \cdot \prod_{i=1}^{M} \frac{f_X(w_i|\mu, \sigma)}{\int_{-\infty}^{x_0} f_X(w < x_0|\mu, \sigma)dw} \tag{8-44}$$

Inserting the normal density function into Equation 8-44 yields:

$$L = \left\{\frac{1}{\sigma\sqrt{2\pi}}\exp\left[-\frac{1}{2}\left(\frac{x-\mu}{\sigma}\right)^2\right]\right\}^N \cdot \left\{\frac{\frac{1}{\sigma\sqrt{2\pi}}\exp\left[-\frac{1}{2}\left(\frac{y-\mu}{\sigma}\right)^2\right]}{\int_{x_0}^{\infty} f_X(y|\mu, \sigma)dy}\right\}^Z \cdot \left\{\frac{\frac{1}{\sigma\sqrt{2\pi}}\exp\left[-\frac{1}{2}\left(\frac{w-\mu}{\sigma}\right)^2\right]}{\int_{-\infty}^{x_0} f_X(w|\mu, \sigma)dw}\right\}^M \tag{8-45}$$

The integrals of Equation 8-45 can be expressed in terms of the cumulative normal distribution, $F_X(x)$:

$$L = \left\{\frac{1}{\sigma\sqrt{2\pi}}\exp\left[-\frac{1}{2}\left(\frac{x-\mu}{\sigma}\right)^2\right]\right\}^N \cdot \left\{\frac{1}{\sigma\sqrt{2\pi}}\exp\left[-\frac{1}{2}\left(\frac{y-\mu}{\sigma}\right)^2\right]\right\}^Z$$
$$\cdot \left\{\frac{1}{\sigma\sqrt{2\pi}}\exp\left[-\frac{1}{2}\left(\frac{w-\mu}{\sigma}\right)^2\right]\right\}^M \Big/ \{[1-F_X(x_0)]^Z[F_X(x_0)]^M\} \tag{8-46}$$

Collecting terms of Equation 8-46 yields:

$$L = \left(\frac{1}{\sigma\sqrt{2\pi}}\right)^H \left\{\exp\left[-\frac{1}{2}\sum_{i=1}^{N}\left(\frac{x-\mu}{\sigma}\right)^2\right] \cdot \exp\left[-\frac{1}{2}\sum_{i=1}^{Z}\left(\frac{y-\mu}{\sigma}\right)^2\right]\right\}$$
$$\cdot \exp\left[-\frac{1}{2}\sum_{i=1}^{M}\left(\frac{w-\mu}{\sigma}\right)^2\right] \Big/ \{[1-F_X(x_0)]^Z[F_X(x_0)]^M\} \tag{8-47}$$

The solution is simplified by using the fact that the maximum of the logarithm of the likelihood function L will occur for the same values of the population parameters as the maximum of the likelihood function itself. Taking the logarithm of the likelihood function yields:

$$\ln L = -H\ln\sqrt{2\pi} - H\ln\sigma - \frac{1}{2\sigma^2}\left[\sum_{i=1}^{N}(x_i-\mu)^2 + \sum_{i=1}^{Z}(y_i-\mu)^2 + \sum_{i=1}^{M}(w_i-\mu)^2\right]$$
$$-Z\ln[1-F_X(x_0)] - M\ln[F_X(x_0)] \tag{8-48}$$

Taking the derivatives with respect to the population parameters, μ and σ, and setting them to zero yield:

$$\frac{\partial(\ln L)}{\partial\mu} = 0 = \frac{1}{\sigma^2}\left[\sum_{i=1}^{N}(x_i-\mu) + \sum_{i=1}^{Z}(y_i-\mu) + \sum_{i=1}^{M}(w_i-\mu)\right]$$
$$+\left[\frac{Z}{1-F_X(x_0)}\right]\frac{\partial F_X(x_0)}{\partial\mu} - \left[\frac{M}{F_X(x_0)}\right]\frac{\partial F_X(x_0)}{\partial\mu} \tag{8-49}$$

$$\frac{\partial(\ln L)}{\partial\sigma} = 0 = -\frac{H}{\sigma} + \frac{1}{\sigma^3}\left[\sum_{i=1}^{N}(x_i-\mu)^2 + \sum_{i=1}^{Z}(y_i-\mu)^2 + \sum_{i=1}^{M}(w_i-\mu)^2\right]$$
$$+\frac{\partial F_X(x)}{\partial\sigma}\left[\frac{Z}{1-F_X(x_0)}\right] - \left[\frac{M}{F_X(x_0)}\right] \tag{8-50}$$

If x_0 is large relative to μ, then the derivatives $\frac{\partial F_X(x_0)}{\partial \mu}$ and $\frac{\partial F_X(x_0)}{\partial \sigma}$ will be small, and the terms that contain the derivatives can be ignored. Thus, Equation 8-49 reduces to:

$$\begin{aligned} 0 &= \sum_{i=1}^{N}(x_i-\mu)+\sum_{i=1}^{Z}(y_i-\mu)+\sum_{i=1}^{M}(w_i-\mu) \\ &= \sum_{i=1}^{N}x_i+\sum_{i=1}^{Z}y_i+\sum_{i=1}^{M}w_i-H\mu \end{aligned} \tag{8-51}$$

and Equation 8-50 reduces to:

$$0 = -H+\frac{1}{\sigma^2}\left[\sum_{i=1}^{N}(x_i-\mu)^2+\sum_{i=1}^{Z}(y_i-\mu)^2+\sum_{i=1}^{M}(w_i-\mu)^2\right] \tag{8-52}$$

The assumptions were made that the unmeasured values w_i are (1) from the same population as the x_i, and (2) less than the threshold x_0. Thus, the values of w_i can be represented by the x_i values, with the adjustment made for the difference in record lengths (i.e., N vs. M); Equations 8-51 and 8-52 can then be solved for the population parameters:

$$\begin{aligned} \mu &= \frac{1}{H}\left[\sum_{i=1}^{N}x_i+\sum_{i=1}^{Z}y_i+\frac{M}{N}\sum_{i=1}^{M}x_i\right] \\ &= \frac{1}{H}\left[\frac{H-Z}{N}\sum_{i=1}^{N}x_i+\sum_{i=1}^{Z}y_i\right] \end{aligned} \tag{8-53}$$

$$\begin{aligned} \sigma^2 &= \frac{1}{H-1}\left[\sum_{i=1}^{N}(x_i-\mu)^2+\sum_{i=1}^{Z}(y_i-\mu)^2+\frac{M}{N}\sum_{i=1}^{M}(x_i-\mu)^2\right] \\ &= \frac{1}{H-1}\left[\frac{H-Z}{N}\sum_{i=1}^{N}(x_i-\mu)^2+\sum_{i=1}^{Z}(y_i-\mu)^2\right] \end{aligned} \tag{8-54}$$

Equations 8-53 and 8-54 are based on the following assumptions:

1. All historic events are greater than the threshold x_0.
2. All unmeasured values in the historic period, H–N, are less than the threshold.
3. All values are from a normal population with location and scale parameters of μ and σ, respectively.
4. The rates of change of the cumulative normal distribution $F_X(x)$ with respect to the population parameters μ and σ are very small at the threshold.

Equations 8-53 and 8-54 can be applied to the logarithms of measured values, with X and Y being the logarithms of the systematic and historic events, respectively. Equations 8-53 and 8-54 are the equations used to compute the weighted mean and variance.

8.5 PROBLEMS

8-1. With respect to collecting samples of streamwater to identify the concentration of dissolved oxygen, discuss the concepts of samples and populations.

8-2. The predicted ($\hat{Y}$) and measured (Y) values of a random variable are:

$\hat{Y}$	6.1	8.3	4.7	5.2	7.6
Y	5.7	7.5	4.4	4.7	7.1

Discuss the model from the standpoint of prediction bias.

8-3. Discuss the concept of bias in terms of an archer shooting arrows at a target.

8-4. If we know that the true specific weight of a dry sand was 100 lb/ft^3, which of the following two methods of estimation provides more precise estimates:

Method 1: 102.4, 101.9, 102.6, 102.1, 101.7
Method 2: 99.7, 101.6, 98.5, 102.3, 100.9

Which method has the lesser bias?

8-5. Discuss the concept of precision in terms of an archer shooting arrows at a target.

8-6. Of the two methods in Problem 8-4, which is more accurate? Explain.

8-7. Discuss the concept of accuracy in terms of an archer shooting arrows at a target.

8-8. Find the mean, variance, and standard deviation for the following values of a discrete random variable X and the corresponding probability mass function ($p_X(x)$):

x	0	1	2	3	4
$p_X(x)$	0.35	0.39	0.19	0.06	0.01

8-9. Given the values of the cumulative mass function $F_X(x)$ for the discrete random variable X, compute the mean, variance, and standard deviation:

x	0	1	2	3	4
$F_X(x)$	0.55	0.88	0.97	0.99	1.00

8-10. Compute the mean and standard deviation for the following pH measurements: 6.4, 5.7, 6.8, 7.3, 6.3.

8-11. Find the mean, variance, and standard deviation of the eight largest values of sediment yield given in Example 9-8.

8-12. Show the calculations of the mean and standard deviation for the data of Equation 9-40a.

8-13. A continuous random variable has the following density function: $f_X(x) = 3x^2/26$ for the interval $(1 \le x \le 3)$. Find the mean, variance, and standard deviation of x.

8-14. Compute the mean, variance, and standard deviation for a continuous random variable x with the density function $f_X(x) = 2x/9$ over the range from 0 to 3.

8-15. Use the method of moments to derive an estimator of b for the following density function:

$$f_X(x) = \begin{cases} 2x/b^2 & \text{for } 0 \le x \le b \\ 0 & \text{otherwise} \end{cases}$$

8-16. Using the histogram for the daily evaporation data of Example 9-2, evaluate the parameters of a population having a uniform distribution using the method of moments ($\overline{X} = 0.1387$, and $S = 0.0935$).

8-17. Use the method of moments to derive estimators of x_0 and y_0 for the density function:

$$f_X(x) = \begin{cases} y_0 & \text{for } 0 \le x \le 1 \\ y_0(x_0 - x)/(x_0 - 1) & \text{for } 1 \le x \le x_0 \\ 0 & \text{otherwise} \end{cases}$$

8.18. Determine the maximum-likelihood estimators of the parameters of (a) the normal distribution, and (b) the lognormal distribution using an approach similar to Example 8-5.

8-19. A random variable has a population mean of 45 and a standard deviation of 10. What is the probability that a sample mean computed from a sample of 15 drawn from the population will be greater than 47.5?

8-20. The population standard deviation of the random variable x is equal to 4. Show how the sample mean varies as the sample size varies from 10 to 50 using a population mean of 25.

8-21. A normal population is known to have a mean of 10. (a) If the standard deviation of the population is 5, what is the probability that a random sample of size 25 will have a mean of 12 or higher? (b) What is the probability in (a) if $n = 100$? (c) If the standard deviation of the population is 10, what is the probability that a random sample of size 25 will have a mean of 12 or higher? (d) What is the probability in (c) for a mean of 6 or lower?

8-22. Write and execute a computer program to derive an approximation of the sampling distribution of the mean. Use a random-number generator that generates standard normal deviates with a mean of zero and a variance of one.

Generate 1000 samples of size n and compute the mean of each sample. Develop a histogram using the 1000 values and plot both the histogram and the normal distribution. Determine the number of sample means that lie in each tail for probabilities of 0.01, 0.025, 0.05, and 0.10. Execute the program for sample sizes of 5, 10, and 25.

8-23. A random variable X has a normal population with a mean μ and a variance of 3. (a) If samples of 10 are taken from the population, sketch the distribution of the variance. (b) Approximately what proportion of 10,000 samples, each with a sample size of 10, would have sample variances greater than 5.6?

8-24. A random variable X has a standard normal distribution. If many samples of 5 are drawn from the population, what sample variance will be exceeded 5% of the time? How does this compare with the 40 samples in Table 8-3c?

8-25. Write and execute a computer program to derive an approximation of the sampling distribution of the variance. Use a random-number generator that generates standard normal deviates with a mean of zero and a variance of one. Generate 1000 samples of size n and compute the variance of each sample. Develop a histogram using the 1000 values and plot both the histogram and the chi-square distribution for $(n - 1)$ degrees of freedom. Determine the number of sample variances that lie in each tail for probabilities of 0.01, 0.025, 0.05, and 0.10. Execute the program for sample sizes of 5, 10, and 25.

8.6. SIMULATION PROJECTS

This section provides a research-oriented simulation project that is not related to the problem statements of the four simulation projects introduced at the end of previous chapters.

The objectives of this project are to use simulation for the following: (1) to show the effect of the sample size on the sampling distribution of the mean; (2) to show the effect of knowing the population standard deviation on the distribution of the mean; and (3) to show the effect of the sampling distribution of the random variable on the sampling distribution of the mean. The simulation analysis needs to be repeated (1) for sample sizes (n) of 5, 10, and 25; (2) using an assumed σ rather than the standard deviation S of each sample of size n to compute the test statistic; and (3) to generate samples assuming that the data are from a uniform, normal, or exponential population. A sample size of 25 will be used for parts 2 and 3. A normal distribution will be used for parts 1 and 2. The sample standard deviation will be used for parts 1 and 3. The simulation procedure is as follows:

1. Set the simulation size N_S as 10^5 (i.e., the number of samples of size n used in each analysis).
2. Compute the mean and standard deviation of the Y variable from Table C-1, C-2, or C-3. These are used as the population parameters μ and σ, respectively.
3. Establish 20 cell bounds for each of three histograms: one for the sample means, a second one for the z statistic $[z = (\bar{X} - \mu)/(\sigma/\sqrt{n})]$, and a third one for the t statistic $[t = (\bar{X} - \mu)/(S/\sqrt{n})]$.

4. Iterate N_S times and perform the following steps: (a) Generate a sample of size n for the assumed distribution. (b) Compute the sample mean ($\overline{X}$) and the standard deviation (S). (c) Compute the test statistics t and z. (d) For each of the three histograms, increase the frequency count by 1 for the cell in which $\overline{X}$, t, and z occur. (*Note:* It is not necessary to store the values of $\overline{X}$, t, and z; only store the frequency counts.)
5. Divide each cell of each frequency histogram by N_S to obtain relative frequency histograms.
6. Compute the theoretical distributions and compare them with the relative frequency histograms of step 5.

Repeat the procedure, as necessary, to meet the three stated objectives. Summarize the results in a way that conclusions can be made about the three objectives.

Chapter 9
Hypothesis Testing

TABLE OF CONTENTS

9.1. INTRODUCTION

Hypothesis testing is a class of statistical analysis that is widely used because it encourages systematic decision making about problems that involve considerable uncertainty. It enables inferences to be made in such a way that sample data can be combined with statitistical theory. It supposedly removes the effect of the biases of the individual, which leads to more rational and accurate decision making.

9.2. GENERAL PROCEDURE

Hypothesis testing is the formal procedure for using statistical concepts and measures in performing decision making. The following six steps can be used to make a statistical analysis of a hypothesis:

1. Formulate hypotheses.
2. Select the appropriate statistical model (theorem) that identifies the test statistic.
3. Specify the level of significance, which is a measure of risk.
4. Collect a sample of data and compute an estimate of the test statistic.
5. Define the region of rejection for the test statistic.
6. Select the appropriate hypothesis.

These six steps are discussed in detail in the following sections.

9.2.1. Step 1: Formulation of Hypotheses

Hypothesis testing represents a class of statistical techniques designed to extrapolate information from samples of data to make inferences about populations for the purpose of decision making. The first step is to formulate two or more hypotheses for testing. The hypotheses are usually statements indicating that a random variable has some specific distribution or that a population parameter has some specific value. If the objective is to compare two or more specific parameters, such as the means of two populations, the hypotheses are then statements formulated to indicate the absence or presence of differences. It is important to note that the hypotheses are composed of statements involving either population distributions or parameters; hypotheses should not be expressed in terms of sample statistics.

The first hypothesis is called the *null hypothesis*, is denoted by H_0, and is formulated as an equality, thus indicating that a difference does not exist. The second hypothesis, which is called the *alternative hypothesis*, is formulated to indicate that a difference does exist. The alternative hypothesis is denoted by either H_1 or H_A. The null and alternative hypotheses should be expressed both grammatically and in mathematical terms and should represent mutually exclusive conditions. Thus, when a statistical analysis of sampled data suggests that the null hypothesis should be rejected, the alternative hypothesis must be accepted. In some cases, the users may hope that the experiment leads to a difference; thus, there may be a desire to present the null hypothesis in a way that would suggest that a difference exists. This practice should not be followed. The null hypothesis must be a statement of equality, not inequality.

9.2.2. Step 2: The Test Statistic and Its Sampling Distribution

The alternative hypothesis of step 1 provides for a difference between specified populations or parameters. To test the two hypotheses, it is necessary to develop a test statistic that reflects the difference suggested by the alternative hypothesis. The computed value of a test statistic varies from one sample to the next; therefore, the test statistic is a random variable and has a sampling distribution. A hypothesis test should be based on a theoretical model that defines the distribution function of the test statistic and the parameters of the sampling distribution. Based on the distributions of the test statistic, probability statements about computed values may be made.

Theoretical models are available for all of the more frequently used hypothesis tests. In cases where theoretical models are not available, approximations have usually been developed. In any case, a model or theorem that specifies the test statistic, its distribution, and its parameters must be found. The test statistic reflects the hypotheses and the data that are usually available. Also, the test statistic is a random variable, thus it has a distribution function that is defined by a functional form and one or more parameters.

9.2.3. Step 3: The Level of Significance

A set of hypotheses was formulated in step 1. In step 2, a test statistic and its distribution were selected to reflect the problem for which the hypotheses were formulated. In step 4, data will be collected to test the hypotheses. Before the data are collected, however, it is necessary to provide a probabilistic framework for accepting or rejecting the null hypothesis and, subsequently, making a decision. The framework should reflect the allowance to be made for the chance variation that can be expected in a sample of data. This chance variation is referred to as *sampling variation*.

Table 9-1 shows the available situations and the potential decisions involved in a hypothesis test. The decision table suggests two types of error:

TABLE 9-1 Decision Table for Hypothesis Testing

	Situation	
Decision	**H_0 is true**	**H_0 is not true**
Accept H_0	Correct decision	Incorrect decision: type II error
Reject H_0	Incorrect decision: type I error	Correct decision

1. Type I error: reject H_0 when, in fact, H_0 is true.
2. Type II error: accept H_0 when, in fact, H_0 is false.

These two types of incorrect decisions are not independent; however, the decision process is most often discussed with reference only to the type I error.

The level of significance, which is a primary element of the decision process in hypothesis testing, represents the probability of making a type I error and is denoted by α. The probability of a type II error is denoted by β. The two possible incorrect decisions are not independent. The level of significance should not be made exceptionally small, because the probability of making a type II error will then be increased. Selection of the level of significance should, therefore, be based on a rational analysis of the effect on decisions, and it should be selected prior to the collection and analysis of the sample data. Specifically, one would expect the level of significance to be different when considering a case involving the loss of human life and a case involving minor property damage. However, the value chosen for α is often based on convention and the availability of statistical tables, with values for α of 0.05 and 0.01 being selected frequently. This traditional means of specifying α should be understood.

9.2.4. Step 4: Data Analysis

Given the significance level α, it is possible to determine the sample size required to meet any rejection criterion. The selection of a sample size is discussed in Section 11.2. After obtaining the necessary data, the sample is used to compute an estimate of the test statistic. In most cases, the data are also used to provide estimates of other parameters required to define the sampling distribution of the test statistic.

9.2.5. Step 5: The Region of Rejection

The region of rejection consists of those values of the test statistic that would be unlikely to occur when the null hypothesis is, in fact, true. Conversely, the region of acceptance consists of those values of the test statistic that would be expected when the null hypothesis is, in fact, true. Extreme values of the test statistic are least likely to occur when the null hypothesis is true. Thus, the region of rejection is usually represented by one or both tails of the distribution of the test statistic.

The critical value of the test statistic is defined as that value which separates the region of rejection from the region of acceptance. The critical value of the test statistic depends on (1) the statement of the alternative hypothesis, (2) the distribution of the test statistic, (3) the level of significance, and (4) characteristics of the sample or data. These four components represent the first four steps of a hypothesis test.

Depending on the statement of the alternative hypothesis, the region of rejection may consist of values associated with either one or both tails of the distribution of the test statistic. This may best be illustrated with examples. Consider the case of a manufacturer of soft drinks. It must be recognized that the accuracy of the bottling process is not sufficient to assure that every bottle will contain exactly 12 oz; some bottles will contain less, whereas others will contain more. If the label on the bottle indicates that the bottle contains 12 oz, the manufacturer wants to be assured that, on the average, each bottle contains at least that amount; otherwise, the manufacturer may be subject to a lawsuit for false advertising. But, if the bottles contain, on the average, more than 12 oz, the manufacturer may be losing money. Thus, the manufacturer is interested in both extremely large and extremely small deviations from 12 oz and would be interested in the following two-tailed test:

$$H_0\text{: } \mu = \mu_0 \tag{9-1a}$$

$$H_A\text{: } \mu \neq \mu_0 \tag{9-1b}$$

where μ is the population mean, and μ_0 is the standard of comparison (i.e., 12 oz). In this case, the region of rejection will consist of values in both tails. This is illustrated in Figure 9-1a, where Z is the test statistic that has the continuous probability density function shown in the figure. In this case, the test is called a *two-tailed test.*

In other cases, the region of rejection may consist of values in only one tail of the distribution. For example, if the mean concentration of a pollutant from ten samples is used as an indicator of water quality, a regulatory agency would only be

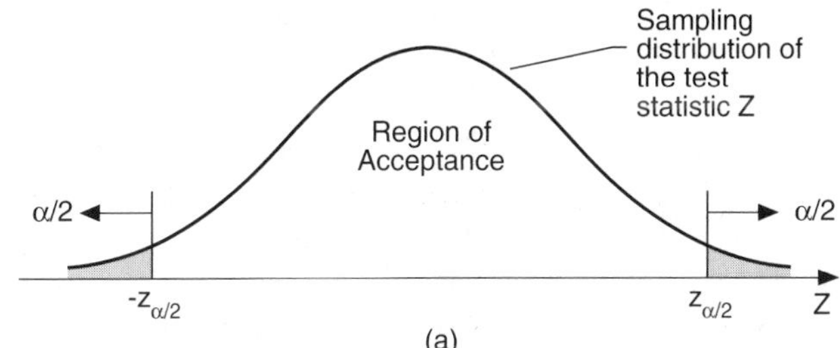

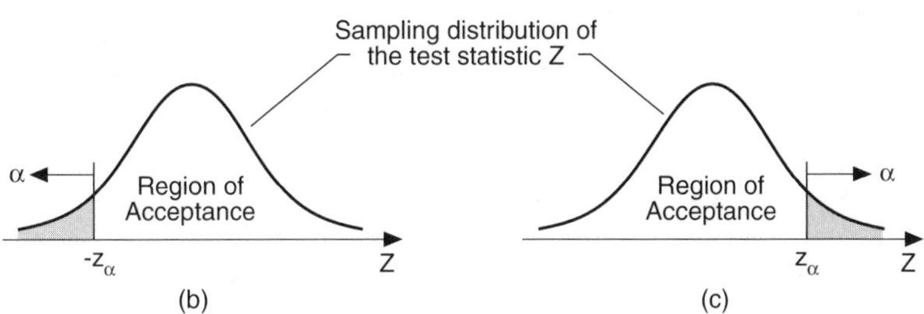

FIGURE 9-1 Representation of the region of rejection (cross-hatched area), region of acceptance, and critical value (z_α): (a) two-sided test, (b) lower tail one-sided test, and (c) upper tail one-sided test.

interested in whether or not the state water-quality standard for the pollutant is met. Specifically, if a manufacturing plant is discharging waste that contains the pollutant, a state may fine the manufacturer or close down the plant if the effluent causes the concentration to exceed the state standard. Thus, the regulatory agency is only interested in whether or not the quality is greater than the standard limit. It is not concerned with the degree to which the concentration is below the standard. This situation represents a one-tailed hypothesis test because the test is directional. The following hypotheses reflect this problem:

$$H_0: \mu = \mu_0 \tag{9-2a}$$

$$H_A: \mu > \mu_0 \tag{9-2b}$$

In this case, the region of rejection is associated with values in only the upper tail of the distribution, as illustrated in Figure 9-1c. The test is called a *one-tailed test.*

Other engineering situations involve concern over the lower tail. Engineers design concrete structures such that a minimum concrete strength is achieved. Strengths higher than a standard or specified value are acceptable assuming the ductility of the structure is adequate. During construction of the structure, samples of the concrete are collected and then tested after they have cured for a required period of time, and the mean strength is computed. This mean is compared with the standard or specified value (μ_0). This situation leads to the following hypotheses:

$$H_0: \mu = \mu_0 \tag{9-2c}$$

$$H_A: \mu < \mu_0 \tag{9-2d}$$

The test is called a *one-tailed, lower test* because the interest is in strengths significantly below the standard. In this case, the region of rejection is associated with values in only the lower tail of the distribution, as illustrated in Figure 9-1b. Rejection of the null hypothesis would suggest that the strength is not adequate. This may necessitate a decision to retrofit the structure to ensure adequate strength.

Although the region of rejection should be defined in terms of values of the test statistic, it is often pictorially associated with an area of the sampling distribution that is equal to the level of significance. The region of rejection, region of acceptance, and the critical value are shown in Figure 9-1 for the two-tailed and both one-tailed tests.

9.2.6. Step 6: Select the Appropriate Hypothesis

A decision on whether or not to accept the null hypothesis depends on a comparison of the computed value of the test statistic and the critical value. The null hypothesis is rejected when the computed value lies in the region of rejection. Rejection of the null hypothesis implies acceptance of the alternative hypothesis. When a computed value of the test statistic lies in the region of rejection, there are two possible explanations. First, the sampling procedure may have resulted in an extreme value purely by chance; although this is a very unlikely event, this corresponds to the type I error of Table 9-1. Because the probability of this event is relatively small, this explanation is most often rejected. Second, the extreme value of the test statistic may have occurred because the null hypothesis is false; this is the explanation most often accepted and forms the basis for statistical inference.

9.3. HYPOTHESIS TESTS OF MEANS

Because the mean of a population is often a primary element of statistical decision making, hypothesis tests of the mean are frequently used. The six-step procedure for testing hypotheses is applicable for testing hypotheses concerning the population mean μ.

9.3.1. Test of the Mean with Known Population Variance

When the standard deviation of the population is known, the procedure for testing the mean is as follows:

Step 1: Formulate hypotheses — The null and alternative hypotheses must be stated in terms of the population parameter μ and the value selected for comparison, which may be denoted as μ_0. The null hypothesis should state that the mean of the population equals a preselected standard value. Acceptance of the null hypothesis implies that it is not significantly different from μ_0. Mathematically, the null hypothesis could be stated as:

$$H_0\text{: } \mu = \mu_0 \tag{9-3}$$

One of three alternative hypotheses may be selected:

$$H_{A1}\text{: } \mu < \mu_0 \quad \text{lower one-tailed test} \tag{9-4a}$$

$$H_{A2}\text{: } \mu > \mu_0 \quad \text{upper one-tailed test} \tag{9-4b}$$

$$H_{A3}\text{: } \mu \neq \mu_0 \quad \text{two-tailed test} \tag{9-4c}$$

Each of the alternative hypotheses indicates that a significant difference exists between the population mean and the standard value. The selected alternative hypothesis depends on the statement of the problem.

Consider, for example, the problem of defining the yield strength of a shipment of steel. A study may be proposed to decide whether or not a shipment of steel meets the required yield strength of 3250 kgf (kilogram force). In this case, the shipment would be rejected only if the mean yield strength of the specimens in the sample was significantly less than 3250 kgf. Whether or not the strength is greater than 3250 kgf is not important because the strength of the structure is assumed to be adequate as long as it is at least 3250 kgf. Thus, a one-sided hypothesis test would be appropriate with the following hypotheses:

$$H_0\text{: } \mu = 3250 \text{ kgf} \tag{9-5a}$$

$$H_A\text{: } \mu < 3250 \text{ kgf} \tag{9-5b}$$

This represents the one-tailed (lower) test.

Step 2: Select the appropriate model — The mean $\overline{X}$ of a random sample is used in testing hypotheses about the population mean μ; $\overline{X}$ is itself a random variable. If the population from which the random sample is drawn has mean μ and variance σ^2, the distribution of the random variable $\overline{X}$ has mean μ and variance σ^2/n for samples from infinite populations. For samples from finite populations of size N, the variance is $[\sigma^2(N - n)]/[n(N - 1)]$.

For a random sample of size n, the sample mean $\overline{X}$ can be used in calculating the value of the test statistic z as:

$$z = \frac{\overline{X} - \mu}{\sigma/\sqrt{n}} \tag{9-6}$$

in which z is the value of a random variable whose distribution function is a standard normal.

Step 3: Select the level of significance — A level of significance of 1% is selected for demonstration of this hypothesis test; however, in actual practice the level selected for use should vary with the problem being studied and the impact of making an incorrect decision.

Step 4: Compute estimate of the test statistic — A random sample consisting of 100 specimens is selected, with a computed mean of 3190 kgf. The standard deviation of the population is 160 kgf. The value of the test statistic of Equation 9-6 is:

$$z = \frac{3190 - 3250}{160/\sqrt{100}} = -3.750 \tag{9-7}$$

Step 5: Define the region of rejection — For the standard normal distribution, the level of significance is the only characteristic required to determine the critical value of the test statistic. The region of rejection depends on the statement of the alternative hypothesis:

if H_A is	then the region of rejection is	
$\mu < \mu_0$	$z < -z_\alpha$	(9-8)
$\mu > \mu_0$	$z > z_\alpha$	
$\mu \neq \mu_0$	$z < -z_{\alpha/2}$ or $z > z_{\alpha/2}$	

For the case study, because the statement of the alternative hypothesis indicates a one-tailed test, the critical value of z for a 1% level of significance (α) can be obtained from Table A-1 as:

$$-z_{\alpha} = -\Phi^{-1}(1-\alpha) = -2.326 \tag{9-9}$$

Thus, the region of rejection consists of all values of Z less than –2.326.

Step 6: Select the appropriate hypothesis — Because the computed statistic lies in the region of rejection, the null hypothesis must be rejected. That is, there is no statistical basis for assuming the population mean to be equal to, or greater than, the selected value of 3250 kgf. If steel having a yield strength of at least 3250 kgf was required, then the sampled data suggest that the shipment from which the sample was obtained is inadequate.

Summary — The decision criterion specified in step 3 was limited to the specification of the level of significance. Because the null hypothesis was rejected for a 1% level of significance, there is a 1% chance of making a type I error; that is, there is a chance of 1 in 100 of returning the shipment of steel when, in fact, it was of adequate strength. The decision criterion of step 3 did not take into account the possibility of a type II error (β). The results of a type II error would be the use of steel of inadequate strength. Obviously, the consequences of a type II error (e.g., structural failure) are probably more severe than the consequences of a type I error (e.g., economic loss). However, it is easier and more direct to specify a value for α than to specify a value for β.

Example 9-1: Hypothesis Testing for the Mean

Assume that the population mean is 7, with a variance of 5.833. Also, assume that a sample size of n = 15 has a mean of 7.2. Based on this, the test statistic of Equation 9-6 is:

$$z = \frac{7.2 - 7.0}{2.416/\sqrt{15}} = 0.321 \tag{9-10}$$

For a level of significance of 5% and a two-tailed test, the critical value of the standard normal distribution is ±1.96. Because the computed value of 0.321 is within the region of acceptance, the null hypothesis H_0: $\mu = 7.0$ is accepted.

Example 9-2: Daily Evaporation Measurements

A mean pan evaporation of 0.1387 in./day is computed from a sample of 354 measurements. Is the mean estimate a biased estimate if the long-term records indicate an average value of 0.1575 in./day and a standard deviation of σ = 0.1

in./day? The hypothesis test for the mean with σ known can be used to test the significance of the bias. The null and alternative hypotheses are:

$$H_0\colon \mu = 0.1575 \text{ in./day} \tag{9-11a}$$

$$H_A\colon \mu \neq 0.1575 \text{ in./day} \tag{9-11b}$$

The computed value of the test statistic is:

$$z = \frac{0.1387 - 0.1575}{0.1/(\sqrt{354})} = -3.537 \tag{9-11c}$$

For a level of significance of 5%, the region of rejection consists of all values of z less than –1.96 and all values of z greater than 1.96. Thus, the null hypothesis must be rejected. This suggests that the sample is biased. Because the sample mean is less than the population mean, one might expect that the sample included more days during cold months.

9.3.2. Test of the Mean with Unknown Population Variance

When the population variance is unknown, the theorem used in the preceding section is not applicable, even though the null and alternative hypotheses and the steps are the same. In such cases, a different theorem is used for testing a hypothesis about a mean. Specifically, for a random sample of size n, the sample mean $\overline{X}$ and standard deviation S can be used in calculating the value of the test statistic t:

$$t = \frac{\overline{X} - \mu}{S/\sqrt{n}} \tag{9-12}$$

The test statistic, t, is the value of a random variable having the Student's t distribution with the degree of freedom $\nu = n - 1$. This statistic requires that the sample be drawn from a normal population. The region of rejection depends on the level of significance, the degrees of freedom, and the statement of the alternative hypothesis:

if H_A is	then reject H_0 if
$\mu < \mu_0$	$t < -t_\alpha$
$\mu > \mu_0$	$t > t_\alpha$
$\mu \neq \mu_0$	$t < -t_{\alpha/2}$ or $t > t_{\alpha/2}$

(9-13)

When the standard deviation of the population is unknown but the sample is large (say, greater than 30), the Z statistic of the preceding section can be used as an approximation of the t statistic. For a sample size of 30, the Z values for 5 and 1%

are obtained from Table A-1 to be 1.645 and 2.324, respectively; the corresponding t values from Table A-2 are 1.699 and 2.462, respectively. The differences are 3.3 and 5.8%, respectively. The decision on whether or not to use the Z statistic in place of the t statistic should be based on the consequences of erroneous decisions.

Example 9-3: Dissolved Oxygen Concentration

The case of sampling the dissolved oxygen concentration of effluent from a manufacturing plant can be used to illustrate the use of this test statistic. If a water-quality regulation required a minimum dissolved oxygen concentration of three parts per million (3 ppm), the following procedure may be used in testing the quality of the effluent:

1. H_0: $\mu = 3$ ppm.
   H_A: $\mu < 3$ ppm.
2. Theorem: t statistic of Equation 9-12 with $\nu = n - 1$.
3. $\alpha = 5\%$.
4. $n = 5$.
   $\bar{X} = 2.8$ ppm.
   $S = 0.32$ ppm.
   $$t = \frac{\bar{X} - \mu_0}{S/\sqrt{n}} = \frac{2.8 - 3.0}{0.32/\sqrt{5}} = -1.398.$$
   $\nu = n - 1 = 4$.
5. For $\nu = 4$, $\alpha = 5\%$, and a one-tailed lower test and using Table A-2, $t_\alpha = -2.132$.
6. Because $t > t_\alpha$, then accept H_0.

Because there is no statistical basis for rejecting the null hypothesis, one must assume that the effluent meets the water-quality regulations of 3 ppm, even though the sample mean is less than the standard of 3 ppm. The fact that H_0 cannot be rejected even though the sample mean is less than the standard of 3 ppm reflects the tolerance that must be permitted, because decision making based only on the sample represents decision making under conditions of uncertainty.

9.3.3. Hypothesis Test of Two Means

The previous hypothesis tests have involved a single random variable. In some cases, samples are obtained from two different populations, and it is of interest to determine if the population means are equal. For example, two manufacturers may advertise that their tires will last for 30,000 miles; samples of tires may be used to test whether or not the means of the two populations are equal. Similarly, tests could be conducted on engineering products to determine whether or not the means are equal. The fraction of downtime for two computer types could be tested to decide whether or not the mean times differ.

A number of tests can be used to test a pair of means. The method presented here should be used to test the means of two independent samples. This test is frequently of interest in engineering research when the investigator is interested in comparing an experimental group to a control group. For example, an environmental engineer might be interested in comparing the mean growth rates of microorganisms in polluted and natural environments. The procedure presented in this section can be used to make the test.

Step 1: Formulate hypotheses — If we denote the means of two populations as μ_1 and μ_2, the null hypothesis for a test on two independent means would be:

$$H_0\text{: the means of two populations are equal} \tag{9-14a}$$

Mathematically, this is:

$$H_0\text{: } \mu_1 = \mu_2 \tag{9-14b}$$

Both one-sided and two-sided alternatives can be used:

$$H_{A1}\text{: } \mu_1 < \mu_2 \tag{9-15a}$$

$$H_{A2}\text{: } \mu_1 > \mu_2 \tag{9-15b}$$

$$H_{A3}\text{: } \mu_1 \neq \mu_2 \tag{9-15c}$$

The selection of the alternative hypotheses should depend on the statement of the problem.

Step 2: Select the appropriate model — For the case of two independent samples, the hypotheses of step 1 can be tested using the following test statistic:

$$t = \frac{\bar{X}_1 - \bar{X}_2}{S_p\left(\dfrac{1}{n_1} + \dfrac{1}{n_2}\right)^{0.5}} \tag{9-16}$$

in which $\bar{X}_1$ and $\bar{X}_2$ are the means of the samples from populations 1 and 2, respectively; n_1 and n_2 are the sample sizes from populations 1 and 2, respectively; t is the value of a random variable having a t distribution with degrees of freedom of ($\nu = n_1 + n_2 - 2$); and S_p is the square root of the pooled variance given by:

$$S_p^2 = \frac{(n_1 - 1)S_1^2 + (n_2 - 1)S_2^2}{n_1 + n_2 - 2} \tag{9-17}$$

in which S_1^2 and S_2^2 are the variances of the samples from population 1 and 2, respectively. This test statistic assumes that the variances of the two populations are equal, but unknown.

Step 3: Select the level of significance — As usual, the level of significance should be selected on the basis of the problem; however, values of either 5% or 1% are used most frequently.

Step 4: Compute an estimate of test statistic — Samples are drawn from the two populations, and the sample means and variances computed. Equation 9-16 can be computed to test the null hypothesis of Equation 9-14.

Step 5: Define the region of rejection — The region of rejection is a function of the degrees of freedom ($\nu = n_1 + n_2 - 2$), the level of significance (α), and the statement of the alternative hypotheses. The regions of rejection for the alternative hypotheses are as follows:

if H_A is	then reject H_0 if	
$\mu_1 < \mu_2$	$t < -t_\alpha$	(9-18)
$\mu_1 > \mu_2$	$t > t_\alpha$	
$\mu_1 \neq \mu_2$	$t < -t_{\alpha/2}$ or $t > t_{\alpha/2}$	

Step 6: Select the appropriate hypothesis — The sample estimate of the t statistic from step 4 can be compared with the critical value, which is based on either t_α or $t_{\alpha/2}$ obtained from step 5. If the sample value lies in the region of rejection, then the null hypothesis should be rejected.

Example 9-4: Effect of Development on Total Nitrogen

A study was made to measure the effect of suburban development on total nitrogen levels in small streams. A decision was made to use the mean concentrations before and after the development as the criterion. Eleven measurements of the total nitrogen (mg/l) were taken prior to the development with a mean of 0.78 mg/l and a standard deviation of 0.36 mg/l. Fourteen measurements were taken after the development with a mean of 1.37 mg/l and a standard deviation of 0.87 mg/l. The data are assumed to test the null hypothesis that the population means are equal against the alternative hypothesis that the urban development increased total nitrogen levels; this requires the following one-tailed test:

$$H_0\!: \mu_b = \mu_a \tag{9-19a}$$

$$H_A\!: \mu_b < \mu_a \tag{9-19b}$$

where μ_b and μ_a are the before and after development means, respectively. Rejection of H_0 would show that the nitrogen levels after development exceed the nitrogen levels before development, with the implication that the development might have caused the increase.

Based on the sample data, the pooled variance of Equation 9-17 is:

$$S_P^2 = \frac{(11-1)(0.36)^2 + (14-1)(0.87)^2}{11+14-2} = 0.4842 \tag{9-20}$$

The computed value of the test statistic is:

$$t = \frac{0.78 - 1.37}{\sqrt{0.4842\left(\frac{1}{11} + \frac{1}{14}\right)}} = -2.104 \tag{9-21}$$

which has $\nu = 11 + 14 - 2 = 23$ degrees of freedom. From Table A-2, with a 5% level of significance and 23 degrees of freedom, the critical value of t is –1.714. Thus, the null hypothesis is rejected, with the implication that the development caused a significant change in nitrogen level. However, for a 1% significance level, the critical value is –2.500, which leads to the decision that the increase is not significant. This shows the importance of selecting the level of significance before analyzing the data.

9.3.4. Summary

Three hypothesis tests were introduced. Each test can be conducted using the six steps that were identified in Section 9.2. In applying a hypothesis test, the important ingredients are the test statistic, the level of significance, the degrees of freedom, and the critical value of a test statistic. Table 9-2 includes a convenient summary of the three statistical tests introduced in this section and other important tests.

9.4. HYPOTHESIS TESTS OF VARIANCES

The variance of a random sample is a measure of the dispersion of the observations about the sample mean. Although the variance is used to indicate the degree of variation about the mean, it is an important statistic in its own right. Large variation in engineering systems reflects instability or nonuniformity, both of which can be considered nonoptimal.

9.4.1. One-Sample Chi-Square Test

Consider, for example, the case of water-distribution systems used for irrigation. They should be designed to distribute water uniformly over an area, such as a lawn or an agricultural field. Failure to provide a uniform application of water over the

TABLE 9-2 Summary of Hypothesis Tests

H_0	Test statistic	H_A	Region of rejection
$\mu = \mu_0$ (σ known)	$z = \dfrac{\bar{X} - \mu}{\sigma/\sqrt{n}}$	$\mu < \mu_0$ $\mu > \mu_0$ $\mu \neq \mu_0$	$z < -z_\alpha$ $z > z_\alpha$ $z < -z_{\alpha/2}$ or $z > z_{\alpha/2}$
$\mu = \mu_0$ (σ unknown)	$t = \dfrac{\bar{X} - \mu}{S/\sqrt{n}}$ $\nu = n - 1$	$\mu < \mu_0$ $\mu > \mu_0$ $\mu \neq \mu_0$	$t < -t_\alpha$ $t > t_\alpha$ $t < -t_{\alpha/2}$ or $t > t_{\alpha/2}$
$\mu_1 = \mu_2$ ($\sigma_1^2 = \sigma_2^2$, but unknown)	$t = \dfrac{\bar{X}_1 - \bar{X}_2}{S_p\left(\dfrac{1}{n_1} + \dfrac{1}{n_2}\right)^{0.5}}$ $\nu = n_1 + n_2 - 2$ $S_p^2 = \dfrac{(n_1 - 1)S_1^2 + (n_2 - 1)S_2^2}{n_1 + n_2 - 2}$	$\mu < \mu_0$ $\mu > \mu_0$ $\mu \neq \mu_0$	$t < -t_\alpha$ $t > t_\alpha$ $t < -t_{\alpha/2}$ or $t > t_{\alpha/2}$
$\sigma^2 = \sigma_0^2$	$\chi^2 = \dfrac{(n-1)S^2}{\sigma_0^2}$ $\nu = n - 1$	$\sigma^2 < \sigma_0^2$ $\sigma^2 > \sigma_0^2$ $\sigma^2 \neq \sigma_0^2$	$\chi^2 < \chi^2_{\alpha-1}$ $\chi^2 < \chi^2_\alpha$ $\chi^2 < \chi^2_{1-\alpha/2}$ or $\chi^2 < \chi^2_{\alpha/2}$
$\sigma_1^2 = \sigma_2^2$ (assuming $\sigma_1^2 > \sigma_2^2$)	$F = \dfrac{S_1^2}{S_2^2}$ $\nu_1 = n_1 - 1$ $\nu_2 = n_2 - 1$	$\sigma_1^2 \neq \sigma_2^2$	$F > F_{\alpha/2}$
Goodness of fit, $X \sim f_X(x,p)$	$\chi^2 = \sum_{i=1}^{k} \dfrac{(O_i - E_i)^2}{E_i}$ $\nu = k - 1 -$ (number of parameters estimated using sample)	$X \neq f_X(x,p)$	$\chi^2 > \chi^2_\alpha$

area may lead to nonoptimum grass or crop output; thus, equipment that does not apply water uniformly would probably not be purchased. A company that manufactures irrigation distribution systems wishes to check whether or not a new system increases the uniformity of water application in comparison with existing models. The variance of depths of water measured at different locations in a field would serve as a measure of uniformity of water application. The following procedure is used to test for a statistical difference in the uniformity of application rates (i.e., a test on the variance of a random variable).

Step 1: Formulate hypotheses — To investigate the possibility of a significant difference existing between the variance of a population σ^2 and the preselected standard variance value (σ_0^2), the following null hypothesis can be used:

H_0: There is no significant difference between the variance of the population σ^2 and the preselected standard value for the variance (σ_0^2) (9-22a)

Mathematically, the null hypothesis is:

$$H_0\text{: } \sigma^2 = \sigma_0^2 \tag{9-22b}$$

The null hypothesis can be tested against either a one-tailed or two-tailed alternative hypothesis as follows:

$$H_{A1}\text{: } \sigma^2 < \sigma_0^2 \tag{9-23a}$$

$$H_{A2}\text{: } \sigma^2 > \sigma_0^2 \tag{9-23b}$$

$$H_{A3}\text{: } \sigma^2 \neq \sigma_0^2 \tag{9-23c}$$

For the irrigation distribution system case, assume that the standard deviation of an existing system is 0.1 cm/hr. The following hypotheses are examined by a manufacturing company wishing to determine if their new system provides a more uniform distribution of water:

$$H_0\text{: } \sigma^2 = (0.1 \text{ cm/hr})^2 \tag{9-24a}$$

$$H_A\text{: } \sigma^2 < (0.1 \text{ cm/hr})^2 \tag{9-24b}$$

Step 2: Select the appropriate model — The variance S^2 of a random sample is a random variable itself and is used in testing the hypotheses about the variance of a population, σ^2. The sampling distribution of the estimated variance of a random sample that is drawn from a normal population has a chi-square distribution. The test statistic for testing the hypotheses is:

$$\chi^2 = \frac{(n-1)S^2}{\sigma_0^2} \tag{9-25}$$

in which χ^2 is the value of a random variable that has a chi-square distribution with $\nu = n - 1$ degrees of freedom, and n is the sample size used in computing the sample variance S^2.

Step 3: Select the level of significance — For the case study, a level of significance of 2.5% is selected.

Step 4: Compute estimate of test statistic — To test the uniformity of application of water for the new irrigation system, the amount of water in each of 25 randomly placed recording devices was observed after 1 hr. The mean and standard deviation of the random sample were 0.31 and 0.063 cm/hr, respectively. The computed test statistic is:

$$\chi^2 = \frac{(25-1)(0.063)^2}{(0.1)^2} = 9.526 \tag{9-26}$$

Step 5: Define the region of rejection — The region of rejection for a test statistic having a chi-square distribution is a function of the level of significance, the statement of the alternative hypothesis, and the degrees of freedom. The regions of rejection for the alternative hypotheses are as follows:

if H_A is	then reject H_0 if
H_{A1}: $\sigma^2 < \sigma_0^2$	$\chi^2 < \chi^2_{1-\alpha}$
H_{A2}: $\sigma^2 > \sigma_0^2$	$\chi^2 > \chi^2_{\alpha}$
H_{A3}: $\sigma^2 \neq \sigma_0^2$	$\chi^2 < \chi^2_{1-\alpha/2}$ or $\chi^2 > \chi^2_{\alpha/2}$

(9-27)

For values of $\alpha = 0.025$ and $\nu = k = 24$, the critical value was obtained from Table A-3 to be 12.397. Thus, the region of rejection consists of all values of the test statistic χ^2 that are less than 12.397.

Step 6: Select the appropriate hypothesis — The computed value of the test statistic is less than the critical value; thus, the null hypothesis must be rejected. This implies that the new irrigation equipment applies water with more uniformity than existing irrigation systems. It is unlikely that the difference between the sample variance (0.004) and the variance of existing irrigation equipment (0.01) is due solely to chance variation within the sample; it is probably due to a real difference in the equipment.

Rationale of test statistic — If the sample variance was exactly equal to the population variance specified in the null hypothesis (i.e., $S^2 = \sigma_0^2$), then the value of the test statistic χ^2 will equal $n - 1$. For the two-tailed alternative, a significant difference exists if the computed value of χ^2 is either significantly larger or smaller than $n - 1$. For the one-tailed alternative, a significant difference exists only when the computed value of χ^2 is much less than $n - 1$; in the case of the irrigation system with $n - 1 = 24$, the computed value of 9.526 is considered to be significant because it is less than the critical value. For the upper one-tailed alternative, a significant difference exists only when the computed value of χ^2 is much greater than $n - 1$.

Example 9-5: Variance of Test Measurements

Assume that a research project requires new specimens to be made when the standard deviation of the crushing strength of five concrete cylinders exceeds 0.75 ksi. The reason is that, for variations greater than this, the specimens are not sufficiently homogeneous for making decisions about the crushing strength. The test procedure is acceptable for small variations, so it would be appropriate to test the following hypotheses:

$$H_0\colon \sigma^2 = (0.75)^2 = 0.5625 \text{ (ksi)}^2$$

$$H_A\colon \sigma^2 > 0.5625 \text{ (ksi)}^2$$

One set of tests yields the following strength measurements:

$$X = \{5.3,\ 6.8,\ 4.4,\ 7.4,\ 5.2\} \text{ ksi}$$

For these measurements, the sample standard deviation is 1.238 ksi. Using Equation 9-25, the computed value of the χ^2 test statistic is:

$$\chi^2 = \frac{(5-1)(1.238)^2}{(0.75)^2} = 10.89$$

which would have 4 degrees of freedom. For 4 degrees of freedom, the critical values for 5% and 2.5% levels of significance are 9.488 and 11.143, respectively. Thus, at a 5% level of significance, the null hypothesis would be rejected; however, at 2.5%, the decision would be to accept the null hypothesis. Thus, the rejection probability is about 3%; therefore, unless specific evidence exists that the 2.5% level of significance is appropriate, it may be better to make new specimens that show greater homogeneity in crushing strength.

9.4.2. Two-Sample *F* Test

For comparing the variances of two random samples, several strategies have been recommended, with each strategy valid when the underlying assumptions hold. One of these strategies is presented here.

For a two-tailed test, an F-ratio is formed as the ratio of the larger sample variance to the smaller sample variance as follows:

$$F = \frac{S_1^2}{S_2^2} \tag{9-28}$$

with degrees of freedom of $\nu_1 = n_1 - 1$ for the numerator and $\nu_2 = n_2 - 1$ for the denominator, where n_1 and n_2 are the sample sizes for the samples used to compute

S_1^2 and S_2^2, respectively. The computed F is compared with the tabulated values for the F probability distribution provided in Table A-4, and the null hypothesis of equal variances (H_0: $\sigma_1^2 = \sigma_2^2$) is accepted if the computed F is less than the tabulated F value for $k = \nu_1$, $u = \nu_2$, and α. If the computed F is greater than the tabulated F value, then the null hypothesis is rejected in favor of the alternative hypothesis (H_A: $\sigma_1^2 \neq \sigma_2^2$). An important note for this two-tailed test is that the level of significance is twice the value from which the tabulated F value was obtained; for example, if the 5% F table is used to obtain the critical F statistic, then the decision to accept or reject the null hypothesis is being made at a 10% level of significance. This is the price paid for using the sample knowledge that one sample has the larger variance.

For a one-tailed test, it is necessary to specify which of the two samples is expected to have the larger population variance. This must be specified prior to collecting the data. The computed F statistic is the ratio of the sample variance of the group expected to have the larger population variance to the sample variance from the second group. If it turns out that the sample variance of the group expected to have the larger variance is smaller than that of the group expected to have the smaller variance, then the computed F statistic will be less than 1. For a test with the level of significance equal to that shown on the table, the null hypothesis is rejected if the computed F is greater than the critical F. Because the direction is specified, the null hypothesis is accepted when the computed F is less than the critical F; the null hypothesis is rejected when the computed F is greater than the critical F.

Example 9-6: Accuracy of Laboratory Measurements

Two pieces of equipment are available to measure the value of a water-quality parameter. A laboratory assistant makes up ten samples of a known concentration (20 ppb) and randomly selects five samples to be tested with each piece of equipment. The objective is to decide whether or not the variations of the two pieces of equipment can be considered equal. This would reflect equal precision. Thus, a two-tailed test would be appropriate.

Measurements were made for both pieces of equipment as follows:

$$E1: \{16.8, 21.3, 18.5, 20.7, 17.6\}$$

$$E2: \{23.6, 18.4, 22.3, 17.9, 19.1\}$$

These measurements yield sample variances of $S_1^2 = 3.807$ ppb and $S_2^2 = 6.423$. Equation 9-28 yields a computed F of:

$$F = \frac{6.423}{3.807} = 1.687$$

which has $v_1 = 4$ and $v_2 = 4$. The 5% and 1% critical F values are 6.39 and 15.98, respectively. Thus, the sample value is not near the region of rejection, so it appears safe to accept the null hypothesis and conclude that the two pieces of equipment have equal precision.

9.4.3. Summary

Two hypothesis tests for the variance were introduced. Table 9-2 includes a summary of these tests.

9.5. TESTS OF DISTRIBUTIONS

In addition to tests on means and variances, it is frequently necessary to decide whether or not it is reasonable to conclude that a specific random variable is from a certain population distribution function. For example, the one-sample t test assumes that the data are from a normal population. The test may lead to an incorrect decision if the test is applied to data that are not normally distributed. Thus, a test on a distribution would be useful prior to using a specific test to decide whether or not the normality assumption is met.

9.5.1. Chi-Square Test for Goodness of Fit

Data analysts are often interested in identifying the density function of a random variable so that the population can be used to make probability statements about the likelihood of occurrence of values of the random variable. Very often, a graphical analysis such as a histogram of the data is used to suggest a likely candidate for the population density function. However, the graphical approach involves a subjective assessment and lacks the validity accorded to systematic analysis such as hypothesis tests. This has led to several tests that can be applied to evaluate the likelihood that a sample is from an assumed population. The chi-square test for goodness of fit is used to assess whether or not a sample of measurements from a random variable can be represented by a selected theoretical probability density function. The following paragraphs provide the six-step framework for performing the chi-square test. To demonstrate the quantitative evaluation, the distribution of arrival times for automobiles at an intersection during a non-peak-volume period is used to illustrate these steps.

Step 1: Formulate hypothesis — The first step is to formulate both the null and the alternative hypotheses that reflect the theoretical density function (continuous random variables) or mass function (discrete random variables). Because a function is not completely defined without the specification of its parameters, the hypothesis must also include specific values for the parameters of the function. For example, if the population is hypothesized to be normal, values for μ and σ must be specified; if the hypotheses deal with the uniform distribution, values for a and b must be specified. Estimates of the parameters may be obtained either empirically or from external conditions. If estimates of the parameters are obtained from the data set used in testing the hypotheses, the degrees of freedom in this test must be modified to reflect this. The following null and alternative hypotheses are typical examples:

H_0: The time between arrivals of automobiles at the specific intersection under investigation (*X*) can be characterized by a uniform density function with a location parameter of 0 and a scale parameter of 40 sec.
H_A: The time between arrivals of automobiles at a specific intersection is not uniformly distributed with parameters of 0 and 40 sec.

Mathematically, these hypotheses are:

$$H_0\text{: } X \sim U\ (a = 0,\ b = 40)$$

$$H_A\text{: } X \neq U\ (0, 40)$$

Rejection of the null hypothesis would not necessarily imply that the arrival time is not uniformly distributed. While rejection of the null hypothesis may occur because the random variable is not uniformly distributed, it may also be rejected because one or both of the parameters, in this case 0 and 40, are incorrect.

The chi-square goodness-of-fit test is always a one-tailed test because the structure of the hypotheses is unidirectional; that is, the random variable is either distributed as specified or it is not.

Step 2: Select the appropriate model — To test the hypotheses formulated in step 1, the chi-square test is based on a comparison of the observed frequencies of sample values with frequencies expected from the population density function that is specified in the null hypothesis. The test statistic, which is a random variable, is a function of the observed and expected frequencies, which are also random variables. The test statistic is:

$$\chi^2 = \sum_{i=1}^{k} \frac{(O_i - E_i)^2}{E_i} \tag{9-29}$$

where χ^2 is the computed value of a random variable having a chi-square distribution with ν degrees of freedom; O_i and E_i are the observed and expected frequencies, respectively, in cell (or interval) *i*; and *k* is the number of discrete cells (intervals) into which the data were separated. The random variable χ^2 has a sampling distribution that can be approximated by the chi-square distribution with $(k - j)$ degrees of freedom, where *j* is the number of quantities that are estimated from the sample of data for use in calculating the expected frequencies. Specifically, because the total number of observations *n* is used to compute the expected frequencies, one degree of freedom is lost. If the mean and standard deviation of the sample are needed to compute the expected frequencies, then two additional degrees of freedom are subtracted (i.e., $k - 3$). However, if the mean and standard deviation are obtained from past experience or other data sources, then the number of degrees of freedom for the test statistic remains $(k - 1)$.

Step 3: Select the level of significance — If the decision is not considered critical, a level of significance of 5% may be considered appropriate, because of convention. For the test of the hypotheses given previously, a value of 5% is used for illustration purposes.

Step 4: Compute estimate of test statistic — The observed data are usually used to form a histogram that shows the observed frequencies in a series of k cells (or intervals). The cell bounds are usually selected such that the width for each cell is the same; however, unequal cell widths are a practical alternative to ensure more uniform values of the observed and expected frequencies across cells. Having selected the cell bounds and counted the observed frequencies (O_i) in the k cells, the expected frequencies (E_i) for the cells can be computed using the density function of the population specified in the hypotheses of step 1. To compute the expected frequencies, the expected probability for each cell is determined and multiplied by the sample size. The expected probability for cell i (p_i), is either the area under the density function between the cell bounds for that cell for a continuous random variable or the sum of the probabilities from the probability mass function for the values within the cell bounds for a discrete random variable. The sum of the expected frequencies must equal the sample size n. The problem is best summarized using the general structure shown in Figure 9-2.

The value of the test statistic of Equation 9-29 is obtained from the cell frequencies of Figure 9-3. The range of the random variable was separated into four equal

Cell bound	$-\infty$			. . .	∞
Cell number i	1	2	3	. . .	k
Observed frequency (O_i)	O_1	O_2	O_3	. . .	O_k
Expected frequency (E_i)	E_1	E_2	E_3	. . .	E_k
$(O_i - E_i)^2/E_i$	$\frac{(O_1 - E_1)^2}{E_1}$	$\frac{(O_2 - E_2)^2}{E_2}$	$\frac{(O_3 - E_3)^2}{E_3}$	. . .	$\frac{(O_k - E_k)^2}{E_k}$

FIGURE 9-2 Cell structure of arrival times for chi-square goodness-of-fit test.

Cell bound	0 sec		10 sec		20 sec		30 sec		40 sec
Cell number		1		2		3		4	
Observed frequency (O_i)		18		19		25		18	
Expected frequency (E_i)		20		20		20		20	
$(O_i - E_i)^2/E_i$		0.20		0.05		1.25		0.20	

FIGURE 9-3 Cell structure of arrival times for chi-square goodness-of-fit test.

intervals of 10 sec. Thus, the expected probability for each cell is 0.25 (because the random variable is assumed to have a uniform distribution and the width of each of the four cells is the same). For a sample size of 80, the expected frequency for each of the four cells is 20 (i.e., the expected probability times the total number of observations). If observed frequencies of 18, 19, 25, and 18 are determined from the sample, then the cell structure would appear as shown in Figure 9-3. The computed χ^2 statistic is, therefore, 1.70, which is obtained by adding up 0.2, 0.05, 1.25, and 0.2. Because the sample size was used in computing the expected frequencies, the number of degrees of freedom is given by $k - 1$, or $4 - 1 = 3$.

Step 5: Define the region of rejection — According to the underlying theorem of step 2, the test statistic has a chi-square distribution with degrees of freedom that depend on the number of cells and the number of pieces of information obtained from the sample. In this case, only the sample size was used; thus, there are 3 degrees of freedom. For a level of significance of 5%, the critical value of the test statistic from Table A-3 is 7.81. Thus, the region of rejection consists of all values of the test statistic greater than 7.81.

Step 6: Select the appropriate hypothesis — Because the computed value of the test statistic (1.70) is less than the critical value (i.e., it is not within the region of rejection), there is no statistical basis for rejecting the null hypothesis based on the location and scale parameters of 0 and 40, respectively; thus, $U(0,40)$ may be used to represent the distribution of arrival times.

Summary — In summary, the chi-square test for goodness of fit provides the means for comparing the observed distribution of a random variable with a theoretical probability mass or density function. An additional point concerning the use of the chi-square test should be noted. The effectiveness of the test is diminished if the expected frequency in any cell is less than five. When this condition occurs, both the expected and observed frequencies of the appropriate cell should be combined with the values of an adjacent cell; the value of k should be reduced to reflect the number of cells in computing the test statistic. Also, it is desirable to have a value for k larger than three. The chi-square test is included in the summary of Table 9-2.

Example 9-7: Chi-Square Test for a Normal Distribution

To illustrate the use of the chi-square test for the normal distribution, a sample of 84 beams was tested to failure. The histogram is shown in Figure 9-4. In general, the normal distribution has a bell shape, but in this case there is one large cell in the lower tail. The sample mean and standard deviation of the failure loads are 10,100 and 780 lb, respectively. A null hypothesis is proposed that the failure loads are normally distributed with a mean and standard deviation of 10,100 and 780 lb, respectively. Table 9-3 gives the standardized variates z_i for the bounds of each interval, the probability that the variate Z is less than z_i, the expected probabilities for each interval, the expected and observed frequencies, and the chi-square values. The test statistic has a computed value of 10.209. Note that because the expected

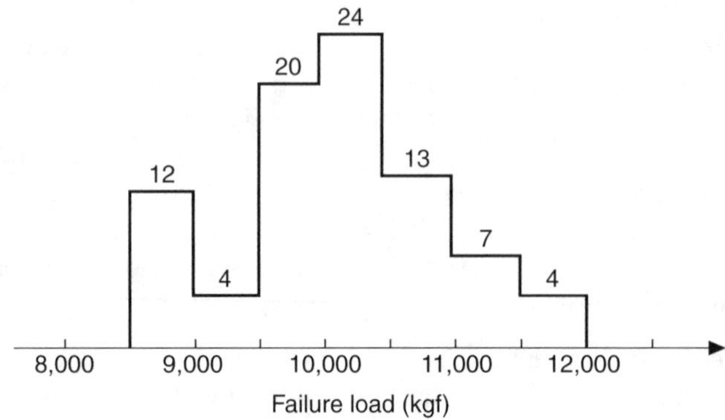

FIGURE 9-4 Frequency distribution of failure loads.

frequency for the seventh interval was less than 5, both the observed and expected frequencies were combined with those of the sixth cell. Thus, $k = 6$ is used to compute the degrees of freedom. Because the sample mean and standard deviation were used to define the hypotheses and compute the expected frequencies, 2 degrees of freedom are lost. Also, 1 degree of freedom is lost because of the sample size; thus, the remaining degrees of freedom are 3 for the test statistic. If past evidence had indicated a mean of 10,000 lb and a standard deviation of 1000 lb, and these statistics were used in Table 9-3 for computing the expected probabilities, then 5 degrees of freedom would be used. For a level of significance of 5% and 3 degrees of freedom, the critical χ^2_α value is 7.817. The null hypothesis is, therefore, rejected because the computed value is greater than the critical value. One may conclude that failure loading is not normally distributed with $\mu = 10{,}100$ lb and $\sigma = 780$ lb. The reason for the rejection of the null hypothesis may be due to one or more of the following: (1) failure loading is not normally distributed, (2) $\mu \neq 10{,}100$ lb, or (3) $\sigma \neq 780$ lb.

Example 9-8: Chi-Square Test for an Exponential Distribution — Case 1

Large volumes of sediment yield are less likely than small volumes. Thus, it is not unreasonable to assume that sediment yields might follow an exponential distribution, which has the following density $f_X(x)$ and cumulative functions $F_X(x)$, respectively:

$$f_X(x) = \lambda e^{-\lambda x} \tag{9-30}$$

and

$$F_X(x) = 1 - e^{-\lambda x} \tag{9-31}$$

TABLE 9-3 Computations for Failure Load Example

Cell i	z_i	$P(Z < z_i)$	Expected probability	Expected frequency (E_i)	Observed frequency (O_i)	$\frac{(O_i - E_i)^2}{E_i}$
Cell 1: $X \le 9000$	$\frac{9000 - 10{,}100}{780} = -1.41$	0.0793	0.0793	6.66	12	$\frac{(12 - 6.66)^2}{6.66} = 4.282$
Cell 2: $9000 < X \le 9500$	$\frac{9500 - 10{,}100}{780} = -0.77$	0.2206	0.1413	11.87	4	$\frac{(4 - 11.87)^2}{11.87} = 5.218$
Cell 3: $9500 < X \le 10{,}000$	$\frac{10{,}000 - 10{,}100}{780} = -0.13$	0.4483	0.2277	19.13	20	$\frac{(20 - 19.13)^2}{19.13} = 0.040$
Cell 4: $10{,}000 < X \le 10{,}500$	$\frac{10{,}500 - 10{,}100}{780} = 0.52$	0.6985	0.2502	21.02	24	$\frac{(24 - 21.02)^2}{21.02} = 0.422$
Cell 5: $10{,}500 < X \le 11{,}000$	$\frac{11{,}000 - 10{,}100}{780} = 1.15$	0.8749	0.1764	14.82	13	$\frac{(13 - 14.82)^2}{14.82} = 0.224$
Cell 6: $11{,}000 < X \le 11{,}500$	$\frac{11{,}500 - 10{,}100}{780} = 1.79$	0.9633	0.0884	7.42 } 10.50	11	$\frac{(11 - 10.50)^2}{10.50} = 0.024$
Cell 7: $11{,}500 < X$	∞	1.0000	0.0367	3.08 }		
Total			1.0000	84	84	10.209

TABLE 9-4 Data for Sediment Yield for Example 9-8 (in acre-feet per square mile per year)

2.67	1.55	0.61	0.21	0.16	0.076
2.65	1.42	0.51	0.21	0.15	0.070
2.37	0.99	0.37	0.20	0.14	0.040
2.31	0.69	0.35	0.18	0.14	0.036
2.20	0.66	0.25	0.17	0.12	0.030
1.65	0.64	0.22	0.17	0.09	0.020
					0.020

TABLE 9-5 Goodness-of-Fit Test for Exponential Density Function — Case 1

Cell i (1)	$F_X(x)$ (2)	x (3)	Expected frequency (E_i) (4)	Observed frequency (O_i) (5)	$\frac{(O_i - E_i)^2}{E_i}$ (6)
1	0.14	0.099	5.18	8	1.53
2	0.28	0.216	5.18	11	6.54
3	0.42	0.358	5.18	3	0.92
4	0.56	0.540	5.18	2	1.95
5	0.70	0.792	5.18	4	0.27
6	0.84	1.206	5.18	1	3.37
7	1.00	∞	5.92	8	0.73
Total			37.00	37	15.32

Note: Column 2, $F_X(x) = 1 - \exp(-\lambda x)$; column 3, $x = -\ln[1 - F_X(x)]/\lambda$; and column 4, $E_i = n[F_X(x_i) - F_X(x_{i-1})]$.

in which λ is a scale parameter. The method of moments yields the reciprocal of a sample mean as an estimator of λ; that is, $\lambda = 1/\overline{X}$.

The sample shown in Table 9-4 includes measured rates of sediment yield in acre-feet per square mile per year. Because the sample is small ($n = 37$) and five or more expected frequencies are needed per cell, the bounds of the cells will be set to get the maximum number of cells possible, which is seven. Using an increment of 0.14 for the cumulative probability provides an expected frequency of 5.18, with one cell having a probability of 0.16 and $E_i = 5.92$, as shown in Table 9-5. The value of λ is computed using the mean of the 37 sample values, which is 0.6579; thus, the estimate of λ is 1.52. Therefore, the hypotheses are:

H_0: Sediment yield has the following density function: $f_X(x) = 1.52 \exp(-1.52x)$.

H_A: $f_X(x)$ cannot be used to represent the sediment yield.

The calculation of the computed chi-square value is shown in Table 9-5. The computed value is 15.32. There are seven cells, but 2 degrees of freedom were lost, one for using n and the second for using the mean. Thus, there is a total of 5 degrees of freedom. Critical values of the χ^2 statistic are obtained from Table A-3. For 5, 1, and 0.5%, the critical values are 11.070, 15.086, and 16.75, respectively. Thus, the null hypothesis would be rejected for 5 and 1%, but not for 0.5%.

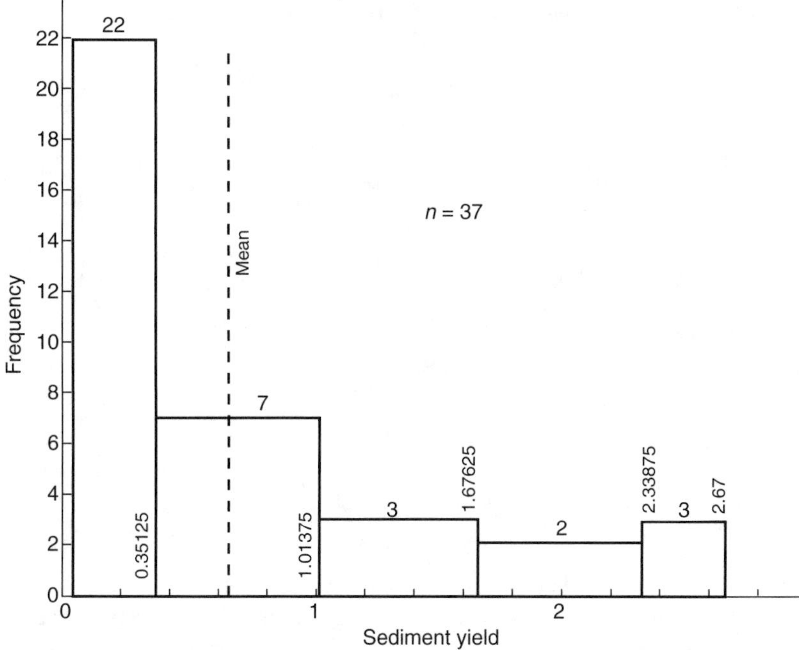

FIGURE 9-5 Histogram of sediment yield data.

Example 9-9: Chi-Square Test for an Exponential Distribution — Case 2

In this example, the sediment yield as discussed in Example 9-8 is considered. In this case, the data are not given, but a histogram is provided as shown in Figure 9-5, which has the shape of an exponential decay function. This example illustrates the effect of not having the original data (i.e., having only the histogram) on the results of the hypothesis testing.

The exponential density function is given in Equation 9-30, and the cumulative function is given in Equation 9-31. The method-of-moments estimator of λ is the reciprocal of the mean (i.e., $\lambda = 1/\overline{X}$). Probabilities can be evaluated by integrating the density function $f_X(x)$ between the upper and lower bounds of each interval.

Using the sediment yield X data and the histogram of Figure 9-5, a test of hypothesis was made for the following hypotheses:

H_0: sediment yield has the following density function: $f_X(x) = 1.52\exp(-1.52x)$.

H_A: $f_X(x)$ cannot be used to represent the sediment yield.

In this case, the calculation of the computed chi-square value is shown in Table 9-6. Although the histogram included five cells, the last three cells were combined

TABLE 9-6 Goodness-of-Fit Test for Exponential Density Function — Case 2

Cell i	$\int_{x_{lower}}^{x_{upper}} f_X(x)dx$	Expected frequency (E_i)		Observed frequency (O_i)		$\frac{(O_i - E_i)^2}{E_i}$
Cell 1: $0 \le X \le 0.35125$	0.4137	15.30		22		$\frac{(22-15.30)^2}{15.30} = 2.93$
Cell 2: $0.35125 < X \le 1.01375$	0.3721	13.77		7		$\frac{(7-13.77)^2}{13.77} = 3.33$
Cell 3: $1.01375 < X \le 1.67625$	0.1359	5.03		3		
Cell 4: $1.67625 < X \le 2.33875$	0.0497	1.84	7.93	2	8	$\frac{(8-7.93)^2}{7.93} = 0.00$
Cell 5: $2.33875 < X$	0.0286	1.06		3		
Total	1.0000	37.00		37		6.263

to ensure that all cells would have an expected frequency greater than 5. The computed chi-square statistic is 6.263. Two degrees of freedom are lost because n and $\overline{X}$ were used to compute the expected frequencies; therefore, with only three cells, there is only one degree of freedom. For levels of significance of 5 and 1% and one degree of freedom, the critical values are 3.841 and 6.635, respectively. Thus, the null hypothesis would be rejected for a 5% level of significance but accepted for 1%. This illustrates the importance of selecting the level of significance on the basis of a rational analysis of the importance of type I and II errors.

9.5.2. Kolmogorov–Smirnov One-Sample Test

A frequent problem in data analysis is verifying that the population can be represented by some specified probability density function. The chi-square goodness-of-fit test was introduced as one possible statistical test, but the chi-square test requires at least a moderate sample size of about 15, with much larger sample sizes preferred. The Kolmogorov–Smirnov one-sample test was also developed for verifying a population distribution and can be used with small samples.

The objective of the Kolmogorov–Smirnov one-sample test is to test the null hypothesis that the sample values of a random variable are likely to have been sampled from a specified probability function; the null hypothesis must specify both the population distribution function and its parameters. The alternative hypothesis is accepted if the distribution function is unlikely to be the underlying function; this may be indicated if either the density function or the specified parameters are incorrect.

The test statistic, which is denoted as KS, is the maximum absolute difference between the values of the cumulative distributions of a random sample and the probability distribution function specified in the null hypothesis. Critical values of the test statistic are available only for selected values of the level of significance as shown in Appendix A, Table A-7.

The Kolmogorov–Smirnov one-sample test may be used for small samples and is generally more efficient than the chi-square goodness-of-fit test when the sample size is small. The test requires data on at least an ordinal scale but is applicable for comparisons with continuous distributions. The Kolmogorov–Smirnov test uses the following six steps:

1. State the null and alternative hypotheses in terms of the proposed probability density function and its parameters.
2. The test statistic, *KS*, is the maximum absolute difference between the cumulative function of the sample (F_S) and the cumulative function of the probability function specified in the null hypothesis (F_X). The cumulative function for the sample (F_S) can be determined by rank-ordering the sample values from the smallest (x_1) to the largest (x_n) for a sample of size n. The sample cumulative function can be computed as:

$$F_S(x) = \begin{cases} 0 & \text{for } x < x_1 \\ \dfrac{i}{n} & \text{for } x_i \le x < x_{i+1} \\ 1 & \text{for } x \ge x_n \end{cases} \tag{9-32}$$

   The resulting step function should be used to determine the maximum absolute difference (*KS*) with the computed $F_X(x)$ at any value of x. The maximum difference can be computed as (with $F_S(x_0) = 0$ and $F_S(x_n) = 1$):

$$KS = \max\begin{Bmatrix} |F_X(x_1) - F_S(x_0)|,\ |F_X(x_1) - F_S(x_1)|,\ |F_X(x_2) - F_S(x_1)|, \\ |F_X(x_2) - F_S(x_2)|,\ \ldots, |F_X(x_i) - F_S(x_{i-1})|,\ |F_X(x_i) - F_S(x_i)|, \\ |F_X(x_n) - F_S(x_{n-1})|,\ |F_X(x_n) - F_S(x_n)| \end{Bmatrix} \tag{9-33}$$

3. The level of significance should be set; values of 0.05 and 0.01 are usually used.
4. A random sample should be obtained and the sample cumulative probability function computed. After computing the cumulative probability function, the value of the test statistic should be computed.
5. The critical value, KS_α, of the test statistic is a function of α and the sample size, n. Table A-7 in Appendix A shows the critical values that can be used with this test.
6. If the computed value *KS* is greater than KS_α, the null hypothesis should be rejected.

Example 9-10: Water Quality

The following is a set of 13 measurements of a water-quality parameter in ppm: {47, 53, 61, 57, 65, 44, 56, 52, 63, 58, 49, 51, 54}. If a level of significance of 5% is used, would it be safe to conclude that the sample is from a population that is normally distributed with a mean of 54.6 ppm and a standard deviation of 6.2 ppm?

TABLE 9-7 Example Using the Kolmogorov–Smirnov One-Sample Test

Rank i	x_i	$F_S(x_i)$	Standard normal variate	$F_X(x_i)$	$\lvert F_X(x_i) - F_S(x_{i-1})\rvert$	$\lvert F_X(x_i) - F_S(x_i)\rvert$
1	44	0.076923	−1.7097	0.043741	0.043740884	0.03318219
2	47	0.153846	−1.2258	0.110118	0.033194626	0.04372845
3	49	0.230769	−0.9032	0.183011	0.029165271	0.04775781
4	51	0.307692	−0.5806	0.280285	0.049516107	0.02740697
5	52	0.384615	−0.4194	0.336871	0.029179031	0.04774405
6	53	0.461538	−0.2581	0.397416	0.012801112	0.06412196
7	54	0.538462	−0.0968	0.460544	0.000994266	0.07791734
8	56	0.615385	0.2258	0.588190	0.049728426	0.02719465
9	57	0.692308	0.3871	0.649462	0.034077401	0.04284568
10	58	0.769231	0.5484	0.707070	0.014762421	0.06216066
11	61	0.846154	1.0323	0.847973	0.078741806	0.00181873
12	63	0.923077	1.3548	0.911449	0.065295000	0.01162808
13	65	1	1.6717	0.952707	0.029629859	0.04729322

Table 9-7 shows the calculations for computing the maximum difference. The cumulative probabilities $F_S(x)$ and $F_X(x)$ are shown in Figure 9-6. The sample mean ($\overline{X}$) and the sample standard deviation (S) were computed to be 54.6 and 6.2, respectively. From Table 9-7 the value of the test statistic, which equals the largest absolute difference, is 0.078742. For a 5% level of significance, the critical value is KS_α = 0.368. Because the test statistic KS is less than KS_α, the null hypothesis cannot be rejected; that is, a normal probability distribution with a mean of 54.6 ppm and standard deviation of 6.2 ppm can be used as a model.

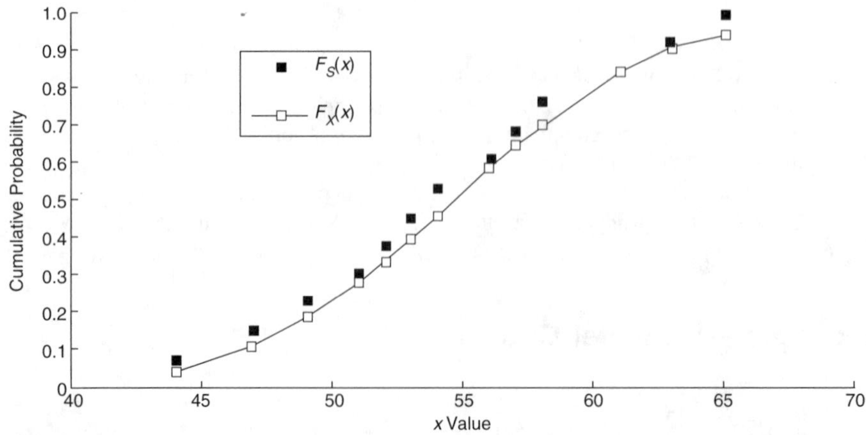

FIGURE 9-6 Example using the Kolmogorov–Smirnov one-sample test.

9.6. APPLICATIONS

Nine examples are used to illustrate the six-step procedure for testing hypotheses. In each case, a statistical analysis of engineering data is used to make inferences about the populations. Statistical inferences about the mean of a random variable depend on knowledge of the variance of the population; a hypothesis test concerning a mean value is illustrated in the first example. The second and third examples deal with a comparison of two population means rather than a test of a single population statistic. The fourth and fifth examples illustrate a hypothesis test on the variance of a population. The sixth, seventh, eighth, and ninth examples illustrate the comparison of two sample variances.

9.6.1. Test of Mean Strength of Steel Prestressing Cables

Samples of steel prestressing cables to be used in a curved gravity dam are tested prior to construction to ensure that the shipment meets the design capacity. Design plans call for 100-wire cables with a mean capacity of 900 kips. Eight samples were randomly selected and tested, with the following capacities: 825, 900, 915, 850, 870, 930, 835, and 885 kips. If the cables selected for testing are representative of the cables supplied by the manufacturer, can we conclude that the design capacity is met?

The design capacity of 900 kips provides the null hypothesis, H_0: $\mu = 900$ kips. Because capacities greater than 900 kips would be acceptable, the problem demands a one-tailed test with the alternative hypothesis, H_A: $\mu < 900$ kips. Therefore, the two hypothesis are:

$$H_0\text{: } \mu = 900 \text{ kips} \tag{9-34a}$$

$$H_A\text{: } \mu < 900 \text{ kips} \tag{9-34b}$$

If the null hypothesis is accepted, then the shipment of 100-wire cables is considered acceptable. If the null hypothesis is rejected, then the engineer should make adjustments, either in the form of a new shipment of cables or redesigning to allow for lower-capacity cables.

Because the standard deviation of the capacity is not known from other studies, the t test should be applied. Analysis of the eight measurements yields a mean and standard deviation of 876.2 and 38.0 kips, respectively. Thus, the computed value of the t statistic is:

$$t = \frac{\bar{X} - \mu_0}{S/\sqrt{n}} = \frac{876.2 - 900}{38/\sqrt{8}} = -1.771 \tag{9-35}$$

A sample of eight has 7 degrees of freedom. Figure 9-7 shows the distribution of the test statistic and the region of rejection for levels of significance of 10, 5, and 1%. For a 5% level of significance and a critical value of –1.895, according to Table

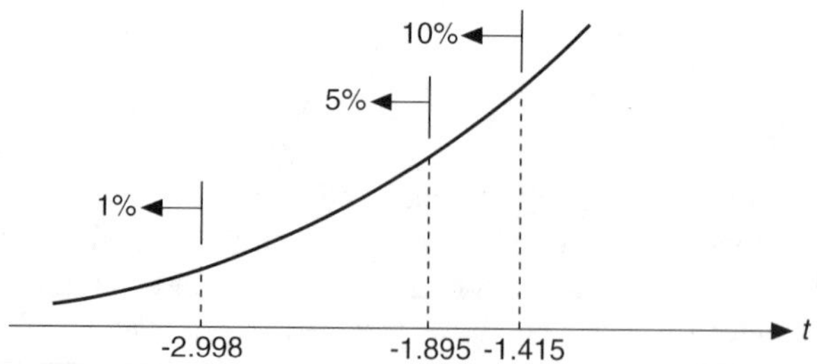

FIGURE 9-7 Distribution of test statistics.

A-2 the null hypothesis would be accepted. There is approximately a 6% chance of making a type I error. If a level of significance greater than about 6% were used, then the null hypothesis would have to be rejected.

9.6.2. Test of Mean Water Use of Hotels

Water use in hotels is measured on a basis of gallons per day per room (gpdr). Using annual data from 12 hotels in Baltimore, the mean and standard deviation are 225 and 35 gpdr, respectively. Data from seven hotels in Las Vegas give a mean and standard deviation of 420 and 90 gpdr, respectively. Based on these data, a test is made to assess whether or not average water use in western cities is greater than that in eastern cities:

$$H_0\text{: } \mu_E = \mu_W \tag{9-36a}$$

$$H_A\text{: } \mu_E < \mu_W \tag{9-36b}$$

The one-sided test on two means can be conducted using the test statistic of Equation 9-16 and the pooled variance of Equation 9-17 which is given by:

$$S_p^2 = \frac{(12-1)(35)^2 + (7-1)(90)^2}{12+7-2} = 3651\ (\text{gpdr})^2 \tag{9-37}$$

Thus, the pooled standard deviation is 60.4 gpdr. The t statistic is given by:

$$t = \frac{225-420}{60.43\left(\frac{1}{12}+\frac{1}{7}\right)^{0.5}} = -6.785 \tag{9-38}$$

with 17 degrees of freedom. The critical t values for 5 and 1% levels of significance are –1.74 and –2.567, respectively, using Table A-2. Thus, even at a 1% level, the null hypothesis must be rejected, with the implication that the average water use in the hotels of western cities is greater than that in eastern cities. In assessing the validity of the implication, it is important to know whether or not water use in the two cities is representative of water use in their respective regions.

9.6.3. Test of Two Sample Means for Car Speed

The average speed of cars was measured at a particular location on an interstate highway before and after passage of a law allowing speeds up to 65 mph. One group argued that the average speed in excess of the 65-mph limit would be smaller than the average speed in excess of the 55-mph limit before the law. Thus, the random variable is the difference between the average speed and the speed limit. For group 1, this would be $\overline{X}_1 - 55$; for group 2, this would be $\overline{X}_2 - 65$.

The comparison dictates the following null hypothesis:

$$H_0\colon \mu_1 = \mu_2 \tag{9-39a}$$

where μ is the difference of the population mean from the speed limit. Given that the group believes that the difference will decrease for the 65-mph limit, the alternative hypothesis is:

$$H_A\colon \mu_1 > \mu_2 \tag{9-39b}$$

Thus, it is a one-tailed test with the direction that the difference is greater for the 55-mph limit.

The following 16 speeds were measured for both the 55- and the 65-mph limit:

$$X_1 = \{62, 59, 67, 66, 56, 63, 58, 67, 72, 54, 61, 63, 60, 57, 68, 64\} \tag{9-40a}$$

$$X_1 = \{73, 66, 68, 63, 76, 72, 68, 71, 75, 67, 74, 59, 70, 58, 75, 78\} \tag{9-40b}$$

For groups 1 and 2, the means are 62.31 and 69.56 mph, respectively, and the standard deviations are 4.91 and 5.91 mph, respectively. Using Equation 9-16, the computed value of the test statistic is:

$$t = \frac{(62.31 - 55) - (69.56 - 65)}{S_p\left(\dfrac{1}{16} + \dfrac{1}{16}\right)^{0.5}} \tag{9-41}$$

where the pooled variance is computed from Equation 9-17 as:

$$S_p^2 = \frac{15(4.91)^2 + 15(5.91)^2}{16 + 16 - 2} = 29.52 \tag{9-42}$$

Thus, the pooled standard deviation is 5.433 mph, the computed value of the test statistic (t) is 1.432, and there are 30 degrees of freedom. For a level of significance of 5%, the critical t value for a one-sided test from Table A-2 is 1.697. Thus, the null hypothesis of equal excesses above the speed limit is accepted.

9.6.4. Variation of Effective Cohesion

The effective cohesion (C_e) is a decision element in problems of slope stability. Given the difficulty in determining the value of the effective cohesion at a site and the inherent variability of soil properties, including C_e, factors of safety are applied in slope stability problems. One engineer decides to estimate the effective cohesion from field sampling by taking five field samples and estimating C_e with each sample. However, as a practice, the engineer resamples when the variance of the sample estimates of C_e is significantly greater than 10^4 (psf)^2.

On one project, the engineer uses five samples to estimate the effective cohesion, with the following values: 620, 410, 730, 570, and 380 lb/ft^2. Based on these measurements, which have a standard deviation of 146.5 psf, the engineer tests the following hypotheses:

$$H_0\text{: } \sigma^2 = 10^4 \text{ (psf)}^2 \tag{9-43a}$$

$$H_A\text{: } \sigma^2 > 10^4 \text{ (psf)}^2 \tag{9-43b}$$

The computed value of the test statistic according to Equation 9-25 is:

$$\chi^2 = \frac{(n-1)S^2}{\sigma^2} = \frac{4(146.5)^2}{(100)^2} = 8.59 \tag{9-44}$$

which has $n - 1 = 4$ degrees of freedom. Using a 5% level of significance, the critical value from Table A-3 is 9.488, so the null hypothesis cannot be rejected. While the sample standard deviation is greater than the limiting value of 100 psf, it is not significantly greater. Thus, the engineer decides not to resample.

9.6.5. Uniformity of Illumination

In addition to the mean illumination in an office building, the lighting engineer is concerned with uniformity of lighting. Except for specific-task locations, the light-source arrangement should provide uniform illumination. Existing lighting systems in similar offices provide a standard deviation of 40 lm/ft^2. For a new building, the engineer is considering a new lightning system and decides to test the uniformity prior to ordering the system. In order to make replacement worthwhile, the engineer believes that the uniformity provided by the new system will have to be significantly better; thus, the following hypotheses will govern the decision:

$$H_0\text{: } \sigma^2 = \sigma_0^2 \tag{9-45a}$$

$$H_A\text{: } \sigma^2 < \sigma_0^2 \tag{9-45b}$$

The null hypothesis suggests that the existing lighting system is as good as the new system. If the illumination provided by the new system has a smaller variance than that of the existing system, then the alternative hypothesis is appropriate. If the null hypothesis is rejected, then the new system should be provided.

To test the uniformity provided by the new system, the engineer decides to collect a random sample of eight measurements in an office building that has the new system and use a 1% level of significance to make a decision. The eight measurements are 62, 97, 83, 92, 54, 78, 86, and 66 lm/ft^2, which yield a standard deviation of 15.20 lm/ft^2. The computed value of the χ^2 test statistic according to Equation 9-25 is:

$$\chi^2 = \frac{(n-1)S^2}{\sigma^2} = \frac{7(15.20)^2}{(40)^2} = 1.011 \tag{9-46}$$

For $k = \nu = n - 1 = 7$ degrees of freedom and a 1% level of significance, the critical value according to Table A-3 is 1.239. Because the computed value is less than the critical value, the null hypothesis should be rejected. Thus, the new lighting system will be used.

9.6.6. Variation of Infiltration Capacity

Two sites are being considered for an infiltration-based stormwater management system. In addition to the average infiltration capacity, it is desirable to have a relative uniform variation of capacity over the site where the facility is to be located. Because infiltration capacity is an important factor in siting, soil tests are made at feasible sites. For one project, two sites are being considered. In addition to the mean infiltration capacity, the variation is also computed. Because of other siting criteria, site A is currently favored over site B and will be selected if it meets the condition that the variability of the infiltration capacity at site A is not greater than that at site B. Thus, the following hypotheses are tested:

$$\text{H}_0\text{: } \sigma_A^2 = \sigma_B^2 \tag{9-47a}$$

$$\text{H}_\text{A}\text{: } \sigma_A^2 > \sigma_B^2 \tag{9-47b}$$

If the null hypothesis is accepted, then site A will be used; if it is rejected, then site B will be used.

Measurements are made at the two sites with the following results: A = {0.9, 1.4, 1.0, 1.7} in./hr, and B = {1.4, 0.9, 1.2, 1.4} in./hr. This yields standard deviations of $S_A = 0.3697$ and $S_B = 0.2363$. The computed value of the test statistic is:

$$F = \frac{S_A^2}{S_B^2} = \frac{(0.3697)^2}{(0.2363)^2} = 2.448 \tag{9-48}$$

with degrees of freedom of $k = \nu_A = n_A - 1 = 3$ and $u = \nu_B = n_B - 1 = 3$. The critical F value from Table A-4 is 9.28 for a 5% level of significance. Thus, we cannot reject the null hypothesis, so site A is selected as the better site for the planned stormwater management facility even though $S_A > S_B$.

9.6.7. Test of Two Variances for Car Speed

For the example of Section 9.6.3, the test statistic is based on the assumption that the variances of the two groups are equal. Furthermore, some argue that the higher speeds associated with the higher speed limit of 65 mph lead to a greater spread of speeds and that the accident rate increases as the spread of speeds increases. Thus, there is both a statistical and an engineering reason for testing for equality of variances for the data. The null hypothesis is:

$$H_0\colon \sigma_1^2 = \sigma_2^2 \tag{9-49a}$$

For the statistical assumption, the alternative hypothesis is:

$$H_A\colon \sigma_1^2 \neq \sigma_2^2 \tag{9-49b}$$

The computed F statistic (Equation 9-28) equals the ratio of the larger variance to the smaller variance, which is:

$$F = \frac{S_2^2}{S_1^2} = \frac{(5.910)^2}{(4.909)^2} = 1.449 \tag{9-50}$$

Both degrees of freedom equal 15, so from Table A-4 for a 5% level of significance, the critical F is 2.40; thus, the null hypothesis is accepted for a 10% level of significance. For the two-tailed alternative, the level of significance is twice the value shown on the table.

For the engineering assessment of the null hypothesis of equal variances, the alternative hypothesis is:

$$H_A\colon \sigma_2^2 > \sigma_1^2 \tag{9-51}$$

Therefore, the computed F statistic is the ratio S_2^2/S_1^2, which equals 1.449. The critical value from Table A-4 is 2.40, which means that the null hypothesis is accepted for a 5% level of significance. Thus, the test suggests that the spread of the speeds for the 65-mph limit is not significantly greater than the spread of the speeds for the 55-mph limit.

9.6.8. Test on Variances of Soil Friction Angle

Two methods of estimating the friction angle ϕ of soil are compared. The first method is more time intensive to conduct but is considered to be more precise. The second method is simple to apply, but it is believed to be less precise. A geotechnical engineer wishes to test whether or not the time-intensive method is more precise. From a homogeneous soil, five samples are obtained and analyzed with the first method, with the following estimates of ϕ: 0.34, 0.37, 0.35, 0.34, and 0.33. Five random samples obtained from the same soil and subjected to the second test yield the following estimates: 0.36, 0.32, 0.38, 0.31, and 0.34. The two-sample *F* test can be applied to evaluate the relative precision of the two methods of estimating ϕ. Because the second test is assumed to be less precise, the engineer expects its sample variance to be the greater, so it appears in the numerator of the *F* statistic of Equation 9-28 as follows:

$$F = \frac{S^2 \text{ of test 2}}{S^2 \text{ of test 1}} = \frac{0.00082}{0.00023} = 3.565 \tag{9-52}$$

which would have degrees of freedom of $\nu_1 = \nu_2 = 4$. The critical *F* value from Table A-4 is 6.39 for a 5% level of significance. Because the computed *F* is less than the critical *F*, the null hypothesis cannot be rejected; the engineer concludes that the two tests are equally precise.

9.6.9. Test on Variances of Nitrogen Levels

In Example 9-4, a test on equality of two means was made for samples of nitrogen levels. The test was based on the assumption that the population variances are equal, so there is a valid reason for testing for equality of variances. The test on the two means was made because it was believed that the development would increase the mean nitrogen level. It would also seem reasonable to test whether or not development increases the variance; thus, there is a second reason for testing for equality of variances. Because of the second reason, a one-sided test on the variances can be used as follows:

$$H_0\text{: } \sigma_b^2 = \sigma_a^2 \tag{9-53a}$$

$$H_A\text{: } \sigma_b^2 < \sigma_a^2 \tag{9-53b}$$

where σ_a^2 and σ_b^2 are the variances for the after- and before-development nitrogen levels, respectively. Because the test is one sided (i.e., with a specified direction), the test value of Equation 9-28 becomes:

$$F = \frac{S_a^2}{S_b^2} \tag{9-54}$$

in which S_a^2 and S_b^2 are the sample variances for the after- and before-development nitrogen-level data, respectively. The degrees of freedom are $k = \nu_1 = n_a - 1 = 13$ and $u = \nu_2 = n_b - 1 = 10$. The sample standard deviations are $S_a = 0.87$ mg/l and $S_b = 0.36$ mg/l. Thus, the computed F is:

$$F = \frac{(0.87)^2}{(0.36)^2} = 5.84 \tag{9-55}$$

For levels of significance of 5 and 1%, the critical F values are 2.89 and 4.66, respectively. Thus, the null hypothesis should be rejected at both levels.

The results of this test have two implications. First, it would suggest that development caused the variance of total nitrogen levels to increase. There may be an ecological consequence of this finding. Second, it would suggest that the assumption of equal variances that was used in the two-sample test on means of Section 9.3.3 was violated. Thus, a different test on two means may need to be investigated. Nonparametric test alternatives can be used instead, but these tests are beyond the scope of this book.

9.7. SIMULATION OF HYPOTHESIS TEST ASSUMPTIONS

In Section 6.5.2, the distribution of the correlation coefficient was approximated using simulation. The topics in this chapter stress the value of knowing the distribution of a statistic; specifically, the distribution function is used as the basis of testing hypotheses and in computing confidence limits. Knowledge of the distribution is also necessary to estimate sample size requirements.

For some statistics, such as the mean from a normal population, the distribution is known from theory. In other cases, theory cannot identify the underlying distribution. The correlation coefficient is an example of a statistic with a distribution function that cannot be identified from theory. In some cases, the distribution can be approximated. In other cases, a transformation can be made such that the resulting statistic has a known distribution or can be approximated with a known distribution. The chi-square goodness-of-fit test is another example where the test statistic is approximated by a known distribution function.

Some of the tests introduced in this chapter have a limitation or an assumption. For example, the t test on a single mean with an unknown population variance assumes that the mean was computed from a sample drawn from a normal population. Other tests are based on similar assumptions. The extent to which the assumption is important is rarely identified in textbooks, and conventional wisdom suggests that use of the test when the assumption is violated is wrong. Yet, if we only used statistical methods when all of the assumptions were met, statistical methods would be rarely used. Public opinion polls, with, for example, a stated accuracy of ±3%, are one prime example where the underlying assumptions are likely to have been violated.

In summary, this brief discussion has identified two factors that are relevant to simulation. First, simulation can be a very useful tool for identifying the probability distribution of a test statistic. Second, simulation can be used to test the importance of assumptions that are assumed valid when using a statistical hypotheses test.

Example 9-11: Assumption of Normality with the *t* Test

The t test for the sample mean of a single sample assumes that the sample is from a normal population. How important is this assumption? Would the t statistic provide reliable decisions if the data were samples from a chi-square distribution? Or an extreme-value distribution? Simulation can give some insight into this problem.

To test the assumption of normality, 20,000 samples, each of size 10, were sampled from normal, uniform, and exponential distributions. The data samples were generated using random-number generators for each of the distributions. The sample mean and standard deviation of each set of ten values were computed and used to compute the t value with Equation 9-12. Based on the t values computed for the 20,000 samples from a distribution, three cumulative distributions were developed. Histograms of the 20,000 t values were formed and were used to obtain critical (i.e., significance) values for 10, 5, 2.5, and 1% of the area in the upper tails. The critical values for samples from the three populations are as follows, along with the true t statistic for 9 degrees of freedom (ν):

Level of significance		Critical value for		
α (%)	$t_{\alpha,\nu}$	Normal	Uniform	Exponential
10	1.383	1.384	1.375	1.050
5	1.833	1.852	1.830	1.362
2.5	2.262	2.220	2.286	1.635
1	2.821	2.759	2.910	1.982

The accuracy of the simulated values can be expected to decrease as the level of significance becomes smaller. For a 10% level of significance, approximately 2000 sample values would be in the upper tail; therefore, the sample size is large enough to get a reasonably accurate estimate of the critical value. For the 1% level of significance, only 200 values would be in the tail of the distribution, such that one or two sample values could cause a noticeable change in the simulated critical value. The simulated values from the samples with a normal distribution produced absolute values of the errors of 0.001, 0.019, 0.042, and 0.062 for the 10, 5, 2.5, and 1% levels of significance, respectively. Greater accuracy could be achieved using more than 20,000 samples.

The errors for the statistics computed from the samples generated from a uniform distribution are 0.008, 0.003, 0.024, and 0.089. These errors are comparable to those from the normal-distribution samples. From these results, it appears that the t test would provide reasonable decisions when sampling is done from a uniform distribution. Both the normal and uniform distribution are symmetric. Could the t test be used with other symmetric distributions without severely violating the accuracy of the critical t values? Simulation could be used to answer this question.

The simulated t values computed from samples generated using an exponential population clearly indicate that the t test cannot be used when sampling from an exponential population. The errors are 0.333, 0.471, 0.627, and 0.839 for the four levels of significance, which are much larger than the errors shown for the normal and uniform critical values. The simulated critical values are in error when compared with the t values by amounts much greater than can be expected from sampling error. All of the critical values are much less than the true values. Thus, decisions made with the upper tail of the t test for samples from an exponential population can lead to rejection of the null hypothesis much less frequently than should be expected. From these results, would you expect that the result would generally be true for all nonsymmetric distributions?

9.8. PROBLEMS

9-1. What are the characteristics of a null hypothesis? Alternative hypothesis?

9-2. What factors influence the selection of the level of significance? Discuss each factor.

9-3. A can of paint is advertised to cover an area of 400 ft^2. What are the implications of type I and type II errors?

9-4. Define the region of rejection in terms of (a) values of the test statistic; (b) proportions of the area of the probability density function of the test statistic; (c) the region of acceptance; and (d) the critical value(s) of the test statistic.

9-5. What four factors influence the critical value of a test statistic? Show pictorially how each factor affects the critical value.

9-6. From Table 8-3a, find the sample that has a sample mean that deviates from 0 by the largest amount. Test the mean to determine if it is significantly different from 0. Use a two-sided alternative hypothesis and a 5% level of significance. Assuming the samples are from a $N(0,1)$ population, is it reasonable that the null hypothesis would be rejected? Explain.

9-7. Results of testing for the presence of pollutants in a local stream have a mean of 10 mg/l and a standard deviation of 2 mg/l. Six samples of water collected from the stream result in the following measurements: 12.7, 15.1, 9.5, 13.7, 19.6, and 16.4 mg/l. Does the level of pollutants in the stream exceed the original finding of 10 mg/l?

9-8. Assume that the compressive strength (lb/in.^{2}) of Boston blue clay has a true variance of 15.5 (lb/in.^{2})^2. Can we conclude that the sample of values in Problem 9-37 has a mean that is significantly less than 50 lb/in.^{2}?

9-9. Using the first sample of 5 in Table 8-3a, test the null hypotheses that $\mu = 0$. Use a two-sided alternative hypothesis and a 5% level of significance.

9-10. Two tests were provided for testing the hypothesis of a mean value against a standard, the Z test of Equation 9-6 and the t test of Equation 9-12. What are the differences of the two tests in terms of the purpose, data analysis requirements, underlying assumptions, and the critical value? Under what circumstances can the Z test be used as an alternative to the t test?

9-11. A sample of 20 yields a mean of 32.4. Test the two-sided hypothesis that the sample was drawn from a population with a mean of 35 for the following cases: (a) if the variance of the population is 33; and (b) if the variance of the population is unknown, but the sample variance is 33. Use a level of significance of 5%.

9-12. A random sample of 10 has a mean of 110. Test the null hypothesis that $\mu = 120$ against the alternative hypothesis that $\mu > 120$ at the 5% level of significance for the following cases: (a) assuming that the population standard deviation of 18 is known; and (b) assuming that the population standard deviation is unknown and the sample value is 18.

9-13. A random sample of 25 has a mean of 4.8 with a standard deviation of 0.32. Test the null hypothesis that $\mu = 4.95$ against the alternative hypothesis that $\mu < 4.95$ at the 1% level of significance.

9-14. A random sample of 10 yields a mean and standard deviation, respectively, of 73.6 and 7.9. Assuming a two-sided test, test the hypothesis that the sample is from a population with a mean of 80. Use a level of significance of (a) 5% and (b) 1%.

9-15. A public water supply official claims that the average household water use is greater than 350 gal/day. To test this assertion, a random sample of 200 households were questioned. If the random sample showed an average use of 359 gal/day and a standard deviation of 35 gal/day, would it be safe to conclude at the 1% level of significance that water use is greater than 350 gal/day?

9-16. A random sample of 12 has a mean and standard deviation of 240 and 30, respectively. Test the null hypothesis that $\mu = 215$ against the alternative hypothesis that $\mu > 215$ at a level of significance of (a) 5% and (b) 1%.

9-17. From the samples in Table 8-3, compare the sample means of the last two samples of 5 (1.412 vs. 0.094). Use a two-sided alternative hypothesis and a 5% level of significance. Assume the population is unknown.

9-18. Destructive tests made on 8 samples of wood to determine the failure strength yield a mean of 56.4 and a standard deviation of 5.22. Nondestructive tests are made on 12 samples of the same type of wood, resulting in a mean of 53.7 and a standard deviation of 3.16. Is it reasonable to conclude that the nondestructive method gives significantly lower measures of the strength?

9-19. Three measurements of a pollutant upgradient of a landfill yield a mean of 15.4 mg/l and a standard deviation of 2.8 mg/l. Five measurements made downgradient of the landfill result in a mean of 29.1 mg/l and a standard deviation of 7.4 mg/l. Is it safe to conclude that the landfill contributes to increasing the pollutant level in the groundwater?

9-20. Two competing lighting systems are installed in two adjacent parking lots, with system A being significantly less costly than system B. The manufacturer

TABLE 9-8 Annual Maximum Discharge (q_p)

Rank i	q_p (ft³/sec)	Rank i	q_p (ft³/sec)
1	21,500	30	7600
2	19,300	31	7420
3	17,400	32	7380
4	17,400	33	7190
5	15,200	34	7190
6	14,600	35	7130
7	13,700	36	6970
8	13,500	37	6930
9	13,300	38	6870
10	13,200	39	6750
11	12,900	40	6350
12	11,600	41	6240
13	11,100	42	6200
14	10,400	43	6100
15	10,400	44	5960
16	10,100	45	5590
17	9640	46	5300
18	9560	47	5250
19	9310	48	5150
20	8850	49	5140
21	8690	50	4710
22	8600	51	4680
23	8350	52	4570
24	8110	53	4110
25	8040	54	4010
26	8040	55	4010
27	8040	56	3100
28	8040	57	2990
29	7780	58	2410

of system B argues that its system provides more uniform lighting. Ten random measurements in parking lots A and B yield readings with standard deviations of 12 and 9, respectively. Can manufacturer B legitimately claim to provide more uniform lighting?

9-21. Using the first two samples of Table 8-3, which have means of 0.258 and –0.491, test the null hypothesis that $\mu_1 = \mu_2$ against a two-tailed alternative. Assume the population variances are not known. Use $\alpha = 5\%$.

9-22. Two nonlinear models are fitted to a sample of data of 12 observations. Model 1 has a bias of –0.34 and a standard error of estimate of 0.62. The corresponding values for model 2 are 0.27 and 0.81, respectively. Can we conclude that the biases of the models are not significantly different?

9-23. The standard deviation of the 58 annual maximum discharges for the Piscataquis River (see Table 9-8) is 4128 ft³/sec. Test the hypothesis that the population standard deviation is 3500 ft³/sec. Make a two-tailed test and use a 5% level of significance.

9-24. The standard deviation of the 58 annual maximum discharges for the Piscataquis River (see Table 9-8) is 4128 ft^3/sec. Regional studies indicate that the standard deviation for that watershed should be at least 3500 ft^3/sec. Using a 1% level of significance, study whether or not the sample value of σ can reasonably be at least 3500 ft^3/sec.

9-25. Using the data from Table 8-3, test the significance of the two sample variances for the last two of the 40 samples (0.421 vs. 1.192). Use a 10% level of significance.

9-26. Using the sample from Table 8-3 with the largest sample variance, test the null hypothesis that the variance is equal to 1 against the two-sided alternative. Use a 5% level of significance. What is the approximate rejection probability?

9-27. Using the sample from Table 8-3 with the smallest sample variance, test the null hypothesis that the variance is equal to 1 against the one-sided lower alternative. Use a 1% level of significance. Is it reasonable for the null hypothesis to be rejected?

9-28. Two irrigation systems are compared on their ability to provide uniform water distribution over a field. Twelve measurements with system X yield a standard deviation of 0.81 cm/hr. Nine measurements with system Y yield a standard deviation of 0.35 cm/hr. Can the conclusion be drawn that the systems provide equally uniform distributions of water?

9-29. Using the sample from Table 8-3 with the largest sample variance, test the null hypothesis that the variance is equal to the smallest sample variance of Table 8-3. Use a 1% level of significance.

9-30. Use the data in Table 9-8 for the Piscataquis River with the sample mean and standard deviation (i.e., parameters) of 8620 and 4128 ft^3/sec, respectively, to test for the goodness of fit of the normal distribution. Use the chi-square test.

9-31. Use the data in Table 9-8 for the Piscataquis River with the sample log-mean (to the base 10) and log-standard deviation of 3.8894 and 0.2031 ft^3/sec, respectively, to test for the goodness of fit of the lognormal distribution. Use the chi-square test. Note that sample log-mean is the mean of the transformed x sample values using $\log(x)$, and sample log-standard deviation is the standard deviation of the transformed x sample values using $\log(x)$.

9-32. The histogram of annual maximum discharges for the Piscataquis River is shown in Table 9-8 and Figure 9-8. Test whether or not it can be assumed that this random variable has a lognormal distribution with (i.e., parameters) log-$\mu = 3.8$ and log-$\sigma = 0.25$. Assume a 1% level of significance and use the chi-square test.

9-33. Using the sediment yield data presented in the histogram of Figure 9-5, test whether or not the sediment yield can be represented by a uniform distribution with $a = 0.0$ and $b = 2.5$. Use a level of significance of 5% and the chi-square test.

9-34. The histogram of annual maximum discharges for the Piscataquis River is shown in Table 9-8 and Figure 9-8. Test whether or not it can be assumed that this random variable has a normal distribution with $\mu = 8620$ and $\sigma = 4128$. Assume a 10% level of significance and use the chi-square test.

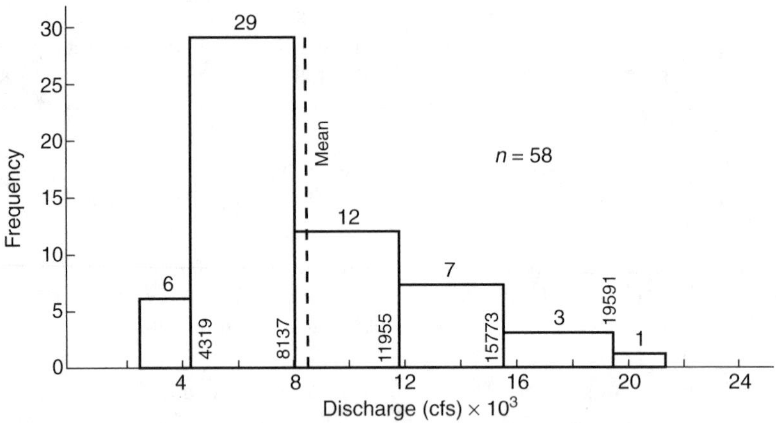

FIGURE 9-8 Histogram for annual maximum discharge in cubic feet per second for the Piscataquis River.

9-35. Use the data in Table 9-8 for the Piscataquis River with the sample mean and standard deviation of 8620 and 4128 ft^3/sec, respectively, to test for the goodness of fit of the normal distribution. Use the Kolmogorov–Smirnov test.

9-36. Use the data in Table 9-8 for the Piscataquis River with the sample log-mean and log-standard deviation (i.e., parameters) of 3.8894 and 0.2031, respectively, to test for the goodness of fit of the lognormal distribution. Use the Kolmogorov–Smirnov test.

9-37. Nine measurements of the compressive strength of a Boston blue clay are 41.6, 48.7, 45.4, 44.0, 46.1, 44.8, 47.7, 45.5, and 42.9 lb/in.^{2}. Using a 5% level of significance and the Kolmogorov–Smirnov one-sample test, decide whether or not the values are normally distributed.

9-38. The time for collecting tolls at a tollbooth on an interstate was measured for a short period, with the following results: 17.2, 23.4, 16.7, 19.0, 21.2, 20.8, 18.7, 20.4, 18.0, 22.1, 17.9, and 19.3 sec. Can the traffic authority responsible for staffing the tollbooth legitimately claim that the service time is normally distributed with a mean of 20 sec and a standard deviation of 1.5 sec? Use the Kolmogorov–Smirnov one-sample test.

9-39. Write a pseudocode that could be used to simulate the critical values for the normal, uniform, and exponential distributions presented in Example 9-11.

9-40. Simulate the critical values for the lower tail of the t statistic for the normal, uniform, and exponential distributions following the procedure of Example 9-11. Use a sample size of 10.

9-41. Gamma-distributed random numbers g_i for scale parameter b and shape parameter c can be generated from uniform variates u_i as follows:

$$g_i = b\ln\left(\prod_{i=1}^{c} u_i\right)$$

Using a sample size of 10, simulate the 10 and 5% critical values of the t statistic $t = (\bar{X} - \mu)/(S/\sqrt{n})$ where $\mu = bc$. With $b = 1$, perform the analyses for $c = 3$, 6, and 10. Assess the applicability of the test for each of these populations. Would it be appropriate to use the t test for any of these populations? Explain.

9.9. SIMULATION PROJECTS

Use one of the data sets of Appendix C (the same data set used for the simulation project in Section 8.6). Repeat the steps of the simulation project in Section 8.6 except in this case set up the histograms for t and z such that they cover the upper and lower tails of the distribution of t and z. This approach will make it possible to verify the tabulated z and t values for levels of significance of 10%, 5%, and 1%. Note any differences and discuss the accuracy of the generation process and sampling variables of computed distributions.

Chapter 10
Analysis of Variance

TABLE OF CONTENTS

10.1. INTRODUCTION

Consider the following situations:

1. The students in a course on introductory statistics are divided into five groups, each of which is assigned a different textbook. The students attend the same lectures and are tested using the same homework assignments and examinations. The average final grade, from 0 to 100, for each of the five groups is computed. If the average for one group is considerably higher than that of the other four groups, can we conclude that the textbook used by this group is a better learning tool?
2. In an effort to reduce traffic accidents in a state, the state highway department examines the number of accidents at sites with four types of traffic control: (1) two-way stop signs, (2) four-way stop signs, (3) flashing red or amber traffic lights, and (4) controlled timing red, yellow, and green traffic lights. The average number of accidents at a number of sites is computed for each of the four control types.

Can we conclude that the traffic-control method having the lowest average accident rate is the most effective?

3. Irrigation equipment is designed to give a uniform spread of water over a field. A nonuniform distribution will result in less than optimum crop production because some areas of the field will receive too much water, while other parts will not receive enough water. A farmer is considering the purchase of a new type of irrigation equipment that is advertised to give a more uniform distribution of water than the equipment that he currently owns. To test the uniformity of the distribution of water, the farmer tries each piece of equipment on his field for 4 hours and collects samples of the depth of water supplied to 25 points throughout the field for both pieces of equipment. If the variance of the 25 measurements for the new piece of equipment is less than the variance for the farmer's existing equipment, can the farmer conclude that the new equipment provides water more uniformly over the field?

In each of these situations, a decision must be made. Should the textbook used by the students having the highest average grade be used by all students in the future? Should all intersections be equipped with the type of traffic control that has the lowest accident rate, as indicated by the sample information? Should the farmer expend funds to buy the new equipment? The decision maker must use sample information to make inferences about the population; therefore, the situations appear to warrant a hypothesis test on characteristics of two or more groups. The textbook decision involved the mean value for five groups. The traffic-control problem required a decision concerning the mean of four groups. The farmer is confronted with a decision involving two group variances.

Decisions such as those outlined can be made with the help of the analysis of variance (ANOVA), which can be classified as a statistical method for comparing two or more populations or treatments. It can be used for testing hypotheses involving two or more means. It can also be used for testing the equality of variances. Because of the frequency with which such problems arise in engineering analysis, the ANOVA is probably the second most widely used statistical technique in engineering analysis; regression is used most frequently. Regression and the ANOVA have much in common, including the assumption of a linear model. Also, the ANOVA test is useful in testing for the significance of the slope coefficient of a regression equation.

10.2. TEST OF POPULATION MEANS

In its simplest form, the ANOVA can be viewed as a hypothesis test of group means. Although it is important to understand the underlying assumptions, it is worthwhile to view the ANOVA in terms of the six steps of a hypothesis test, after which the method can be viewed in more general terms.

10.2.1. Steps in the ANOVA

In Chapter 8, the six hypothesis steps were discussed. Using these same six steps, the ANOVA can be solved computationally as follows.

Step 1: Formulation of hypotheses — If a problem involves k groups, the following hypotheses are appropriate for comparing k group means:

H_0: $\mu_1 = \mu_2 = \ldots = \mu_k$ (10-1a)

H_A: at least one pair of group means are not equal(10-1b)

The test compares the means, but if the null hypothesis is rejected, the following five steps do not identify which pair or pairs of means are not equal; this problem is covered later. The group means are expressed as population values, not sample values.

Step 2: Define the test statistic and its distribution — The hypotheses of step 1 can be tested using the following test statistic:

$$F = \frac{MS_b}{MS_w} \tag{10-2}$$

in which MS_b and MS_w are the mean squares between and within variations, respectively, and F is the value of a random variable having an F distribution with degrees of freedom of $(k - 1, N - k)$. The mean squares are computed as shown in Table 10-1.

Step 3: Specify the level of significance — The level of significance is used in the same way that it is for other tests of hypotheses. Tables of the F distribution are usually available only for levels of significance of 5 and 1%; however, these are somewhat arbitrary values and may not have physical significance.

Step 4: Collect data and compute test statistic — The data should be collected and used to compute the value of the test statistic (F) of Equation 10-2. The data are best portrayed as a matrix as shown in Table 10-2. There are k columns, with each column representing a group. The jth column contains n_j values. The total number of observations N is given by:

$$N = \sum_{j=1}^{k} n_j \tag{10-3}$$

All n_j do not have to be equal.

TABLE 10-1 Computation Table for the Analysis of Variance Test

Source of variation	Sum of squares	Degrees of freedom	Mean square
Between	$SS_b = \sum_{j=1}^{k} n_j(X_j - \bar{X})^2$	$k - 1$	$MS_b = SS_b/(k-1)$
Within	$SS_w = \sum_{j=1}^{k}\sum_{i=1}^{n_j}(X_{ij} - \bar{X}_j)^2$	$N - k$	$MS_w = SS_w/(N-k)$
Total	$SS_t = \sum_{j=1}^{k}\sum_{i=1}^{n_j}(X_{ij} - \bar{X})^2$	$N - 1$	—

TABLE 10-2 Data Matrix for the Analysis of Variance

	Observations for group				
	1	**2**	**3**	**...**	***k***
	X_{11}	X_{12}	X_{13}		X_{1k}
	X_{21}	X_{22}	X_{23}		X_{2k}
	X_{31}	X_{32}	X_{33}		X_{3k}
	.	.	.		.
	.	.	.		.
	.	.	.		.
	$X_{n_1 1}$	$X_{n_2 2}$	$X_{n_3 3}$		$X_{n_k k}$
Mean	$\bar{X}_1$	$\bar{X}_2$	$\bar{X}_3$	—	$\bar{X}_k$

Step 5: Determine the critical value of the test statistic — The critical value of the F statistic is obtained from tables of the F distribution (see Table A-4); it is a function of the level of significance and the degrees of freedom. The region of rejection consists of all values of F greater than the critical F value. If the computed value of step 4 is greater than the critical value, the null hypothesis of Equation 10-1a should be rejected and the alternative hypothesis of Equation 10-lb accepted.

Step 6: Make a decision — The computed and table values should be compared in order to select the appropriate hypothesis.

Example 10-1: Effectiveness of Textbooks

Before examining either the rationale for the ANOVA test or the details of computation, it may be instructive to consider a simple example. The experiment described previously about the use of five different textbooks for a statistics class can be used to illustrate the ANOVA test. The distribution of final grades is given in Table 10-3. The class of 20 was divided into five groups of four students each. The individual grades ranged from 66 to 91, and the group means ranged from 74 to 82. The grand

TABLE 10-3 Example: Test Scores for the ANOVA

	Grades for group				
	1	**2**	**3**	**4**	**5**
	82	67	91	66	82
	75	79	82	73	71
	87	77	76	89	67
	76	81	79	84	76
Mean	80	76	82	78	74

TABLE 10-4 Example: ANOVA Table

Source of variation	Degrees of freedom	Sum of squares	Mean square
Between	4	160	40.00
Within	15	788	52.53
Total	19	948	—

mean, $\overline{X}$, equals 78. The equations in Table 10-1 were used to compute the values of the ANOVA table of Table 10-4. The computed F value of Equation 10-2 equals 0.76. For a 5% level of significance and degrees of freedom (4, 15), the critical F value is 3.06. Because the computed value is less than the critical value, the null hypothesis cannot be rejected. Thus, even though the group means had a range of 8, the evidence is not sufficient to reject the null hypothesis of equal means. This implies that the textbook is not responsible for the differences in the group means. One must conclude that the differences are due to "random" factors; that is, factors that were not controlled, such as students' past experiences and native intelligence, are responsible for the observed variation of the scores.

10.2.2. Rationale of the ANOVA Test

The ANOVA test is a comparison of means. The example demonstrated that the null hypothesis can be accepted even when the means are not equal, but how much variation can be permitted? First, it must be recognized that two sources of variation must be considered: variation within each group and variation between group means. This leads to the values for the "sources of variation" column in the ANOVA table. The "sums of squares between" entry is intended to reflect the variation between the group means. It has the general form of a variance calculation; that is, the value of SS_b involves the square of the difference between a value and a mean. In the case of the "sums of squares between," each squared difference is multiplied by the number of elements in the group, n_j; the value of n_j serves as a weight that is attached to the squared difference. The larger the value of n_j, the more weight given to the squared difference. The "sums of squares within" entry is intended to reflect the variation of the scores, X_{ij}, within the group. It also has the general form of a variance calculation; however, in this case, the difference being squared depends on the sample score, X_{ij}, and the group mean. The F statistic is then a comparison of the two variance measurements. A large value of F results when the variation between groups is large in comparison to the variation within groups. The sums of squares are divided by the degrees of freedom, which are measures of the number of independent values used in calculating the sums of squares. The greater the degrees of freedom, the more confidence one can have in a computed mean square. Chapter 11 shows that the confidence in a mean value improves as the sample size increases; the width of a confidence interval on the mean decreases inversely as the square

TABLE 10-5 Equality of Group Means

	Scores for group		
	1	**2**	**3**
	66	72	80
	73	81	85
	89	84	70
	84	75	77
Mean	78	78	78

TABLE 10-6 ANOVA Table for Equality of Group Means

Source of variation	Degrees of freedom	Sum of squares	Mean square
Between	2	0	0
Within	9	534	59.33
Total	11	534	—

TABLE 10-7 Inequality of Group Means

	Scores for group		
	1	**2**	**3**
	78	82	74
	78	82	74
	78	82	74
	78	82	74
Mean	78	82	74

root of the sample size in accordance to the sampling distribution of the mean. The same concept applies here.

Two simple examples may help illustrate the rationale of the F statistic. Table 10-5 shows the scores for three groups, with each group having four values. The scores are such that there is no variation in the group means; that is, the group means for the sample data are equal, so there would be no reason to expect the population means to be unequal. The ANOVA table for the data of Table 10-5 is shown in Table 10-6. Because all of the group means equal the grand mean of 78, the "between sums of squares" is zero. Thus, the computed F statistic equals zero. As expected, the null hypothesis is accepted.

The scores for the second example are given in Table 10-7. In this case, no variation within the groups is found. Thus, one would intuitively feel that the sample mean for a group is a good indicator of the population mean of the group. After all, the standard error of the mean, $S/\sqrt{n}$, for each group is zero. In this case, the ANOVA table indicates that both the "within sum of squares" and "mean square within" equal zero. Thus, the computed F statistic equals infinity (Table 10-8), and it would be larger than the critical value.

TABLE 10-8 ANOVA Table for Inequality of Group Means

Source of variation	Degrees of freedom	Sum of squares	Mean square
Between	2	128	64
Within	9	0	0
Total	11	128	—

In summary, these two extreme examples were designed to show that the F statistic is a comparison of two variances, and the ANOVA test is designed to systematically weigh the variation within a sample against the variation between the sample means. The critical F statistic is a measure of the variation that would result if the variation was due totally to random factors.

A second factor of importance is that the ANOVA tests involve a linear separation of the total variation into two parts. The equation for computing the total sum of squares is nothing more than the summation used in computing the variance of the scores. It is linearly divided into two parts. The first part is the experimental part; that is, the "between sum of squares" is intended to measure the variation that is due to the factor that led to the experiment in the first place. The second part is the confusion part; that is, the "within sum of squares" is the random variation due to other factors that were not controlled. Thus, the total sum of squares is separated into parts reflecting the experimental effect and the random effect.

The null hypothesis of Equation 10-la indicates that the ANOVA test is a test of k population means. This process can be represented schematically as shown in Figure 10-lb. Samples are drawn from k populations, with each population having mean μ_i, and the samples are subjected to the same treatment. After the treatment has been applied, the subjects in all groups are given the same test. Because the treatment and testing are the same, any difference in sample means can be attributed to differences in the original population means. The ANOVA test allows for random variation when making a decision.

Alternatively, the null hypothesis could be stated as:

$$H_0: \mu_1 = \mu_2 = \ldots = \mu_k = \mu \tag{10-4}$$

in which μ is the mean of a single population from which all the samples are drawn. In this case (shown schematically in Figure 10-1a), the samples are subjected to different treatments. After the treatments have been applied, a common test is given, and the group mean scores are computed. In this case, differences in group means suggest that the treatment was the factor that caused the differences in the sample means.

A simple example illustrates the difference between the two cases of Figure 10-la and b. When the group of 20 seniors was divided into five groups of four students and taught using five different textbooks, we had one population and five treatments; thus, this example is represented by Figure 10-la. If we had a class consisting of four freshmen, four sophomores, four juniors, four seniors, and four graduate students and the students were taught in the same classroom and used the same textbook, we would

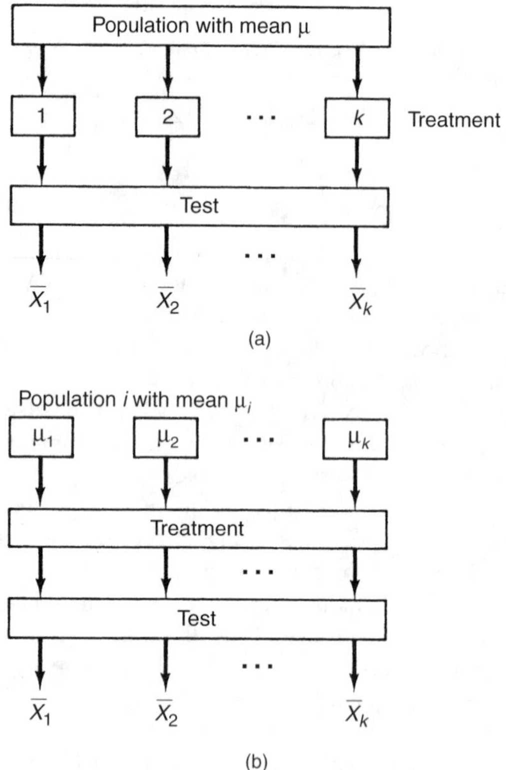

FIGURE 10-1 Schematic of ANOVA test. (a) One population and k treatments; (b) k populations with one treatment.

be dealing with five populations, with one common treatment and one test; thus, this example is represented by Figure 10-lb. Although the ANOVA test procedure is the same, the distinction is important in understanding the application of the ANOVA test.

10.2.3. Linear Separation of the Total Variation

It may be useful to examine the ANOVA test from the standpoint of a linear separation of the total variation. Using the notation discussed previously, an observation, or score, X_{ij}, can be represented as:

$$X_{ij} = \mu + (\mu_j - \mu) + (X_{ij} - \mu_j) \tag{10-5}$$

The term $(\mu_j - \mu)$ represents the variation in X_{ij} that can be attributed to the difference between its group mean and the grand mean. The term $(X_{ij} - \mu_j)$ represents the variation of the observation from its group mean. If we let $\beta_j = \mu_j - \mu$ and $\epsilon_{ij} = (X_{ij} - \mu_j)$ and subtract μ from both sides, Equation 10-5 reduces to:

$$X_{ij} - \mu = \beta_j + \epsilon_{ij} \tag{10-6}$$

If the null hypothesis of Equation 10-4 is correct, all $\beta_j = 0$; thus, the hypotheses for the ANOVA test can be rewritten as:

$$H_0\text{: } \beta_j = 0 \text{ for } j = 1, 2, \ldots, k \tag{10-7a}$$

$$H_A\text{: at least one } \beta_j \neq 0 \tag{10-7b}$$

Each of the terms in Equation 10-6 represents a difference between a value and a mean. The accuracy of a value is reflected by the variance of its distribution. For the β_j term, the variance of interest is the variance of the group means, $S_{\bar{X}}^2$:

$$S_{\bar{X}}^2 = \frac{\sum_{j=1}^{k} (\bar{X}_j - \bar{X})^2}{k - 1} \tag{10-8}$$

The error of a population mean is represented by its standard error, which, in general, is equal to the population standard deviation divided by the square root of the sample size. For the k group means, $S_{\bar{X}}^2$ will be an estimate of (σ^2/k) if H_0 is true, where σ^2 is the population variance; therefore, the between-group estimate of the population variance, which will be denoted as $\hat{\sigma}_B^2$, equals $kS_{\bar{X}}^2$. It is also necessary to compute the average of the variances of the groups, which is denoted as $\bar{S}^2$ and is computed by:

$$\bar{S}^2 = \frac{1}{k}\sum_{j=1}^{k} S_j^2 \tag{10-9}$$

in which the group variance S_j^2 is computed by:

$$S_j^2 = \frac{\sum_{i=1}^{n_j} (X_{ij} - \bar{X}_j)^2}{n_j - 1} \tag{10-10}$$

The variance within a group is also an estimate of the variance within the population; that is, S_j^2 is an estimate of σ^2. Thus, $\bar{S}^2$ will also be an estimate of σ^2. If we denote $\bar{S}^2$ by $\hat{\sigma}_W^2$, then both $\hat{\sigma}_B^2$ and $\hat{\sigma}_W^2$ are estimates of σ^2. If the null hypothesis is true, $\hat{\sigma}_B^2$ will equal $\hat{\sigma}_W^2$, and the F statistic will equal unity. If the null hypothesis is false, we would expect $\hat{\sigma}_B^2$ to be significantly greater than $\hat{\sigma}_W^2$, and the computed F statistic would be significantly greater than 1. Thus, the linear separation of Equation 10-5 leads to two estimates of the population error variance, which are then used to test for the equality of group means.

10.2.4. Computational Equations

The equations given in Table 10-1 for determining the sums of squares are of interest primarily because they reflect the definition of the sum of squares. When performing an ANOVA test, other formulas can be used that reduce the computational effort. The computational formulas for the total sum of squares (SS_t), the within-group sum of squares (SS_w), and the between-group sum of squares (SS_b) are:

$$SS_t = \left(\sum_{j=1}^{k}\sum_{i=1}^{n_j} X_{ij}^2\right) - \frac{T^2}{N} \quad (10\text{-}11)$$

$$SS_w = \left(\sum_{j=1}^{k}\sum_{i=1}^{n_j} X_{ij}^2\right) - \sum_{j=1}^{k}\frac{T_j^2}{n_j} \quad (10\text{-}12)$$

$$SS_b = \left(\sum_{j=1}^{k}\frac{T_j^2}{n_j}\right) - \frac{T^2}{N} \quad (10\text{-}13)$$

in which T_j is the total of the scores in group j, and T is the sum of the T_j values; that is, T is the total of all scores in all samples.

Example 10-2: Fertilizer Mixes and Crop Yield

A 20-acre field is divided into 20 1-acre plots that are randomly assigned to four groups, with five plots in each group. Plots within each group are treated with a specific mix of fertilizer. The crop yield from each plot is determined (Table 10-9). This case is an example of the ANOVA test with one population, which is shown schematically in Figure 10-1a. The mean crop yields are determined for each group (Table 10-9) and used to test the null hypothesis of equal means:

$$H_0: \mu_1 = \mu_2 = \mu_3 = \mu_4 = \mu \quad (10\text{-}14)$$

TABLE 10-9 Example: Crop Yields (bushels/acre) for 20 Plots

	Group				
Plot	**1**	**2**	**3**	**4**	**Total**
1	75	103	67	87	
2	93	89	84	79	
3	82	97	73	86	
4	104	108	82	73	
5	96	98	74	80	
Mean	90	98	76	81	
Standard deviation	11.51	7.11	6.96	5.70	
T_j	450	495	380	405	1730
$\sum_{i=1}^{5} X_{ij}^2$	41,030	49,207	29,074	32,935	152,246

TABLE 10-10 ANOVA Table for Crop Yield Example

Source of variation	Degrees of freedom	Sum of squares	Mean square
Between	3	1545	515
Within	16	1056	66
Total	19	2601	—

The alternative hypothesis is that at least one pair of means are not equal. Acceptance of the null hypothesis would imply that the fertilizer mix would not have a significant effect on the crop yield. Rejection of the null hypothesis would imply that the mix does have an effect on crop yield.

The computation formulas of Equations 10-11, 10-12, and 10-13 were used to compute the entries to the ANOVA table of Table 10-10. The computed F statistic of Equation 10-2 equals 7.80. For a level of significance of 5% and degrees of freedom of (3, 16), the critical F value equals 3.24. Thus, the null hypothesis must be rejected because the computed F value is greater than the critical value. This indicates that the fertilizer mix has an effect on the crop yield and that at least one pair of group means are significantly different.

10.3. MULTIPLE COMPARISONS IN THE ANOVA TEST

If the null hypothesis is rejected, we can conclude only that at least one pair of group means are unequal. The test does not specify which pair or pairs of means are unequal. It is usually insufficient to simply determine whether or not the null hypothesis of Equation 10-la is valid; that is, we usually wish to know what differences exist in the group means when the null hypothesis is rejected. Unfortunately, a unique solution to the problem does not exist. Because a unique, theoretically valid solution cannot be obtained and because ANOVA types of problems involve different data characteristics, a number of solutions to these problems have evolved. Although a number of alternatives are available, only two are discussed herein.

10.3.1. Duncan Multiple Range Test

The Duncan multiple range test is applicable to problems for which the sample sizes are equal; the test also assumes that the populations have normal distributions with equal variances and that the random samples are independent. The test is usually applied to each pair of group means. Thus, the null hypothesis for groups i and j would be:

$$H_0\colon \mu_i = \mu_j \tag{10-15}$$

with the alternative hypothesis that the pair of means are not equal. The test statistic (R_m) is:

$$R_m = S_{\bar{X}}(r_m) \tag{10-16}$$

where r_m is a value obtained from Table A-9 and is a function of the level of significance, α; the number of groups being compared, m; and the degrees of freedom $(N - k)$. R_m is the value of a random variable called the least significant range, and $S_{\bar{X}}$ is given by:

$$S_{\bar{X}} = \left(\frac{MS_w}{n_j}\right)^{0.5} = \left[\frac{SS_w}{n_j(N-k)}\right]^{0.5} \tag{10-17}$$

The test is conducted by obtaining MS_w from the ANOVA table to compute $S_{\bar{X}}$. The value of R_m is computed with Equation 10-16. For any pair of sample means, the difference is computed and compared with the least significant range, R_m. If the absolute value of the difference is greater than R_m, the null hypothesis of Equation 10-15 should be rejected. The test can be used on each pair of group means to identify which pairs are significantly different.

Example 10-3: Assessing Effectiveness of Fertilizer Mixes on Crop Yield

The null hypothesis of Equation 10-14 was rejected. Thus, it is of interest to know which group means were unequal so that some assessment of the effectiveness of a fertilizer mix can be made. The Duncan multiple range test was applied to each pair of group means, and the computations are given in Table 10-11. From Table 10-10, the "mean square within" is 66; therefore, $S_{\bar{X}}$ equals 3.63. For $(N - k)$ degrees of freedom, which equals 16, the values of r_m are 3.00 and 4.13 for levels of significance of 5 and 1%, respectively. Thus, the corresponding values of R_m are 10.89 and 14.99, respectively. A computed range that is greater than R_m is considered significant. The decisions to accept or reject H_0 are given in Table 10-11 for both levels of significance. With one exception, the decision is the same for both levels of significance. The means for the following pairs of groups are significantly different: (2 and 3) and (2 and 4). At the 5% level of significance, the means for groups 1 and 3 are also different.

Multiple Group Comparisons

It is also possible to use the Duncan multiple range test to make comparisons of more than two group means. Again, the least significant range is computed with Equation 10-16, but the value of r_m is obtained for m groups and $(N - k)$ degrees of freedom. The computed range is the difference between the largest and the smallest group means for the groups being compared.

For the crop yield example, there are five sets of three or more groups, with one of these sets being composed of all four groups. For four groups, the values of r_m are 3.23 and 4.43 for levels of significance of 5 and 1%, respectively. Given that $S_{\bar{X}}$ equals 3.63, the least significant ranges for levels of significance of 5 and 1% are 11.7 and 16.1, respectively. The range of the four group means is 23; therefore, as indicated in Table 10-12, the computed range exceeds the least significant range, and the null hypothesis of equality of group means is rejected. This conclusion should be expected

TABLE 10-11 Duncan Multiple Range Tests for Pairs of Group Means for Crop Yield Example

Groups	$\overline{X}_i$	$\overline{X}_j$	Range	Decision $\alpha = 0.05$	Decision $\alpha = 0.01$
1, 2	90	99	9	Accept H_0	Accept H_0
1, 3	90	76	14	Reject H_0	Accept H_0
1, 4	90	81	9	Accept H_0	Accept H_0
2, 3	99	76	23	Reject H_0	Reject H_0
2, 4	99	81	18	Reject H_0	Reject H_0
3, 4	76	81	5	Accept H_0	Accept H_0

TABLE 10-12 Duncan Multiple Range Tests for Multiple Comparisons for Crop Yield Example

Groups	Group means	Range	Decision $\alpha = 0.05$	Decision $\alpha = 0.01$
1, 2, 3, 4	99, 90, 81, 76	23	Reject H_0	Reject H_0
1, 2, 3	99, 90, 76	23	Reject H_0	Reject H_0
1, 2, 4	99, 90, 81	18	Reject H_0	Reject H_0
1, 3, 4	90, 81, 76	14	Reject H_0	Accept H_0
2, 3, 4	99, 81, 76	23	Reject H_0	Reject H_0

because the ANOVA test, which used the same null hypothesis, rejected the equality of group means.

For the four sets of three group means, the values of r_m for levels of significance of 5 and 1% are 3.14 and 4.31, respectively; therefore, the corresponding values of the least significant range for R_m are 11.4 and 15.6. The ranges of the sample means varied from 14 to 23 (Table 10-12). Thus, at the 5% level of significance, the null hypothesis of equality of three group means should be rejected for all groups. At the 1% level of significance, the null hypothesis would be rejected for all sets of three groups except for the set consisting of groups 1, 3, and 4.

In summary, it appears that the major difference lies in individual pairs, rather than in sets of three or four groups; as sets of three, differences are observed in all combinations. When the tests were applied to pairs of group means, the primary difference appears to be that the set consisting of groups 1 and 2 differs from the set consisting of groups 3 and 4; however, this conclusion is subjective because there is no significant difference between groups 1 and 4. The difficulty in arriving at a totally defensible conclusion probably lies in the arbitrary selection of the level of significance. Whereas there is probably no difference in the effectiveness of the fertilizer mix of groups 3 and 4, there is probably some difference between groups 2 and 1 and between groups 1 and 3; these two pairs of groups both have the same range (i.e., 9).

10.3.2. Scheffé Test

The Scheffé test is an alternative to the Duncan multiple range test. It does not require samples of equal size, although it is often desirable to have equal sample sizes. In general, the Scheffé test is relatively insensitive to departures from the assumptions of normality and homogeneity of variances; thus, its loss of power is not that great when these assumptions are violated.

The test statistic for comparing two group means $\overline{X}_i$ and $\overline{X}_j$ is:

$$F = \frac{(\overline{X}_i - \overline{X}_j)^2}{MS_w\left(\frac{1}{n_i} + \frac{1}{n_j}\right)(k-1)} \tag{10-18}$$

in which MS_w and k are the values obtained from the ANOVA table; that is, even though the test is only comparing the means for two groups, k is still the number of groups used in computing MS_w. The computed F value is the value of a random variable with degrees of freedom $(k - 1, N - k)$.

The Scheffé test could also be used to compare the mean of group i with the combined mean of two other groups, say groups j and h. The mean of two groups, $\overline{X}_{jh}$, is computed as:

$$\overline{X}_{jh} = \frac{n_j\overline{X}_j + n_h\overline{X}_h}{n_j + n_h} \tag{10-19}$$

In this case, the computed F value is:

$$F = \frac{(\overline{X}_i - \overline{X}_{jh})^2}{MS_w\left(\frac{1}{n_i} + \frac{1}{n_j + n_h}\right)(k-1)} \tag{10-20}$$

For both cases (Equations 10-18 and 10-20), the null hypothesis is rejected when the computed F value is greater than the critical value.

For example, if the mean of group 1 is compared with the mean of groups 3 and 4, then $\overline{X}_{34}$ equals:

$$\overline{X}_{34} = \frac{n_3\overline{X}_3 + n_4\overline{X}_4}{n_3 + n_4} = \frac{5(76) + 5(81)}{5 + 5} = 78.5 \tag{10-21}$$

Thus, the computed F statistic is:

$$F = \frac{(\overline{X}_1 - \overline{X}_{34})^2}{MS_w\left(\frac{1}{n_1} + \frac{1}{n_3 + n_4}\right)(k-1)} = \frac{(90 - 78.5)^2}{66\left(\frac{1}{5} + \frac{1}{10}\right)(3)} = 2.23 \tag{10-22}$$

For degrees of freedom of (3, 16), the critical F value for a 5% level of significance is 3.24; therefore, the Scheffé test would suggest that there is no significant difference between group 1 and groups 3 and 4.

10.4. TEST OF POPULATION VARIANCES

Although variances of random variables are important in comparing population means, they are also important in their own right; that is, many decision situations involve variances. For example, administrators of a university may be considering installing a new security lighting system. The uniformity of the lighting intensity is an important characteristic of a lighting system. Thus, the level of uniformity, which could be measured by the variance of measurements of lighting intensity, may influence the decision of which brand to select. The same characteristic is important in selecting from among alternative irrigation systems; it is desirable to have an irrigation system that provides a uniform distribution of water over the field. Similarly, it would be desirable to have a sprinkling system for fire control that provides a uniform level of water over the area for which the water sprinkler is intended to control any combustion. These are systems for which knowledge of the population variance is important, and decisions depend, in part, on the lack of variance in a specific characteristic.

One assumption of the ANOVA test on the equality of population means is the equality of population variances. For the case of Figure 10-1b, the ANOVA test assumes that the population variances σ_i^2 are equal; for the case of Figure 10-1a, the ANOVA test for equa lity of means assumes that the treatments do not affect the variances in such a way as to cause the variances of the measurements within a group to be different. If the variances are significantly different, the ANOVA test on the means may lead to an incorrect decision; that is, rejection of the null hypothesis of either Equation 10-la or Equation 10-4 may actually be due to an inequality of variances rather than an inequality of means. Thus, there is a good reason to test for equality of population variances before testing for the equality of population means.

In summary, the variances of different populations are important both for making decisions associated with variances of random variables and for testing an assumption of the ANOVA test on population means. Therefore, it is reasonable to test for the equality of population variances.

Although a number of tests are available for testing the equality of variances, Bartlett's test is easy to apply and will be presented here. The hypotheses are:

$$H_0\text{: } \sigma_1^2 = \sigma_2^2 = \ldots = \sigma_k^2 \quad (10\text{-}23a)$$

$$H_A\text{: at least one pair of variances are unequal} \quad (10\text{-}23b)$$

The test statistic, χ^2, is the value of a random variable having a χ^2 distribution with v degrees of freedom, where $v = k - 1$, and the test statistic is:

$$\chi^2 = 2.3026\frac{V_a}{D} \quad (10\text{-}24)$$

$$V_a = (N-k)\log_{10} S_p^2 - \sum_{i=1}^{k} (n_i - 1)\log_{10} S_i^2 \tag{10-25}$$

$$D = 1 + \frac{1}{3(k-1)}\left[\left(\sum_{i=1}^{k}\frac{1}{n_i - 1}\right) - \frac{1}{N-k}\right] \tag{10-26}$$

in which N is the total number of scores, n_i is the number of scores in group i, S_i^2 is the variance of the scores in group i, k is the number of groups, and S_p^2 is given by:

$$S_p^2 = \frac{\sum_{i=1}^{k}(n_i - 1)S_i^2}{\sum_{i=1}^{k}(n_i - 1)} \tag{10-27}$$

For a given level of significance the critical value is obtained from the χ^2 distribution for $(k - 1)$ degrees of freedom. The null hypothesis is rejected if the computed value is greater than the critical value.

The test can be applied to the scores of Table 10-9. The standard deviations of Table 10-9 can be used in testing the null hypothesis that the treatments (i.e., the various fertilizer applications) do not cause differences in the sample variances. The value of the pooled variance of Equation 10-27 is:

$$S_p^2 = \frac{4(11.51^2 + 7.11^2 + 6.96^2 + 5.70^2)}{4+4+4+4} = 65.99 \tag{10-28}$$

The value for V_a is:

$$V_a = (20-4)\log_{10}(65.99) - 4\begin{bmatrix}\log_{10}(11.51^2) + \log_{10}(7.11^2) \\ + \log_{10}(6.96^2) + \log_{10}(5.70^2)\end{bmatrix} = 1.02 \tag{10-29}$$

Using Equation 10-26, the value of n_k is:

$$D = 1 + \frac{1}{3(4-1)}\left(\frac{1}{4} + \frac{1}{4} + \frac{1}{4} + \frac{1}{4} - \frac{1}{20-4}\right) = 1.104 \tag{10-30}$$

The computed value of the test statistic is:

$$\chi^2 = 2.3026\frac{1.02}{1.104} = 2.127 \tag{10-31}$$

For 3 degrees of freedom and a level of significance of 5%, the critical value is 7.815; therefore, we have no reason to reject the null hypothesis of equal variances, as the computed value is much less than the critical value.

It is of interest to examine the rationale of the test statistic of Equation 10-24. If all sample variances were equal, V_a of Equation 10-25 would equal zero and the computed χ^2 value would equal zero. A computed value of χ^2 of zero is obviously less than the critical value, so the null hypothesis would be accepted with the implication that population variances are equal; if all of the sample variances are equal, there is no reason to believe that the population variances are not equal.

For given values of N and k, the value of D will be greatest when all of the groups except one have sample sizes (n_i) of 2; the other group would have a sample size of $N - 2\ (k - 1)$. For this case, the summation:

$$\sum_{i=1}^{k} \frac{1}{n_i - 1}$$

would equal

$$(k-1) + \frac{1}{N - 2(k-1)}$$

which is the maximum possible value for the summation. Thus, because the other terms in Equation 10-26 are constant for given values of N and k, D will be the maximum. This produces the smallest possible value of χ^2 and implies that the computed value of χ^2 will be smallest when the sample sizes of $(k - 1)$ groups are a minimum (i.e., 2). This should be expected because, as the sample size decreases, the accuracy of an estimate of a variance decreases, thus the smaller value of χ^2 and the smaller chance of rejecting the null hypothesis.

10.5. RANDOMIZED BLOCK DESIGN

To illustrate the one-way ANOVA of Figure 10-1, an example was used in which five groups of students within a class each used a different textbook. Any difference in the group mean grade was attributed to the textbook. A bias might be introduced if the students were permitted to select the group to which they would belong and if students having high ability selected one group while students with less native intellect selected another group. Then, differences in the mean grades of the groups may be due more to intellectual ability than to the textbook.

To eliminate such sources of bias, the students would have to be assigned to the groups with due consideration to their intellectual ability. If some measure of the students' scholastic ability was available, such as grade point averages (GPAs), that measure could be used to assign students to groups such that the average intellectual ability of each group was the same as that of the other groups. Then, a one-way ANOVA test could be used to evaluate the effectiveness of the textbooks without an inherent bias.

Randomized block design is a way of conducting a one-way analysis of variance while accounting for variation in an independent variable. More specifically, a randomized block design is a one-way ANOVA with (1) an independent variable that is measured on an ordinal scale, (2) random assignment of subjects within a given level of the independent variable, and (3) systematic assignment of subjects between levels of the independent variable. For the textbook example, the five students having the highest GPAs would be identified, and one would be randomly assigned to each group. The five students having the next highest GPA level would be identified, with one student being randomly assigned to each group. This systematic separation on the basis of GPA and random separation within the GPA level would be continued until all students were assigned to groups. Their final grades could then be analyzed using a one-way ANOVA for equality of means. The randomized block design is a method of assigning subjects to groups; it is not a new analysis of variance test. However, when performing the ANOVA test, the total variation is separated differently.

10.5.1. Randomized Block Design Model

The total variation of the one-way ANOVA problem was separated into the "between" and the "within" variations. For a randomized block design problem, the total variation is separated into three parts: the treatment, the block, and the error variations. Thus, a score X_{ij} can be represented by the following linear model:

$$X_{ij} = \mu + (\mu_{i\bullet} - \mu) + (\mu_{\bullet j} - \mu) + (X_{ij} - \mu_{i\bullet} - \mu_{\bullet j} + \mu) \tag{10-32}$$

in which μ is the grand mean; $\mu_{i\bullet}$ is the mean of the ith block, with $i = 1, 2, \ldots, b$; $\mu_{\bullet j}$ is the mean of the jth treatment, with $j = 1, 2, \ldots, k$; k is the number of treatments; and b is the number of blocks. Letting α_i represent the variation of the ith block mean about μ (i.e., the effect of block i) and β_j represent the variation of the jth treatment mean about μ (i.e., the effect of the jth treatment), Equation 10-32 can be rewritten as:

$$X_{ij} = \mu + \alpha_i + \beta_j + \varepsilon_{ij} \tag{10-33}$$

in which ε_{ij} is the random variation — that is, the variation that is not accounted for by the treatment and block variables; thus, ε_{ij} is given by:

$$\varepsilon_{ij} = X_{ij} - \mu_{i\bullet} - \mu_{\bullet j} + \mu \tag{10-34}$$

The means are given by:

$$\mu = \frac{1}{kb}\sum_{j=1}^{k}\sum_{i=1}^{b} X_{ij} \tag{10-35}$$

$$\mu_{i\bullet} = \frac{1}{k}\sum_{j=1}^{k} X_{ij} \tag{10-36}$$

$$\mu_{\bullet j} = \frac{1}{b}\sum_{i=1}^{b} X_{ij} \tag{10-37}$$

The randomized block analysis involves testing two sets of hypotheses. First, one is interested in the equality of block means:

$$H_0: \mu_{1\bullet} = \mu_{2\bullet} = \ldots = \mu_{b\bullet} \quad (10\text{-}38a)$$

$$H_A: \text{all block means are not equal} \quad (10\text{-}38b)$$

Because the model of Equation 10-33 indicates that the separation of scores includes a treatment effect, the equality of treatment means is also of interest:

$$H_0: \mu_{\bullet 1} = \mu_{\bullet 2} = \ldots = \mu_{\bullet k} \quad (10\text{-}39a)$$

$$H_A: \text{all treatment means are not equal} \quad (10\text{-}39b)$$

If the null hypothesis of Equation 10-39a is rejected, one can conclude that the treatments have different effects on the criterion variable X_{ij}. Similarly, if the null hypothesis of Equation 10-38a is rejected, one can conclude that the blocked variable has a significant effect on the criterion variable.

To test the hypotheses of Equations 10-38 and 10-39, the total variation, which is evident if μ is subtracted from both sides of Equation 10-32, can be separated into the three parts: treatment, block, and error. In terms of sample means (rather than population means), Equation 10-32 can be written as a linear separation of variation:

$$\sum_{i=1}^{b}\sum_{j=1}^{k}(X_{ij}-\bar{X})^2 = k\sum_{i=1}^{b}(\bar{X}_{i\bullet}-\bar{X})^2 + b\sum_{j=1}^{k}(\bar{X}_{\bullet j}-\bar{X})^2 + \sum_{i=1}^{b}\sum_{j=1}^{k}(X_{ij}-X_{i\bullet}-X_{\bullet j}+\bar{X})^2 \quad (10\text{-}40)$$

or, in terms of variation,

$$\text{total} = \text{treatment} + \text{block} + \text{error} \quad (10\text{-}41a)$$

$$SS_t = SS_a + SS_b + SS_e \quad (10\text{-}41b)$$

The summations of Equation 10-40 can be used to compute the terms of the linear separation. Alternatively, the following equations can be used to compute the sums of squares:

$$T = \sum_{j=1}^{k}\sum_{i=1}^{b} X_{ij} \quad (10\text{-}42)$$

$$SS_t = \sum_{j=1}^{k}\sum_{i=1}^{b} X_{ij}^2 - \frac{T^2}{bk} \quad (10\text{-}43)$$

$$T_{i\bullet} = \sum_{j=1}^{k} X_{ij} \quad (10\text{-}44)$$

$$SS_a = \frac{1}{b}\sum_{j=1}^{k} T_{\bullet j}^2 - \frac{T^2}{bk} \tag{10-45}$$

$$T_{\bullet j} = \sum_{i=1}^{b} X_{ij} \tag{10-46}$$

$$SS_b = \frac{1}{k}\sum_{i=1}^{b} T_{i\bullet}^2 - \frac{T^2}{bk} \tag{10-47}$$

$$SS_e = SS_t - SS_a - SS_b \tag{10-48}$$

The sums of squares can be entered into a randomized block analysis table, which is shown in Table 10-13. Using the corresponding degrees of freedom, which are shown in Table 10-13, the mean squares are computed as the ratio of the sum of squares to the degrees of freedom. The mean squares are used to compute two F ratios. The F statistic for evaluating the significance of the treatment effect, which is denoted as F_1, is the ratio of the mean square for treatment to the mean square interaction. The F statistic for evaluating the significance of the block effect, which is denoted as F_2, is the ratio of the mean square for the block effect to the mean square interaction. Both F_1 and F_2 are shown in Table 10-13.

The two F statistics, F_1 and F_2, are used to test the hypotheses of Equations 10-38 and 10-39. If the computed value of F_1 is greater than the critical value for the desired level of significance, the null hypothesis of Equation 10-39a should be rejected; this implies that the treatment effect is significant. If the computed value of F_2 is greater than the critical value for the desired level of significance, the null hypothesis of Equation 10-38a should be rejected; this implies that the block effect is significant.

Rationale — Two simple examples will be used to illustrate the randomized block analysis. In Table 10-14, the scores are listed for a case involving three blocks and four treatments; thus, N equals 12, b equals 3, and k equals 4. In Table 10-14a, the block effect is obvious, with no treatment effect. The scores are the same for each treatment, thus the treatment means are identical, and there is no reason to believe that the treatment has an effect on the criterion variable. The values within each block show no variation, so the standard error of the block mean (i.e., $S_b/\sqrt{k}$)

TABLE 10-13 Randomized Block Analysis Table

Source of variation	Degrees of freedom	Sum of squares	Mean square	Computed F
Treatments	$k - 1$	SS_a	$MS_a = \frac{SS_a}{k-1}$	$F_1 = \frac{MS_a}{MS_e}$
Blocks	$b - 1$	SS_b	$MS_b = \frac{SS_b}{b-1}$	$F_2 = \frac{MS_b}{MS_e}$
Interaction	$(k - 1)(b - 1)$	SS_e	$MS_e = \frac{SS_e}{(k-1)(b-1)}$	—
Total	$kb - 1$	SS_t	—	—

TABLE 10-14 Rationale of the Randomized Block Analysis

(a) Block effect

Block	Treatment 1	Treatment 2	Treatment 3	Treatment 4
1	1	1	1	1
2	2	2	2	2
3	3	3	3	3

(b) Treatment effect

Block	Treatment 1	Treatment 2	Treatment 3	Treatment 4
1	1	2	3	4
2	1	2	3	4
3	1	2	3	4

equals zero. With high confidence in the block mean, one would expect the difference between the block means to be significant; thus a block effect would be expected. The ANOVA table of Table 10-15a indicates a significant F for the block effect, whereas the F value for the treatment effect is zero and not significant.

In Table 10-14b, the treatment effect is obvious. Because both the values in each block and the block means are identical, there is no reason to expect the block F to be significant. In each treatment group, there is no variation; that is, the standard deviations of the scores within each treatment (and therefore the standard errors) equal zero. Thus, the differences between each pair of treatment means must be considered significant, and one would expect the treatment effect to be significant. The ANOVA table of Table 10-15b supports these findings. The F for the block variance is zero, while the F for treatment variation equals ∞; therefore, the block effect is not statistically significant, whereas the treatment effect is highly significant. (The reader should perform the summations that are necessary to derive Table 10-15 from the scores of Table 10-14.)

TABLE 10-15 Randomized Block Tables for Experiments of Table 10-14

(a) Block effect

Source of variation	Degrees of freedom	SS^*	MS	F
Treatment	3	0	0	0.00
Block	2	8	4	∞
Interaction	6	0	0	—
Total	11	8	—	—

(b) Treatment effect

Source of variation	Degrees of freedom	SS^{**}	MS	F
Treatment	3	15	5	∞
Block	2	0	0	0
Interaction	6	0	0	—
Total	11	15	—	—

$$^{*}\quad SS_b = \frac{4^2}{4} + \frac{8^2}{4} + \frac{12^2}{4} - \frac{24^2}{12} = 8$$

$$SS_a = \frac{6^2}{3} + \frac{6^2}{3} + \frac{6^2}{3} + \frac{6^2}{3} - \frac{24^2}{12} = 0$$

$$SS_t = 1^2 + 1^2 + 1^2 + 1^2 + 2^2 + \ldots + 3^2 - (24)^2/12 = 8$$

$$^{**}\quad SS_b = \frac{10^2}{4} + \frac{10^2}{4} + \frac{10^2}{4} - \frac{30^2}{12} = 0$$

$$SS_a = \frac{3^2}{3} + \frac{6^2}{3} + \frac{9^2}{3} + \frac{12^2}{3} - \frac{30^2}{12} = 15$$

$$SS_t = 1^2 + 1^2 + 1^2 + 2^2 + \ldots + 4^2 - (30)^2/15 = 15$$

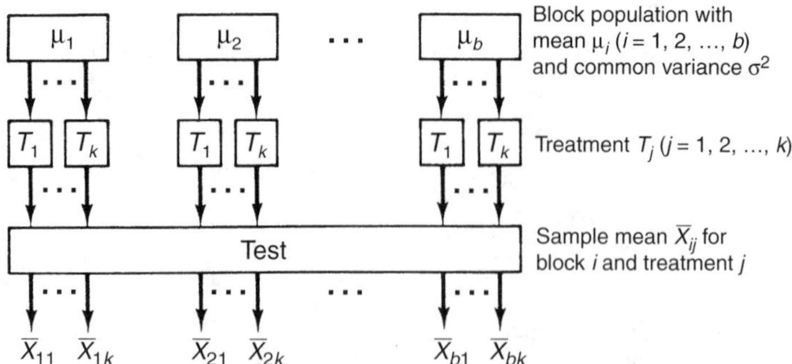

FIGURE 10-2 Schematic of ANOVA test for randomized block designs.

The randomized block design can be represented schematically as shown in Figure 10-2. The system consists of b normal populations that have a common variance σ^2. Each block has a mean value μ_i, with i = 1, 2, ..., b. If the blocked factor does not really influence the value of the criterion variable, the block means will be equal; otherwise, the ANOVA test that is used for the randomized design case will identify a significant difference among the block means. Samples from each block are subjected to the treatments. If the treatment does not have a significant effect on the criterion variable score, the treatment means will not be statistically different. As indicated in Figure 10-2, the sample means are used to draw inferences about the population mean.

Example 10-4: Effectiveness of Fertilizer Mixes and Soil Texture

Consider the case in which one wants to examine the effect of the fertilizer mix on crop yield but the fields on which the experiments are conducted vary in soil characteristics. On the basis of soil texture, the soils can be separated into six groups ranging from well drained to poorly drained. Thus, the fertilizer mix is the treatment variable and the drainage characteristic of the soil is the block variable. If four fertilizer mixes are to be investigated, k equals 4. For the six soil drainage classes, b equals 6. If 24 plots are available, with four in each drainage class, each combination of fertilizer mix and soil drainage type can be tested on only one plot. The resulting crop yields are shown in Table 10-16. The ANOVA table is shown in Table 10-17 (the reader may wish to verify the entries in Table 10-17 using Equations 10-42 to 10-48). The computed F value for the treatment F equals 17.17. For a 5% level of significance and degrees of freedom of (3, 15), the critical value is 3.29. Thus, the computed F is significant and the null hypothesis of equal treatment means should be rejected; this would imply that the fertilizer mix does have a significant effect. For degrees of freedom of (5, 15) and a 5% level of significance, the critical F value is 2.90. Thus, the computed value is significant, and the null hypothesis of equal block means should be rejected; this implies that the drainage characteristics of the soil have an effect on the crop yield.

TABLE 10-16 Crop Yields (bushels/acre) for Four Fertilizer Mixes (Treatments) and Six Drainage Classes (Blocks)

	Fertilizer mix			
Drainage class	**1**	**2**	**3**	**4**
1. Well drained	103	97	108	92
2.	94	89	102	87
3.	97	94	99	96
4.	88	85	96	91
5.	81	72	82	75
6. Poorly drained	74	67	79	64

TABLE 10-17 Randomized Block Design Analysis for Crop Yield Example

Source of variation	Degrees of freedom	Sum of squares	Mean square	F
Treatment	3	438.33	146.11	17.17
Block	5	2578.00	515.6	60.58
Interaction	15	127.67	8.51	—
Total	23	3144.00	—	—

10.6. TWO-WAY ANALYSIS OF VARIANCE

Before introducing another ANOVA model structure, it may be worthwhile to summarize the methods examined to this point, which would also be a useful way of placing the two-way ANOVA, or ANOVA2, in perspective. Usually, analysis of variance methods are used for testing hypotheses concerning mean values, a problem that has been discussed in Chapter 9. The first tests on mean values were the one-sample z and t tests. These two tests were concerned with one sample of measurements on a criterion variable measured on a continuous scale. The null hypothesis equates the population mean μ with a standard of comparison μ_0. The second test for means was the two-sample t test, for which there are several alternatives. For the two-sample test, the comparison of the null hypothesis involves the means for two populations, μ_1 and μ_2; in this case, the values of the criterion variable are measured on a continuous scale. The two-sample t test is a special case of the one-way ANOVA, which is a test of hypothesis for equal means of two or more populations; a sample corresponds to each population, and the criterion variable is measured on a continuous scale. The randomized block design is similar to the one-way ANOVA but with the addition of an independent variable that is measured on a discrete scale. The one-way ANOVA problem is viewed as a matrix of scores on the criterion variable, with the scores in a column representing the result of one treatment. The randomized block design can be viewed in the same way, but with the added characteristic that the scores in the different rows represent the effect of an independent variable that is measured on a discrete scale. These four analysis problems are summarized in Table 10-18.

TABLE 10-18 Summary of Hypothesis Tests on Means

Test	Null hypothesis	Number of discrete independent variables	Test statistic	Data
t test (or z) on one criterion variable	$\mu = \mu_0$	0	$t = \dfrac{\overline{X}_0 - \mu_0}{S/\sqrt{n}}$	Vector $[x_1 x_2 \cdots x_n]$
Two-sample t test	$\mu_1 = \mu_2$	0	$t = \dfrac{\overline{X}_1 - \overline{X}_2}{S_p\sqrt{\dfrac{1}{n_1} + \dfrac{1}{n_2}}}$	Matrix $(n \times 2)$ $\begin{bmatrix} x_{11} & x_{12} \\ x_{21} & x_{22} \\ \vdots & \vdots \\ x_{n1} & x_{n2} \end{bmatrix}$
One-way ANOVA	$\mu_1 = \mu_2 = \ldots = \mu_k$	0	$F = \dfrac{MS_{\text{between}}}{MS_{\text{within}}}$	Matrix $(n \times k)$ $\begin{bmatrix} x_{11} & x_{12} & \ldots & x_{1k} \\ \vdots & \vdots & \vdots & \vdots \\ x_{n1} & x_{n2} & \ldots & x_{nk} \end{bmatrix}$
ANOVA with randomized block design	$\mu_1 = \mu_2 = \ldots = \mu_k$ $\mu_1' = \mu_2' = \ldots = \mu_b'$	1	$F_1 = \dfrac{MS_{\text{treatment}}}{MS_{\text{error}}}$ $F_2 = \dfrac{MS_{\text{block}}}{MS_{\text{error}}}$	Treatment Block 1 2 … k 1 2 ⋮ b
Two-way ANOVA	$\mu_1 = \mu_2 = \ldots = \mu_{rc}$ $\mu_1' = \mu_2' = \ldots = \mu_r'$ $\mu_1'' = \mu_2'' = \ldots = \mu_c''$	2	$F_{rc} = \dfrac{MS_{rc}}{MS_{\text{error}}}$ $F_{\text{row}} = \dfrac{MS_r}{MS_{\text{error}}}$ $F_{\text{column}} = \dfrac{MS_c}{MS_{\text{error}}}$	Column variable 1 2 … n_c Row variable 1 2 ⋮ n_r

TABLE 10-19 Interaction Effect between Independent Variables

	Fertilizer application rate			
Level of irrigation	**None**	**Low**	**Medium**	**High**
None	36	39	35	40
Low	42	45	51	56
Medium	53	61	74	86
High	72	83	95	109

A fifth ANOVA method, two-way analysis of variance, extends the analysis to the case where two discrete independent variables are involved. The ANOVA2 problem can involve interaction between the two independent variables; this characteristic is a distinguishing factor between the ANOVA with randomized block design and the ANOVA2. The distinction is important because the significance of the interaction must be examined before the effects of the row and column variables are studied. The data arrangement for ANOVA2 is shown in Table 10-18.

Two-way ANOVA designs are applicable to many engineering problems. For example, if the criterion variable is traffic accidents at intersections, one may wish to examine whether or not accident rates are affected by (1) the degree of lighting at the intersection (none, poor, good); (2) the degree of traffic control (none, stop sign, traffic light); or (3) the degree of interaction between lighting and traffic control. As another example, crop yield may be the criterion variable of interest. Crop yield is affected by the levels of irrigation and fertilizer. Because these two factors may interact, the ANOVA2 procedure would be applicable. That is, the fertilizer may not be effective unless sufficient water is provided by irrigation. Another way of viewing interaction is in terms of the variation of the effect of one independent variable as the value of the second independent variable changes; for example, if there is no variation in the mean values of the crop yield for different levels of fertilizer application at the low level of water application but a significant difference in mean values at high levels of water application, the two independent variables are interacting. This concept is easily illustrated with the data of Table 10-19. For the case of no irrigation, the crop yield does not change significantly; however, for high levels of irrigation, the crop yield changes significantly as the fertilizer application rate increases. For low and medium levels of irrigation, the effect of the fertilizer application rate is greater than for the case of no irrigation but less than that for the high level of irrigation. The interaction between the independent variables must be accounted for if proper decisions are to be made.

10.6.1. ANOVA2 Model

ANOVA2 analysis involves three null hypotheses, one for each of the independent variables and one for the interaction between the two independent variables; both null hypotheses for the independent variables test for equality of means:

H_0: column means are not significantly different $(\mu_1' = \mu_2' = \ldots = \mu_{nc}')$ (10-49a)

H_A: at least one pair of columns means are different (10-49b)

and

H_0'' : row means are not significantly different $(\mu_1'' = \mu_2'' = \ldots = \mu_{nc}'')$ (10-50a)

H_A'' : at least one pair of row means are different (10-50b)

in which the subscripts *nc* and *nr* are the number of columns and rows, respectively. The hypotheses for testing for a significant interaction between the two independent variables are:

H_0: the interaction between the independent variables is not significant (10-51a)

H_A: the interaction between the independent variables is significant (10-51b)

The hypotheses of Equation 10-51 must be tested before those of Equations 10-49 and 10-50.

Just as the other ANOVA models involved a separation of variation into its component parts, the ANOVA2 model separates the variation in a way that the hypotheses of Equations 10-49, 10-50, and 10-51 can be tested. The ANOVA2 model is:

$$X_{ijk} = \mu + (X_{ijk} - \mu_{jk}) + (\mu_j - \mu) + (\mu_k - \mu) + (\mu_{jk} - \mu_j - \mu_k + \mu) \quad (10\text{-}52a)$$

$$= \mu + \varepsilon_{ijk} + \beta_j + \gamma_k + \alpha_{jk} \quad (10\text{-}52b)$$

in which X_{ijk} is the *i*th value of the criterion variable for levels of the independent variables *j* and *k*; μ_{jk} is the mean for cell *jk*; μ_j and μ_k are the means for independent variables *j* and *k*, respectively; μ is the grand mean; ε_{ijk} is the deviation of the *i*th value of the criterion variable from the mean of cell *jk*; β_j and γ_k are the deviations of the column and row means, respectively, from the grand mean; and α_{jk} is the cell deviations.

The sampling procedure for a two-way ANOVA is similar to that for a one-way ANOVA. One can either draw a number of samples from the same population and subject each to a different experimental analysis or draw the samples from different populations and subject each to the same experimental analysis. The values of the criterion variable can be placed into a double-entry table, such as that given in Table 10-20; the table consists of *nr* rows and *nc* columns that reflect the different levels of the independent variables. Each cell of the double-entry table should contain at least one value of the criterion variable.

Once the experiment has been designed and the level of significance selected, the experimental analysis is made, and the values of the criterion variable are computed and placed into the double-entry table. The sums of squares can be computed using the following equations:

TABLE 10-20 Double-Entry Table for ANOVA2

Level of independent variable 2	Level of independent variable 1			
	1	2	. . .	*nc*
1				
2				
.				
.				
.				
nr				

1. The between-columns sum of squares:

$$SS_c = \sum_{k=1}^{nc} \frac{T_{\cdot k}^2}{n_k} - \frac{T^2}{N} \tag{10-53}$$

2. The between-rows sum of squares:

$$SS_r = \sum_{j=1}^{nr} n_j(\bar{X}_{j\cdot} - \bar{X})^2 \tag{10-54a}$$

$$= \sum_{j=1}^{nr} \frac{T_{j\cdot}^2}{n_j} - \frac{T^2}{N} \tag{10-54b}$$

3. The between-cells sum of squares:

$$SS_b = \sum_{j=1}^{nr} \sum_{k=1}^{nc} \frac{T_{jk}^2}{n_{jk}} - \frac{T^2}{N} \tag{10-55}$$

4. The interaction sum of squares:

$$SS_{rc} = SS_b - SS_c - SS_r \tag{10-56}$$

5. The total sum of squares:

$$SS_t = \sum_{j=1}^{nr} \sum_{k=1}^{nc} \sum_{i=1}^{n_{jk}} X_{ijk}^2 - \frac{T^2}{N} \tag{10-57}$$

6. The within-cells sum of squares:

$$SS_w = SS_t - SS_b \tag{10-58}$$

in which $T_{j\bullet}$ and $T_{\bullet k}$ are the total of the scores in row j and column k, respectively; n_j and n_k are the number of scores in row j and column k, respectively; n_{jk} is the number of scores in cell jk; T is the total of all the scores in the double-entry table; and N is the total number of scores and is equal to:

$$N = \sum_{j=1}^{nr} n_j = \sum_{k=1}^{nc} n_k = \sum_{j=1}^{nr}\sum_{k=1}^{nc} n_{jk} \tag{10-59}$$

The sums of squares can be placed as input into an ANOVA2 table, as shown in Table 10-21. The three F values are computed by dividing the corresponding mean square by the mean square within, MS_w. The degrees of freedom v for each sum of squares are listed in Table 10-21.

After the entries for the ANOVA2 table are computed, the first step is to test the significance of the interaction, which corresponds to the null hypothesis of Equation 10-51. The critical value F_{rc} is obtained from a table of values for the F distribution; the critical value is selected based on the level of significance α and the degrees of freedom (v_{rc}, v_w). If the computed value F_{rc} is greater than the table F value, the null hypothesis is rejected; this implies that the interaction is significant. If the interaction is significant, the effects of the row variable may not be the same for each column, and the effects of the column variable may not be the same for each row. If the interaction is not significant (i.e., $F_{rc} < F_\alpha$), the row and column variables may be analyzed independently. If the interaction is significant, one-way ANOVAs should be performed separately for the row and column variables. If the interaction is not significant, the values of F_r and F_c computed in Table 10-21 can be used to test the significance of the null hypotheses of Equations 10-50 and 10-49, respectively. The decision procedure can be summarized as:

1. Compute all entries to the ANOVA2 table (Table 10-21).
2. Using the level of significance and degrees of freedom (v_{rc}, v_w), find the critical F value (say, F_α).

TABLE 10-21 ANOVA2 Table

Source of variation	Degrees of freedom	Sum of squares	Mean square	Computed F
Between cells	—	SS_b	—	—
Between columns	$v_c = nc - 1$	SS_c	MS_c	$F_c = \dfrac{MS_c}{MS_w}$
Between rows	$v_r = nr - 1$	SS_r	MS_r	$F_r = \dfrac{MS_r}{MS_w}$
Interaction	$v_{rc} = (nc - 1)(nr - 1)$	SS_{rc}	MS_{rc}	$F_{rc} = \dfrac{MS_{rc}}{MS_w}$
Within cells	$v_w = N - nc(nr)$	SS_w	MS_w	—
Total	$v_t = N - 1$	SS_t	—	—

3. If $F_{rc} > F_{\alpha}$, reject H_0 of Equation 10-51 and perform one-way ANOVA for the row and column variables.
4. If $F_{rc} < F_{\alpha}$, use F_c and F_r of Table 10-21 to test the null hypotheses of Equations 10-49 and 10-50, respectively; use degrees of freedom (v_c, v_w) and (v_r, v_w) to test for significant column and row effects, respectively.

10.6.2. Computational Examples

The model of Equation 10-52 indicates that variations from the grand mean can be the result of deviations from the row mean, the column mean, the cell mean, and interaction. The task of an ANOVA2 is to separate the variation that exists in a set of values on the criterion variable into its component parts, so that the significance of the effects can be assessed. The model of Equation 10-52b could be expressed as:

$$X_{ijk} = \mu + \varepsilon_{ijk} + \beta_j + \gamma_k + \alpha_{jk}$$

$$X_{ijk} = \text{grand mean} + \text{error effect} + \text{column effect} + \text{row effect} + \text{interaction} \quad (10\text{-}60)$$

Equation 10-60 implies that given the five components of the righthand side, the value of the criterion variable can be computed as the sum of the grand mean and the four effects; of course, the purpose of ANOVA2 is the opposite — that is, to break an observed value into the individual effects — but it can be instructive to show cases where the values of the criterion variable are formed from given values of the grand mean and the four effects. Table 10-22 uses a grand mean of 10 to show seven cases involving various combinations of the four effects. The values of the criterion variable, which are the sum of the individual parts of Equation 10-60, are shown in the double-entry table on the right side of Table 10-22.

A two-way ANOVA was performed for each of the seven double-entry tables of Table 10-22; the resulting ANOVA2 tables are given in Table 10-23. If we define zero divided by zero to be zero and any positive value divided by zero to be infinity, the computed F values of Table 10-23 reflect the effect that was used to generate the values of the criterion variable that are shown in the double-entry tables of Table 10-22. Although there are no degrees of freedom for the MS_w in six of the cases, we must also assume that an F value of infinity is significant.

Case 1: row effect. The first double-entry table was generated using a grand mean of 10 and row effects of +2 and –2. Thus, the row means differ, but the column means equal the grand mean. The ANOVA2 table indicates that both the column effect and the interaction are not significant. The row effect is significant. Because each cell has only one value, the within-cell variation is zero. Thus, the F values must be either zero or infinity.

Case 2: column effect. The second double-entry table of Table 10-22 was generated using a grand mean of 10 and column effects of –4, 1, and 3. The sum of the column effects is zero. Whereas the column means differ, the row means equal the grand mean; therefore, one would expect the column effect to be significant because

TABLE 10-22 Formulation of Examples To Illustrate Row (R), Column (C), Interaction (I), and Error (E) Effects

Effect	Grand mean	Effect for row 1	Effect for row 2	Effect for column 1	Effect for column 2	Effect for column 3	Interaction effects	Error effects	Double-entry table with mean
R	10	2	–2	0	0	0	0 0 0 0 0 0	0 0 0 0 0 0	12 12 12 12 8 8 8 8 10 10 10 10
C	10	0	0	–4	1	3	0 0 0 0 0 0	0 0 0 0 0 0	6 11 13 10 6 11 13 10 6 11 13 10
R, C	10	2	–2	–4	1	3	0 0 0 0 0 0	0 0 0 0 0 0	8 13 15 12 4 9 11 8 6 11 13 10
I	10	0	0	0	0	0	–2 5 –3 2 –5 3	0 0 0 0 0 0	8 15 7 10 12 5 13 10 10 10 10 10

TABLE 10-22 Formulation of Examples To Illustrate Row (*R*), Column (*C*), Interaction (*I*), and Error (*E*) Effects (cont.)

Effect	Grand mean	Effect for row		Effect for column			Interaction effects	Error effects	Double-entry table with mean
		1	2	1	2	3			
R, I	10	2	–2	0	0	0	–2 | 5 | –3 2 | –5 | 3	0 | 0 | 0 0 | 0 | 0	10 | 17 | 9 | 12 10 | 3 | 11 | 8 10 | 10 | 10 | 10
R, C, I	10	2	–2	–4	1	3	–2 | 5 | –3 2 | –5 | 3	0 | 0 | 0 0 | 0 | 0	6 | 18 | 12 | 12 6 | 4 | 14 | 8 6 | 11 | 13 | 10
R, C, I, E	10	2	–2	–4	1	3	–2 | 5 | –3 2 | –5 | 3	±2 | 0 | ±1 ±1 | ±2 | 0	4 8 | 18 18 | 11 13 | 12 5 7 | 2 6 | 14 14 | 8 6 | 11 | 13 | 10

TABLE 10-23 ANOVA2 Tables for Examples Illustrating Row (*R*), Column (*C*), Interaction (*I*), and Error (*E*) Effects[a]

	Effect: *R*				Effect: *C*				Effect: *R, C*				Effect: *I*			
Source of variation	v	*SS*	*MS*	*F*	v	*SS*	*MS*	*F*	v	*SS*	*MS*	*F*	v	*SS*	*MS*	*F*
Between cells	x	24	x	x	x	52	x	x	x	76	x	x	x	76	x	x
Between columns	2	0	0	0	2	52	26	∞	2	52	26	∞	2	0	0	0
Between rows	1	24	24	∞	1	0	0	0	1	24	24	∞	1	0	0	0
Interaction	2	0	0	0	2	0	0	0	2	0	0	0	2	76	38	∞
Within	0	0	0	x	0	0	0	x	0	0	0	x	0	0	0	x
Total	5	24	x	x	5	52	x	x	5	76	x	x	5	76	x	x

	Effect: *R, I*				Effect: *R, C, I*				Effect: *R, C, I, E*			
	v	*SS*	*MS*	*F*	v	*SS*	*MS*	*F*	v	*SS*	*MS*	*F*
Between cells	x	100	x	x	x	152	x	x	x	304	x	x
Between columns	2	0	0	0	2	52	26	∞	2	104	52	15.6
Between rows	1	24	24	∞	1	24	24	∞	1	48	48	14.4
Interaction	2	76	38	∞	2	76	38	∞	2	152	76	22.8
Within	0	0	0	x	0	0	0	x	6	20	10/3	x
Total	5	100	x	x	5	152	x	x	11	324	x	x

[a] v = degrees of freedom; *SS* = sums of squares; *MS* = mean squares; *F* = computed *F* statistic.

the within-cell variation is zero. The row effect and the interaction are not significant, as is evident from the zero values of the corresponding F statistics.

Case 3: row and column effects. The third double-entry table of Table 10-22 was generated using a grand mean of 10 and the row and column effects of the previous two cases; specifically, the row effect is generated using values of +2 and –2, while the column effect is generated with values of –4, 1, and 3. Thus, both the row and column means differ from the grand mean. The i row means are the same as the values for case 1, and the column means are the same as the values for case 2. The F values of the ANOVA2 table (Table 10-23) indicate that the interaction is not significant, but the row and column effects are significant. As the number of effects increases, the variation of the cell values increases; this makes it more difficult to identify visually what effects are present and significant.

Case 4: interaction. The fourth double-entry table in Table 10-22 was generated using a grand mean of 10 and an interaction effect. Row and column effects are not present; therefore, the row and column means equal the grand mean even though the values within a row or column vary. The interaction effects sum to zero for each row and column. The computed F values show that the interaction is significant. Because the interaction is significant, it would be necessary to perform a separate one-way ANOVA for the row and column variables; it is simple to show that the F values will equal zero for the one-way ANOVAs.

Case 5: row and interaction effects. The values in the fifth double-entry table of Table 10-22 were generated using a grand mean of 10 and both a row and interaction effect. Because there is no column effect, the column means equal the grand mean. Also, because the interaction effects are such that they sum to zero in each row, the row means are the same as for case 1. The values in each row differ from the values of case 1 because of the interaction effect. The F value for interaction is significant, as expected; therefore, a one-way ANOVA would be necessary to show that the row effect is significant.

Case 6: row, column, and interaction effects. The values in the sixth double-entry table of Table 10-22 were generated using a grand mean of 10 and the row, column, and interaction effects used in the previous cases. Both the row and column means differ from the grand mean but equal the means for the cases where only row or column effects were present. The computed F values for row, column, and interaction effects indicate that the effects are significant. Because the interaction F is significant, it would be necessary to perform one-way ANOVAs to test the significance of the row and column effects.

Case 7: row, column, interaction, and error effects. The seventh double-entry table is the only one that contains more than one value per cell. This occurs because an error effect is included. As a result, the F values of Table 10-23 for this case will not be zero or infinity. The row and column means still equal the values for the cases where there were row and column effects only; this results because the interaction effects sum to zero in each row and column, and the error effects sum to zero within each cell. The critical F values for the interaction effects would be

TABLE 10-24 ANOVA2 Example: Coefficient of Friction

	Percentage of lead and antimony					
Machining	**High**		**Medium**		**Low**	
Highly polished	0.0005	0.0011	0.0017	0.0019	0.0031	0.0031
Moderately polished	0.0022	0.0022	0.0027	0.0031	0.0023	0.0025
No machining	0.0039	0.0043	0.0036	0.0038	0.0033	0.0037

determined using degrees of freedom of (2, 6). For the 5% level of significance, the critical value is 5.14; therefore, the interaction effect is significant. Thus, it is necessary to perform a one-way ANOVA for both the row and column effects. The F values from the one-way ANOVAs are 1.74 and 2.13 for the row and column effects, respectively. For a 5% level of significance and 1 and 10 degrees of freedom, the critical F value for testing the row effects is 4.96; therefore, the row effect is not statistically significant. For a 5% level of significance and degrees of freedom (2, 9), the critical F value for testing the significance of the column effects is 4.26. Thus, the column effect is also not significant. This case shows that even though some variation is attributed to the rows (±2) and columns (–4, 1, 3), the variation is not significant when compared with the within-cell variation. Additionally, the case illustrates that the F values in the ANOVA2 table for the row and column effects can lead to erroneous conclusions when interaction effects are present.

Example 10-5: ANOVA2 of Coefficient of Friction

A series of experiments is conducted in which the coefficient of friction is measured between a shaft and the bearing. Two factors that potentially affect the coefficient are the degree of machining of the bearing (highly polished, moderately polished, and no machining) and the percentage of lead and antimony in the shaft (high, moderate, low). The experiments produced the coefficients given in Table 10-24. The row means are 0.0019, 0.0025, and 0.0038; the column means are 0.0024, 0.0028, and 0.0030. The grand mean is 0.0027.

Because the two factors, machining and the composition of the alloy used in making the shaft, have the potential for interaction, a two-way ANOVA should be used to test the significance of the factors. Table 10-25 shows the ANOVA2 table for the problem. Because all of the F values are large, it would seem that all effects are significant. But the ANOVA2 procedure requires that the F for interaction be tested first for significance. For degrees of freedom of (4, 9) and a 5% level of significance, the critical value is 3.63; therefore, the computed value of 23.24 is significant. Because the computed F for the interaction effect is significant, a separate one-way ANOVA should be made for both of the independent factors. The total and between-row sums of squares from Table 10-25 were entered into the one-way ANOVA table for row effects shown in Table 10-26. These were used to compute the within sums of squares.

TABLE 10-25 ANOVA2 Test for Example

Source of variation	Degrees of freedom	Sum of squares	Mean square	Computed F
Between cells	x	17.1111×10^{-6}	x	x
Between columns	2	1.2578×10^{-6}	0.6289×10^{-6}	11.80
Between rows	2	10.8978×10^{-6}	5.4489×10^{-6}	102.23
Interaction	4	4.9555×10^{-6}	1.2389×10^{-6}	23.24
Within	9	0.4800×10^{-6}	0.0533×10^{-6}	x
Total	17	17.5911×10^{-6}	x	x

TABLE 10-26 One-Way ANOVA Table for Row Effect

Source of variation	Degrees of freedom	Sum of squares	Mean square	Computed F
Between	2	10.8978×10^{-6}	5.4489×10^{-6}	12.21
Within	15	6.6933×10^{-6}	0.4462×10^{-6}	—
Total	17	17.5911×10^{-6}	—	—

TABLE 10-27 One-Way ANOVA Table for Column Effect

Source of variation	Degrees of freedom	Sum of squares	Mean square	Computed F
Between	2	1.2578×10^{-6}	0.6289×10^{-6}	0.58
Within	15	16.3333×10^{-6}	1.0889×10^{-6}	—
Total	17	17.5911×10^{-6}	—	—

The ratio of the mean squares results in an F value of 12.21. For degrees of freedom of (2, 15) and a 5% level of significance, the critical F value is 3.68; therefore, the row effect is significant. This indicates that the level of machining influences the coefficient of friction. The total and between-columns sums of squares from Table 10-25 were entered into the one-way ANOVA table for column effects shown in Table 10-27. The within sums of squares and degrees of freedom were then computed. The F ratio was computed using the mean squares. For a 5% level of significance and degrees of freedom of (2, 15), the critical F value is 3.68; therefore, the computed value of 0.58 is not statistically significant. Thus, the percentage of lead and antimony in the shaft does not influence the coefficient of friction as a separate variable; it does, however, interact with the degree of machining to affect the coefficient of friction.

10.6.3. Discussion

Methods involving the analysis of variance have been introduced as a set of hypothesis tests. The one-way and two-way ANOVA are used to test hypotheses of equality of means. Bartlett's test is used to detect differences in the variances of a set of variables. Although it is important to recognize the role of such tests in data analysis, it is equally important to be aware of the similarities between the analysis of variance and regression analysis. Both methods use the linear model relating the criterion variable and the other variables, or factors. Whereas regression separates the total variance in a way that the regression coefficients reflect the rate of change of the criterion variable (y) with respect to change in the predictor variable (x), the analysis of variance provides a "yes/no" statement of the effect of x on y. In fact, the analysis of variance will be introduced as a means of testing the significance of the regression slope coefficient (see Chapter 12). Although regression is usually applied to variables measured on an interval scale, the factors used in the analysis of variance can be measured on nominal and ordinal scales. Another closely related analysis of variance method, which is called the analysis of covariance, uses regression to hold the effect of a variable that is measured on the interval measurement scale constant when performing the test of hypothesis.

Another analysis of variance test that is becoming more widely used is the multivariate alternative. In this case, the criterion variable is a vector variable, such as when one is interested in the effect of different factors on several interdependent criterion variables. For example, one may wish to detect a significant difference between various models that were designed to estimate the same criterion variable on the basis of bias, precision, and the cost of application. One may expect that as the cost of application increases, the bias will decrease and the precision will improve, thus the three criterion variables are not independent and cannot be analyzed separately. The multivariate analysis of variance can be used to detect changes when all variables are considered simultaneously.

10.7. APPLICATIONS

10.7.1 Steel Corrosion

A structural engineer measures the corrosion rate (μm) for four types of steel (carbon, C; copper, Cu; A588; and A242) after 8 years of exposure in a rural environment. The number of specimens available differed with steel type. Given the corrosion rates provided below, do the four types of steel exhibit different rates of corrosion? If so, which ones differ?

C: {110, 165, 95, 145, 130}

Cu: {80, 115, 105, 75}

A588: {45, 80, 75, 60, 45, 40}

A242: {25, 55, 30}

Solution: The means ($\overline{X}$) and standard deviations (S) for the four samples are:

	C	**Cu**	**A588**	**A242**
Mean	129	93.8	57.5	36.7
Standard deviation	27.7	19.3	17.0	16.1

The samples were analyzed using a one-way ANOVA with the following ANOVA table:

Variation	ν	Sum of squares	Mean square	F
Between	3	21,364	7121	16.23
Within	14	6143	438.8	—
Total	17	27,507	—	—

The critical F values for 5% and 1% significance levels are 3.34 and 5.56, respectively; therefore, the computed F of 16.23 is very significant, with an extremely small rejection probability. A Scheffé test was made for each pair of steel types. The same critical F values for 5% and 1% apply. The resulting computed F values are:

Steel 1	Steel 2	F	5% decision	1% decision
C	Cu	2.10	Accept H_0	Accept H_0
C	A588	10.59	Reject H_0	Reject H_0
C	A242	12.14	Reject H_0	Reject H_0
Cu	A588	2.40	Accept H_0	Accept H_0
Cu	A242	4.24	Reject H_0	Accept H_0
A588	A242	0.66	Accept H_0	Accept H_0

The differences in means are not significant for C vs. Cu, Cu vs. A588, and A588 vs. A242. The differences are significant for C vs. A588 and C vs. A242. The difference for Cu vs. A242 is inconclusive, unless clear evidence is available that would lead to a specific level of significance.

10.7.2. Sheet Erosion

A soil scientist is investigating the development of sheet erosion. On several experimental plots with similar topographic characteristics, the scientist uses an artificial rain generator to simulate a 1-hour rainstorm of constant intensity. Nineteen plots are available. The plots differ in soil type, each having one of four types: sandy loam, sandy clay loam, loam, and clay loam. After subjecting each plot to the simulated rain, the hydraulic radius of the largest rill is computed, with the following results:

Soil type	Measured hydraulic radius (mm)
Sandy loam (S_1)	13, 10, 9, 12
Sandy clay loam (S_2)	11, 8, 12, 8, 9
Loam (S_3)	10, 11, 6, 7
Clay loam (S_4)	8, 5, 7, 8, 6, 5

The soil scientist is interested in knowing if the soil types differ significantly in their development of rills. Do the four soil types differ? If so, which ones are different?

Solution: The means and standard deviations are:

	S_1	S_2	S_3	S_4
Mean	11.00	9.60	8.50	6.50
Standard deviation	1.83	1.82	2.38	1.38

The samples were analyzed using a one-way ANOVA with the following results:

Variation	v	Sum of squares	Mean square	F
Between	3	54.41	18.14	5.47
Within	15	49.70	3.31	—
Total	18	104.11	—	—

The 5% and 1% critical F values for 3 and 15 degrees of freedom are 3.29 and 5.42, respectively. Thus, the null hypothesis of equal means is rejected for both levels of significance; however, the rejection probability is only slightly smaller than 1%. Scheffé tests were applied to each pair of soil types to decide which soils showed significant differences. The resulting computed values are:

Soil 1	Soil 2	F	5% decision	1% decision
S_1	S_2	0.44	Accept H_0	Accept H_0
S_1	S_3	1.26	Accept H_0	Accept H_0
S_1	S_4	4.89	Reject H_0	Accept H_0
S_2	S_3	0.27	Accept H_0	Accept H_0
S_2	S_4	2.64	Accept H_0	Accept H_0
S_3	S_4	0.97	Accept H_0	Accept H_0

The only apparent difference is between the sandy loam and the clay loam, and this difference is only detected for a 5% level of significance. At a 1% level of significance, none of the means is significantly different. The small samples are likely responsible for the difference between the ANOVA1 decision to reject and the decision to accept most differences for the Scheffé test. Accurate decisions are difficult to achieve with small samples.

10.8. PROBLEMS

10-1. For the six measurements on the two variables X_1 and X_2, conduct both a two-sample t test (assume σ_1 and σ_2 are unknown but equal) and an ANOVA test for comparing means. Perform the tests for (a) $Y_1 = Y_2 = 10$, and (b) $Y_1 = Y_2 = 15$. Use a level of significance of 5%. Explain the difference in the results of the two tests. What is the relationship between the t statistic ($t_{\alpha/2}$) for $v = n_1 + n_2 - 2$ and the F_α statistic for $v_1 = 1$ and $v_2 = n_1 + n_2 - 2$?

X_1	7	8	8	10	11	Y_1
X_2	9	11	12	15	15	Y_2

10-2. Four sections of the same statistics course are taught by four teachers. Using the data given below and a level of significance of 5%, are the average grades of the four classes significantly different?

Class 1	78	83	65	74	91	83	—	—
Class 2	92	81	87	76	94	85	90	—
Class 3	63	71	65	68	83	—	—	—
Class 4	94	87	89	92	90	89	85	93

10-3. A series of experiments is conducted, and the coefficient of friction μ between a shaft and bearing is computed. Tests are run using six different lubricants, with four measurements of μ for each lubricant. At the 1% level of significance, can we conclude that any of the lubricants are more effective in reducing friction? If so, which ones?

Lubricant 1	0.0105	0.0082	0.0076	0.0093
Lubricant 2	0.0074	0.0110	0.0086	0.0077
Lubricant 3	0.0062	0.0073	0.0059	0.0068
Lubricant 4	0.0107	0.0113	0.0094	0.0098
Lubricant 5	0.0052	0.0075	0.0064	0.0071
Lubricant 6	0.0083	0.0066	0.0080	0.0058

10-4. Four electric resistance furnaces are tested to determine the temperature produced in each one. The four furnaces differ only in the composition of the coil of wire that is wound on the refractory material. A number of measurements of temperature (°C) are made for each furnace and are given below. Use a 5% level of significance in determining whether or not the mean temperatures produced in the furnaces differ. If an overall difference is found, determine the furnace or furnaces that differ significantly.

Furnace 1	1250	1175	1275	1200
Furnace 2	1400	1375	1525	1475
Furnace 3	1350	1375	1325	1425
Furnace 4	1625	1575	1650	1675

10-5. Three different acoustical systems are proposed for controlling noise intensity in an office. Measurements of noise intensity (decibels) that simulate heavy street traffic are made for all three acoustical systems at various locations in the office. Using a 5% level of significance, test whether or not the mean noise intensity for the three systems differs.

System 1	22	28	26	25	23
System 2	31	34	29	36	34
System 3	19	23	22	25	21

10-6. Using the Scheffé test and the database for Example 10-2, determine why the null hypothesis of equal means was rejected.

10-7. Test the null hypothesis of equality of group variances for both of the sets of three groups given below. Use a level of significance of 1%. What do the two tests show?

	Set 1				Set 2			
Group 1	1	2	3	4	1	2	3	4
Group 2	2	3	4	5	11	12	13	14
Group 3	3	4	5	6	21	22	23	24

10-8. Test the null hypothesis of equal variances in the population for the database given below. Use a 5% level of significance.

Group 1	2	3	4	5
Group 2	–1	0	7	8
Group 3	–4	–3	10	11

10-9. Measurements in lumens per square foot are made for testing the uniformity of lighting using four lighting systems (X_1 to X_4). Using a 5% level of significance and the following data, determine whether or not the four lighting systems provide equal uniformity in lighting intensity:

X_1	56	63	49	58	65
X_2	47	52	48	56	51
X_3	62	65	56	59	60
X_4	48	62	55	47	64

10-10. For the data of Problem 10-5, determine whether or not the noise intensity is uniform throughout the office. Use a 5% level of significance.

10-11. Measurements of the percentage soil moisture are made on the top of a sloping field (X_1), the side of the slope (X_2), and at the bottom of the slope (X_3). Using a 1% level of significance, determine whether or not the soil moisture characteristics are constant over the field.

X_1	9.6	10.3	8.8	8.5	10.4
X_2	15.2	19.7	14.4	11.6	17.0
X_3	16.5	18.3	16.9	20.2	17.7

10-12. A 20-acre field is divided into l-acre plots, with the plots randomly assigned to one of four groups. For each plot fertilizer is either applied (F) or not applied (NF). For each plot, water is either applied (W) or not applied (NW). Thus, there are five plots for each of the four treatments: (F, W), (F, NW), (NF, W), and (NF, NW). The crop yields in bushels per acre are shown below. Using a 5% level of significance, determine if the within-treatment variances are significantly different. If the null hypothesis of equal variances is accepted, test for a significant difference between group means. If the null hypothesis of equal means is rejected, use the Duncan multiple range to determine if: (a) the application of water causes a difference: (F, W) and (NF, W) vs. (F, NW) and (NF, NW); (b) if the application of fertilizer causes a difference: (F, W) and (F, NW) vs. (NF, W) and (NF, NW); (c) if both fertilizer and water have a significant effect on crop yield: (F, W) vs. (F, NW), (NF, W), and (NF, NW).

(F, W)	(NF, W)	(F, NW)	(NF, NW)
123	117	104	98
118	110	97	95
121	113	112	105
113	99	106	89
126	108	103	100

10-13. For the given data set, which consists of three treatments and five blocks, use a 5% level of significance to test the significance of the treatment and block effects.

	Treatment		
Block	**1**	**2**	**3**
1	0.070	0.081	0.075
2	0.065	0.074	0.068
3	0.066	0.068	0.071
4	0.062	0.071	0.066
5	0.059	0.067	0.066

10-14. For the given data set, which consists of four treatments and four blocks, use a 1% level of significance to test the significance of the treatment and block variables.

	Treatment			
Block	**1**	**2**	**3**	**4**
1	15	18	12	10
2	13	19	11	13
3	17	22	8	10
4	16	17	14	8

10-15. A series of experiments is conducted, and the coefficient of friction μ between a shaft and a bearing is computed. Tests are made using four different lubricants and three levels of machining of the bearings (highly polished, little polishing, no polishing). Using a 1% level of significance, determine whether or not the type of lubricant has an effect and whether or not the degree of machining has an effect on the coefficient of friction. The coefficients of friction are:

Machining	Lubricant 1	2	3	4
High	0.0045	0.0058	0.0061	0.0048
Little	0.0051	0.0063	0.0064	0.0053
None	0.0068	0.0067	0.0073	0.0066

10-16. Using the data below, perform a two-way ANOVA. Test the hypothesis of interaction at the 1% level of significance. Also, use a 1% level of significance to test the null hypotheses of equal column and equal row means.

Factor	Level 1			Level 2		
Level A	14	16	18	12	16	16
Level B	10	12	16	12	12	14

10-17. Given the data below, perform a two-way ANOVA. Use a level of significance of 5% to test (a) the hypothesis of interaction, (b) the null hypothesis of equal column means, and (c) the null hypothesis of equal row means.

Factor	Level 1				Level 2			
Level A	14	16	16	18	12	14	14	16
Level B	10	12	14	16	10	12	12	14

10-18. Given the data below, perform a two-way ANOVA. Use a level of significance of 1% to test (a) the hypothesis of interaction; (b) the null hypothesis of equal column means; and (c) the null hypothesis of equal row means.

Factor	Level A		Level B		Level C	
Level 1	27	29	18	24	13	15
Level 2	17	21	21	21	12	16
Level 3	18	20	19	23	13	15

10-19. A 240-acre field is divided into 24 10-acre plots. The plots are randomly assigned to 12 groups, with the two fields in each group receiving the same treatment of fertilizer and irrigation. Three levels of fertilizer are applied (low, medium, and high), and four levels of water application are used (none, low, medium, and high). The crop yield in bushels per acre is determined at the end of the season. For the yields shown in the data table opposite, determine (a) if the interaction between irrigation and fertilizer application is significant; (b) if the mean crop yield is affected by fertilizer application; and (c) if the mean crop yield is affected by irrigation rates. Use a 5% level of significance.

Fertilizer	Water application: None		Low		Medium		High	
Low	69	75	81	89	102	106	109	113
Medium	89	95	96	100	107	111	110	116
High	105	113	106	110	99	105	95	99

10.9. SIMULATION PROJECTS

10-1. The analysis of variance for means assumes that the data are drawn from normal populations. To test the sensitivity of the test to this assumption, generate 10^6 samples of size 10 for three populations: normal (mean = standard deviation = 1.464), uniform (lower bound = –1.072, upper bound = 4), and exponential (parameter = 1.464). Thus, all three distributions have the same mean and variance. While the mean and variance of each population are the same (1.464), the shapes of the distributions differ. For each run, generate three samples, compute the three sample means, and perform the ANOVA test. Count the number of times that the null hypothesis is rejected at 5% and at 1%. Compare the proportion of the rejections for the three distributions and discuss the implications of the results.

10-2. The ANOVA test on means is insensitive to slight departures from the assumption of equal population variances when the sample sizes are equal. To test the sensitivity of the ANOVA F test to unequal sample sizes, generate 10^6 tests using one sample of 5 and a second sample of 5, both from normal populations with a mean of 5 and a standard deviation of 1. Count the number of rejections, and compute the proportion of rejections. Repeat the simulation when the second sample is of size 25, then 50, then 100. Assess the sensitivity to unequal sample sizes.

Chapter 11
Confidence Intervals and Sample Size Determination

TABLE OF CONTENTS

11.1. INTRODUCTION

From a sample we obtain single-valued estimates such as the mean, the variance, a correlation coefficient, or a regression coefficient. These single-valued estimates represent our best estimate of the population values, but they are only estimates of random variables, and we know that they probably do not equal the corresponding true values. Thus, we should be interested in the accuracy of these sample estimates.

If we are only interested in whether or not an estimate of a random variable is significantly different from a standard of comparison, we can use a hypothesis test. However, the hypothesis test only gives us a "yes" or "no" answer and not a statement of the accuracy of an estimate of a random variable that may be the object of our attention.

In Example 9-3, a water-quality standard of 3 ppm was introduced for illustration purposes. The hypothesis test showed that the sample mean of 2.8 ppm was not significantly different from 3 ppm. The question arises: Just what is the true mean? Although the best estimate (i.e., expected value) is 2.8 ppm, values of 2.75 or 3.25 ppm could not be ruled out. Is the true value between 2 and 4 ppm, or is it within the range 2.75 to 3.25 ppm? The smaller range would suggest that we are more sure of the population value; that is, the smaller range indicates a higher level of accuracy. The higher level of accuracy makes for better decision making, and this is the reason for examining confidence intervals as a statistical tool.

Confidence intervals represent a means of providing a range of values in which the true value can be expected to lie. Confidence intervals have the additional advantage, compared with hypothesis tests, of providing a probability statement about the likelihood of correctness.

11.2. GENERAL PROCEDURE

The theoretical basis of a confidence interval is the same as that for the theorem for the corresponding hypothesis test: The theorem used in making a statistical hypothesis test is also used to derive the confidence interval. Just as hypothesis tests can be one or two sided, confidence intervals can be computed as either one or two sided.

Many two-sided confidence intervals have the following general form:

$$K_{est} \pm F_d D \tag{11-1}$$

in which K_{est} is the estimated value of the statistic K, F_d is a distribution factor, and D is a measure of the dispersion of the sampling distribution of K_{est}. For one-sided confidence intervals, the ± is replaced by either + or –, depending on the intent (i.e., upper or lower interval, respectively).

The following general procedure can be used to compute a confidence interval:

1. Specify whether a one-sided or two-sided confidence interval is of interest.
2. Identify the theorem that specifies the sampling distribution and distribution factor F_d of the statistic of interest.
3. State the desired level of confidence, $\gamma = 1 - \alpha$, where α is the level of significance used in hypothesis tests.
4. Collect the sample and compute the necessary statistics (e.g., K_{est} and D of Equation 11-1).
5. Determine the value of the distribution factor F_d based on the confidence level and possibly the sample size.
6. Compute the confidence interval.

The six steps of this procedure correspond directly to the six steps of the hypothesis test. It is important to notice that the theoretical model is selected prior to obtaining the sample. This corresponds to making the statement of the hypothesis prior to obtaining the data to test the hypothesis.

11.3. CONFIDENCE INTERVALS ON SAMPLE STATISTICS

11.3.1. Confidence Interval for the Mean

The same theorems that were used for testing hypotheses on the mean are used in computing confidence intervals. In testing a hypothesis for the mean, the choice of test statistic depends on whether or not the standard deviation of the population, σ, is known; this is also true in computing confidence intervals. The theorem for the case where σ is known specifies a *Z* statistic, whereas the *t* statistic is used when σ is unknown; the theorems are not repeated here.

For the case where σ is known, the theorem is given in Section 9.3.1. Confidence intervals on the population mean are given by:

$$\bar{X} - Z_{\alpha/2}\left(\frac{\sigma}{\sqrt{n}}\right) \le \mu \le \bar{X} + Z_{\alpha/2}\left(\frac{\sigma}{\sqrt{n}}\right) \qquad \text{Two-sided interval} \qquad (11\text{-}2)$$

$$\bar{X} - Z_{\alpha}\left(\frac{\sigma}{\sqrt{n}}\right) \le \mu \le \infty \qquad \text{Lower one-sided interval} \qquad (11\text{-}3)$$

$$-\infty \le \mu \le \bar{X} + Z_{\alpha}\left(\frac{\sigma}{\sqrt{n}}\right) \qquad \text{Upper one-sided interval} \qquad (11\text{-}4)$$

in which $\bar{X}$ is the sample mean; *n* is the sample size; Z_α and $Z_{\alpha/2}$ are values of random variables having the standard normal distribution and cutting off (1 – γ) or (1 – γ/2) percent in the tail of the distribution, respectively; and α is the level of significance and equals 1 – γ. The value of $\bar{X}$ corresponds to K_{est} of Equation 11-1, while Z_α and $Z_{\alpha/2}$ correspond to the distribution factor F_d. The measure of dispersion is given by $\sigma\sqrt{n}$, as $\sigma\sqrt{n}$ is the standard error of the mean. Equation 11-2 is a two-sided confidence interval, while Equations 11-3 and 11-4 are one sided. Equation 11-3 gives a lower confidence limit, with no limit on the upper side of the mean; similarly, Equation 11-4 gives an upper limit, with no lower limit.

For the case where σ is unknown, the theorem is given in Section 9.3.2. Confidence intervals on the population mean are given by:

$$\bar{X} - t_{\alpha/2}\left(\frac{S}{\sqrt{n}}\right) \le \mu \le \bar{X} + t_{\alpha/2}\left(\frac{S}{\sqrt{n}}\right) \qquad \text{Two-sided interval} \qquad (11\text{-}5)$$

$$\bar{X} - t_{\alpha}\left(\frac{S}{\sqrt{n}}\right) \le \mu \le \infty \qquad \text{Lower one-sided interval} \qquad (11\text{-}6)$$

$$-\infty \le \mu \le \bar{X} + t_{\alpha}\left(\frac{S}{\sqrt{n}}\right) \qquad \text{Upper one-sided interval} \qquad (11\text{-}7)$$

in which *S* is the sample standard deviation, and t_α and $t_{\alpha/2}$ are values of random variables having a *t* distribution with $\nu = n - 1$. The significance level (α) is used for a one-sided confidence interval, and $\alpha/2$ is used for a two-sided confidence interval. The value of $\overline{X}$ corresponds to K_{est} of Equation 11-1, while t_α and $t_{\alpha/2}$ are the distribution factors (F_d). Because $S/(\sqrt{n})$ is the sample estimate of the standard error of the mean, it is the measure of dispersion (D). Equation 11-5 gives a two-sided confidence interval on the mean, while Equations 11-6 and 11-7 give one-sided lower and upper confidence intervals, respectively. For sample sizes greater than 30, Equations 11-2, 11-3, and 11-4 can be used in place of Equations 11-5, 11-6, and 11-7, respectively.

Example 11-1: Daily Evaporation Measurements

Using a sample of 354 daily evaporation measurements, the mean evaporation was computed to be 0.1387 in./day. The long-term mean and standard deviation for that location were 0.1575 and 0.1 in./day, respectively. The test of hypothesis in Example 9-2 indicated that the sample mean was significantly different from the long-term mean value. The confidence interval would give a better indication of the difference between $\overline{X}$ and the long-term average. The confidence interval can be computed using Equation 11-2 because the sample size is greater than 30. The two-sided, 95% confidence interval is given by:

$$0.1387 \pm \frac{1.96(0.1)}{\sqrt{354}} \tag{11-8}$$

or

$$0.1283 \le \mu \le 0.1491 \text{ in./day} \tag{11-9}$$

Because the upper limit of the 95% confidence interval is not near the long-term mean, one must conclude that the sample is definitely biased; that is, the sample of evaporation data is not representative of the long-term evaporation characteristics at the site.

Example 11-2: Water Quality

Example 9-3 showed that the mean water-quality parameter of 2.8 ppm was not significantly different from the standard of 3 ppm. The accuracy of the mean value is usually more important than a simple "yes/no" decision; therefore, a confidence interval on the mean would be useful in decision making. Because the standard deviation of the population is not known, the confidence interval can be computed using Equation 11-6; the one-sided lower interval is selected because the upper limit is of no physical importance. For a 95% confidence (i.e., $\alpha = 0.05$), the distribution factor was obtained from Table A-2 to be −1.833 for 9 degrees of freedom. The confidence interval is:

$$2.80 - 1.833(0.4)/\sqrt{10} \le \mu \le \infty \tag{11-10a}$$

$$2.568 \text{ ppm } \le \mu \le \infty \tag{11-10b}$$

Thus, we are 95% confident that the true mean is at least 2.568 ppm. Note that if the sample statistics (i.e., $\bar{X}$ and S) were based on a sample size of 20 rather than 10, the lower limit would be:

$$2.80 - 1.729(0.4)/\sqrt{20} \leq \mu \leq \infty \tag{11-11a}$$

$$2.645 \text{ ppm } \leq \mu \leq \infty \tag{11-11b}$$

The higher value of the lower limit reflects the increased confidence that results from the larger sample size; as n is increased, both the standard error of the mean $S/\sqrt{n}$ and the distribution factor decrease.

11.3.2. Factors Affecting a Confidence Interval

It is of interest to notice that a confidence interval is a function of the level of confidence, the theoretical sampling distribution, and the sample characteristic. The level of confidence, γ, is a user-supplied measure of the desired accuracy. It equals 1 minus the level of significance (i.e., $\gamma = 1 - \alpha$). For a given set of sample statistics, the width of the confidence interval increases as the confidence level is increased. The increased width is the price that must be paid for the increased level of confidence (i.e., increased accuracy).

The size of each sample is an important factor that affects the width of the confidence interval. It was shown that the standard error of the mean is inversely proportional to the square root of the sample size. Because D of Equation 11-1 is a measure of the sampling variation of the statistic, one would expect the width of the confidence interval to decrease as the sample size is increased.

11.3.3. Confidence Interval for the Variance

The confidence interval on the population variance (σ^2) can be computed using the same theorem that was used in testing a hypothesis for the variance (see Section 9.4.1). The two-sided and one-sided confidence intervals are:

$$\frac{(n-1)S^2}{\chi^2_{\alpha/2}} \leq \sigma^2 \leq \frac{(n-1)S^2}{\chi^2_{1-\alpha/2}} \quad \text{Two-sided interval} \tag{11-12}$$

$$\frac{(n-1)S^2}{\chi^2_{\alpha}} \leq \sigma^2 \leq \infty \quad \text{Lower one-sided interval} \tag{11-13}$$

$$0 \leq \sigma^2 \leq \frac{(n-1)S^2}{\chi^2_{1-\alpha}} \quad \text{Upper one-sided interval} \tag{11-14}$$

in which $\chi^2_{\alpha/2}$ and χ^2_{α} are values of a random variable having a chi-square distribution that cuts $\alpha/2$ and α percent of the right tail of the distribution, respectively; similarly, $\chi^2_{1-\alpha/2}$ and $\chi^2_{1-\alpha}$ are values of a random variable having a chi-square distribution with cuts at $1 - \alpha/2$ and $1 - \alpha$, respectively. The confidence interval provides an interval in which we are $100(1 - \alpha)$ percent confident that the population value lies within the interval. Although Equations 11-12, 11-13, and 11-14 do not have the same form as Equation 11-1, S^2 is the estimated value of the statistic, and the χ^2 value represents the distribution factor. Although the general form is not the same, the six-step procedure for confidence intervals is directly applicable.

Example 11-3: Uniformity of Irrigation

In Section 9.4.1, the example centered on the variance of water depths resulting from irrigation. Twenty-five measurements produced a variance of 0.003969 $(\text{cm/hr})^2$, or $S = 0.063$ cm/hr. In this case, the greater the variance, the poorer the equipment, so an upper limit would be of interest. Thus, an upper confidence limit is computed using Equation 11-14. For a 95% level of confidence ($\alpha = 0.05$) and 24 degrees of freedom, the critical χ^2 value is 13.848. Thus, the upper limit on the interval is:

$$\frac{(n-1)S^2}{\chi^2_{1-\alpha}} = \frac{24(0.003969)}{13.848} = 0.006879 \tag{11-15a}$$

The confidence interval on the standard deviation is:

$$0 \le \sigma \le 0.0829 \text{ cm/hr} \tag{11-15b}$$

Because the standard deviation of 0.01 cm/hr in Section 9.4.1 falls outside this limit, the null hypothesis of equality is rejected.

It is important to note that $\chi^2_{1-\alpha}$, which equals 13.848, is used for computing the upper limit. If χ^2_{α} were used, which equals 36.415, then the upper limit would actually be smaller than the sample value; thus, using χ^2_{α} would be incorrect.

11.4. SAMPLE-SIZE DETERMINATION

The selection of a sample size is a first step in performing statistical analysis. The selection of the sample size should be based on a clearly stated objective. For example, if we are interested in estimating the mean value of concrete strength that is received at a construction site, then the sample size should be determined based on the objective of estimating the mean concrete strength with specified confidence level (α) and desired width for the corresponding confidence interval. A different sample size would be required if our objective was the development of a regression model with concrete strength as one of the variables. The development of methods for determining the sample size is based on the theorems and methods presented for hypothesis testing and confidence intervals.

11.4.1. Sample Size for Estimating a Mean Value with Known Population Variance

The methods for determining the sample size for the purpose of estimating the mean value depend on whether or not the variance of the population is known or unknown. If the population variance is known, Equation 11-2 can be used to solve for half the width of the confidence interval (*H*) as:

$$H = Z_{\alpha/2}\frac{\sigma}{\sqrt{n}} \tag{11-16}$$

Equation 11-16 can be written as:

$$\frac{H}{\sigma} = \frac{Z_{\alpha/2}}{\sqrt{n}} \tag{11-17}$$

The righthand size of Equation 11-17 can be viewed as the half-width of the desired confidence interval for the parameter of interest expressed in units of standard deviations. This quantity is called the *normalized half-width* H_n; that is,

$$H_n = \frac{H}{\sigma} \tag{11-18}$$

Values for H_n in the range of 0.01 to 1 are commonly used. This range will result in the half-width for the desired confidence interval of 0.01 to 1 standard deviation, which is sufficient for most engineering applications. Therefore, a decision needs to be made about the value of H_n; this should be based on the importance of the decision and practical aspects of the engineering problem under consideration. The smaller the value of H_n, the larger the sample size, which would reflect our interest in a greater certainty level in the mean and, thus, a smaller confidence interval. Substituting Equation 11-18 into Equation 11-17 and solving for the sample size *n* produces:

$$n = \left(\frac{Z_{\alpha/2}}{H_n}\right)^2 \tag{11-19a}$$

or

$$n = \left(\frac{Z_{\alpha/2}}{\frac{H}{\sigma}}\right)^2 = \left(\frac{Z_{\alpha/2}\sigma}{H}\right)^2 \tag{11-19b}$$

Example 11-4: Sample Size for Estimating Daily Evaporation

In Examples 9-2 and 11-1, daily evaporation measurements were discussed. The information in these examples is used to illustrate the determination of the sample

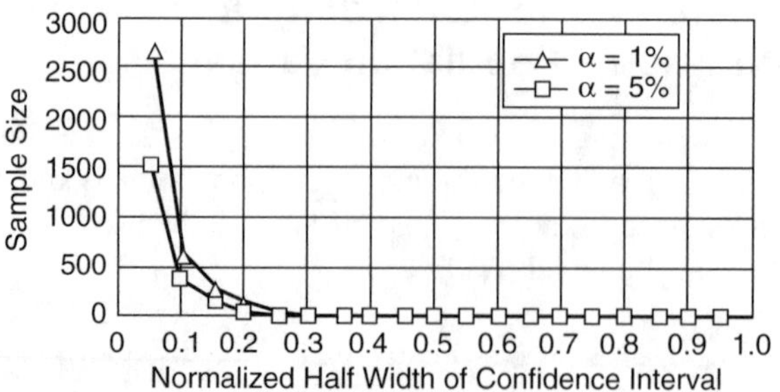

FIGURE 11-1 Sample size with known variance.

size in the case of known population variance. Assuming that we are interested in a normalized half-width of 0.1 and α of 5%, the sample size using Equation 11-19a is:

$$n = \left(\frac{1.96}{0.1}\right)^2 = 384.2 \tag{11-20}$$

Therefore, a sample size of 385 measurements is selected. Figure 11-1 shows the sample size as a function of the normalized half-width of the confidence interval for two α values of 5 and 1%.

11.4.2. Sample Size for Estimating a Mean Value with Unknown Population Variance

If the population variance is unknown, Equation 11-5 can be used to solve for the half-width of the confidence interval (H) as:

$$H = t_{\alpha/2}\frac{S}{\sqrt{n}} \tag{11-21}$$

Equation 11-21 can be written as:

$$\frac{H}{S} = \frac{t_{\alpha/2}}{\sqrt{n}} \tag{11-22}$$

The normalized half-width H_n is:

$$H_n = \frac{H}{S} \tag{11-23}$$

Therefore, the sample size n is:

$$n = \left(\frac{t_{\alpha/2}}{H_n}\right)^2 \tag{11-24a}$$

or

$$n = \left(\frac{t_{\alpha/2}}{\frac{H}{S}}\right)^2 = \left(\frac{t_{\alpha/2}S}{H}\right)^2 \tag{11-24b}$$

In Examples 9-3 and 11-2, water-quality data were used to illustrate one-sided hypothesis testing and confidence interval estimation. Assuming that we are interested in a normalized half-width of 0.5 and α of 5% for the one-sided interval, the sample size using Equation 11-24a can be utilized with t_α instead of $t_{\alpha/2}$. In order to look up a value for t_α, a sample size is needed. Therefore, the solution procedure is iterative and begins by assuming a sample size, looking up t_α from Table A-2, computing the sample size (using Equation 11-24b), and then repeating the process.

Example 11-5: Sample Size for Water Quality

This example uses the same problem presented in Examples 9-3 and 11-2 in which water-quality data were used to illustrate one-sided hypothesis testing and confidence interval estimation. For a normalized half-width of 0.5 and α of 5% for the one-sided interval, a sample size is assumed to be 10; therefore, from Table A-2, $t_\alpha = 1.833$. The resulting sample size is:

$$n = \left(\frac{1.833}{0.5}\right)^2 = 13.4 \tag{11-25a}$$

Therefore, a sample size of 14 measurements is obtained. A revised estimate of t_α is obtained from Table A-2 to be 1.771, thus the revised sample size is:

$$n = \left(\frac{1.771}{0.5}\right)^2 = 12.55 \tag{11-25b}$$

and a sample size of 13 can be used.

11.5. APPLICATIONS

11.5.1. Accuracy of Principal Stress

Six random samples of a sand from a deposit are taken and their principal stresses measured: {5.6, 4.9, 5.3, 4.8, 4.6, 5.5} tons/ft^2. What is the two-sided 80% confidence interval on the mean?

Solution: The sample of 6 yields a mean and standard deviation of 5.117 and 0.407 tons/ft^2, respectively. A sample of size 6 yields 5 degrees of freedom and a t value of 1.476 for 10% in each tail. Thus, the two-sided confidence interval is:

$$5.117 - 1.476(0.407)/\sqrt{6} \leq \mu \leq 5.117 + 1.476(0.407)/\sqrt{6}$$

$$4.872 \leq \mu \leq 5.362 \text{ tons/ft}^2$$

Thus, the true mean is expected to lie within these bounds with 80% certainty.

11.5.2. Compression of Steel

Eight specimens of a particular carbon steel are tested for compression on the vertical plane, with the following results: {5270, 4720, 5110, 4800, 4650, 4940, 5320, 5080} lb/in.^{2}. Because a low mean is undesirable, what is the 97.5% one-sided lower confidence interval?

Solution: The sample of 8 yields a mean and standard deviation of 4986 and 249.8 lb/in.^{2}, respectively. For 7 degrees of freedom and 2.5% in the tail, the t statistic is –2.365. Thus, the lower confidence interval is:

$$4986 - 2.365(249.8)/\sqrt{8} \leq \mu$$

$$4777 \leq \mu$$

Thus, we are 97.5% confident that the true mean is greater than 4777 lb/in.^{2}.

11.5.3. Sample Size of Organic Carbon

A state water-quality agency wants to begin certifying laboratories that are used to analyze water-quality samples. For total organic carbon, their detailed studies suggest that assessments using the prescribed procedure have a population standard deviation of 1.5 mg/l at a mean level of 20 mg/l. To test the accuracy of measurements made by laboratories wishing to be certified, the state plans on sending them periodic samples of a known concentration. How many samples do they need to send to be 90% sure that the laboratory is accurate to within 0.5 mg/l on the mean?

Solution: Because errors greater than and less than the mean are expected, a two-tailed analysis is used. With 5% in each tail, the z statistic is ±1.645 from Table A-1. The z statistic is used rather than the t statistic because the population standard deviation is assumed to be known. Therefore, the required sample size is:

$$n = \left(\frac{z\sigma}{H}\right)^2 = \left[\frac{1.645(1.5)}{0.5}\right]^2 = 24.3 \quad \text{(use 25)}$$

If a sample of 25 is used, the state can be more than 90% assured of an accuracy of at least 0.5 mg/l.

11.5.4. Resampling for Greater Accuracy

In Section 11.5.2, the users of the steel believe that the confidence interval on the mean is too wide for their application. They believe that the mean should be accurate to 150 psi. How many additional specimens must they analyze to be 97.5% confident of their decision?

Solution: It is necessary to assume that the standard deviation of the existing sample of 8 specimens is an accurate estimate of the variation that will result from additional testing; therefore, the relationship for computing the required sample size is:

$$n = (ts/H)^2 = (249.8t/150)^2 = 2.773t^2$$

Because t depends on n and n depends on t, the solution is iterative. Assuming the t value is obtained for 2.5% in one tail, $\nu = n - 1$, and the iteration starts with the existing sample size of 8, the solution is:

n used	ν	$t_{.025}$	Computed *n*
8	7	2.365	15.5
16	15	2.131	12.6
13	12	2.179	13.2
14	13	2.160	12.9
13	12	2.179	13.2

Therefore, a sample size of 14 is needed, so an additional 6 specimens must be tested.

11.6. PROBLEMS

11-1. Provide a definition of a confidence interval and discuss how the confidence interval is used in decision making.

11-2. Compare the six steps of hypothesis testing with the six steps used to compute a confidence interval. Note the similarities and differences between the two procedures.

11-3. An engineering firm manufactures a missile component that will have a life of use that is approximately normally distributed with a population standard deviation of 3 hr. If a random sample of ten components has mean life of use of 13.4 hr, find a two-sided 99.5% confidence interval for the population mean of all components manufactured by this firm.

11-4. The mean life of a random sample of 20 tires of a certain brand is 27,500 mi. (a) If the standard deviation of the population is known to be 3500 mi, what is the two-sided 90% confidence on the mean? (b) If the variance of the population is not known and the standard deviation of the random sample is 3500 mi, what is the two-sided 90% confidence interval?

11-5. For the conditions of Example 11-2, show the variation of the confidence interval with change in the confidence coefficient. Assume $\nu = 9$.

11-6. Five cores are removed from a new section of an asphalt highway. The following percentages of air voids were obtained from laboratory analyses: 3.7, 4.5, 4.1, 4.7, 3.9. Past studies have suggested that the standard deviation of the air void content is 0.5%. Compute the 95%, two-sided confidence interval on the mean.

11-7. Twenty-five engineering students independently measured the melting point of lead. The mean and standard deviation of the 25 measurements were 329.6 and 5.1°C, respectively. Using a level of significance of 10%, what is the error that can be expected if the sample mean is used as an estimate of the true melting point of lead?

11-8. A random sample of test scores of the GRE exam by 25 students produces a mean of 575 and standard deviation of 60. (a) Determine the 99% confidence interval on the mean score for all students at the university. (b) What assertion can be made about the magnitude of the error, if we state that the mean score of all students at the university is 575 and assume that we wish to be 99% confident in our assertion?

11-9. A laboratory testing of five samples on the content of a certain water-quality indicator results in the following values: 3.41, 3.06, 2.94, 3.27, and 3.32 mg/l. (a) Find the 95% confidence interval for the mean content of this indicator. (b) What assumptions are made?

11-10. Write and execute a computer program that generates 1000 samples from a standard normal distribution, with each sample of size n. For each sample compute the mean, standard deviation, and two-sided confidence intervals for confidence levels of 90 and 99%. For each confidence level, determine the number of samples for which the confidence intervals do not cover the true mean. Execute the program for sample sizes of 10, 25, and 50.

11-11. Identify the four factors that influence the width of a confidence interval on the mean. Provide general statements about the effect of each factor on the width of the interval.

11-12. Using the first sample of 5 in Table 8-3, compute a two-sided, 95% confidence interval for the variance for the first sample ($S^2 = 1.764$). Does the interval cover the true variance in this case?

11-13. Using the sample standard deviation of the data of Problem 11-6, compute a one-sided upper 90% confidence interval on the variance. Does the interval cover the true variance of $0.25(\%)^2$?

11-14. The corrosion depth of four samples taken from a steel I beam are 72, 217, 145, and 259 mils. Assuming an interest in being 90% certain, use a confidence interval to decide whether or not the true variance could be as small as 2500 $(\text{mils})^2$.

11-15. A laboratory believes that its equipment needs adjustment when the standard deviation of five samples of a known concentration exceeds 4 mg/l. During one test, the following sample measurements were recorded: 50.7, 58.2, 46.3, 53.0, and 62.4 mg/l. Does a confidence interval suggest that the equipment requires adjustment? Explain.

11-16. A lab technician wishes to determine the mean time to perform a lab experiment. What sample size will be needed to be 96% confident that the sample mean will be within 10 min of the true mean? Previous studies have indicated that $\sigma = 25$ min.

11-17. An efficiency expert is interested in knowing the mean number of days of sick leave taken by workers in the auto industry. How many workers' sick-leave records would have to be checked to be able to state that the sample mean will be within 3.0 days of the true mean? Assume a 5% level of significance and a value of the population standard deviation of 4.7 days.

11-18. A manufacturer produces a stereo component that will have a length of life that is normally distributed with a population standard deviation of 70 hr. (a) Show the change in the width of the 90% confidence interval for the population mean of all components produced by this manufacturer as a function of sample size. (b) What sample size is needed to ensure that the confidence interval is no greater than 40 hr?

11-19. Six-inch cores are to be sampled from a section of asphalt concrete pavement and tested for a mechanical property. Accuracy requirements indicate that a normalized half-width of 0.33 is needed. Regulations dictate a 90% level of confidence. What sample size should be collected?

11-20. Measurements of the thickness of the corrosion layer on a steel I beam are needed for a corrosion study. The researcher wishes to be 99% confident in the estimate of the means, with a normalized half-width of 0.75. What sample size will be needed?

11-21. For the conditions of Problem 11-17, compute the sample size that would be needed if the variance of the population is not known but the sample standard deviation would be approximately 4.7 days.

11.7. SIMULATION PROJECTS

Using the simulations of the project from Chapter 9, show that the confidence interval on the mean for a normal population and sample size of 10 is accurately represented by the theory of Section 11.3.1.

Chapter 12
Regression Analysis

TABLE OF CONTENTS

12.1. INTRODUCTION

In analyzing single random variables, it is important to identify the underlying distribution of the random variable and make estimates of its parameters. Additionally, decision making requires an analysis of the uncertainty in the knowledge of the population. Hypothesis testing and interval estimation (i.e., confidence intervals) are means of assessing the significance of and quantifying this uncertainty.

In many cases, variation in the value of a random variable is associated with variation in one or more additional variables. By establishing the relationship between two or more variables, it may be possible to reduce the uncertainty in an estimate of the random variable of interest. Thus, before making a decision based on the analysis of observations on a random variable alone, it is wise to consider the possibility of systematic associations between the random variable of interest and other variables that have a causal relationship with it. These associations among variables can be understood using correlation and regression analyses.

To summarize these concepts, the variation in the value of a variable may be either random or systematic in its relationship with another variable. Random variation represents uncertainty. If the variation is systematically associated with one or more other variables, the uncertainty in estimating the value of a variable can be reduced by identifying the underlying relationship. Correlation and regression analyses are important statistical methods to achieve these objectives. They should be preceded by a graphical analysis to determine (1) if the relationship is linear or nonlinear, (2) if the relationship is direct or indirect, and (3) if any extreme events might control the relationship.

12.2. CORRELATION ANALYSIS

Correlation analysis provides a means of drawing inferences about the strength of the relationship between two or more variables. That is, it is a measure of the degree to which the values of these variables vary in a systematic manner. Thus, it provides a quantitative index of the degree to which one or more variables can be used to predict the values of another variable.

It is just as important to understand what correlation analysis cannot do as it is to know what it can do. Correlation analysis does not provide an equation for predicting the value of a variable. Also, correlation analysis does not indicate whether a relationship is causal; that is, it is necessary for the investigator to determine whether there is a cause-and-effect relationship between the variables. Correlation analysis only indicates whether the degree of common variation is significant.

Correlation analyses are most often performed after an equation relating one variable to another has been developed. This happens because the correlation coefficient, which is a quantitative index of the degree of common linear variation, is used as a measure of the goodness of fit between the prediction equation and the data sample used to derive the prediction equation, but correlation analyses can be very useful in model formulation, and every model development should include a correlation analysis.

12.2.1. Graphical Analysis

The first step in examining the relationship between variables is to perform a graphical analysis. Visual inspection of the data can identify the following information:

1. Degree of common variation, which is an indication of the degree to which the two variables are related
2. Range and distribution of the sample data points
3. Presence of extreme events
4. Form of the relationship between the two variables
5. Type of relationship

Each of these factors is of importance in the statistical analysis of sample data and in decision making.

When variables show a high degree of association, we assume that a causal relationship exists. If there is a physical reason to suspect that a causal relationship exists, the association demonstrated by the sample data provides empirical support for the assumed relationship. Common variation implies that, when the value of one of the random variables is changed, the value of the other variable will change in a systematic manner. For example, an increase in the value of one variable occurs when the value of the other variable increases. If the change in the one variable is highly predictable from a given change in the other variable, there is a high degree of common variation. Figure 12-1 shows graphs of different samples of data for two variables having different degrees of common variation. The common variation is measured in the figure using the correlation coefficient (R). In Figures 12-1a and 12-1e the degree of common variation is very high; thus, the variables are said to be correlated. In Figure 12-1c there is no correlation between the two variables because, as the value of X is increased, it is not certain whether Y will increase or decrease. In Figures 12-1b and 12-1d, the degree of correlation is moderate; in Figure 12-1b, evidently Y will increase as X is increased, but the exact change in Y for a change in X is difficult to estimate just by looking at the graph. A more quantitative description of the concept of common variation is needed.

It is important to use a graphical analysis to identify the range and distribution of the sample data points, so that the stability of the relationship can be assessed and so that we can assess the ability of the data sample to represent the distribution of the population. If the range of the data is limited, a computed relationship may not be stable; that is, it may not apply to the distribution of the population. The data set of Figure 12-2 shows a case where the range of the sample is much smaller than the

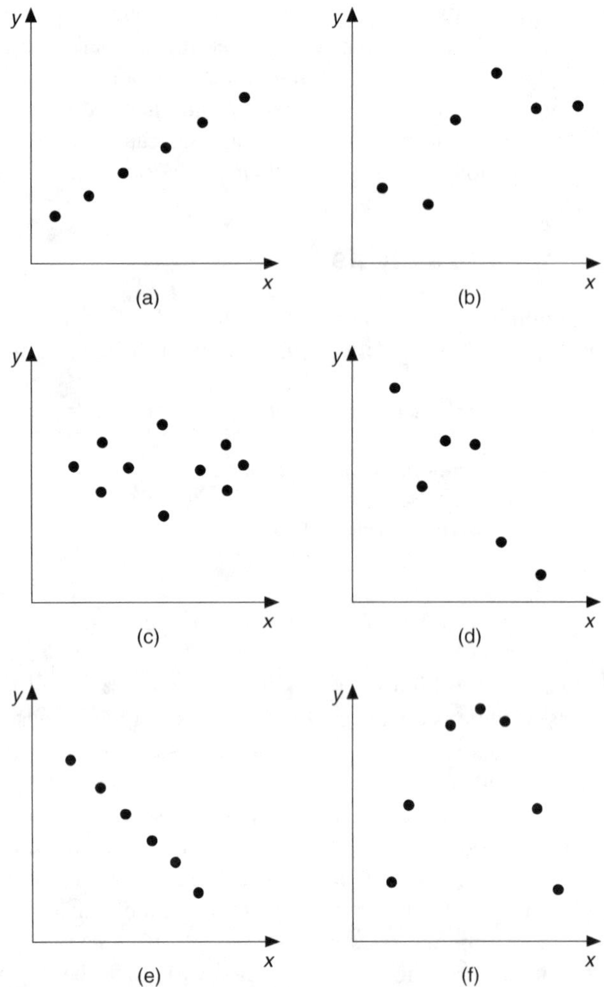

FIGURE 12-1 Different degrees of correlation between the variables X and Y: (a) $R = 1.0$; (b) $R = 0.5$; (c) $R = 0.0$; (d) $R = -0.5$; (e) $R = -1.0$; (f) $R = 0.3$.

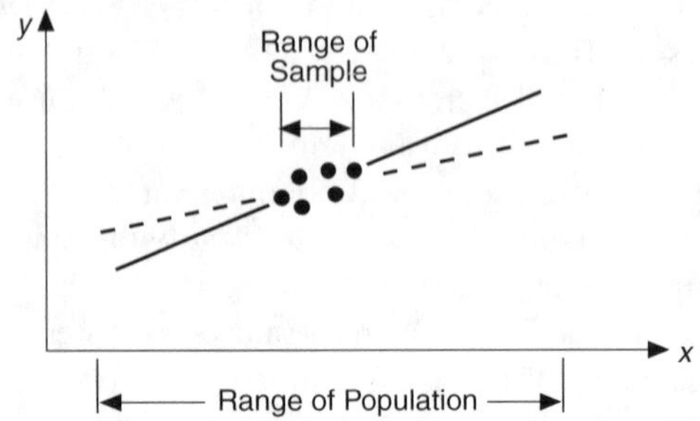

FIGURE 12-2 Instability in the relationship between random variables.

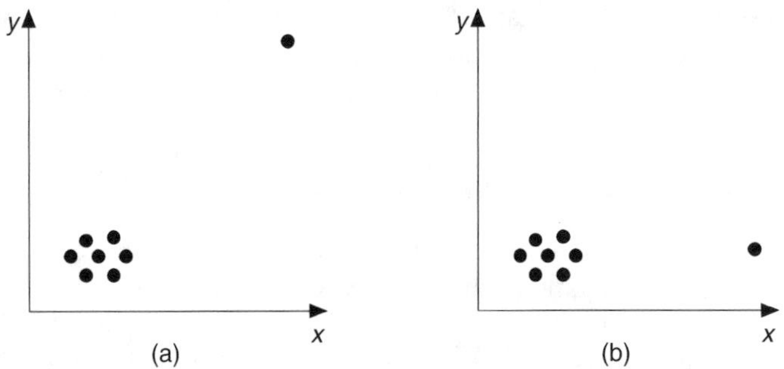

FIGURE 12-3 Effect of an extreme event in a data sample on the correlation: (a) high correlation; (b) low correlation.

expected range of the population. If an attempt is made to use the sample to project the relationship between the two random variables, a small change in the slope of the relationship causes a large change in the predicted estimate of *Y* for values of *X* at the extremes of the range of the population. A graph of two random variables might alert the investigator to a sample in which the range of the sample data may cause stability problems in a derived relationship between two random variables, especially when the relationship is projected beyond the range of the sample data.

It is important to identify extreme events in a sample of data for several reasons. First, extreme events can dominate a computed relationship between two variables. For example, in Figure 12-3a, the extreme point suggests a high correlation between *X* and *Y*; in this case, the cluster of points acts like a single observation. In Figure 12-3b, the extreme point causes a poor correlation between the two random variables; since the mean of the cluster of points is nearly the same as the value of *Y* of the extreme point, the data of Figure 12-3b suggest that a change in *X* is not associated with a change in *Y*. A correlation coefficient is more sensitive to an extreme point when the sample size is small. An extreme event may be due to (1) errors in recording or plotting the data, or (2) a legitimate observation in the tail of the distribution. Therefore, an extreme event must be identified and the reason for the event determined. Otherwise, it is not possible to properly interpret the results of the correlation analysis.

Relationships can be linear or nonlinear. Because the statistical methods to be used for the two forms of a relationship differ, it is important to identify the form. Additionally, the most frequently used correlation coefficient depends on a linear relationship existing between the two random variables; thus, low correlation may result for a nonlinear relationship even when the relationship is obvious. For example, the bivariate relationship of Figure 12-1f suggests a very predictable trend in the relationship between *Y* and *X*, but the correlation is poor and is certainly not as good as that in Figure 12-1a, even though the two plots suggest equal levels of predictability.

Graphs relating pairs of variables can be used to identify the type of the relationship. Linear trends can be either direct or indirect, with an indirect relationship indicating a decrease in *Y* as *X* increases. This information is useful for checking the rationality of the relationship, especially when dealing with data sets that include more than two variables. A variable that is not dominant in the physical relationship may demonstrate a physically irrational relationship with another variable because of the values of the other variable affecting the physical relationship.

12.2.2. Bivariate Correlation

Correlation is the degree of association between the elements of two samples of data — that is, between observations on two variables. Correlation coefficients provide a quantitative index of the degree of linear association. Examples of variables that are assumed to have a causal relationship and significant correlation are (1) the cost of living and wages, and (2) the volumes of rainfall and flood runoff. However, examples that have no cause-and-effect relationship but may have significant correlation include (1) the crime rate and the sale of chewing gum over the past two decades, and (2) annual population growth rates in 19th-century France and annual cancer death rates in the U.S. in the 20th century.

Many indexes of correlation exist. The method that is used most frequently is the Pearson product-moment correlation coefficient. Nonparametric correlation indexes include the contingency coefficient, the Spearman rank correlation coefficient, and the Kendall rank correlation coefficient. Only the Pearson correlation coefficient is considered in this chapter.

12.2.3. Separation of Variation

A set of observations on a random variable Y has a certain amount of variation, which may be characterized by the variance of the sample. The variance equals the sum of squares of the deviations of the observations from the mean of the observations divided by the degrees of freedom. Ignoring the degrees of freedom, the variation in the numerator can be separated into two parts: (1) variation associated with a second variable X, and (2) variation not associated with X. That is, the total variation (TV), which equals the sum of the squares of the sample data points about the mean of the data points, is separated into the variation that is explained by variation in the second variable (EV) and the variation that is not explained — that is, the unexplained variation (UV). Thus, TV can be expressed as:

$$\text{TV} = \text{EV} + \text{UV} \tag{12-1}$$

Using the general form of the variation of a random variable, each of the three terms in this equation can be represented by a sum of squares as follows:

$$\sum_{i=1}^{n}(y_i - \overline{Y})^2 = \sum_{i=1}^{n}(\hat{y}_i - \overline{Y})^2 + \sum_{i=1}^{n}(y_i - \hat{y}_i)^2 \tag{12-2}$$

where y_i is an observation on the random variable, $\hat{y}_i$ is the value of Y estimated from the best relationship with the second variable X, and $\overline{Y}$ is the mean of the observations on Y. These variation components are shown in Figure 12-4; the dashed

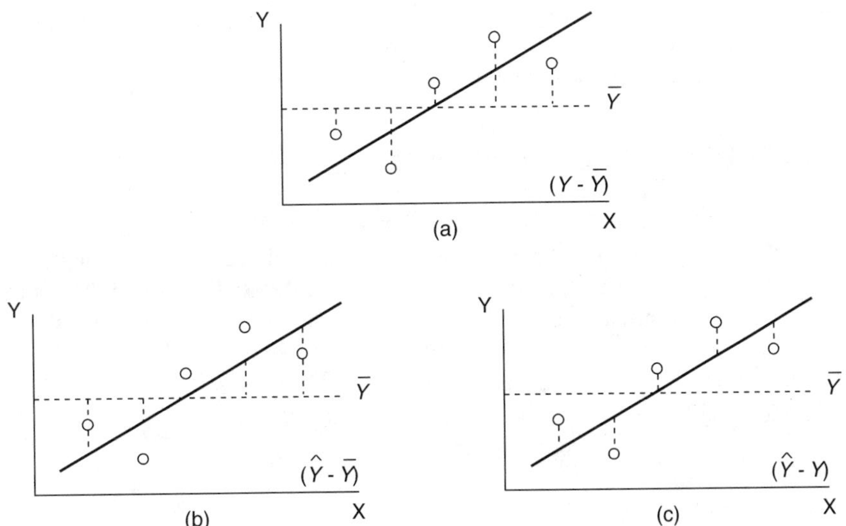

FIGURE 12-4 Separation of variation: (a) total variation; (b) explained variation; (c) unexplained variation.

lines illustrate the deviations of Equation 12-2. Figure 12-4a shows the variation of the observations about the mean value; this represents the variation that exists in the sample and that may potentially be explained by the variable X, and it reflects the left side of Equation 12-2. The variation in Figure 12-4b represents the variation of the line about the mean and corresponds to the first term of the right side of Equation 12-2. If all of the sample points fall on the line, the explained variation will equal the total variation. The variation in Figure 12-4c represents the variation of the points about the line; therefore, it corresponds to the second term on the right side of Equation 12-2. This is the variation that is not explained by the relationship between Y and X. If this unexplained variation equals the total variation, the line does not explain any of the variation in the sample, and the relationship between X and Y is not significant.

12.2.4. Correlation: Fraction of Explained Variation

The separation of variation concept is useful for quantifying the correlation coefficient by Pearson. Specifically, dividing both sides of Equation 12-1 by the total variation TV gives:

$$1 = \frac{\text{EV}}{\text{TV}} + \frac{\text{UV}}{\text{TV}} \tag{12-3}$$

The ratio EV/TV represents the fraction of the total variation that is explained by the linear relationship between Y and X; this is called the *coefficient of determination* and is given by:

$$R^2 = \frac{\text{EV}}{\text{TV}} = \frac{\sum_{i=1}^{n} (\hat{y}_i - \bar{Y})^2}{\sum_{i=1}^{n} (y_i - \bar{Y})^2} \tag{12-4}$$

The square root of the ratio is the correlation coefficient, R. If the explained variation equals the total variation, the correlation coefficient will equal 1. If the relationship between X and Y is inverse, as in Figure 12-1e, and the explained variation equals the total variation in magnitude, R will equal –1. These represent the extremes, but both values indicate a perfect association, with the sign only indicating the direction of the relationship. If the explained variation equals zero, R equals zero. Thus, a correlation coefficient of zero, which is sometimes called the *null correlation*, indicates no linear association between the two variables X and Y.

12.2.5. Computational Form for Correlation Coefficient

While Equation 12-4 provides the means to compute a value of the correlation coefficient, it can be shown that Equation 12-4 can be rearranged to the following form using a linear model for $\hat{Y}$ that minimizes the unexplained variation (UV):

$$R = \frac{\sum_{i=1}^{n} x_i y_i - \frac{1}{n}\left(\sum_{i=1}^{n} x_i\right)\left(\sum_{i=1}^{n} y_i\right)}{\sqrt{\sum_{i=1}^{n} x_i^2 - \frac{1}{n}\left(\sum_{i=1}^{n} x_i\right)^2}\sqrt{\sum_{i=1}^{n} y_i^2 - \frac{1}{n}\left(\sum_{i=1}^{n} y_i\right)^2}} \tag{12-5}$$

This equation is used most often because it does not require prior computation of the means, thus the computational algorithm is easily programmed for a computer. The linear model $\hat{Y}$ that minimizes UV is based on the principle of least squares, as discussed in Section 12.4.

12.2.6. Distribution of the Correlation Coefficient

It should be recognized that the estimated correlation coefficient is a random variable, and each random variable must have a distribution function. The distribution of R is a function of the sample size n and the population value of the correlation coefficient ρ. Unfortunately, the correlation coefficient does not have a recognizable probability distribution.

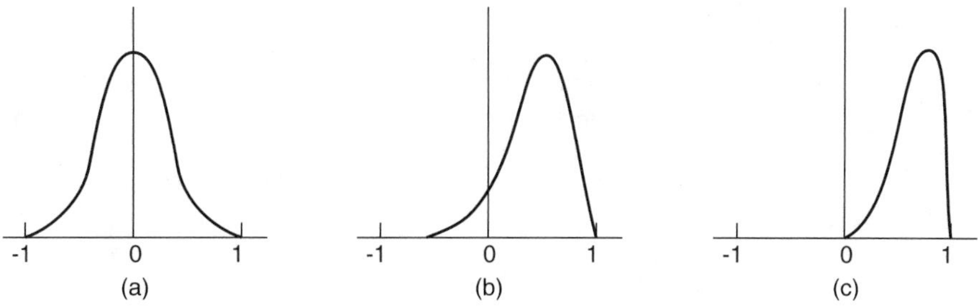

FIGURE 12-5 Effect of ρ (n = 10) on the distribution of R: (a) ρ = 0; (b) ρ = 0.5; (c) ρ = 0.9.

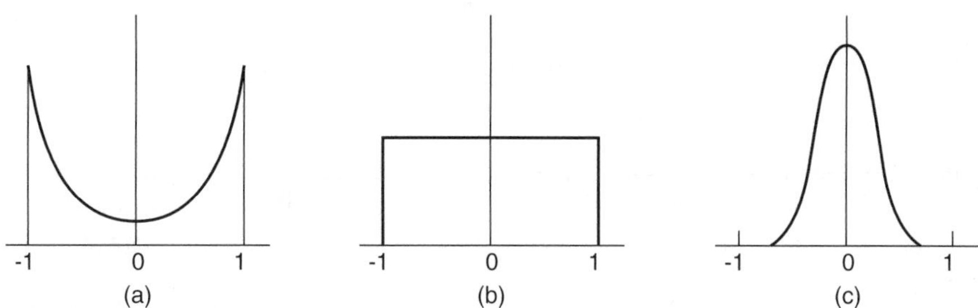

FIGURE 12-6 Effect of n (ρ = 0) on the distribution of R: (a) n = 3; (b) n = 4; (c) n = 25.

12.2.6.1. Effect of ρ (n Constant)

Figure 12-5 shows the distribution of R for three values of ρ, with n held constant at 10. The most likely sample value is the population value. For ρ = 0, a sample value near zero (i.e., ρ = 0) is most likely. For ρ = 0.5, the probability of a sample value at or near zero is considerably smaller than when ρ = 0. For ρ = 0.9, there is very little chance of obtaining a sample estimate that is near zero, but a value near 0.9 is highly likely.

12.2.6.2. Effect of n (ρ Constant)

Figure 12-6 shows the distribution of R for sample sizes of 3, 4, and 25, with ρ held constant at zero. Figure 12-6a shows that for a sample size of 3, a large sample estimate of the correlation coefficient is more likely than a value near zero even when the population value equals zero. For a sample size of 4 (Figure 12-6b), all values of R are equally likely. As n increases to 10 (Figure 12-5a) and 25 (Figure 12-6c), sample estimates near zero become more likely. In summary, it is important to recognize that for small sample sizes, large sample estimates of R are very likely to occur and that even high correlation may not represent a significant relationship between the two random variables.

12.2.7. Hypothesis Testing

A correlation coefficient of zero indicates that there is no linear relationship between the two variables. Thus, it would be of interest to make a two-sided test of the hypotheses:

$$H_0\colon \rho = 0 \tag{12-6}$$

$$H_A\colon \rho \neq 0 \tag{12-7}$$

It can be shown that for $n > 2$ these hypotheses can be tested using a t statistic that is given by:

$$t = \frac{R}{\sqrt{\dfrac{1-R^2}{n-2}}} \tag{12-8}$$

where t is the value of a random variable that has approximately a t distribution with $(n-2)$ degrees of freedom. The correlation coefficient is considered statistically significant if the computed t value is greater in absolute value than a critical value having a t distribution and a level of significance of α; that is, the null hypothesis is rejected when the absolute value of the computed t statistic is greater than the critical value. One-sided tests of hypotheses are also possible using the test statistic above. Table A-5 provides the critical values of R for both one-sided and two-sided tests. A value of R greater than the critical value is considered significant (i.e., the null hypothesis of Equation 12-6 should be rejected). For negative values of R, the absolute value of the critical value in Table A-5 should be used.

In many cases, it is of interest to examine whether or not one can assume that the population has a value of the correlation coefficient ρ_0 other than zero. The two-sided hypotheses for this test are:

$$H_0\colon \rho = \rho_0 \tag{12-9a}$$

$$H_A\colon \rho \neq \rho_0 \tag{12-9b}$$

One-sided tests can also be performed using the same procedure as the two-sided alternative. The test statistic for the hypotheses of Equation 12-9 is:

$$R' = 0.5\ln\left(\frac{1+R}{1-R}\right) \tag{12-10}$$

in which R is the sample correlation coefficient. The random variable R' has a mean $\bar{R}'$ and standard deviation $S_{R'}$ given by:

$$\bar{R}' = 0.5\ln\left(\frac{1+\rho_0}{1-\rho_0}\right) \tag{12-11}$$

$$S_{R'} = \frac{1}{\sqrt{n-3}} \tag{12-12}$$

in which n is the sample size. A standard normal transformation yields:

$$Z = \frac{R' - \bar{R}'}{S_{R'}} \tag{12-13}$$

in which Z is the value of a random variable having a standard normal distribution with a mean of zero and a standard deviation of 1. If either the sample (R) or population (ρ_0) correlation coefficient is negative, the absolute value should be used. The null hypothesis is rejected where the absolute value of the computed Z coefficient of Equation 12-13 is greater than the critical value for the selected level of significance.

Example 12-1: Hypothesis Testing of Correlation Coefficients

The objective of this example is to illustrate both the hypothesis tests for correlation coefficients and the problem of using small sample sizes to estimate the correlation between two random variables. Assume that a sample of five yields a correlation coefficient of 0.87 (i.e., $R = 0.87$). The null hypothesis of Equation 12-6 is tested using the t statistic of Equation 12-8 as follows:

$$t = \frac{0.87}{\sqrt{\frac{1-0.87^2}{5-2}}} = 3.056 \tag{12-14}$$

For a 5% level of significance, the critical t value for 3 degrees of freedom and $\alpha/2 = 0.025$ obtained from Table A-2 is 3.182; therefore, the null hypothesis should be accepted, and there is no reason to believe that the sample does not come from a population having a correlation coefficient of zero.

The same sample can be used to test the hypotheses of Equation 12-9 with $\rho_0 = 0.99$. The computations for Equations 12-10 to 12-13 are:

$$R' = 0.5\ln\left(\frac{1+0.87}{1-0.87}\right) = 1.3331 \tag{12-15}$$

$$\bar{R}' = 0.5\ln\left(\frac{1+0.99}{1-0.99}\right) = 2.6467 \tag{12-16}$$

$$S_{R'} = \frac{1}{\sqrt{5-3}} = 0.7071 \tag{12-17}$$

$$Z = \frac{1.3331 - 2.6467}{0.7071} = -1.8577 \tag{12-18}$$

For a 5%, two-sided level of significance, the critical z value obtained from Table A-1 is 1.96; therefore, the null hypothesis that ρ equals 0.99 should be accepted.

In summary, the sample of five was used to illustrate the computation of hypothesis tests on the population correlation coefficient. However, it is more important to note important characteristics of hypothesis testing for correlation coefficients. First, estimates of R obtained from small samples may not be close to the population value; in this case, the sample value of 0.87 was not significantly different from either 0.0 or 0.99. This suggests that with a small sample, an engineer could prove almost anything, including an opinion based on personal bias. Second, the 5% level of significance may not be the appropriate level. Third, it is inadequate just to investigate the null hypothesis for null correlation; confidence intervals on ρ should be calculated.

Example 12-2: Computational Illustration

A sample of five observations, given as the values of X and Y in Table 12-1, is used to illustrate the computational methods outlined herein. The last row of Table 12-1 contains column summations. Using Equation 12-4, the correlation is:

$$R = \sqrt{\frac{8.10}{10}} = 0.9 \tag{12-19}$$

The predicted values, $\hat{y}_i$, were obtained by fitting a regression equation to the data points; the regression method is discussed in the remaining sections of this chapter. Therefore, Equation 12-5 yields:

$$R = \frac{54 - \frac{15(15)}{5}}{\sqrt{55 - \frac{15(15)}{5}}\sqrt{55 - \frac{15(15)}{5}}} = 0.9 \tag{12-20}$$

Thus, 81% of the total variation in Y is explained by variation in X.

12.3. INTRODUCTION TO REGRESSION

In many engineering projects, estimates of random variables are required. For example, in the design of irrigation projects, it is necessary to provide estimates of evaporation. If we are fortunate to have a past record of daily evaporation rates from a pan, the mean of these observations multiplied by some proportionality constant (a pan coefficient) may be our best estimate. However, the error in such an estimate can be significant. Because the standard deviation of the observations is a measure

TABLE 12-1 Calculation of the Correlation Coefficient

	x_i	y_i	$(x_i - \overline{X})$	$(y_i - \overline{Y})$	$(x_i - \overline{X}) \cdot (y_i - \overline{Y})$	$\hat{y}_i$	$\hat{y}_i - \overline{Y}$	$(\hat{y}_i - \overline{Y})^2$	$(y_i - \overline{Y})^2$	$x_i y_i$	x_i^2	y_i^2
	1	2	–2	–1	2	1.2	–1.8	3.24	1	2	1	4
	2	1	–1	–2	2	2.1	–0.9	0.81	4	2	4	1
	3	3	0	0	0	3.0	0.0	0.00	0	9	9	9
	4	4	1	1	1	3.9	0.9	0.81	1	16	16	16
	5	5	2	2	4	4.8	1.8	3.24	4	25	25	25
Sum	15	15			9			8.10	10	54	55	55

of the variation in the past, it is our best estimate of the future variation and the error in a single measurement. The error in a mean value is also a function of the standard deviation.

If the random variable can be related to other variables, it may be possible to reduce the error in a future estimate. For example, evaporation is a function of the temperature and humidity of the air mass that is over the water body. Thus, if measurements of the air temperature and the relative humidity are also available, a relationship, or model, can be developed. The relationship may provide a more accurate estimate of evaporation for the conditions that may exist in the future.

The process of deriving a relationship between a random variable and measured values of other variables is called *optimization*, or model calibration, and makes use of the fundamental concepts of calculus and numerical analysis. The objective of optimization is to find the values of a vector of unknowns that provides the minimum or maximum value of some function. Before examining regression analysis, it is necessary to discuss the fundamentals of statistical optimization.

12.3.1. Elements of Statistical Optimization

In statistical optimization, the function to be optimized is called the *objective function*, which is an explicit mathematical function that describes what is considered the optimal solution. A mathematical model relates a random variable, called the *criterion* or dependent variable, to a vector of unknowns, called *regression coefficients*, and a vector of predictor variables, which are sometimes called *independent variables*. The predictor variables are usually variables that have a causal relationship with the criterion variable. For example, if we are interested in using measurements of air temperature and humidity to predict evaporation rates, the evaporation would be the criterion variable and the temperature and relative humidity would be the predictor variables. While the objective function of statistical optimization corresponds to the explicit function, the vector of unknowns in statistical optimization corresponds to the unknowns in the explicit function. The unknowns are the values that are necessary to transform values of the predictor variable(s) into a predicted value of the criterion variable. It is important to note that the objective function and the mathematical model are two separate explicit functions. The third element of statistical optimization is a data set. The data set consists of measured values of the criterion variable and the predictor variable(s). In summary, statistical optimization requires: (1) an objective function, which defines what is meant by best fit; (2) a mathematical model, which is an explicit function relating a criterion variable to vectors of unknowns and predictor variable(s); and (3) a matrix of measured data.

As an example, we may attempt to relate evaporation (E), the criterion variable, to temperature (T), which is the predictor variable, using the equation:

$$\hat{E} = b_0 + b_1 T \tag{12-21}$$

in which b_0 and b_1 are the unknown coefficients, and $\hat{E}$ is the predicted value of E. To evaluate the unknowns b_0 and b_1, a set of simultaneous measurements on E and T would be made. For example, if we were interested in daily evaporation rates, we may measure both the total evaporation for each day in a year and the corresponding mean daily temperature; this would give us 365 observations from which we could estimate the two regression coefficients. An objective function would have to be selected to evaluate the unknowns; for example, regression minimizes the sum of the squares of the differences between the predicted and measured values of the criterion variable. Other criterion functions, however, may be used.

12.3.2. Zero-Intercept Model

Assume that the criterion variable Y is related to a predictor variable X using the linear model:

$$\hat{Y} = bX \tag{12-22}$$

in which b is the unknown, and $\hat{Y}$ is the predicted value of Y. The objective function, F, is defined as minimizing the sum of the squares of the differences (e_i) between the predicted values ($\hat{y}_i$) and the measured values (y_i) of Y as follows:

$$F = \min \sum_{i=1}^{n} e_i^2 = \min \sum_{i=1}^{n} (\hat{y}_i - y_i)^2 \tag{12-23}$$

where n is the sample size.

To derive a value for b, differentiate $\sum_{i=1}^{n} e_i^2$ with respect to the unknown, set the derivative equal to zero, and solve for the unknown. Substituting Equation 12-22 into Equation 12-23 yields:

$$F = \min \sum_{i=1}^{n} (\hat{y}_i - y_i)^2 = \min \sum_{i=1}^{n} (bx_i - y_i)^2 \tag{12-24}$$

The derivative of $\sum_{i=1}^{n} e_i^2$ with respect to the unknown b is:

$$\frac{\partial \sum_{i=1}^{n} e_i^2}{\partial b} = 2\sum_{i=1}^{n}(bx_i - y_i)(x_i) \tag{12-25}$$

Setting Equation 12-25 to 0 and solving for b yields:

$$b = \frac{\sum_{i=1}^{n} x_i y_i}{\sum_{i=1}^{n} x_i^2} \tag{12-26}$$

The value of b can be computed using the two summations, $\sum_{i=1}^{n} x_i y_i$ and $\sum_{i=1}^{n} x_i^2$.

Example 12-3: Fitting a Zero-Intercept Model

The following table gives the data that can be used for fitting the value of b:

i	x_i	y_i
1	3	2
2	5	3
3	6	4

The following table shows the necessary computations:

i	x_i	x_i^2	y_i	$x_i y_i$
1	3	9	2	6
2	5	25	3	15
3	6	36	4	24
Column summation		70		45

Therefore, the value of b equals (45/70) or 0.64286. The following equation could then be used to derive a future estimate of Y for a future value of X:

$$\hat{Y} = 0.6428X \tag{12-27}$$

It is important to note that Equation 12-26 is valid only for the linear model of Equation 12-22 that does not include an intercept; Equation 12-26 would not be valid for Equation 12-21, as Equation 12-21 includes an intercept coefficient (b_0).

12.3.3. Regression Definitions

The objective of regression is to evaluate the coefficients of an equation relating the criterion variable to one or more other variables, which are called the *predictor variables*. The predictor variables are variables for which their variation is believed to cause or agree with variation in the criterion variable. A predictor variable is often called an *independent variable*; this is a misnomer in that independent variables are usually neither independent of the criterion variable nor independent of other predictor variables.

The most frequently used linear model relates a criterion variable Y to a single predictor variable X by the equation:

$$\hat{Y} = b_0 + b_1 X \tag{12-28}$$

in which b_0 is the intercept coefficient and b_1 is the slope coefficient; b_0 and b_1 are called *regression coefficients* because they are obtained from a regression analysis. Because Equation 12-28 involves two variables, Y and X, it is sometimes referred to as the *bivariate model*. The intercept coefficient represents the value of Y when X equals zero. The slope coefficient represents the rate of change in Y with respect to change in X. Whereas b_0 has the same dimensions as Y, the dimensions of b_1 equal the ratio of the dimensions of Y to X.

The linear multivariate model relates a criterion variable to two or more predictor variables:

$$\hat{Y} = b_0 + b_1 X_1 + b_2 X_2 + \ldots + b_p X_p \tag{12-29}$$

in which p is the number of predictor variables, X_i is the ith predictor variable, b_i is the ith slope coefficient, and b_0 is the intercept coefficient, where $i = 1, 2, \ldots, p$. The coefficients b_i are often called *partial regression coefficients* and have dimensions equal to the ratio of the dimensions of Y to X_i. Equation 12-28 is a special case of Equation 12-29. This model is discussed in Chapter 13.

12.4. PRINCIPLE OF LEAST SQUARES

12.4.1. Definitions

The values of the slope and intercept coefficients of Equations 12-28 and 12-29 can be computed using the principle of least squares. The principle of least squares is a process of obtaining *best* estimates of the coefficients and is referred to as a *regression method*. Regression is the tendency for the expected value of one of two jointly correlated random variables to approach more closely the mean value of its set than any other. The principle of least squares is used to regress Y on either X or the X_i values of Equations 12-28 and 12-29, respectively. To express the principle of least squares, it is important to define the error, e, or residual, as the difference between the predicted and measured values of the criterion variable:

$$e_i = \hat{y}_i - y_i \tag{12-30}$$

in which $\hat{y}$ is the ith predicted value of the criterion variable, y_i is the ith measured value of Y, and e_i is the ith error. It is important to note that the error is defined as the measured value of Y subtracted from the predicted value. Some computer programs use the measured value minus the predicted value. This definition is avoided because it indicates that a positive residual implies underprediction. With Equation 12-30, a positive residual indicates overprediction, while a negative residual indicates underprediction. The objective function for the principle of least squares is to minimize the sum of the squares of the errors:

$$F = \min \sum_{i=1}^{n} (\hat{y}_i - y_i)^2 \tag{12-31}$$

in which n is the number of observations on the criterion variable (i.e., the sample size).

12.4.2. Solution Procedure

After inserting the model of $\hat{Y}$ in Equation 12-31, the objective function of Equation 12-31 can be minimized by taking the derivatives with respect to each unknown, setting the derivatives equal to zero, and then solving for the unknowns. The solution requires the model for predicting y_i to be substituted into the objective function. It is important to note that the derivatives are taken with respect to the unknown b_i and not the predictor variables X_i.

To illustrate the solution procedure, the model of Equation 12-28 is substituted into the objective function of Equation 12-31, which yields:

$$F = \min \sum_{i=1}^{n} (b_0 + b_1 x_i - y_i)^2 \tag{12-32}$$

The derivatives of the sum of the squares of the errors with respect to the unknowns b_0 and b_1 are, respectively,

$$\frac{\partial F}{\partial b_0} = 2\sum_{i=1}^{n} (b_0 + b_1 x_i - y_i) = 0 \tag{12-33a}$$

$$\frac{\partial F}{\partial b_1} = 2\sum_{i=1}^{n} (b_0 + b_1 x_i - y_i)x_i = 0 \tag{12-33b}$$

Dividing each equation by 2, separating the terms in the summations, and rearranging yield the set of normal equations:

$$nb_0 + b_1 \sum_{i=1}^{n} x_i = \sum_{i=1}^{n} y_i \tag{12-34a}$$

$$b_0 \sum_{i=1}^{n} x_i + b_1 \sum_{i=1}^{n} x_i^2 = \sum_{i=1}^{n} x_i y_i \tag{12-34b}$$

All the summations in Equation 12-34 are calculated over all values of the sample. Sometimes the index values of the summation are omitted when they refer to summation over all elements in a sample of size n. Also, the subscripts for X and Y can be omitted but are inferred. The two unknowns b_0 and b_1 can be evaluated by solving the following two simultaneous equations:

$$b_1 = \frac{\sum_{i=1}^{n} x_i y_i - \frac{1}{n} \sum_{i=1}^{n} x_i \sum_{i=1}^{n} y_i}{\sum_{i=1}^{n} x_i^2 - \frac{1}{n}\left(\sum_{i=1}^{n} x_i\right)^2} \tag{12-35a}$$

$$b_0 = \bar{Y} - b_1 \bar{X} = \frac{\sum_{i=1}^{n} y_i}{n} - \frac{b_1 \sum_{i=1}^{n} x_i}{n} \tag{12-35b}$$

Example 12-4: Least-Squares Analysis

The data of Table 12-2 can be used to illustrate the solution of the least-squares principle for a data set of five observations on two variables. Substituting the sums of Table 12-2 into the normal equations of Equation 12-34 gives:

$$5b_0 + 50b_1 = 10 \tag{12-36a}$$

$$50b_0 + 510b_1 = 109 \tag{12-36b}$$

Solving Equations 12-36a and 12-36b for the unknowns yields the model:

$$\hat{Y} = -7 + 0.9X \tag{12-37}$$

The errors, which are shown in Table 12-2, can be computed by substituting the measured values of X into Equation 12-37 to obtain predicted values that are used in Equation 12-30 to obtain the errors. In solving a linear model using the least-squares principle, the sum of the errors always equals zero, as shown in Table 12-2. The sum of the squares of the errors equals 1.9, which is a minimum.

TABLE 12-2 Least-Squares Computations

i	x_i	y_i	x_i^2	y_i^2	$x_i y_i$	e_i	e_i^2
1	12	4	144	16	48	–0.2	0.04
2	11	3	121	9	33	–0.1	0.01
3	10	2	100	4	20	0.0	0.00
4	9	0	81	0	0	1.1	1.21
5	8	1	64	1	8	–0.8	0.64
Column summation	50	10	510	30	109	0.0	1.90

Note that if a line is passed through the first three observations of Table 12-2 the error for each of these three observations is zero. Such a model would be:

$$\hat{Y} = -8 + 1.0X \tag{12-38}$$

While three of the residuals are zero, the sum of the squares of the errors for Equation 12-38 equals 2.0, which is greater than the corresponding sum for Equation 12-37. Thus, the least-squares solution always provides the straight line with the smallest sum of squares of the errors.

12.5. RELIABILITY OF THE REGRESSION EQUATION

Having evaluated the coefficients of the regression equation, it is of interest to evaluate the reliability of the regression equation. The following criteria should be assessed in evaluating the model: (1) the correlation coefficient; (2) the standard error of estimate; (3) the F statistic for the analysis of variance; (4) the rationality of the coefficients and the relative importance of the predictor variables, both of which can be assessed using the standardized partial regression coefficients; and (5) the degree to which the underlying assumptions of the regression model are met.

12.5.1. Correlation Coefficient

As suggested in Section 12.2, the correlation coefficient (R) is an index of the degree of linear association between two random variables. The magnitude of R indicates whether the regression provides accurate predictions of the criterion variable. Thus, R is often computed before the regression analysis is performed in order to determine whether it is worth the effort to perform the regression. However, R is always computed after the regression analysis because it is an index of the goodness of fit. The correlation coefficient measures the degree to which the measured and predicted values agree and is used as a measure of the accuracy of future predictions. It must be recognized that if the measured data are not representative of the population (that is, data that will be observed in the future) the correlation coefficient cannot be indicative of the accuracy of future predictions.

The square of the correlation coefficient (R^2) equals the percentage of the variance in the criterion variable explained by the predictor variable. Because of this physical interpretation, R^2 is a meaningful indicator of the accuracy of predictions.

12.5.2. Standard Error of Estimate

In the absence of additional information, the mean is the best estimate of the criterion variable; the standard deviation S_Y of Y is an indication of the accuracy of prediction. If Y is related to one or more predictor variables, the error of prediction is reduced from S_Y to the standard error of estimate, S_e. Mathematically, the standard error of estimate equals the standard deviation of the errors, has the same units as Y, and is given by:

$$S_e = \sqrt{\frac{1}{\nu}\sum_{i=1}^{n}(\hat{y}_i - y_i)^2} \tag{12-39}$$

in which ν is the degree of freedom, which equals the sample size minus the number of unknowns estimated by the regression procedure. For the bivariate model of Equation 12-28, $p = 1$ and $\nu = n - 2$. For the general linear model with an intercept, Equation 12-29, there are $(p + 1)$ unknowns, thus $\nu = n - p - 1$. In terms of the separation-of-variation concept discussed previously, the standard error of estimate equals the square root of the ratio of the unexplained variation to the degrees of freedom. It is important to note that S_e is based on $(n - p - 1)$ degrees of freedom, while the value, S_Y, is based on $(n - 1)$ degrees of freedom. Thus, in some cases, S_e may be greater than S_Y. To assess the reliability of the regression equation, S_e should be compared with the bounds of zero and S_Y. If S_e is near S_Y, the regression has not been successful. If S_e is much smaller than S_Y and is near zero, the regression analysis has improved the reliability of prediction.

The standard error of estimate is sometimes computed using the following relationship:

$$S_e = S_Y\sqrt{1 - R^2} \tag{12-40}$$

Equation 12-40 must be considered as only an approximation to Equation 12-39 because R is based on n degrees of freedom and S_Y is based on $(n - 1)$ degrees of freedom.

Using the separation-of-variation concept, the exact relationship between S_e, S_Y, and R can be computed as follows:

$$\text{TV} = \text{EV} + \text{UV} \tag{12-41}$$

The total variation is related to the variance of Y by:

$$S_Y^2 = \frac{\text{TV}}{n - 1} \tag{12-42}$$

The square of the correlation coefficient is the ratio of the explained variation (EV) to the total variation (TV):

$$R^2 = \frac{\text{EV}}{\text{TV}} \tag{12-43}$$

The standard error of estimate is related to the unexplained variation (UV) by:

$$S_e^2 = \frac{\text{UV}}{n - p - 1} \tag{12-44}$$

Equation 12-42 can be solved for TV, which can then be substituted into both Equations 12-41 and 12-43. Equation 12-44 can be solved for UV, which is also substituted into Equation 12-41. Solving for S_e^2 yields:

$$S_e^2 = \left(\frac{n-1}{n-p-1}\right) S_Y^2 (1 - R^2) \tag{12-45}$$

Thus, Equation 12-45 is a more exact relationship than Equation 12-40. However, for large sample sizes the difference between the two estimates is small.

Although S_e may actually be greater than S_Y, in general S_e is within the range from zero to S_Y. When a regression equation fits the data points exactly, S_e equals zero; this corresponds to a correlation coefficient of 1. When the correlation coefficient equals zero, S_e equals S_Y. As was indicated previously, S_e may actually exceed S_Y when the degrees of freedom have a significant effect. The standard error of estimate is often preferred to the correlation coefficient because S_e has the same units as the criterion variable and its magnitude is a physical indicator of the error; the correlation coefficient is only a standardized index and does not properly account for degrees of freedom lost because of the regression coefficients.

12.5.3. Analysis of Variance

The regression coefficient (b_1) for the predictor variable X in Equation 12-28 is a measure of the change in Y that results from a change in X. Recognizing that the regression coefficient b_1 is dependent on the units of both Y and X, it is reasonable to ask whether or not a change in X causes a significant change in Y. This is easily placed in the form of the following statistical hypothesis test:

$$\text{H}_0\text{: } \beta_1 = 0 \tag{12-46a}$$

$$\text{H}_\text{A}\text{: } \beta_1 \neq 0 \tag{12-46b}$$

in which β_1 is the population value of the slope coefficient. The hypotheses are designed to test whether or not the relationship between Y and X is significant; this is exactly the same as testing the hypothesis that the population correlation coefficient equals zero (i.e., Equations 12-6 and 12-7).

The hypothesis test based on Equations 12-46a and 12-46b relies on the computations involved in the separation of variation concept, thus it represents an analysis of variance (ANOVA), as discussed in Chapter 10. The total variation is represented by the total sum of squares SS_T:

$$SS_T = \sum_{i=1}^{n} y_i^2 - \frac{\left(\sum_{i=1}^{n} y_i\right)^2}{n} \tag{12-47}$$

The total sum of squares is separated into the regression (SS_R), or the explained variation, and the error (SS_E) sums of squares, or the unexplained variation, as follows:

$$SS_R = b_1\left(\sum_{i=1}^{n} x_i y_i - \frac{\sum_{i=1}^{n} x_i \sum_{i=1}^{n} y_i}{n}\right) \tag{12-48}$$

$$SS_E = SS_T - SS_R \tag{12-49}$$

in which b_1 is the computed value of the slope coefficient. The mean squares, which correspond to variances, equal the sums of squares divided by the degrees of freedom, which are shown in Table 12-3.

The null hypothesis is tested for significance using the ratio of the mean square for regression to the mean square for error, as follows:

$$F = \frac{MS_R}{MS_E} \tag{12-50}$$

in which F is the computed value of a random variable having an F distribution with 1 and $(n - 2)$ degrees of freedom. For a stated level of significance, α, the null hypothesis is accepted if F is less than the critical value F_α obtained from Table A-4 of Appendix A. The region of rejection consists of all values of F greater than F_α.

TABLE 12-3 ANOVA Table for a Bivariate Regression

Souce of variation	Sum of squares (*SS*)	Degrees of freedom	Mean square (*MS*)	*F*
Regression	SS_R	1	$MS_R = \frac{SS_R}{1}$	$\frac{MS_R}{MS_E}$
Error	SS_E	$n - 2$	$MS_E = \frac{SS_E}{n-2}$	—
Total	SS_T	$n - 1$	—	—

It is of interest to compare R, S_e, and the F from the ANOVA. The value of R is based on the ratio of EV to TV. The value of S_e is based on UV. The computed value of F is based on EV and UV. Thus, the three indices that are used to examine the goodness of fit of the data to the linear model are based on the individual elements of the separation of variation concept (see Section 12.2.3).

12.5.4. Standardized Partial Regression Coefficients

Because the partial regression coefficient, b_1 of Equation 12-28, is dependent on the units of both Y and X, it is often difficult to use its magnitude to measure the rationality of the model. The partial regression coefficient can be standardized by:

$$t = \frac{b_1 S_X}{S_Y} \tag{12-51}$$

in which t is called the *standardized partial regression coefficient*, and S_X and S_Y are the standard deviations of the predictor and criterion variables, respectively. Because the units of b_1 are equal to the ratio of the units of Y to the units of X, t is dimensionless. For the bivariate regression model, t equals the correlation coefficient of the regression of the standardized value of Y on the standardized value of X. This suggests that t is a measure of the relative importance of the corresponding predictor variable. The value of t has the same sign as b_1. Therefore, if the sign of t is irrational, we can conclude that the model is not rational. Although it is more difficult to assess the rationality of the magnitude of a regression coefficient than it is to assess the rationality of the sign, the magnitude of the standardized partial regression coefficient can be used to assess the rationality of the model. For rational models, t must vary in the range $-1 \leq t \leq 1$. Because t equals R for the bivariate model of Equation 12-28, values of t outside this range should not occur; this is not true for multivariate models, such as Equation 12-29.

Example 12-5: Errors in Regression Models

The data of Table 12-2 were previously used to compute the standard error of estimate and the correlation coefficient; they can also be used to illustrate the use of the analysis of variance (ANOVA) for the analysis of the bivariate regression model. Table 12-4 shows the necessary ANOVA computations. For 1 and 3 degrees of freedom, the critical F values for levels of significance of 0.05 and 0.01 from Table A-4 are 10.13 and 34.12, respectively. Because the computed value of 12.8 is larger than the F value for $\alpha = 0.05$, the null hypothesis can be rejected at the 5% level. However, in comparing the computed F with the critical F for $\alpha = 1\%$, the null hypothesis must be accepted at the 1% level of significance. This example illustrates that it is important to evaluate the consequences of selecting a level of significance arbitrarily.

TABLE 12-4 Example of ANOVA Computations

Souce of variation	Sum of squares	Degrees of freedom	Mean square	F
Regression	$0.9\left[109 - \frac{(50)(10)}{5}\right] = 8.1$	1	8.1	12.8
Error	$10 - 8.1 = 1.9$	3	0.633	—
Total	$30 - \frac{(10)^2}{5} = 10$	4	—	—

The standardized partial regression coefficient, which is computed using Equation 12-51, equals:

$$t = 0.9\sqrt{\frac{510 - \frac{(50)^2}{5}}{30 - \frac{(10)^2}{5}}} = 0.9 \tag{12-52}$$

Using the equations given previously for computing the correlation coefficient, it is simple to show that t equals R.

12.5.5. Assumptions Underlying the Regression Model

The principle of least squares assumes that the errors, that is, the differences between the predicted and measured values of the criterion variable, (1) are independent of each other, (2) have zero mean, (3) have a constant variance across all values of the predictor variables, and (4) are normally distributed. If any of these assumptions are violated, we must assume that the model structure is not correct. Violations of these assumptions are easily identified using statistical analyses of the residuals.

When the sum of the residuals does not equal zero, it reflects a bias in the model. The regression approach, when applied analytically, requires the sum of the residuals for a linear model to equal 0. For a nonlinear model, the sum of the errors may not equal 0, which suggests bias. When an inadequate number of significant digits is computed for the regression coefficients, the sum of the residuals may not equal 0, even for a linear model. However, a model may be biased even when the sum of the residuals equals zero. For example, Figure 12-7 shows an X–Y plot in which the trend of the data is noticeably nonlinear, with the linear regression line also shown. If the errors e_i are computed and plotted vs. the corresponding values of the predictor variable, a noticeable trend appears (Figure 12-8). The errors are positive for both low and high values of X, while the errors show a negative bias of prediction for intermediate values of X. While the sum of the residuals equals 0, the trends in the residuals suggest a biased model that should be replaced by a model that has a different structure.

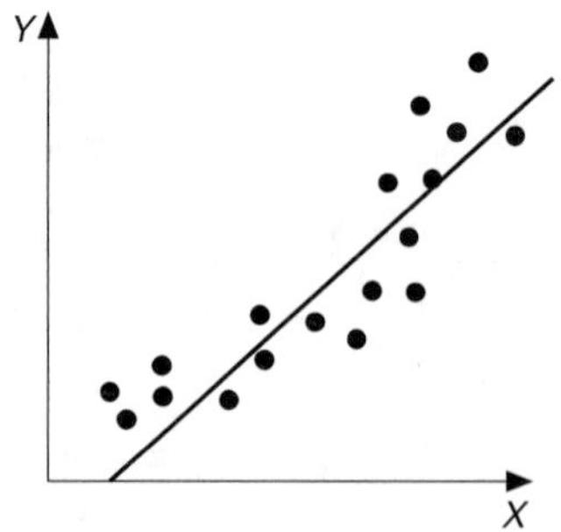

FIGURE 12-7 Biased linear regression model.

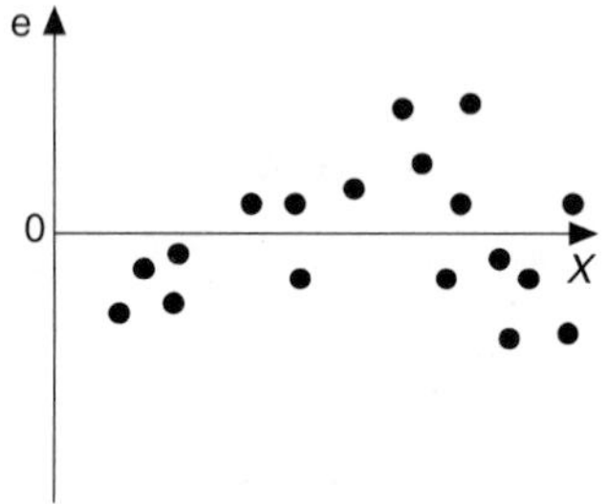

FIGURE 12-8 Residual plot for a biased linear regression model.

The population model for linear regression has the form:

$$Y = \beta_0 + \beta_1 X + \varepsilon \tag{12-53}$$

in which β_0 and β_1 are the population statistics for the intercept and slope coefficients, respectively, and ε is a random variable having a normal distribution with zero mean and a constant variance. The variance of ε equals the square of the standard error of estimate. Therefore, the population of the residuals is defined as:

$$\varepsilon \sim N(0, \sigma_e^2 = \text{constant}) \tag{12-54}$$

If the error variance (σ_e^2) is not constant across all values of the predictor and criterion variable, the underlying model assumptions are not valid and a more accurate model structure should be identified. The data shown in Figure 12-9 reveal a trend in the error variance; specifically, as X increases, the error variance increases. Thus, the following relationship is apparent:

$$\varepsilon \sim N[0, \sigma_e^2 = f(X)] \tag{12-55}$$

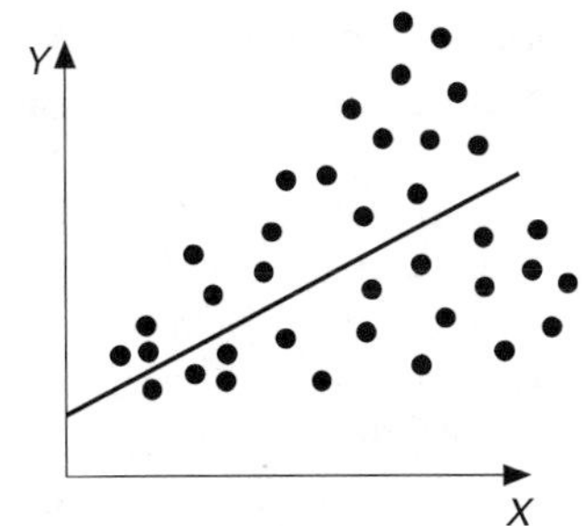

FIGURE 12-9 Model with a nonconstant error variance.

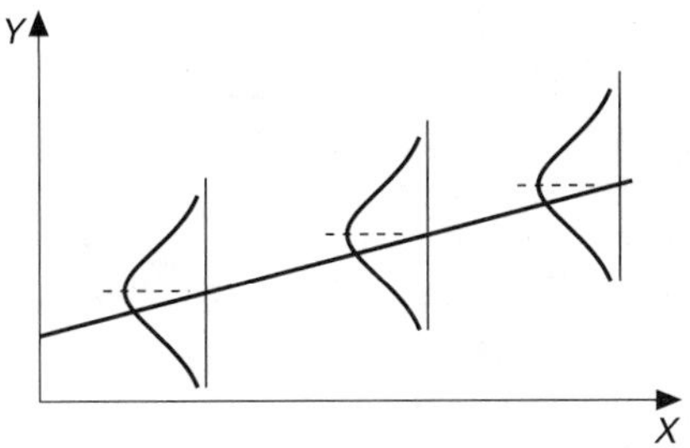

FIGURE 12-10 Model in which $\varepsilon \sim N(0, \text{constant } \sigma_e^2)$.

Although an exact test does not exist for testing a data set for a nonconstant error variance, it is not unreasonable to separate the errors into two or more sets based on the value of X and use a hypothesis test for comparing variances to detect a significant difference. A systematic method for selecting either the number of groups or the points at which the separation will be made does not exist. Therefore, one should attempt to have large subsamples and a rational explanation for choosing the points of separation.

The linear regression procedure also assumes that the errors are normally distributed across the range of the values of the predictor variables. The normality assumption is illustrated in Figure 12-10. If the errors are not normally distributed, the model structure should be examined. An exact procedure for testing for normality does not exist. However, in a manner similar to the test for constant variance, the residuals could be separated into groups that have similar values of X, and such tests as a rank-order frequency analysis, the chi-square goodness-of-fit test, or the Kolmogorov–Smirnov test could be performed to study the normality question.

The fourth assumption deals with the independence of the observations. This is not usually a problem except when time or space is the predictor variable. In such cases, the measurement of Y for one time period may not be completely independent of the measurement for an adjacent time period. The runs test and/or the serial correlation coefficient may be used to check for statistical independence. Statistical independence should not be used in place of an examination of the physical independence of the data.

12.6. RELIABILITY OF POINT ESTIMATES OF THE REGRESSION COEFFICIENTS

The regression coefficients are important because of their physical significance; the slope coefficient represents the effect of change in the predictor variable on the criterion variable, and the intercept coefficient equals the value of the criterion variable when the predictor variable equals zero. Because of their physical significance, it is important to examine the reliability of the coefficients. Confidence intervals and tests of hypotheses are useful for indicating the quality of the regression statistics.

12.6.1. Sampling Distributions of the Regression Coefficients

The computed coefficients b_0 and b_1 are sample estimates of the population parameters β_0 and β_1, respectively. Recognizing that b_0 and b_1 are random variables, it is important to examine their probability density functions. Assuming that the true error variance σ_e^2 is not known, it could be shown that the coefficients are random variables having t distributions with $(n - 2)$ degrees of freedom, mean estimates equal to the estimated values of b_0 and b_1, and the following error variances:

$$S_{e,b_0}^2 = \frac{S_e^2 \sum_{i=1}^{n} x_i^2}{n \sum_{i=1}^{n} (x_i - \bar{X})^2} \tag{12-56}$$

and

$$S_{e,b_1}^2 = \frac{S_e^2}{\sum_{i=1}^{n} (x_i - \bar{X})^2} \tag{12-57}$$

Therefore, two-sided confidence intervals for the coefficients are:

$$b_0 \pm t_{\alpha/2, n-2} S_{e,b_0} \tag{12-58}$$

and

$$b_1 \pm t_{\alpha/2, n-2} S_{e,b_1} \tag{12-59}$$

in which α is the level of significance.

Example 12-6: Confidence Intervals for Regression Coefficients

The data of Table 12-2 resulted in the regression equation of Equation 12-37, with slope and intercept coefficients of 0.9 and –7.0, respectively. Because the sum of squares equals 1.9, the standard error of estimate is given by:

$$S_e = \sqrt{\frac{1.9}{5-2}} = 0.7958 \tag{12-60}$$

Therefore, the error variances of the regression coefficients are:

$$S^2_{e,b_0} = \frac{(0.7958)^2(510)}{5(10)} = 6.4596 \tag{12-61}$$

and

$$S^2_{e,b_1} = \frac{(0.7958)^2}{10} = 0.06333 \tag{12-62}$$

The width of the confidence intervals is a function of the level of significance; for a level of significance of 5%, the two-sided confidence intervals on b_0 and b_1 are:

$$-7.0 - 3.182\sqrt{6.4596} \le \beta_0 \le -7.0 + 3.182\sqrt{6.4596} \tag{12-63a}$$

$$-15.087 \le \beta_0 \le 1.0873 \tag{12-63b}$$

and

$$0.9 - 3.182\sqrt{0.06333} \le \beta_1 \le 0.9 + 3.182\sqrt{0.06333} \tag{12-64a}$$

$$0.09924 \le \beta_1 \le 1.7008 \tag{12-64b}$$

One can conclude with 95% confidence that the true value is within the bounds set by Equations 12-63 and 12-64; however, because of the small sample size, the confidence intervals are extremely wide.

12.6.2. Hypothesis Test on the Slope Coefficient

In addition to interval estimates of a statistic, tests of hypotheses are also of value in decision making. If one is interested in a particular value of β_1, say, $\tilde{\beta}_1$, the following null hypothesis would be of interest:

$$H_0\text{:} \quad \beta_1 = \tilde{\beta}_1 \tag{12-65}$$

This can be tested against either one- or two-sided alternative hypotheses. The null hypothesis of Equation 12-65 can be tested with the following test statistic:

$$t = \frac{b_1 - \tilde{\beta}_1}{S_{e,b_1}} \tag{12-66}$$

in which t is the value of a random variable having a t distribution with $(n - 2)$ degrees of freedom. For a given level of significance, the following summarizes the decision process for the three possible alternative hypotheses:

If H_A is:	Reject H_0 if:
$\beta_1 < \tilde{\beta}_1$	$t < -t_\alpha$
$\beta_1 > \tilde{\beta}_1$	$t > t_\alpha$
$\beta_1 \neq \tilde{\beta}_1$	$t < -t_{\alpha/2}$ or $t > t_{\alpha/2}$

The selection of a one- or two-sided test depends on the problem. The selection of the level of significance should also be selected by considering the physical problem and the implications of making errors in the decision. The computed value can be compared with the critical value according to the decision criteria given in the tabulation above.

If $\tilde{\beta}_1$ equals zero, one may conclude that variation in X does not cause variation in Y; that is, X and Y are independent. Thus, the following hypothesis may be of special interest:

$$H_0\colon \quad \beta_1 = 0 \tag{12-67}$$

This can be tested against either one- or two-sided alternative hypotheses by setting $\tilde{\beta}_1 = 0$. This null hypothesis corresponds exactly to the null hypothesis that $\rho = 0$. This is equivalent to the test of Section 12.5.3.

Example 12-7: Hypothesis Testing of Slope

While the sample estimate of β_1 in Example 12-6 was 0.9, three of the five points in the sample discussed previously fell on a straight line defined by Equation 12-38. Thus, it may be of interest to test the null hypothesis that $\beta_1 = 1$ against the two-sided alternative hypothesis that $\beta_1 \neq 1$. The computed value of the test statistic would be:

$$t = \frac{0.9 - 1.0}{\sqrt{0.06333}} = -0.397 \tag{12-68}$$

For a level of significance of 10% and 3 degrees of freedom, the critical t value from Table A-2 is 2.353. For the two-sided alternative, the computed t value lies in the region of acceptance. Therefore, one could not reject the null hypothesis that the true value of β_1 equals 1.0. However, the value of 0.9 is still the best estimate.

12.6.3. Hypothesis Test on the Intercept Coefficient

Because of the physical significance of the intercept, it may be of value to test the hypothesis that the population intercept β_0 equals some specific value, $\tilde{\beta}_0$:

$$H_0: \quad \beta_0 = \tilde{\beta}_0 \tag{12-69}$$

The null hypothesis can be tested against either a one- or two-sided alternative hypothesis using the following test statistic:

$$t = \frac{b_0 - \tilde{\beta}_0}{S_{e,b_0}} \tag{12-70}$$

in which t is the value of a random variable having a t distribution with $(n - 2)$ degrees of freedom. For a given level of significance, the following summarizes the decision process for the three possible alternative hypotheses:

If H_A is:	Reject H_0 if:
$\beta_0 < \tilde{\beta}_0$	$t < -t_\alpha$
$\beta_0 > \tilde{\beta}_0$	$t > t_\alpha$
$\beta_0 \neq \tilde{\beta}_0$	$t < -t_{\alpha/2}$ or $t > t_{\alpha/2}$

12.7. CONFIDENCE INTERVALS OF THE REGRESSION EQUATION

Confidence intervals provide a useful way for assessing the quality of a point estimate. Whereas a hypothesis test provides a "yes/no" response to a hypothesis, a confidence interval is an interval within which we believe a given population parameter is located. In regression analysis, several confidence intervals are of interest, including: (1) a confidence interval for the line as a whole, (2) a confidence interval for a single point on the line, and (3) a confidence interval for a single future value of Y corresponding to a chosen value of X.

12.7.1. Confidence Interval for a Line as the Whole

When the entire line is of interest, such as when an experiment is repeated many times and a regression equation is computed for each sample, it may be of interest to compute a confidence interval on the line as a whole; such a confidence interval permits one to simultaneously make confidence statements about estimates of Y for a number of values of the predictor variable. For a specified level of confidence, m points x_{ai} (i = 1, 2, …, m) are selected such that the points adequately cover the range of interest of the predictor variable; usually, seven to ten values are adequate for delineating the confidence interval. The F statistic is obtained for (2, n – 2) degrees of freedom and a level of significance of $\alpha = 1 - \gamma$. For each value of x_{ai} selected, the estimated values of the criterion variable $\hat{y}_{ai}$ and the confidence limits $\hat{y}_{ci}$ are computed as follows:

$$\hat{y}_{ai} = \bar{Y} + b_1(x_{ai} - \bar{X}) \qquad (12\text{-}71)$$

$$\hat{y}_{ci} = \hat{y}_{ai} \pm S_e \sqrt{2F\left[\frac{1}{n} + \frac{(x_{ai} - \bar{X})^2}{\sum_{i=1}^{n}(x_i - \bar{X})^2}\right]} \qquad (12\text{-}72)$$

The regression line and the confidence interval can then be plotted for each value of x_{ai}.

12.7.2. Confidence Interval Estimate for a Single Point on the Line

If a particular value of the predictor variable (say, x_0) is of special importance, a confidence interval on the value of the criterion variable (i.e., average Y at x_0) corresponding to x_0 may be of interest. In such a case, the two-sided, γ-percent confidence interval on the estimated value of Y corresponding to x_0 can be computed as:

$$\hat{y} \pm S_e t_{\alpha/2} \sqrt{\frac{1}{n} + \frac{(x_0 - \bar{X})^2}{\sum_{i=1}^{n}(x_i - \bar{X})^2}} \qquad (12\text{-}73)$$

in which $t_{\alpha/2}$ is the value of a random variable having a t distribution with (n – 2) degrees of freedom with a level of significance α, which equals $1 - \gamma$. The estimated value of Y, which is denoted by $\hat{y}$, is computed by:

$$\hat{y} = \bar{Y} + b_1(x_0 - \bar{X}) \qquad (12\text{-}74)$$

If the experiment was repeated many times, γ percent of the computed intervals would cover the true value of Y corresponding to x_0. It is important to note that this confidence interval is concerned with a mean value; that is, the line represents the mean value of the distribution of Y values for a given x_0 and the confidence interval of Equation 12-73 is for the line at the point x_0 only. The interval is computed only for the point x_0.

12.7.3. Confidence Interval Estimate for a Future Value

A third confidence interval of interest can be used to evaluate the accuracy of a single (future) value of Y corresponding to a chosen value of X (say, x_f). The two-sided, γ-percent confidence interval for an estimated value $\hat{y}$, which can be computed from Equation 12-74 with $x_0 = x_f$, is given by:

$$\hat{y} \pm S_e t_{\alpha/2} \sqrt{1 + \frac{1}{n} + \frac{(x_f - \bar{X})^2}{\sum_{i=1}^{n} (x_i - \bar{X})^2}} \tag{12-75}$$

It is of interest to compare the form of Equations 12-73 and 12-75. They differ by a factor of 1.0 in the square root radical. The larger confidence interval that results from the 1.0 factor reflects the less certainty resulting from the estimation of a single future value of Y rather than the mean value computed for the confidence interval of Equation 12-73. Equation 12-75 provides a confidence interval on $\hat{y}$ only for the single point x_f.

12.7.4. Summary of Confidence Intervals

Confidence intervals provide a measure of reliability. Three interval estimates have been described: (1) the confidence intervals for the line as a whole, (2) the confidence intervals for the mean value of Y corresponding to a specific value x_0, and (3) the confidence interval for a value of Y corresponding to a future estimate of the predictor variable x_f. It should be of interest to distinguish among these three using a very simple example. Assume that the problem involves finding the relationship between the height and weight of graduate students at a university, with weight being the criterion variable. If the problem was to determine the true relationship, a confidence interval for the line as a whole would be appropriate for assessing the overall accuracy. If the mean value of those students who were 6 ft tall was of special interest, we could compute the confidence interval for the point on the line corresponding to $x_0 = 6$ ft. This confidence interval would indicate the accuracy of the estimated value of the weight (Y) for a height (x_0) of 6 ft. If the problem of interest is the weight of the very next 6-ft-tall graduate student who passes by the door, the estimated weight would be the same as that computed for the previous problem.

However, because it is a single value for the weight of a specific person, the confidence interval would be wider than that for the average of all graduate students who are 6 ft tall. This confidence interval would be computed using Equation 12-75.

Example 12-8: Confidence Intervals in Regression

Three types of confidence intervals were described in the previous paragraphs; the selected type of confidence interval depends on the type of decision that must be made. For the data of Table 12-2, the confidence interval on the line as a whole can be computed using Equation 12-72. If it can be assumed that values of the predictor variable X are likely to be in the range from 6 to 16, it is reasonable to compute the confidence interval within these bounds. Values of x_{ai} are selected in increments of two because six values should be sufficient to draw the confidence interval. For a 5% level of significance and degrees of freedom of 2 and 3, the F statistic from Table A-4 is 9.55. For the standard error of 0.7958, the 95% confidence interval for any value of x_{ai} is:

$$\hat{y}_{ai} \pm 0.7958\sqrt{2(9.55)}\left[\frac{1}{5} + \frac{(x_{ai} - 10)^2}{10}\right]^{0.5} \tag{12-76a}$$

$$\hat{y}_{ai} \pm 3.478\left[\frac{1}{5} + \frac{(x_{ai} - 10)^2}{10}\right]^{0.5} \tag{12-76b}$$

The computed values of $\hat{y}_{ai}$ are determined using Equation 12-37 with $X = x_{ai}$. The computations are given in Table 12-5, and the confidence interval is shown in Figure 12-11. The confidence interval is smallest at the mean value of X (i.e., $\bar{X} = 10$) and increases in width as X differs from the mean. This results because of the greater confidence in estimated values near the mean. For values beyond the range of the data, the confidence interval is especially wide. The confidence interval is highly nonlinear because of the small sample size. For large samples, the width of the confidence interval is not characterized by such noticeable nonlinearity. For large samples it is not necessary to use as many values of x_{ai} to get a reasonable delineation of the confidence interval.

TABLE 12-5 Computation of Confidence Interval for the Line as a Whole

x_{ai}	y_{ai}	$\frac{(x_{ai}-10)^2}{10}$	Half-width	Lower limit	Upper limit
6	−1.6	1.6	4.67	−6.3	3.1
8	0.2	0.4	2.69	−2.5	2.9
10	2.0	0.0	1.56	0.4	3.6
12	3.8	0.4	2.69	1.1	6.5
14	5.6	1.6	4.67	0.9	10.3
16	7.4	3.6	6.78	0.6	14.2

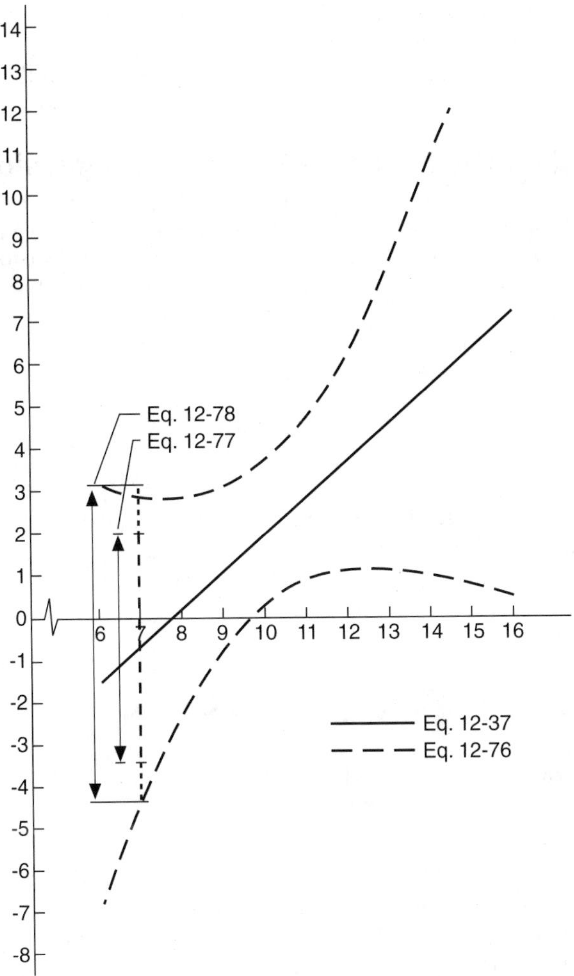

FIGURE 12-11 Confidence intervals in regression analysis.

If a particular value of X and the corresponding mean value of Y are of special interest, the confidence interval of Equation 12-73 can be used to show the accuracy of the value of Y on the regression line that corresponds to the X value. If a value of X of 7 (i.e., $x_0 = 7$) is of special interest, with the data of Table 12-2, the 95% confidence interval at that point is:

$$\hat{y} \pm 3.182(0.7958)\left[\frac{1}{5} + \frac{(7-10)^2}{10}\right]^{0.5} \tag{12-77a}$$

$$\hat{y} \pm 2.66 \tag{12-77b}$$

The predicted value of Y at X equal to 7 is –0.7; therefore, the confidence limits are –3.4 and 2.0 as shown in Figure 12-11. The width of the confidence interval for the mean value is smaller than the width of the confidence interval of the line as a whole at $X = 7$, because we are only interested in a single value of X and not for all values of X.

If a particular occurrence of the variable X and the corresponding value of Y is of special interest, the confidence interval of Equation 12-75 can be used to assess the expected accuracy of the predicted value. If the specific value of X is 7 (i.e., $x_f = 7$), the 95% confidence interval is:

$$\hat{y} \pm 3.182(0.7958)\left[1 + \frac{1}{5} + \frac{(7-10)^2}{10}\right]^{0.5} \tag{12-78a}$$

$$\hat{y} \pm 3.67 \tag{12-78b}$$

Thus, for the predicted value of –0.7, the confidence limits are –4.4 and 3.0, which are shown in Figure 12-11. The confidence interval of Equation 12-78 is wider than that of Equation 12-77 because Equation 12-78 is intended as a measure of the accuracy of the predicted value of Y for a specific occurrence of X, while Equation 12-78 measures the accuracy of the mean value of Y at $X = 7$.

12.8. CORRELATION VS. REGRESSION

Before discussing other topics in regression, it is important to emphasize the differences between correlation and regression. Regression is a means of calibrating the unknown coefficients of a prediction equation, whereas correlation provides a measure of goodness of fit. Thus, regression is a method for model calibration, while correlation would have usefulness in model formulation and model verification. When using regression, it is necessary to specify which variable is the criterion and which is the predictor; when using correlation, it is not necessary to make such a distinction. The distinction is necessary with regression because a regression equation is not transformable unless the correlation coefficient equals 1.0. That is, if Y is the criterion variable when the equation is calibrated, the equation cannot be rearranged algebraically to get an equation for predicting X. Specifically, Equation 12-29 can be rearranged to predict X as:

$$Y = b_0 + b_1 X \tag{12-79a}$$

$$Y - b_0 = b_1 X \tag{12-79b}$$

Therefore,

$$X = \frac{Y - b_0}{b_1} \tag{12-79c}$$

or

$$X = -\frac{b_0}{b_1} + \frac{1}{b_1}Y \tag{12-79d}$$

$$X = \frac{Y - b_0}{b_1} \tag{12-79e}$$

If X is regressed on Y, the resulting coefficients will not be the same as the coefficients obtained by regressing Y on X (Equation 12-79a) and then setting $a_0 = -(b_0/b_1)$ and $a_1 = (1/b_1)$. The correlation for the regression of Y on X is the same as the correlation for the regression of X on Y.

Computationally, the correlation coefficient is a function of the explained and total variation; the slope coefficient of a regression equation is related to the correlation coefficient by:

$$b_1 = \frac{RS_Y}{S_X} \tag{12-80}$$

in which S_X and S_Y are the standard deviations of X and Y, respectively.

After a regression equation is calibrated, it is very important to examine the rationality of the regression coefficients; specifically, is the predicted value of Y rational for all reasonable values of X? Because the slope coefficient of the regression equation represents the rate of change of Y with respect to X, is the effect of X on Y rational? A regression equation that is not considered rational should be used with caution, if at all. In addition to checking for rationality, the goodness-of-fit statistics, R and S_e, should be computed. Also, a hypothesis test, either ANOVA or a test on ρ, should be made. Confidence intervals can be computed on both the regression coefficients and the line and used to assess the accuracy of predictions made with the regression equation. If the expected accuracy is not acceptable, we may elect to collect more data or develop a model that uses other predictor variables.

12.9. APPLICATIONS OF BIVARIATE REGRESSION ANALYSIS

12.9.1. Estimating Trip Rate

A traffic planner needs estimates of trip rates for residential areas. The planner conducts studies in ten residential areas of different densities of development, with the following measured rates:

X	3	4	8	8	13	15	18	22	24	27
Y	4.5	3.3	3.5	2.3	3.8	2.6	2.7	1.6	1.9	1.7

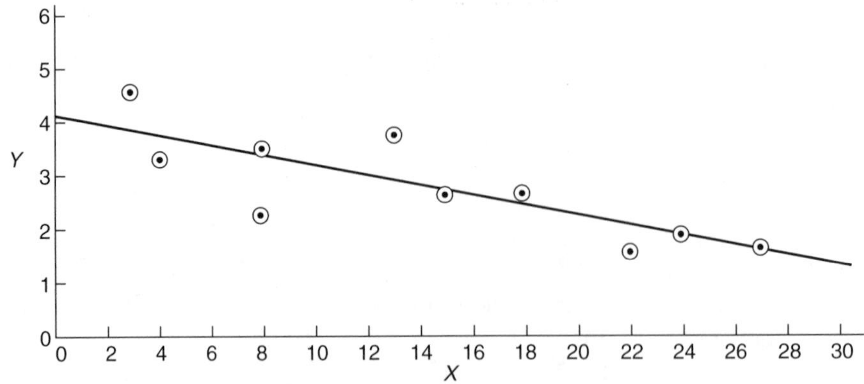

FIGURE 12-12 Bivariate regression of trip rate (*Y*, daily trip per household) on residential density (*X*, households per acre).

where *X* is the residential density (households per acre) and *Y* is the trip rate (daily trips per household). The regression of trip rate on density yielded the following linear equation:

$$\hat{Y} = 4.1000 - 0.09226X \tag{12-81}$$

The negative slope coefficient indicates that the trip rate decreases by 0.092 daily trips per household for each unit increase in the number of households per acre. The standard error of estimate is 0.603, which is an improvement over the standard deviation of 0.965 ($S_e/S_y = 0.625$). The correlation coefficient is –0.808, which yields a normalized explained variance (R^2) of 0.653. The standard error ratio of the slope coefficient ($S_{e,b_1}/b_1$) equals 0.258, which suggests that the coefficient is accurate. Figure 12-12 shows the measured data and the regression line; the moderate variation of the measured points about the line confirms the moderate correlation between residential density and trip rate.

12.9.2. Breakwater Cost

A construction engineering firm has recently had a number of contracts for the construction of breakwaters in coastal environments. To reduce the effort in developing cost estimates for future clients, the company compiled a data record of 14 breakwater construction projects. The data consist of the length of the breakwater and the cost for rock, rock spurs, groins, marsh plants, and beach fill (see Table 12-6). The cost (*Y*) was regressed on the length of the breakwater (*X*):

$$\hat{Y} = 253.4 + 0.08295X \tag{12-82}$$

TABLE 12-6 Bivariate Regression Analysis of Breakwater Cost

X	Y	$\hat{Y}$	e
450	230	291	61
625	490	305	–185
670	150	309	159
730	540	314	–226
810	280	321	41
880	120	326	206
1020	380	338	–42
1050	170	340	170
1100	530	345	–185
1175	290	351	61
1230	460	355	–105
1300	230	361	131
1350	470	365	–105
1510	360	379	19

The standard error of estimate is \$149.1 × 10^3, which is slightly greater than the standard deviation of the measured costs (S_y = \$145.5 × 10^3); that is, $S_e/S_y = 1.024$. Because S_e is greater than S_y, it would be more accurate to use the mean cost (\$335.7 × 10^3) than the value estimated with the regression equation. If we use Equation 12-5 to compute the correlation coefficient, we get a value of 0.176; however, Equation 12-45 shows that we have essentially no reliability in the estimated values. Thus, for practical purposes, we have a correlation of zero. The adjusted correlation computed with Equation 12-45 is a more realistic measure of reliability than is the value of R computed with the unadjusted, but more commonly used, equation (Equation 12-5), because Equation 12-45 accounts for differences in the degrees of freedom. The standard error ratio of the slope coefficient, $S_{e,b_1}/b_1$, is 1.61; this exceptionally large value is another indication of the poor reliability of the linear, bivariate equation.

Why does the computed equation give such poor results? Certainly, breakwater length is a determinant of cost; however, it appears that other factors are more important and necessary to obtain accurate estimates of the cost.

12.9.3. Stress–Strain Analysis

Table 12-7 includes nine measurements of the observed stress and the resulting axial strain for an undisturbed sample of a Boston blue clay. A linear equation was computed by regressing the strain (Y) on the stress (X):

$$\hat{Y} = -8.29 + 0.6257X \tag{12-83}$$

TABLE 12-7 Bivariate Regression Analysis of Stress–Strain Data

X	Y	$\hat{Y}$	e
10.8	0.1	−1.5	−1.6
14.1	0.4	0.5	0.1
16.3	1.1	1.9	0.8
17.5	1.4	2.7	1.3
18.6	2.7	3.4	0.7
19.8	3.6	4.1	0.5
22.1	5.7	5.5	−0.2
23.8	6.8	6.6	−0.2
24.3	8.3	6.9	−1.4

The equation has a standard error of estimate of 1.04%, which is much smaller than the standard deviation of the measured strains, that is, $S_e/S_y = 0.35$. The correlation coefficient of 0.945 also suggests that the linear equation provides a reasonable fit to the measured data. An R^2 of 0.89 indicates that almost 90% of the total variation is explained by the linear regression; the standard error ratio of 0.35 suggests that the fit is not as good as suggested by the R^2. The predicted strains and residuals are given in Table 12-7 and shown in Figure 12-13. While most of the errors are small, there is a noticeable bias in the residuals, with negative residuals for small and large stresses. While the mean bias ($\bar{e}$) is zero, the local biases indicate that a nonlinear structure may be more appropriate. Another problem with the linear model is the negative estimates of strain that occur for stresses of less than 13.244 psi. The negative strain is physically not meaningful, so the model is not rational for small strains.

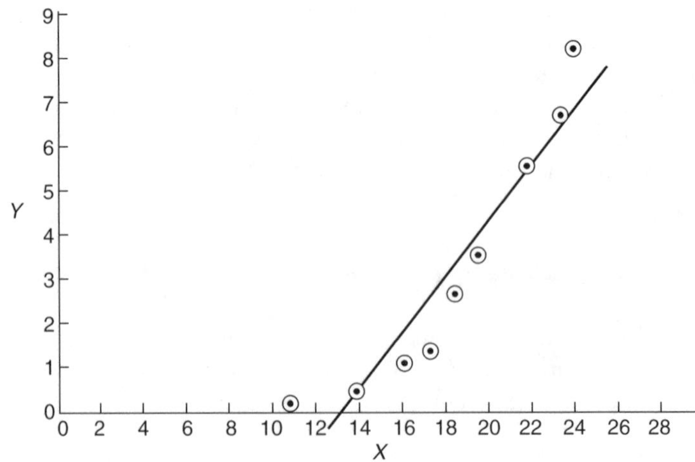

FIGURE 12-13 Axial strain (Y) of a Boston blue clay due to stress (X).

TABLE 12-8 Bivariate Regression Analyses of Project Cost (Y) vs. Time (X)

		Equation 12-84		Equation 12-85	
X	Y	$\hat{Y}$	e	$\hat{Y}$	e
6	1.5	3.03	1.53	5.10	3.60
12	4.3	6.05	1.75	7.84	3.54
18	11.2	9.08	–2.12	10.59	–0.61
24	15.5	12.10	–3.40	13.33	–2.17
30	19.1	15.13	–3.97	16.07	–3.03
36	21.8	18.15	–3.65	18.81	–2.99
42	24.0	21.18	–2.82	21.55	–2.45
48	26.2	24.20	–2.00	24.30	–1.90
54	26.9	27.23	0.33	27.04	0.14
60	27.8	30.25	2.45	29.78	1.98
66	30.1	33.28	3.18	32.52	2.42
72	33.8	36.30	2.50	35.26	1.46

Note: X = time (months) from start of project, Y = value of work completed (\$ × 10^6), $\hat{Y}$ = predicted value, and e = error in predicted value = $\hat{Y} - Y$.

12.9.4. Project Cost vs. Time

For large projects that require years to complete, the construction engineer needs to schedule project finances just as it is necessary to schedule construction activities. While costs vary with time, a construction engineer believes that a linear model can be used to represent the relationship between the value of work completed (Y) and the time since the start of the project (X). Because no work would be completed at the start of the project, a zero-intercept model would seem appropriate. The engineer compiled semiannual costs for a recent 12-year project, and the data are given in Table 12-8. The linear, zero-intercept, bivariate model was fit to the data as follows:

$$\hat{Y} = 0.50421X \tag{12-84}$$

where X is the time in months from the start of project, and $\hat{Y}$ is the predicted value of work in \$ × 10^6. The model is shown in Figure 12-14, and the residuals are given in Table 12-8. The model underpredicts for the middle part of the period while it overpredicts for the early and latter parts of the time period. The standard error of estimate is \$7.73 × 10^6, which gives $S_e/S_y = 0.757$; this indicates that the model does not provide highly accurate estimates of Y. Additionally, the sum of the errors is not zero, so the model is biased. Whereas a linear model with an intercept will provide an unbiased model, a zero-intercept model can be biased, as it is for this example with a mean error of –\$0.52 × 10^6. This means that, on the average, Equation 12-84 tends to underpredict by \$520,000. A correlation coefficient should not be computed for a biased model. The correlation coefficient assumes a linear model with an

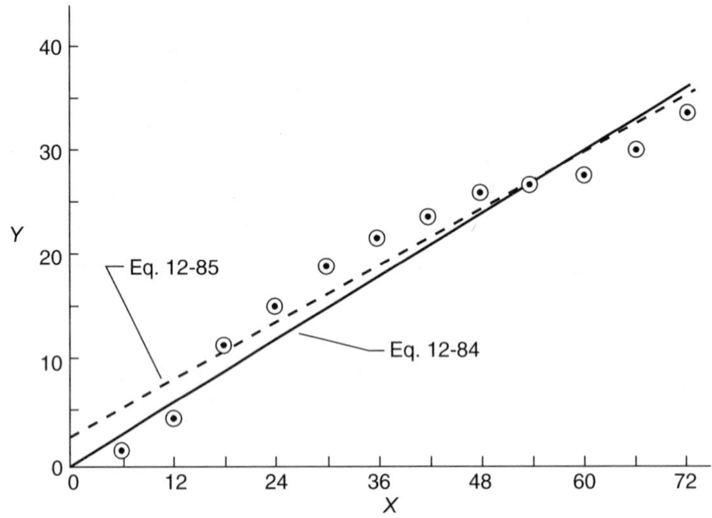

FIGURE 12-14 Regression of project cost (Y, $\$10^6$) on time from start of project (X, months) for zero-intercept (Equation 12-84) and linear (Equation 12-85) models.

intercept, so it would be incorrect to use either Equation 12-5 or 12-45 to compute a correlation coefficient for Equation 12-84.

Because of the poor accuracy of the model of Equation 12-84, the following linear model with an intercept was fit to the data of Table 12-8:

$$\hat{Y} = 2.36 + 0.4570X \tag{12-85}$$

This model is also shown in Figure 12-14 and the residuals are given in Table 12-8. Equation 12-85 is unbiased and has a standard error of $\$2.651 \times 10^6$, which gives $S_e/S_y = 0.26$. The model explains (R^2) 94% of the variation in Y and has a correlation coefficient of 0.969. In spite of the reasonably accurate estimates suggested by the goodness-of-fit statistics, the model still has local biases in the residuals (see Table 12-8), with underpredictions for the range $18 \le X \le 48$ months. However, even with the local biases, the model of Equation 12-85 may be sufficiently accurate for making project-planning estimates. The nonzero intercept of Equation 12-84 suggests a cost of $\$2.36 \times 10^6$ at the time the project is initiated. While a value of \$0 is expected, the nonzero intercept may suggest that the model may not be realistic for the initial part of a project.

In comparing the two models, the standard error ratio would suggest that the zero-intercept model is considerably less accurate than the nonzero-intercept model. However, comparing the residuals of the two models reveals very little difference. The averages of the absolute values of the residuals are 2.5 and 2.2 for Equations 12-84 and 12-85, respectively. The nonzero-intercept model is slightly better, but greater accuracy may be achieved with another model structure.

TABLE 12-9 Regression Analysis

		Equation 12-86		Equation 12-87	
X	Y	$\hat{Y}$	e	$\hat{Y}$	e
640	0.063	0.066	0.003	—	—
1930	0.061	0.050	–0.011	0.058	–0.003
1980	0.057	0.049	–0.008	0.055	–0.002
2010	0.049	0.049	0.000	0.053	0.004
2080	0.043	0.048	0.005	0.049	0.006
2120	0.048	0.048	0.000	0.046	–0.002
2200	0.039	0.047	0.008	0.041	0.002
2250	0.042	0.046	0.004	0.038	–0.004

Note: X = distance from nuclear power facility, Y = concentration of ^{241}Am in soil samples, $\hat{Y}$ = predicted value of Y, and e = residual = $\hat{Y} - Y$.

12.9.5. Effect of Extreme Event

Small samples very often contain an extreme event, which can greatly influence the resulting equation. Table 12-9 contains eight measurements of the concentration of ^{241}Am in soil samples and the distance of the soil sample from the nuclear facility. The eight measurements were used to regress the ^{241}Am concentration (Y) on the distance (X):

$$\hat{Y} = 0.07366 - 0.0000123X \tag{12-86}$$

The negative slope coefficient indicates that the concentration decreases with distance. The standard error of estimate is 0.00694, which gives $S_e/S_y = 0.764$. This suggests poor prediction ability. The correlation coefficient of –0.706 means that 50% of the total variation in Y is explained by the regression equation. The standard error ratio of the slope coefficient, $S_{e,b_1}/b_1$, equals 0.41, which is moderate and suggests a less than adequate model. The predicted values of Y and the errors for Equation 12-86 are shown in Table 12-9. Except for the value for a distance of 640, the predicted values range from 0.046 to 0.050 picocuries per gram (pCi/g); however, the measured values for these distances ranged from 0.039 to 0.061 pCi/g. These predicted values are not much different from the mean of 0.050 pCi/g. The batch of seven points acts as a single point.

A second model was developed using seven of the eight values. By eliminating the extreme event, the following model was fitted:

$$\hat{Y} = 0.1755 - 0.0000611X \tag{12-87}$$

This model has a standard error of 0.00412 pCi/g ($S_e/S_y = 0.51$). The correlation coefficient in this case is –0.885, which means that 78.3% of the total variation in the seven measurements was explained by the regression equation of Equation 12-87.

TABLE 12-10 Bivariate Regression of Fuel Consumption (*Y*) on the Reciprocal of Speed (*X*)

X	W	Y	$\hat{Y}$	e
3.4	0.2940	0.320	0.3190	–0.0010
8.1	0.1230	0.140	0.1570	0.0170
16.4	0.0610	0.130	0.0975	–0.0325
23.7	0.0422	0.094	0.0796	–0.0144
31.3	0.0319	0.062	0.0698	0.0078
38.8	0.0258	0.066	0.0640	–0.0020
42.3	0.0236	0.051	0.0619	0.0109
49.0	0.0204	0.056	0.0588	0.0028
54.6	0.0183	0.055	0.0568	0.0018
58.2	0.0172	0.049	0.0558	0.0068
66.1	0.0151	0.051	0.0538	0.0028

Note: X = fuel consumption (gallons/mile), $W = 1/X$, and Y = mean speed (miles/hour).

The standard error ratio for the slope coefficient, $S_{e,b_1}/b_1$, was 0.235, which is considerably better than the corresponding value for the model of Equation 12-86. The residuals for Equation 12-87 are small, and the predicted values range from 0.038 to 0.058 pCi/g, which is much greater than the range for the model of Equation 12-86. But, for a distance of 640, which is the measured value not included in fitting Equation 12-87, the value predicted with Equation 12-87 is 0.136 pCi/g; this is considerably higher than the measured value of 0.063 pCi/g.

In summary, in spite of the goodness-of-fit statistics for both models, neither model is an accurate representation of the system. This is a result of both the small sample size ($n = 8$) and the distribution of the data. It is undesirable to have a cluster of points and then one extreme event. Such a distribution of values makes it difficult to attain a model that is representative over a wide range of values of the predictor variable.

12.9.6. Variable Transformation

Studies have shown that fuel consumption (Y, gallons per mile) is related to the vehicle velocity (X, miles per hour) by:

$$\hat{Y} = b_0 + \frac{b_1}{X} \tag{12-88}$$

By creating a new variable $W = 1/X$, a relationship between Y and X can be fitted. The data of Table 12-10 can be used to illustrate the fitting of a model that requires a transformation of the predictor variable.

A linear regression analysis of Y on W produces the following equation:

$$Y = 0.03939 + \frac{0.9527}{X} \tag{12-89}$$

The standard error of estimate for Equation 12-89 is 0.0141 gal/mi, which is 17.5% of the standard deviation of Y (i.e., $S_e/S_y = 0.175$). Accurate estimates are also suggested by the correlation coefficient of 0.986 ($R^2 = 0.972$). The standard error ratio for the slope coefficient, $S_{e,b_1}/b_1$, is 0.056, which indicates that the computed value of b_1 is highly accurate. The residuals (see Table 12-10) are small and a trend is not apparent.

In this case, only the predictor variable was transformed, which does not create problems. Problems can arise when it is necessary to transform the criterion variable in order to put the relationship into a linear form.

12.10. SIMULATION AND PREDICTION MODELS

Prediction models based on statistical analysis are important in many engineering design methods. Traditionally, the linear models with an intercept have been used. As computers have enabled more complex models to be developed and fit to data, greater accuracy has been achieved because of the greater flexibility of the complex models. However, one advantage of the linear model is that theory provides the basis for computing confidence intervals and performing tests of hypotheses. Theoretical developments are not available for more complex models. Thus, while the more complex models may provide greater prediction accuracy, the lack of theory that would enable hypothesis tests to be made and confidence intervals to be computed is a drawback.

For complex models where theory does not provide a description of the underlying probability distributions, simulation can be used to derive the distribution and then develop methods for constructing confidence intervals and making hypothesis tests. While the full procedure is beyond the scope of this discussion, the use of simulation for the purpose of understanding the distribution of a regression-model statistic is presented using a simple example.

Example 12-9: Distribution of the Slope Coefficient of the Zero-Intercept Model

Equation 12-57 provides the standard error of the slope coefficient for the bivariate model with an intercept (Equation 12-28). If the zero-intercept model of Equation 12-22 were being used, could the standard error of Equation 12-57 be used to make hypothesis tests or construct confidence intervals on the slope coefficient? The answer is no. Equation 12-57 applies only to the bivariate model of Equation 12-28.

TABLE 12-11 Analysis of Critical Values for the Distribution of the Slope Coefficient of the Zero-Intercept Model

(1) Probability α	(2) Critical value	(3) z_α or $z_{1-\alpha}$	(4) Simulated values $-b + z\, S_e(b)$
0.005	1.171	–2.575	1.171
0.010	1.195	–2.327	1.193
0.050	1.254	–1.645	1.254
0.100	1.286	–1.282	1.286
0.900	1.515	1.282	1.515
0.950	1.547	1.645	1.547
0.990	1.604	2.327	1.608
0.995	1.631	2.575	1.630

To learn more about the distribution of the slope coefficient (b) of the zero-intercept model, the following population model was assumed:

$$\hat{Y} = \beta X + Z\sigma_e \tag{12-90}$$

where β is the slope coefficient, X is the predictor variable, Z is a standard normal deviate, and σ_e is the standard error of the population. For this example, a value of 1.4 was assumed for β. The variable X was assumed to be normally distributed with $\mu = 10$ and $\sigma_x = 2$. The standard error of the population model was assumed to equal 2.857. Sample values of X and Z were generated for sample sizes of 10. The model of Equation 12-90 was used to compute values of Y, and then the simulated values of X and Y were used to fit the slope coefficient b. A total of 100,000 samples (Ns) of a size 10 for each sample (N) were generated using the program shown in Figure 12-15. The mean and standard deviation of the 100,000 values of b were 1.4006 and 0.08912, respectively. The standard deviation is the standard error of the slope coefficient. The computed mean of 1.4006 agrees with the assumed population value. In addition to the two moments, a histogram of the values was computed, and the critical values of the slope coefficient were obtained from the cumulative histogram for selected probabilities, shown in Table 12-11 (column 2). The standard error of b was multiplied by the standard normal deviates (column 3) for the probabilities (column 1) to obtain the simulated values that would be used to compute confidence intervals or make hypothesis tests. A comparison of columns 2 and 4 of Table 12-11 indicates that the normal distribution can be applied to the analysis of the zero-intercept model.

The above analysis is limited in that it considered only one sample size and one standard error. To develop a general expression relating $S_e(b)$ and S_e, it would be necessary to make many analyses, with S_e, n, and the characteristics of X allowed to vary. A general expression could then be identified. It is important to recognize that simulation provides a means for obtaining statistical results that cannot be obtained from theory because of the complexity of the problem.

```
      PROGRAM SIMDIST
C           Simulate Upper Tails of Distribution for Slope Coefficient
C           for Zero-intercept Model
C           N = Sample Size
C           NS = Number of Samples
C           ISEED = Seed Value for Simulation
            DIMENSION B(100000), Y(10), X(10), F(20)
            CHARACTER*75 FILE2
            IAD = 10
            IADS = 100000
            FILE2 = 'OUTPUT'
            OPEN (6,FILE=FILE2,STATUS='NEW')
C           INPUT ISEED, N, NS
            EN = N
C           delete first 10 random numbers
            RN = RNN(ISEED)
            DO 10 I = 1, 10
               J = INT(RN)
               RN = RNN(J)
   10          CONTINUE
C           generate samples
            DO 80 K = 1, NS
               S1 = 0.0
               S2 = 0.0
               DO 20 I = 1, N
                  J = INT(RN)
                  RN = RNN(J)
                  X(I) = 10. + RN * 2.
                  J = INT(RN)
                  RN = RNN(J)
                  Y(I) = 1.4 * X(I) + 2.857 * RN
                  S1 = S1 + X(I) * Y(I)
                  S2 = S2 + X(I) ** 2
C                  WRITE(6,19) K, I, X(I), Y(I), RN
   19             FORMAT(2I5, 3F10.5)
   20             CONTINUE
               B(K) = S1 / S2
C               WRITE(6,21) S1, S2, B(K)
   21          FORMAT(40X, 3F12.5)
   80          CONTINUE
C           Compute Statistics of Coefficient B and Print
C           STAT Subroutine not Provided
            CALL STAT
C           Construct Histogram for Coefficient B and Print
C           FREQ Subroutine not Provided
            CALL FREQ
            CLOSE (6,STATUS='KEEP')
            END

            FUNCTION RNN(K)
C           Generates random normal numbers (not provided)
            FUNCTION RAND(K)
C           Generates random numbers (not provided)
            SUBROUTINE STAT
C           Computes Mean and Standard Deviation for AN ARRAY
            SUBROUTINE FREQ
C           Determines the number of elements in each histogram cell
```

FIGURE 12-15 Outline of a computer program used in Example 12-9.

12.11. PROBLEMS

12-1. Plot the following data, assess the degree of systematic variation, and assess the appropriateness of using a linear model:

X	0.7	1.3	2.8	5.1	7.3	8.9	11.5
Y	1.1	3.8	7.2	7.9	9.7	8.8	9.6

12-2. Plot the following data, and assess the quality of the data for prediction:

X	0.7	0.8	1.3	1.2	1.7	1.9	4.1
Y	1.0	1.8	1.2	2.1	1.0	2.8	5.7

12-3. Expand Equation 12-2 to show that the total variation is the sum of the explained and unexplained variation.

12-4. Using the following values of X and Y, show that the total variation (TV) equals the sum of the explained variation (EV) and the unexplained variation (UV):

X	2	5	7	6	2
Y	1	3	5	7	9

Assume that the regression equation relating Y and X is given by:

$$\hat{Y} = 4.5849 + 0.09434X$$

12-5. Using the following values of X and Y, show that the total variation (TV) equals the sum of the explained variation (EV) and the unexplained variation (UV):

X	1	2	3	4	5	6
Y	2	2	3	5	4	5

Assume that the regression equation relating Y and X is given by:

$$\hat{Y} = 1.2 + 0.657X$$

12-6. For the data in Problem 12-5, compute the correlation coefficient.

12-7. For the following observations on X and Y, compute the correlation coefficient.

X	–3	–2	–1	0	1	2
Y	2	2	3	0	–2	–1

12-8. For the data given in Problem 12-4, compute the correlation coefficient using the standard normal transformation definition of Equation 12-5.

12-9. Derive Equation 12-5 from Equation 12-4.

12-10. Given the following observations on X and Y, (a) construct a graph of X vs. Y, and (b) compute the Pearson correlation coefficient:

X	1	2	3	4	5
Y	1	1	2	4	4

12-11. Compute the correlation coefficient for the data of Problem 12-2.

12-12. Given the following paired observations on variables X and Y, and the values of the standardized variables Z_X and Z_Y, compute the Pearson correlation coefficient between each of the following pairs: (a) X and Y; (b) Z_X and Z_Y; (c) Z_X and X; (d) Z_X and Y; (e) X and $Y-3$; and (f) X and $2Y$. Briefly discuss your results.

X	Y	Z_X	Z_Y
3	4	–1.5	–1.5
5	8	–0.5	0.5
6	7	0.0	0.0
7	6	0.5	–0.5
9	10	1.5	1.5

12-13. Engineering analyses often involve small samples, often fewer than five. Explain why a high sample correlation coefficient obtained with a small sample should not necessarily be taken as an indication of a strong relationship between two variables.

12-14. Seven students in an introduction to statistics course received the following grades on the first two tests. Does the evidence suggest that knowledge of the material covered on the first test helped the students on the second test? Use a 5% level of significance.

Test 1	98	94	93	90	87	85	84
Test 2	96	94	91	85	93	86	90

12-15. A sample of four yields a correlation coefficient of 0.915. Is it safe to conclude that the two variables are related?

12-16. Given the following observations on random variable X and Y, (a) calculate the correlation coefficient; (b) using a level of significance of 5%, test the null hypothesis that the two variables are uncorrelated; (c) using a level of significance of 1%, test the null hypothesis that the population correlation is 0.8.

X	4	3	9	6	8
Y	3	6	8	8	10

12-17. The correlation coefficient computed from a sample of 15 observations equals 0.431. Test the null hypothesis using a level of significance of 5% that the correlation of the population is (a) 0.0; and (b) 0.65.

12-18. A sample of 12 observations on X and Y yields a correlation coefficient of 0.582. Using a level of significance of 5%, test the null hypothesis that the correlation coefficient of the population is (a) 0.0; and (b) –0.1.

12-19. The design engineer believes that a correlation of 0.8 is needed in the population to have the necessary prediction accuracy. A sample of 22 measurements on Y and X yields a sample correlation coefficient of 0.49. Is it safe to assume that the population correlation coefficient is at least 0.8? Use a 1% level of significance.

12-20. What is the smallest sample size for which a correlation coefficient of 0.61 is statistically significant at 5%, with a two-tailed alternative hypothesis?

12-21. Using the data of Problem 12-2, fit the zero-intercept model of Equation 12-22.

12-22. Using the data of Problem 12-14, fit the zero-intercept model of Equation 12-22. Predict values of test 2 using values of test 1.

12-23. Given the following four pairs of observations

X	1	1	2	2	?
Y	1	2	1	2	?

Compute the correlation coefficient and the regression coefficients if (a) the fifth pair is ($Y = 8$, $X = 8$); (b) the fifth pair is ($Y = 2$, $X = 8$). (c) Plot both sets of points and draw the regression lines. What general observations do the results suggest?

12-24. Show that Equation 12-38 results in a larger sum of squares of the errors for the data of Table 12-2 than the value obtained with Equation 12-37.

12-25. Using the following values of X and Y, compute the values of the regression model of Equation 12-28:

X	1	3	5
Y	5	2	1

12-26. Compute the correlation coefficient for the data of Problem 12-25. What is the fraction of explained variation?

12-27. Given the following paired observations on Y and X, (a) graph Y vs. X; (b) calculate the correlation coefficient; (c) determine the slope and intercept coefficients of the linear regression line; (d) show the regression line on the graph of part (a); (e) compute the predicted value of Y for each observed value of X; and (f) calculate $\sum e$ and S_e. (g) What is the predicted value of Y for $X = 5$? $X = 10$?

X	3	5	6	7	9
Y	4	8	7	6	10

12-28. Using the data of Problem 12-12, compute the standard error of estimate S_e and the standard error ratio, S_e/S_y.

12-29. Using the data of Problem 12-12, compare the standard error of estimate based on Equation 12-40 with the correct value of 12-45. Discuss the cause of the difference.

12-30. Using the data of Problem 12-12, show that the SS_T of Equation 12-47 is equal to the total variation of Equation 12-2 and that the SS_R of Equation 12-48 is equal to the explained sum of squares of Equation 12-2.

12-31. Using the data set of Problem 12-16, compute the regression coefficients for the linear bivariate model with X as the predictor variable. Also, determine the standard error of estimate and perform an analysis of variance (use a 5% level of significance).

12-32. Compute the standardized partial regression coefficient t (Equation 12-51) for the data of Problem 12-12, and compare the value with the correlation coefficient computed in Problem 12-12.

12-33. If the correlation coefficient is equal to 0.71 and the slope coefficient b_1 is equal to 46.3, which variable, X or Y, has the larger variance?

12-34. State the four assumptions that underlie the linear regression analysis and identify statistical methods that could be used to test the assumptions.

12-35. Compute the standard errors of the regression coefficients and the two-sided 95% confidence intervals for the regression coefficients of Problem 12-27.

12-36. Using the data of Problem 12-16 and the regression coefficients of Problem 12-31, compute standard errors of the regression coefficients and the two-sided, 90% confidence intervals for the coefficients.

12-37. The ratio of the standard error of the slope coefficient (Equation 12-57) to the slope coefficient b_1 is a measure of the relative accuracy of the coefficient, with a value below 30% indicating good accuracy and a value greater than 50% indicating poor accuracy. Evaluate the relative accuracy of the slope coefficient of Problem 12-25.

12-38. Perform hypothesis tests for the significance of the regression coefficients for Problem 12-27. Use a 5% level of significance and alternative hypothesis that the coefficients are different from zero.

12-39. Perform hypothesis tests for the significance of the regression coefficients for Problem 12-31. Use a 10% level of significance and alternative hypothesis that the coefficients are different from zero.

12-40. Compute a two-sided, 95% confidence interval for the line as a whole using the data for Problem 12-27. Provide values of $\overline{X} - 2S_X$, $\overline{X} - S_X$, $\overline{X} + S_X$, and $\overline{X} + 2S_X$.

12-41. Compute a two-sided, 95% confidence interval for the line as a whole using the data for Problem 12-16. Provide values of $\overline{X} - 2S_X$, $\overline{X} - S_X$, $\overline{X} + S_X$, and $\overline{X} + 2S_X$.

12-42. Using the data from Problem 12-27, compute the two-sided, 95% confidence interval for the mean value of Y at $X = 5$. Also, compute a similar interval at $X = 11$. Compare the two intervals and explain the difference in their widths.

12-43. Using the data from Problem 12-16, compute the two-sided, 95% confidence interval for the mean value of Y at $X = 6$. Also, compute a similar interval at $X = 0$. Compare the two intervals and explain the difference in their widths.

12-44. Compare correlation analysis and regression analysis as tools in decision making.

12-45. Using the data set given below, regress (a) Y on X, and (b) X on Y. (c) Compute the correlation coefficient for each regression; (d) Transform the regression of Y on X into an equation for computing X, and compare these coefficients with the coefficients for the regression of X on Y. (e) Why do the coefficients of parts (b) and (d) differ?

X	2	1	3	2
Y	1	2	4	5

12-46. Regress Y on X_1, Y on X_2, and Y on X_3 with the following data:

X_1	X_2	X_3	Y
1	2	3	1
2	2	1	3
5	5	7	5
6	4	4	7

Then transform the three equations algebraically to solve for X. Regress X_1 on Y, X_2 on Y, and X_3 on Y. Use the equations for regressing the X's on Y and the transformed equation to predict X. Compare the differences in the predictions with the correlation coefficient and develop a general observation on the accuracy of transformed equations.

12-47. The data shown are the results of a tensile test of a steel specimen, where Y is elongation in thousands of an inch that resulted when the tensile force was X thousands of pounds. Fit the data with a linear model and make a complete assessment of the results (use $\alpha = 5\%$ where necessary).

X	0.8	1.6	3.1	4.4	6.3	7.9	9.2
Y	2.8	4.9	6.5	8.1	8.8	9.1	8.9

12-48. The stream reaeration coefficient (Y) is related to the water temperature (T) in °C. Fit the following with a linear model and make a complete assessment of the results (use $\alpha = 10\%$ where necessary):

X	14	23	17	11	20	15	18	11	25	19	13	24	16	23	21
Y	2.89	4.20	4.17	3.69	3.78	3.56	3.35	2.69	3.58	3.06	2.41	3.30	3.05	4.18	4.08

12-49. Use simulation to verify the relationships for computing the standard errors of the two coefficients of the linear bivariate regression equation.

12-50. Use simulation to find the standard error and the critical values for 10 and 5% of the slope coefficient of the linear zero-intercept bivariate regression model ($\hat{Y} = bX$) for a sample size of 10. Assume X to be normally distributed, $X \sim$ N(mean = 20, standard deviation = 4), and generate values of Y based on the generated values of X using the following equation:

$$Y = \beta_1 x + Z\sigma_Y\sqrt{1-\rho^2}$$

where β_1 = population value of b_1, Z is the random normal variate, σ_Y is the population standard deviation of Y, and ρ is the population value of correlation between X and Y. Use $\rho = 0.6$, $n = 10$, $\mu_X = 20$, $\sigma_X = 4$, $\beta_1 = 3$, and $\sigma_Y = 2$.

12.12. SIMULATION PROJECTS

Two cases are provided in this section.

Case 1

Using the data of Table C-1, C-2, or C-3, calibrate bivariate linear regression for Y vs. each X. Make a full analysis of the accuracy of each model and decide which model, if any, should be used to predict values of Y for future analyses.

Case 2

The objective of this project is to simulate the distribution of the correlation coefficient under the condition that the population value of ρ is zero. The Y variable and one X variable of Table C-1, C-2, or C-3 are used to establish the assumed population. The means and standard deviations of Y (μ_Y and σ_Y) and X (μ_X and σ_X) are estimated from the data in the Appendix. Use 10^5 simulations to obtain the distribution of the correlation coefficient, with the following procedure:

1. Generate a sample of 25 values of Y for the assumed normal population $N(\mu_Y, \sigma_Y)$.
2. Generate a sample of 25 values of X for the assumed normal population $N(\mu_X, \sigma_X)$.
3. Compute the correlation coefficient R with the sample of X and Y of steps 2 and 1.
4. Increase the frequency count of the histogram cell corresponding to the sample value of R. (*Note:* The histogram cells can be established from theory.)

Repeat steps 1 to 4 Ns times (where the number of simulation cycles equals Ns), and then divide the cell counts of the frequency histogram by Ns to obtain a relative frequency distribution. Discuss the results and compare them to the expected distribution.

Chapter 13
Multiple and Nonlinear Regression Analysis

TABLE OF CONTENTS

13.1. INTRODUCTION

What options are available to improve the accuracy of predictions if the accuracy from a bivariate regression is still not sufficient for the design problem? Multivariate systems are one possible solution to this problem. Where one predictor variable may not be adequate, several predictor variables may provide sufficient prediction accuracy. Thus, one reason for using multivariate models rather than a bivariate model is to reduce the standard error of estimate.

Multivariate data analyses are also of value when a goal of the analysis is to examine the relative importance of the predictor variables. If a predictor variable is not included in the model, then a regression analysis cannot determine the sensitivity of the criterion variable to variation in that predictor variable. Also, if a predictor variable that has not been included in the model is correlated with another predictor variable that is included in the model, then the regression coefficient for the former variable may reflect the effects of both the former and latter predictor variables. Thus, predictor variables that have only slight correlation with the criterion variable are often included in the model so that a complete analysis may be made.

How does a multivariate analysis compare with a bivariate analysis? Actually, the two are very similar. First, the same least-squares objective function is used to calibrate the regression coefficients. Second, bivariate correlations are still computed. Third, the data should be properly screened using graphical analyses prior to selecting a model structure. The major difference between multivariate and bivariate analyses is necessary to account for the interdependence (i.e., correlation) of the predictor variables.

13.2. CORRELATION ANALYSIS

The first step in a multivariate analysis is to perform a graphical analysis. This includes developing plots of the criterion variable vs. each of the predictor variables in the data set. Also, plots of each pair of predictor variables should be made; such plots will help to identify the characteristics of the interdependence between predictor variables. It will become evident that the interdependence is extremely important in the analysis of multivariate systems. All the plots should be examined for the characteristics identified previously for bivariate analyses — for example, extreme events, nonlinearity, and random scatter.

After a graphical analysis, the bivariate correlation coefficients should be computed for each pair of variables; this includes the correlation between the criterion variable and each predictor variable, as well as the correlation between each pair of predictor variables. The correlations are best presented in matrix form. The correlation matrix is a means of presenting in an organized manner the correlations between pairs of variables in a data set; it appears as:

	X_1	X_2	X_3	...	X_p	Y
X_1	1.0	r_{12}	r_{13}	...	r_{1p}	r_{1Y}
X_2		1.0	r_{23}	...	r_{2p}	r_{2Y}
X_3			1.0	...	r_{3p}	r_{3Y}
$\vdots$				$\ddots$	$\vdots$	$\vdots$
X_p					1.0	r_{pY}
Y						1.0

(13-1)

Note that the correlation matrix, which includes p predictor variables (X_i, $i = 1, 2, \ldots, p$) and the criterion variable (Y), is shown in upper-triangular form because the matrix is symmetric (i.e., $r_{ij} = r_{ji}$); also, the elements on the principal diagonal equal 1.0 because the correlation between a variable and itself is unity. The matrix is $(p + 1) \times (p + 1)$. Whereas the correlation matrix shown above uses the sample correlation coefficients (r_{ij}), the correlation matrix for the population can be indicated using ρ_{ij} as the elements.

Example 13-1: Sediment Yield Data

The sediment database includes four predictors and a criterion variable, the sediment yield. The ultimate objective would be to develop a model for estimating the sediment yield from small watersheds. The predictor variables are (1) the precipitation/temperature ratio (X_1), which reflects the vegetation potential of the area; (2) the average watershed slope (X_2), which reflects the erosion potential due to the momentum of surface runoff; (3) the soil particle size index (X_3), which reflects the coarse-particle composition of the soil; and (4) the soil aggregation index (X_4), which reflects the dispersion characteristics of the small particles. Because vegetation retards erosion, we would expect r_{1Y} to be negative. Similarly, coarse particles are more difficult to transport, so r_{3Y} should also be negative. As the slope increases, the water will move with a greater velocity and more soil can be transported, so r_{2Y} should be positive. Erosion should increase as the dispersion characteristics of small particles increases, so r_{4Y} should be positive.

The correlation matrix for 37 measurements of these five variables is given in Table 13-1. In general, the intercorrelations are low, with the largest in absolute value being for r_{14} (−0.445); this suggests that more than one predictor variable will probably be used to obtain accurate estimates of the criterion variable. Many of the intercorrelations are very low, less than 0.2; this is usually a desirable characteristic. The predictor–criterion correlations are low to moderate; the largest is 0.570. Using the square of the predictor–criterion correlations, the fraction of variance explained ranges from 0.062 to 0.325. These low values would also suggest that more than one predictor variable will be necessary to accurately estimate the sediment yield.

TABLE 13-1 Correlation Matrix of the Sediment Data Base

	X_1	X_2	X_3	X_4	Y
X_1: precipitation/temperature ratio	1.000	0.340	−0.167	−0.445	−0.297
X_2: average slope		1.000	−0.051	−0.185	0.443
X_3: soil particle size			1.000	0.069	−0.253
X_4: soil aggregation index				1.000	0.570
Y: sediment yield					1.000

TABLE 13-2 Correlation Matrix for the Evaporation Data Base

	X_1	X_2	X_3	X_4	Y
X_1: temperature (°F)	1.000	−0.219	0.578	0.821	0.581
X_2: wind speed (miles/day)		1.000	−0.261	−0.304	−0.140
X_3: radiation (equivalent inches)			1.000	0.754	0.578
X_4: vapor pressure deficit				1.000	0.635
Y: pan evaporation (inches)					1.000

Example 13-2: Evaporation Data

The evaporation data set includes four predictor variables and a criterion variable, which is the pan evaporation from a small pond in southeastern Georgia. The predictor variables are (1) the mean daily temperature in degrees Fahrenheit (°F), X_1; (2) the wind speed in miles per day, X_2; (3) the radiation in equivalent inches, X_3; and (4) the vapor pressure deficit, X_4. The wind speed attempts to define the rate at which the drier air is moved over the pond to replace the more moist air. The temperature and the radiation measure the potential energy of the water molecules and their potential for escaping from the water surface. The vapor pressure deficit represents the potential for the overlying air mass to accept water molecules.

The correlation matrix, which was based on 354 daily measurements, is shown in Table 13-2. The correlation matrix provides a sharp contrast to that for the sediment data because the matrix of Table 13-2 is characterized by high intercorrelation and moderate predictor–criterion correlations. The high intercorrelations suggest that only one or two predictor variables will be necessary to estimate evaporation rates. The moderate predictor–criterion correlations suggest that the potential accuracy of predictions is only moderate. The fraction of the variance explained by X_4 is 0.403. Because most of the intercorrelations are high, it is unlikely that the variation explained by the four predictor variables will be much greater than the 40% explained by just one of the predictor variables.

13.3. MULTIPLE REGRESSION ANALYSIS

13.3.1. Calibration of the Multiple Linear Model

The three components of regression analysis are the model, the objective function, and the data set. The data set consists of a set of n observations on p predictor variables and one criterion variable, where n should be, if possible, at least four times greater than p. The data set can be viewed as a matrix having dimensions of n by $(p + 1)$. The principle of least squares is used as the objective function. The model, in raw score form, is:

$$\hat{Y} = b_0 + b_1X_1 + b_2X_2 + \ldots + b_pX_p \tag{13-2}$$

in which X_j $(j = 1, 2, \ldots, p)$ are the predictor variables, b_j $(j = 1, 2, \ldots, p)$ are the partial regression coefficients, b_0 is the intercept coefficient, and Y is the criterion variable. Using the least-squares principle and the model of Equation 13-2, the objective function becomes:

$$F = \min\sum_{i=1}^{n} e_i^2 = \min\sum_{i=1}^{n}\left(b_0 + \sum_{j=1}^{p} b_jx_{ij} - y_i\right)^2 \tag{13-3}$$

in which F is the value of the objective function. It should be noted that the predictor variables include two subscripts, with i indicating the observation and j the specific predictor variable.

The method of solution is to take the $(p + 1)$ derivatives of the objective function, Equation 13-3, with respect to the unknowns, b_j $(j = 0, 1, \ldots, p)$; setting the derivatives equal to zero; and solving for the unknowns. A set of $(p + 1)$ normal equations is an intermediate result of this process.

As an example, consider the case where $p = 2$; thus, Equation 13-2 reduces to:

$$\hat{Y} = b_0 + b_1X_1 + b_2X_2 \tag{13-4}$$

Also, the objective function, Equation 13-3, is given by:

$$F = \min\sum_{i=1}^{n} e_i^2 = \min\sum_{i=1}^{n}(b_0 + b_1x_{i1} + b_2x_{i2} - y_i)^2 \tag{13-5}$$

The resulting derivatives are:

$$\frac{\partial\sum_{i=1}^{n} e_i^2}{\partial b_0} = 2\sum_{i=1}^{n}(b_0 + b_1x_{i1} + b_2x_{i2} - y_i)(1) = 0 \tag{13-6a}$$

$$\frac{\partial\sum_{i=1}^{n} e_i^2}{\partial b_1} = 2\sum_{i=1}^{n}(b_0 + b_1x_{i1} + b_2x_{i2} - y_i)(x_{i1}) = 0 \tag{13-6b}$$

$$\frac{\partial\sum_{i=1}^{n} e_i^2}{\partial b_2} = 2\sum_{i=1}^{n}(b_0 + b_1x_{i1} + b_2x_{i2} - y_i)(x_{i2}) = 0 \tag{13-6c}$$

Rearranging Equations 13-6 yields a set of normal equations:

$$nb_0 + b_1 \sum_{i=1}^{n} x_{i1} + b_2 \sum_{i=1}^{n} x_{i2} = \sum_{i=1}^{n} y_i \tag{13-7a}$$

$$b_0 \sum_{i=1}^{n} x_{i1} + b_1 \sum_{i=1}^{n} x_{i1}^2 + b_2 \sum_{i=1}^{n} x_{i1}x_{i2} = \sum_{i=1}^{n} x_{i1}y_i \tag{13-7b}$$

$$b_0 \sum_{i=1}^{n} x_{i2} + b_1 \sum_{i=1}^{n} x_{i1}x_{i2} + b_2 \sum_{i=1}^{n} x_{i2}^2 = \sum_{i=1}^{n} x_{i2}y_i \tag{13-7c}$$

The solution of the three simultaneous equations yields values of b_0, b_1, and b_2.

13.3.2. Standardized Model

When the approach of the previous section is used and the means and standard deviations of the predictor and criterion variables are significantly different, round-off error, which results from the inability to maintain a sufficient number of significant digits in the computations, may cause the partial regression coefficient to be erroneous. Thus, most multiple regression analyses are made using a standardized model:

$$Z_Y = t_1 Z_1 + t_2 Z_2 + \ldots + t_p Z_p \tag{13-8}$$

in which the t_j ($j = 1, 2, \ldots, p$) are called *standardized partial regression coefficients*, and Z_Y and the Z_j ($j = 1, 2, \ldots, p$) are the *criterion variable* and the *predictor variables*, respectively, expressed in standardized form; specifically, for $i = 1, 2, \ldots, n$, they are computed by:

$$Z_Y = \frac{Y_i - \bar{Y}_i}{S_Y} \tag{13-9}$$

and

$$Z_j = \frac{X_j - \bar{X}_j}{S_j} \tag{13-10}$$

in which S_Y is the standard deviation of the criterion variable and S_j ($j = 1, 2, \ldots, p$) are the standard deviations of the predictor variables. It can be shown that the standardized partial regression coefficients (i.e., the t_j) and the partial regression coefficients (i.e., the b_j) are related by:

$$b_j = \frac{t_j S_Y}{S_j} \tag{13-11}$$

The intercept coefficient can be computed by:

$$b_0 = \bar{Y} - \sum_{j=1}^{p} b_j \bar{X}_j \tag{13-12}$$

Thus, the raw score model of Equation 13-2 can be computed directly from the standardized model, Equation 13-8, and Equations 13-11 and 13-12.

Although the differences between regression and correlation have been emphasized, the correlation matrix can be used in solving for the standardized partial regression coefficients. The solution is represented by:

$$\boldsymbol{R}_{11} \times \boldsymbol{t} = \boldsymbol{R}_{12} \tag{13-13}$$

in which $\boldsymbol{R}_{11}$ is the $p \times p$ matrix of intercorrelations between the predictor variables, $\boldsymbol{t}$ is a $p \times 1$ vector of standardized partial regression coefficients, and $\boldsymbol{R}_{12}$ is the $p \times 1$ vector of predictor–criterion correlation coefficients. Because $\boldsymbol{R}_{11}$ and $\boldsymbol{R}_{12}$ are known while $\boldsymbol{t}$ is unknown, it is necessary to solve the matrix equation, Equation 13-13, for the $\boldsymbol{t}$ vector. This involves premultiplying both sides of Equation 13-13 by $\boldsymbol{R}_{11}^{-1}$ (i.e., the inverse of $\boldsymbol{R}_{11}$) and using the matrix identities, $\boldsymbol{R}_{11}^{-1} \times \boldsymbol{R}_{11} = \boldsymbol{I}$ and $\boldsymbol{I} \times \boldsymbol{t} = \boldsymbol{t}$, where $\boldsymbol{I}$ is the unit matrix, as follows:

$$\boldsymbol{R}_{11}^{-1} \times \boldsymbol{R}_{11} \times \boldsymbol{t} = \boldsymbol{R}_{11}^{-1} \times \boldsymbol{R}_{12} \tag{13-14a}$$

$$\boldsymbol{I} \times \boldsymbol{t} = \boldsymbol{R}_{11}^{-1} \times \boldsymbol{R}_{12} \tag{13-14b}$$

$$\boldsymbol{t} = \boldsymbol{R}_{11}^{-1} \times \boldsymbol{R}_{12} \tag{13-14c}$$

It is important to recognize from Equation 13-14c that the elements of the $\boldsymbol{t}$ vector are a function of both the intercorrelations and the predictor–criterion correlation coefficients. If $\boldsymbol{R}_{11} = \boldsymbol{I}$, then $\boldsymbol{R}_{11}^{-1} = \boldsymbol{I}$, and $\boldsymbol{t} = \boldsymbol{R}_{12}$; this suggests that because the t_j values serve as weights on the standardized predictor variables, the predictor–criterion correlations also reflect the importance (or "weight") that should be given to a predictor variable. However, when $\boldsymbol{R}_{11}$ is very different from $\boldsymbol{I}$ (i.e., when the intercorrelations are significantly different from zero), the t_j values will provide considerably different estimates of the importance of the predictors than would be indicated by the elements of $\boldsymbol{R}_{12}$.

13.3.3. Intercorrelation

It is evident from Equation 13-14c that intercorrelation can have a significant effect on the t_j values and thus on the b_j values. In fact, if the intercorrelations are significant, the t_j values can be irrational. It should be evident that irrational regression coefficients can lead to irrational predictions. Thus, it is important to assess the rationality of the coefficients.

The irrationality results from the difficulty in taking the inverse of the $\boldsymbol{R}_{11}$ matrix that is necessary for Equation 13-14c; this corresponds to the round-off-error problem associated with the solution of the normal equations of Equations 13-7a to 13-7c. A matrix in which the inverse cannot be evaluated is called a *singular matrix*. A near-singular matrix is one in which one or more pairs of the standardized normal equations are nearly identical (i.e., linearly dependent).

The determinant of a square matrix, such as a correlation matrix, can be used as an indication of the degree of intercorrelation. The determinant is a unique scalar value that characterizes the intercorrelation of the $\boldsymbol{R}_{11}$ matrix; that is, the determinant is a good single-valued representation of the degree of linear association between the normal equations.

Consider the following four matrices which differ in their level of intercorrelation:

$$\boldsymbol{A}_1 = \begin{bmatrix} 1 & 0 & 0 \\ 0 & 1 & 0 \\ 0 & 0 & 1 \end{bmatrix} \tag{13-15}$$

$$\boldsymbol{A}_2 = \begin{bmatrix} 1.0 & 0.5 & 0.5 \\ 0.5 & 1.0 & 0.5 \\ 0.5 & 0.5 & 1.0 \end{bmatrix} \tag{13-16}$$

$$\boldsymbol{A}_3 = \begin{bmatrix} 1.0 & 0.9 & 0.9 \\ 0.9 & 1.0 & 0.9 \\ 0.9 & 0.9 & 1.0 \end{bmatrix} \tag{13-17}$$

$$\boldsymbol{A}_4 = \begin{bmatrix} 1.0 & 0.9 & 0.0 \\ 0.9 & 1.0 & 0.0 \\ 0.0 & 0.0 & 1.0 \end{bmatrix} \tag{13-18}$$

The determinants of these matrices are:

$$|\boldsymbol{A}_1| = 1 \tag{13-19a}$$

$$|\boldsymbol{A}_2| = 0.5 \tag{13-19b}$$

$$|\boldsymbol{A}_3| = 0.028 \tag{13-19c}$$

$$|\boldsymbol{A}_4| = 0.19 \tag{13-19d}$$

These values indicate that (1) the determinant of a correlation matrix will lie between 0.0 and 1.0; (2) if the intercorrelations are zero, the determinant of a correlation matrix equals 1.0; (3) as the intercorrelations become more significant, the determinant approaches zero; and (4) when any two rows of a correlation matrix are nearly identical, the determinant approaches zero.

13.3.4. Criteria for Evaluating a Multiple Regression Model

After a multiple regression model has been calibrated, we may ask how well the linear model represents the observed data. The following criteria should be used in answering this question: (1) the rationality of the coefficients; (2) the coefficient of multiple determination (R^2); (3) the standard error of estimate, which is usually compared with S_Y; (4) the relative importance of the predictor variables; and (5) the characteristics of the residuals. The relative importance of each of these five criteria may vary with the problem as well as with the analyst.

The coefficient of multiple determination equals the fraction of the variation in the criterion variable that is explained by the regression equation. Thus, it is the square of the correlation coefficient. Mathematically, it is defined as:

$$R^2 = \frac{\sum_{i=1}^{n}(\hat{y}_i - \bar{Y})^2}{\sum_{i=1}^{n}(y_i - \bar{Y})^2} \tag{13-20}$$

It may be computed by:

$$R^2 = \sum_{J=1}^{P} t_j r_{jY} \tag{13-21}$$

in which r_{jY} is the predictor–criterion correlation for predictor j. The value of R^2 is always in the range from zero to 1.0, with a value of zero indicating that Y is not related to any of the predictor variables. The coefficient of multiple determination must be at least as large as the square of the largest predictor–criterion correlation; if the intercorrelations are high, there will be little difference between the two, which would indicate that including more than one predictor variable in the regression equation does little to improve the accuracy of predictions. If the intercorrelations are low, including more than one predictor variable will probably provide greater accuracy than a bivariate model.

The standard error of estimate (S_e) is defined as the square root of the sum of the squares of the errors divided by the degrees of freedom (ν):

$$S_e = \left[\frac{1}{\nu}\sum_{i=1}^{n}(\hat{y}_i - y_i)^2\right]^{0.5} \tag{13-22}$$

Previously, the degrees of freedom used to compute S_e were $(n - 2)$; however, in general, the degrees of freedom are defined as:

$$\nu = n - q \tag{13-23}$$

in which q is the number of unknowns. For the case where there are p partial regression coefficients and one intercept coefficient, $q = p + 1$ and $\nu = n - p - 1$. The S_e has the same units as the criterion variable and should be compared with S_Y in assessing the accuracy of prediction.

If we use Equation 13-11 to compute t_i for the bivariate case (i.e., one predictor variable), it will be evident that t, the standardized partial regression coefficient, equals the correlation coefficient. Even in the case where $p > 1$, t_i is still a measure of the common variation. However, Equation 13-14c indicates that the t_i values are a function of both the intercorrelation and the predictor–criterion correlations. When the intercorrelations are significant (i.e., $\boldsymbol{R}_{11}$ of Equation 13-14c is significantly different from $\boldsymbol{I}$, the identity matrix), the t_i values may be irrational; thus, they would not be valid measures of the relative importance of the corresponding predictor variable. Because the t value corresponds to a correlation coefficient in the bivariate case, it is not unreasonable to provide a similar interpretation for a multiple regression; thus, t_i values should be considered irrational in magnitude if they are greater than 1.0 in absolute value. Even values of t near 1.0 may indicate irrationality. If a t value is irrational, the corresponding b value must be considered irrational, as b is a scaled value of t.

13.3.5. Analysis of Residuals

The residuals (i.e., $\hat{y}_i - y_i$) were shown to be an important criterion in assessing the validity of a bivariate regression equation; the same concepts apply in the analysis of the residuals for a multiple regression equation. Plots of the residuals may indicate an incorrect model structure, such as a nonlinear relationship between the criterion variable and one of the predictor variables. Additionally, the residuals may suggest a need to consider another predictor variable that was not included in the model. Plots of the residuals should be made to assess the assumptions of zero mean, constant error variation, independence, and normality. A frequency analysis using probability papers of the residuals can be used to check for normality. A plot of the residuals vs. the predicted value of the criterion variable (i.e., e_i vs. $\hat{y}_i$) may identify either a violation of the assumption of constant variance or a lack of independence of the observations. The Durbin–Watson test can be used to test whether or not significant correlation exists among the residuals.

13.3.6. Computational Example

Although we rarely perform a multiple regression without the aid of a computer, it is instructive to see the manual computations at least one time. Table 13-3 provides values for the criterion variable (Y) and two predictor variables (X_1 and X_2); the sample consists of six observations (i.e., $n = 6$). Figure 13-1 shows the

TABLE 13-3 Database for Example of Two-Predictor Model

	Y	X_1	X_2	X_1^2	X_1X_2	X_1Y	X_2^2	X_2Y
	2	1	2	1	2	2	4	4
	2	2	3	4	6	4	9	6
	3	2	1	4	2	6	1	3
	3	5	5	25	25	15	25	15
	5	4	6	16	24	20	36	30
	6	5	4	25	20	30	16	24
Column summation	21	19	21	75	79	77	91	82

graphical relationship between each pair of variables; in general, linear associations are evident, although the scatter about the apparent linear relationships is significant. The correlation between the two predictor variables appears to be high relative to the degree of linearity between the criterion variable and the predictor variables.

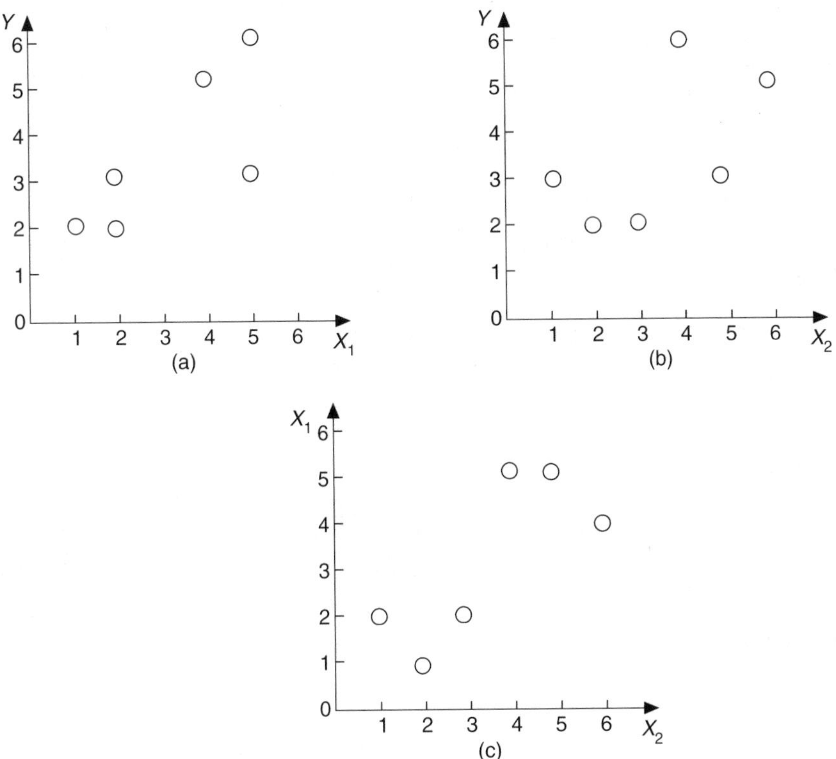

FIGURE 13-1 Graphs of variables for computational example: (a) Y vs. X_1; (b) Y vs. X_2; (c) X_1 vs. X_2.

After examining the plots, the correlation matrix ($\boldsymbol{R}$) should be computed:

$$\boldsymbol{R} = \begin{array}{c|ccc} & X_1 & X_2 & Y \\ \hline X_1 & 1.000 & 0.776 & 0.742 \\ X_2 & & 1.000 & 0.553 \\ Y & & & 1.000 \end{array} \tag{13-24}$$

The correlation coefficient of 0.776 between X_1 and X_2 indicates that the intercorrelation is high. The determinant of the intercorrelation matrix equals 0.398, which should be considered as moderate; that is, it suggests that we should carefully examine the resulting regression equation for rationality.

The summations shown in Table 13-3 can be used to develop the normal equations (Equation 13-7), as follows:

$$19b_0 + 75b_1 + 79b_2 = 77 \tag{13-25a}$$

$$21b_0 + 79b_1 + 91b_2 = 82 \tag{13-25b}$$

$$6b_0 + 19b_1 + 21b_2 = 21 \tag{13-25c}$$

The normal equations can be solved to obtain values for the coefficients of the following equation:

$$\hat{Y} = 1.30 + 0.75X_1 - 0.05X_2 \tag{13-26}$$

Similarly, the correlation matrix could be used to develop the normal equations for the standardized model:

$$\begin{bmatrix} 1.000 & 0.776 \\ 0.776 & 1.000 \end{bmatrix} \begin{bmatrix} t_1 \\ t_2 \end{bmatrix} = \begin{bmatrix} 0.742 \\ 0.553 \end{bmatrix} \tag{13-27}$$

which yields:

$$t_1 + 0.776t_2 = 0.742 \tag{13-28a}$$

$$0.776t_1 + t_2 = 0.553 \tag{13-28b}$$

Solving for t_1 and t_2 yields the following standardized equation:

$$Z_Y = 0.786Z_1 - 0.057Z_2 \tag{13-29}$$

Both the raw score equation (Equation 13-26) and the standardized equation (Equation 13-29) indicate that as X_2 increases, Y decreases; this is in conflict with both the correlation coefficient and the graphical relationship shown in Figure 13-1. This irrationality is the result of the high intercorrelation between the predictor variables.

The correlation coefficient can be computed from either Equation 13-20 or 13-21. Equation 13-21 yields:

$$R^2 = t_1 r_{1Y} + t_2 r_{2Y} \tag{13-30a}$$

$$= 0.786(0.742) + (-0.057)(0.553) = 0.552 \tag{13-30b}$$

Therefore, R equals 0.743. Predictor X_1 explained 55.1% (i.e., 0.742^2) of the variation in Y by itself. Thus, the use of X_2 is not justified because it does not make a significant contribution to the explained variation and, even more important, it results in an irrational model. The standard error of estimate is:

$$S_e = S_Y \sqrt{\left(\frac{n-1}{n-p-1}\right)(1-R^2)} = 1.643\sqrt{\left(\frac{5}{3}\right)(1-0.743^2)} \tag{13-31a}$$

$$S_e = 1.643(0.864) = 1.420 \tag{13-31b}$$

Thus, the two-predictor model reduced the error variance by 25.3%; the standard error of estimate is about 86% of the standard deviation of the criterion variable. For the one-predictor model with X_1 as the predictor variable, the standard error of estimate would be 1.231, which is less than the standard error of estimate for the two-predictor model; this occurs because the ratio $(n - 1)/(n - p - 1)$ is much less for the one-predictor model. This example illustrates the importance of computing the standard error of estimate when the sample size is small. The standard error provides a more realistic measure of the goodness of fit than the correlation coefficient does.

13.4. POLYNOMIAL REGRESSION ANALYSIS

In most empirical analyses, linear models are attempted first because of the relative simplicity of linear analysis. Also, linear models are easily applied, and the statistical reliability is easily assessed.

Linear models may be rejected because of either theoretical considerations or empirical evidence. Specifically, theory may suggest a nonlinear relationship between a criterion variable and one or more predictor variables; for example, biological growth curves used by environmental engineers are characterized by nonlinear forms. When a model structure cannot be identified by theoretical considerations, empirical evidence can be used in model formulation and may suggest a nonlinear form. For example, the hydrologic relationship between peak discharge and the drainage area of a watershed has been found to be best represented by a log–log equation, which is frequently referred to as a *power model*. A nonlinear form may also be suggested by the residuals that result from a linear analysis; that is, if a linear model produces nonrandomly distributed residuals (i.e., a bias is apparent), the underlying assumptions are violated. A nonlinear functional form may produce residuals that satisfy the assumptions that underlie the principle of least squares.

One very common use of nonlinear models is the fitting of existing design curves that have significant nonlinearity. The equations are then used as part of computer software that performs a design procedure that, in the past, was solved manually.

13.4.1. Structure of Polynomial Models

Linear models were separated into bivariate and multivariate; the same separation is applicable to nonlinear models. It is also necessary to separate nonlinear models on the basis of the functional form. Although polynomial and power models are the most frequently used nonlinear forms, it is important to recognize that other model structures are available and may actually provide the correct structure. In addition to the power and polynomial structures, forms such as a square root, exponential, and logarithmic may provide the best fit to a set of data.

A bivariate polynomial of order p has the following form:

$$\hat{Y} = b_0 + b_1 X + b_2 X^2 + \ldots + b_p X^p \qquad (13\text{-}32)$$

A second-order polynomial with two predictor variables X_1 and X_2 has the form:

$$\hat{Y} = b_0 + b_1 X_1 + b_2 X_2 + b_3 X_1^2 + b_4 X_2^2 + b_5 X_1 X_2 \qquad (13\text{-}33)$$

In these two relationships, Y is the criterion variable, X is the predictor variable in the bivariate case, X_i is the ith predictor variable in the multivariate case, b_j ($j = 0, 1, \ldots, p$) is the jth regression coefficient, and p is the order of the polynomial. Other polynomial forms of higher order or with more predictor variables can be similarly developed.

13.4.2. Calibration of Polynomial Models

Polynomial models are widely used nonlinear forms because they can be transformed in a way that makes it possible to use the principle of least squares. Although the transformation to a linear structure is desirable from the standpoint of calibration, it has important consequences in terms of assessing the goodness-of-fit statistics.

The bivariate polynomial of Equation 13-32 can be calibrated by forming a new set of predictor variables:

$$W_i = X^i \quad \text{for } i = 1, 2, \ldots, p \qquad (13\text{-}34)$$

This results in the model:

$$\hat{Y} = b_0 + b_1 W_1 + b_2 W_2 + \ldots + b_p W_p \tag{13-35}$$

The coefficients b_j can be estimated using a standard linear multiple regression analysis.

The multivariate polynomial of Equation 13-33 can also be solved using a multiple regression analysis for a set of transformed variables. For the model given by Equation 13-33, the predictor variables are transformed as follows:

$$W_1 = X_1 \tag{13-36a}$$

$$W_2 = X_2 \tag{13-36b}$$

$$W_3 = X_1^2 \tag{13-36c}$$

$$W_4 = X_2^2 \tag{13-36d}$$

$$W_5 = X_1 X_2 \tag{13-36e}$$

The revised model has the form:

$$\hat{Y} = b_0 + \sum_{j=1}^{5} b_j W_j \tag{13-36f}$$

It is important to note that the polynomial models do not require a transformation of the criterion variable.

The model given by Equation 13-33 has only two predictor variables (i.e., X_1 and X_2) and is a second-order equation. In practice, a model may have more predictor variable and may be of higher order. In some cases, the interaction terms (i.e., $X_i X_j$) may be omitted to decrease the number of coefficients that must be calibrated. However, if the interaction terms are omitted when they are actually significant, the goodness-of-fit statistics may suffer, unless the variation is explained by the other terms in the model; when this occurs, the coefficients may lack physical significance.

13.4.3. Analysis of Variance for Polynomial Models

Just as the analysis of variance (ANOVA) is used to examine the statistical adequacy of linear models, it can be used to test whether or not the polynomial models are adequate. Both the total and partial F statistics can be used. Table 13-4 provides a summary of the ANOVA test for first-, second-, and third-order polynomials that include a single predictor variable. However, the formulas are easily adapted for polynomial models involving more than one predictor variable, including analyses involving interaction terms. Although the partial F test can be used sequentially to select terms for the final polynomial model, it is always important to remember that the selection of a level of significance is a somewhat arbitrary decision that can affect the resulting model.

TABLE 13-4 ANOVA Table for Polynomial Analysis

Model	Correlation coefficient	F test	H_0	Sum of squares	Degrees of freedom	Test statistic: F
First-order (linear) $Y = a + b_1X$	R_1	Total and partial	$\beta_1 = 0$	Regression: R_1^2 Error: $1 - R_1^2$	1 $n - 2$	$F = \dfrac{R_1^2}{\left(\dfrac{1 - R_1^2}{n - 2}\right)}$
Second-order (quadratic) $Y = a + b_1X + b_2X^2$	R_2	Total	$\beta_1 = \beta_2 = 0$	Regression: R_2^2 Error: $1 - R_2^2$	2 $n - 3$	$F = \dfrac{\frac{1}{2}R_2^2}{\left(\dfrac{1 - R_2^2}{n - 3}\right)}$
		Partial	$\beta_2 = 0$	Regression: $R_2^2 - R_1^2$ Error: $1 - R_2^2$	1 $n - 3$	$F = \dfrac{R_2^2 - R_1^2}{\left(\dfrac{1 - R_2^2}{n - 3}\right)}$
Third-order (cubic) $Y = a + b_1X + b_2X^2 + b_3X^3$	R_3	Total	$\beta_1 = \beta_2 = \beta_3 = 0$	Regression: R_3^2 Error: $1 - R_3^2$	3 $n - 4$	$F = \dfrac{\frac{1}{3}R_3^2}{\left(\dfrac{1 - R_3^2}{n - 4}\right)}$
		Partial	$\beta_3 = 0$	Regression: $R_3^2 - R_2^2$ Error: $1 - R_3^2$	1 $n - 4$	$F = \dfrac{R_3^2 - R_2^2}{\left(\dfrac{1 - R_3^2}{n - 4}\right)}$

Example 13-3: One-Predictor Polynomial of Sediment Yield vs. Slope

A computer program was used to derive first-, second-, and third-order polynomials that regress sediment yield on watershed slope. The correlation matrix is given in Table 13-5. It is evident that the intercorrelation coefficients are very significant and should lead to irrational regression coefficients; furthermore, because of the high intercorrelation it is unlikely that the explained variance will improve significantly as the order of the equation increases. Table 13-6 provides a summary of the correlation coefficients, the standardized partial regression coefficients, and the partial regression coefficients. The second- and third-order models do not provide a significant increase in the explained variation when compared with the linear model.

The ANOVA results are given in Tables 13-7 and 13-8. The total F tests indicate that all three models are statistically significant at the 5% level of significance. However, the partial F tests indicate that including the quadratic term in the equation does not improve the statistical goodness of fit, as the computed F value is less than the critical value.

TABLE 13-5 Correlation Matrix: Sediment Yield (Y) vs. Slope (S)

	S	S^2	S^3	Y
S	1.000	0.908	0.791	0.443
S^2		1.000	0.972	0.443
S^3			1.000	0.428
Y				1.000

TABLE 13-6 Summary of Polynomial Analyses for Sediment Yield vs. Slope Models

		Intercept	Partial regression coefficients			Standardized partial regression coefficients		
Model	R	b_0	b_1	b_2	b_3	t_1	t_2	t_3
First-order	0.443	0.220	0.0263	—	—	0.443	—	—
Second-order	0.454	0.323	0.0136	2.31×10^{-4}	—	0.229	0.235	—
Third-order	0.499	0.042	0.0865	–0.0031	3.43×10^{-5}	1.456	–3.164	2.35

TABLE 13-7 ANOVA: Total F Test

Model	R^2	$1 - R^2$	First degrees of freedom	Second degrees of freedom	Total F	$F_{0.05}$	Decision
First-order	0.196	0.804	1	35	$\frac{0.196/1}{0.804/35} = 8.53$	4.12	Reject H_0
Second-order	0.206	0.794	2	34	$\frac{0.206/2}{0.794/34} = 4.41$	3.28	Reject H_0
Third-order	0.249	0.751	3	33	$\frac{0.249/3}{0.751/33} = 3.65$	2.90	Reject H_0

TABLE 13-8 ANOVA: Partial F Tests

Model	ΔR^2	$1 - R^2$	First degrees of freedom	Second degrees of freedom	Partial F	$F_{0.05}$	Decision
First-order	0.196	0.804	1	35	$\frac{0.196/1}{0.804/35} = 8.53$	4.12	Reject H_0
Second-order	0.010	0.794	1	34	$\frac{0.01/1}{0.794/34} = 0.43$	4.13	Accept H_0
Third-order	0.043	0.751	1	33	$\frac{0.043/1}{0.751/33} = 1.89$	4.14	Accept H_0

13.5. REGRESSION ANALYSIS OF POWER MODELS

The power model is a widely used nonlinear model. The bivariate model between criterion variable Y and predictor variable X is:

$$\hat{y}_i = b_0 X_i^{b_1} \tag{13-37}$$

in which $\hat{y}_i$ is the predicted value of Y for the value of x_i, b_0 is the intercept coefficient, and b_1 is the power coefficient. When Y is affected by more than one predictor variable, the multiple-predictor power model is used:

$$\hat{y}_i = b_0 X_{i1}^{b_1} X_{i2}^{b_2} \ldots X_{ip}^{b_p} \tag{13-38}$$

in which p is the number of predictor variables, and b_j is the slope coefficient for predictor x_{ij}.

13.5.1. Fitting a Power Model

The bivariate power model of Equation 13-37 can be calibrated by forming the following set of variables:

$$\hat{Z} = \ln \hat{Y} \tag{13-39a}$$

$$c = \ln b_0 \tag{13-39b}$$

$$W = \ln X \tag{13-39c}$$

These transformed variables form the following linear equation, which can be calibrated using bivariate linear regression analysis:

$$\hat{Z} = c + b_1 W \tag{13-40}$$

After values of c and b_1 are obtained, the coefficient b_0 can be determined using:

$$b_0 = e^c \tag{13-41}$$

Base 10 logarithms can also be used; in such a case, Equations 13-39a and 13-39c will use the base 10 logarithm and Equation 13-41 will use a base 10 rather than e.

The multivariate power model of Equation 13-38 can be evaluated by making a logarithmic transformation of both the criterion and the predictor variables:

$$\hat{Z} = \ln \hat{Y} \tag{13-42}$$

$$c = \ln b_0 \tag{13-43}$$

$$W_i = \ln X_i \quad \text{for } i = 1, 2, \ldots, p \tag{13-44}$$

The resulting model has the form:

$$\hat{Z} = c + \sum_{j=1}^{p} b_j W_j \tag{13-45}$$

The coefficients of Equation 13-45 can be evaluated using a multiple regression analysis. The value of b_0 can be determined by making the transformation of Equation 13-41. Again, a base 10 transformation can be used rather than a natural log transformation.

13.5.2. Goodness of Fit

Because most computer programs that perform multiple regression analyses include goodness-of-fit statistics as part of the output, it is important to recognize the meaning of these statistics. For nonlinear models in which the criterion variable $\hat{Y}$ is not transformed, the goodness-of-fit statistics are valid indicators of the reliability of the model; however, when the criterion variable is transformed, such as is necessary for the power model form, the principle of least squares is applied in the log–log space. As a result, the residuals that are used to compute the standard error of estimate, and therefore the correlation coefficient, are measured in the domain of the logarithm of $\hat{Y}$ and not the $\hat{Y}$ domain. Therefore, the goodness-of-fit statistics are not necessarily a reliable indicator of model reliability, especially as engineering decisions are made to predict Y and not $\log(Y)$. Therefore, when a model requires a transformation of the criterion variable $\hat{Y}$ in order to calibrate the coefficients, the goodness-of-fit statistics that are included with the multiple regression output should not be used as measures of reliability. Instead, values for the goodness-of-fit statistics should be recomputed using values of $\hat{Y}$ rather than $\log(\hat{Y})$. The correlation coefficient and the standard error of estimate should be computed using Equations 13-20 and 13-22, respectively. In summary, when an equation is calibrated with the criterion variable transformed, such as the log–log space, the least-squares concepts apply only to the transformed space, not to the measurement nontransformed space. The sum of the residuals based on the logarithms of Y will equal zero; however, the sum of the residuals of the transformed values of Y will not equal zero, and the sum of the squares of the errors may not be a minimum in the Y domain even though it is in the log(Y) domain. Furthermore, the assumption of a constant variance may also not be valid. Because these basic assumptions of regression are not valid in the X–Y space, many practitioners object to data transformations. However, in spite of these theoretical considerations, the transformations may provide reasonable estimates in the X–Y space.

13.5.3. Additional Model Forms

In addition to the polynomial and power models, other forms can provide good approximations to the underlying population relationship between two or more variables. An exponential model has the form:

$$\hat{Y} = b_0 e^{b_1 X} \tag{13-46}$$

in which b_0 and b_1 are coefficients requiring values by fitting to data. Values for b_0 and b_1 can be obtained by linear bivariate regression after taking the logarithms of both sides of Equation 13-46 as follows:

$$\ln \hat{Y} = \ln b_0 + b_1 X \tag{13-47}$$

In this case, $\hat{Y}$ is transformed; thus, the correlation coefficient and standard error of estimate apply to the $\ln \hat{Y}$ and not to $\hat{Y}$. Also, the intercept coefficient obtained from the regression analysis for Equation 13-47 must be transformed for use with Equation 13-46.

A logarithmic curve can also be used to fit values of Y and X:

$$\hat{Y} = b_0 + b_1 \ln X \tag{13-48}$$

In this case, $\hat{Y}$ is not transformed, so the values of the correlation coefficient and the standard error of estimate are valid indicators of the goodness of fit in the $\hat{Y}$ space.

13.6. APPLICATIONS

13.6.1. One-Predictor Polynomial of Evaporation vs. Temperature

The evaporation data were used to derive polynomials for the regression of evaporation on temperature. The correlation matrix, which is shown in Table 13-9, shows very high intercorrelation. Table 13-10 provides a summary of the regression analyses.

TABLE 13-9 Correlation Matrix: Evaporation (*E*) vs. Temperature (*T*)

	T	T^2	T^3	E
T	1.000	0.995	0.981	0.581
T^2		1.000	0.996	0.602
T^3			1.000	0.627
E				1.000

TABLE 13-10 Summary of Regression Analysis

Model	R	S_e^*	t_1	t_2	t_3	b_0	b_1	b_2	b_3
First-order	0.581	0.0763	0.581	—	—	–0.114	0.00383	—	—
Second-order	0.648	0.0715	2.160	2.76	—	0.425	–0.01420	0.000143	—
Third-order	0.654	0.0711	3.110	–8.44	5.98	–0.239	0.02050	–0.000439	3.14×10^{-6}

* S_y = 0.0936 in./day.

Whereas the linear model explains approximately 34% of the variation in evaporation, the higher-order terms provide only a marginal increase in the explained variation. The quadratic model increases the explained variance by about 8%. The marginal improvement is also evident from the small decrease in the standard error of estimate. In judging the rationality of a model, it is of value to examine the predicted values of evaporation for values of the predictor variable that are likely to be observed. The three models were used to estimate the daily evaporation depth for selected temperatures (Table 13-11). Because water freezes at 32°F, evaporation rates near zero are expected; only the linear model gives a rational estimate at 32°F. Estimates made using the second-order model decrease with increases in temperature for temperatures up to about 50°F. Both the second- and third-order models provide especially high rates at 100°F. In summary, the high intercorrelations lead to irrational estimates when the higher-order forms are used.

Tables 13-12 and 13-13 summarize the ANOVA. The total F test (Table 13-12) indicates that all models are statistically significant. Similarly, the partial F tests (Table 13-13) indicate that each term is significant, and thus the third-order equation should be selected.

This example illustrates that different criteria for model selection may conflict — for example, the partial F tests and the standardized partial regression coefficients.

TABLE 13-11 Predicted Evaporation Rates (in./day)

	Temperature (°F)			
Model	**32**	**50**	**75**	**100**
First-order	0.009	0.078	0.173	0.269
Second-order	0.117	0.073	0.164	0.435
Third-order	0.070	0.081	0.154	0.561

TABLE 13-12 ANOVA: Total F Tests

Model	R^2	$1 - R^2$	First degrees of freedom	Second degrees of freedom	Total F	$F_{0.05}$	Decision
First-order	0.338	0.662	1	352	$\frac{0.338/1}{0.662/352} = 179.5$	3.86	Reject H_0
Second-order	0.420	0.580	2	351	$\frac{0.420/2}{0.580/351} = 127.1$	3.02	Reject H_0
Third-order	0.428	0.572	3	350	$\frac{0.428/3}{0.572/350} = 87.1$	2.62	Reject H_0

TABLE 13-13 ANOVA: Partial F Tests

Model	ΔR^2	$1 - R^2$	First degrees of freedom	Second degrees of freedom	Partial F	$F_{0.05}$	Decision
First-order	0.338	0.662	1	352	$\frac{0.338/1}{0.662/352} = 179.7$	3.86	Reject H_0
Second-order	0.082	0.580	1	351	$\frac{0.082/1}{0.580/351} = 49.6$	3.86	Reject H_0
Third-order	0.008	0.572	1	350	$\frac{0.008/1}{0.572/350} = 4.9$	3.86	Reject H_0

For this example, it appears that the ANOVA is an inadequate criterion for model selection. Quite possibly, the 5% level of significance is not an adequate criterion; after all, use of the 5% level is based solely on convention and not on a rational analysis of the relationship between the level of significance, the sample size, and the physical separation criterion.

13.6.2. Single-Predictor Power Model

A data set that consists of five observations will be used to illustrate the fitting of a power model that includes a single predictor variable. The data set is given in Table 13-14. The values were converted to natural logarithms, and the coefficients of Equation 13-40 were calibrated using least squares:

$$\ln \hat{Y} = -0.1 + 1.9 \ln X \tag{13-49}$$

A correlation coefficient of 0.9851 was computed for Equation 13-49, which can be retransformed to the X–Y coordinate space:

$$\hat{Y} = 0.9048 X^{1.9} \tag{13-50}$$

TABLE 13-14 Data Base for Single-Predictor Power Model

				Using Equation 13-51		Using Equation 13-50	
X	Y	ln$X$	ln$Y$	$\hat{Y}$	e	ln $\hat{Y}$	e_e
2.718	7.389	1	2	–214.54	–221.929	6.048	–1.341
7.389	20.086	2	3	53.18	33.095	40.445	20.359
20.086	403.429	3	6	780.93	377.499	270.427	–133.002
54.598	2980.960	4	8	2759.03	–221.927	1807.960	–1173.000
148.413	8103.080	5	9	8136.00	32.923	12087.400	3984.310
					$\sum e = 0$		$\sum e_e \neq 0$

A linear analysis of the values of X and Y provides a correlation coefficient of 0.9975 and the following regression equation:

$$\hat{Y} = 370.3301 + 57.3164X \tag{13-51}$$

The residuals for Equations 13-50 and 13-51 are given in Table 13-14. Whereas the residuals for Equation 13-51 have a sum of zero, the sum of the residuals for Equation 13-50 is not equal to zero; however, the sum of the residuals for Equation 13-49 will equal zero in the natural-log space. The nonzero sum of the residuals indicates that the model is biased.

The correlation coefficient and standard error of estimate that are computed for the model in the natural-log space are not valid indicators of the accuracy of the estimates in the domain of the nontransformed criterion variable. For example, the correlation coefficient of 0.9851 computed with Equation 13-49 is not a measure of the accuracy of Equation 13-50. Values of Y that were estimated with Equation 13-50 are given in Table 13-14, and the resulting residuals are denoted as e_e. If the separation of variation concept is applied to Equation 13-50, the components are:

$$\sum_{i=1}^{n}(y_i - \bar{Y})^2 = \sum_{i=1}^{n}(y_i - \hat{y}_i)^2 + \sum_{i=1}^{n}(\hat{y}_i - \bar{Y})^2 \tag{13-52}$$

$$48{,}190{,}499.4 \neq 17{,}276{,}427.6 + 100{,}525{,}199.1 \tag{13-53}$$

The separation of variation is not valid for Equation 13-50 because the model was calibrated by taking logarithms. The separation of variation does, however, apply to the logarithms. It is evident that if the correlation coefficient was computed as the ratio of the explained variation to the total variation (Equation 13-20), then from the values of Equation 13-53 the correlation coefficient would be greater than 1.0, which is not rational. Thus, the standard error of estimate, rather than the correlation coefficient, should be used as a measure of accuracy:

$$S_e = \left(\frac{17{,}276{,}427.6}{5-2}\right)^{0.5} = 2399.75 \tag{13-54}$$

The standard deviation of the criterion variable is 3470.97; therefore, the power model is not highly accurate, and the computed correlation for the log model (i.e., 0.9851) certainly does not reflect the accuracy of the model for predicting Y.

13.6.3. Multivariate Power Model

Sediment yield data will be used to illustrate a power model that includes more than one predictor variable. The small particle size index cannot be included as a predictor variable because some of the observations equal zero; thus, the model regresses sediment yield (Y) on three predictor variables: the precipitation/temperature ratio (P), the slope (S), and the soil index (I). The data are given in Table 13-15. The resulting regression equation in power form is:

$$\ln \hat{Y} = -4.563 - 1.123 \ln P + 0.8638 \ln S - 0.1089 \ln I \tag{13-55}$$

Transforming Equation 13-55 to the X–Y space yields:

$$\hat{Y} = 0.01043 P^{-1.123} S^{0.8638} I^{-0.1089} \tag{13-56}$$

The computer program provided a correlation coefficient of 0.725 and a standard error estimate of 1.019; however, these are for Equation 13-55 and may not be accurate indications of the accuracy of Equation 13-56. The standard error of estimate is expressed in logarithmic units. The standard error of estimate as measured in the X–Y space was computed as 0.595, which can be compared with a standard deviation of 0.817 for the 37 observations of sediment yield.

For a comparison, the following linear model was calibrated using the same three predictor variables:

$$\hat{Y}_{\text{Linear}} = 0.8920 - 1.428P + 0.03664S - 0.01308I \tag{13-57}$$

This resulted in a multiple correlation coefficient of 0.720 and a standard error of estimate of 0.592. Thus, for this case, there is no significant difference in the expected accuracy of the linear and power models.

13.6.4. Estimating Breakwater Costs

In Section 12.9.2, a bivariate model between the construction cost of a breakwater and length of the breakwater was developed. It was found to be inaccurate, with the suggestion given that other variables associated with cost were probably better determinants of the cost.

Table 13-16 includes the original measurements of length and cost along with a second predictor variable, the average depth of the breakwater. The correlation matrix is:

	X_1	X_2	Y
X_1: length	1.000	–0.229	0.176
X_2: depth		1.000	0.905
Y: cost			1.000

While length and depth are poorly correlated, depth and cost have a high correlation, with R^2 equal to 0.819.

TABLE 13-15 Sediment Yield Data*

P	*S*	*I*	*Y*
0.135	4.000	40.	0.140
0.135	1.600	40.	0.120
0.101	1.900	22.	0.210
0.353	17.600	2.	0.150
0.353	20.600	1.	0.076
0.492	31.000	0.	0.660
0.466	70.000	23.	2.670
0.466	32.200	57.	0.610
0.833	41.400	4.	0.690
0.085	14.900	22.	2.310
0.085	17.200	2.	2.650
0.085	30.500	15.	2.370
0.193	6.300	1.	0.510
0.167	14.000	3.	1.420
0.235	2.400	1.	1.650
0.448	19.700	3.	0.140
0.329	5.600	27.	0.070
0.428	4.600	12.	0.220
0.133	9.100	58.	0.200
0.149	1.600	28.	0.180
0.266	1.200	15.	0.020
0.324	4.000	48.	0.040
0.133	19.100	44.	0.990
0.133	9.800	32.	0.250
0.133	18.300	19.	2.200
0.356	13.500	58.	0.020
0.536	24.700	62.	0.030
0.155	2.200	27.	0.036
0.168	3.500	24.	0.160
0.673	27.900	22.	0.370
0.725	31.200	10.	0.210
0.275	21.000	64.	0.350
0.150	14.700	37.	0.640
0.150	24.400	29.	1.550
1.140	15.200	11.	0.090
1.428	14.300	1.	0.170
1.126	24.700	44.	0.170

* Y = sediment yield, X_1 = climate variable, X_2 = watershed slope variable, and X_3 = soil aggregation variable.

Source: Flakman, E.M., *J. Hydraulic Div. ASCE*, 98(HY12), 2073, 1972.

TABLE 13-16 Data for Breakwater Cost Estimation

Length, X_1 (ft)	Depth, X_2 (ft)	Cost, Y ($\$ \times 10^3$)
450	6.3	230
625	9.6	490
670	3.1	150
730	10.1	540
810	5.4	280
880	3.5	120
1020	6.5	380
1050	2.9	170
1100	8.8	530
1175	3.9	290
1230	7.4	460
1300	3.3	230
1350	7.2	470
1510	4.6	360

A multiple regression analysis yields the following equation:

$$\hat{Y} = -199.8 + 0.1904X_1 + 58.72X_2 \tag{13-58}$$

The multiple correlation coefficient equals 0.987, thus, $R^2 = 0.974$, which indicates that the two-predictor model provides accurate estimates of the cost. The standard error of estimate is $\$25.3 \times 10^3$, with $S_e/S_y = 0.174$. The F statistic of 210 is exceptionally large and well beyond any F values for commonly used level of significance. The standard error ratios for the regression coefficients are small: $S_{e,b_1}/b_1 = 0.122$ and $S_{e,b_2}/b_2 = 0.050$. Overall, the goodness-of-fit statistics suggest an accurate model.

The standardized partial regression coefficients are $t_1 = 0.405$ and $t_2 = 0.998$. While t_2 is below 1, it is still very large.

The intercept is quite large in magnitude. While none of the predicted costs is negative, it is still possible because the intercept is negative. For example, at a length of 450 ft and depth of 2 ft, the estimated cost would be negative. Equation 13-58 should be used cautiously beyond the range of the data: $450 < X_1 < 1510$ ft and $2.9 < X_2 < 10.1$ ft.

It is interesting to compare the two-predictor multiple regression model with both one-predictor models. The length vs. cost model was discussed in Section 12.9.2. It provided very poor accuracy, $R^2 = 0.031$. The depth vs. cost model would explain 81.9% of the variation in cost (0.905^2). Because the multiple regression model explains 97.4%, adding the length to the model increases the explained variance by 15.5% (97.4 − 81.9); this is a significant increase, so the two-predictor model is preferable to either of the one-predictor models. Also, when the depth is added to the equation, the importance of the length increases. The t for the bivariate model is 0.176, and $t_1 = 0.405$ for the multiple regression model.

TABLE 13-17 Regression Analysis

X_1	X_2	Y	$\hat{Y}$	e
1	0	0.9	0.42	–0.48
2	0	1.8	1.87	0.07
3	0	2.4	3.32	0.92
4	0	4.3	4.76	0.46
5+	0	6.6	6.21	–0.39
1	1	1.9	1.13	–0.77
2	1	2.7	2.58	–0.12
3	1	3.6	4.03	0.43
4	1	5.4	5.47	0.07
5+	1	7.7	6.92	–0.78
1	2+	2.1	1.84	–0.26
2	2+	3.1	3.29	0.19
3	2+	3.7	4.74	1.04
4	2+	5.9	6.18	0.28
5+	2+	8.3	7.63	–0.67

13.6.5. Trip Generation Model

For a regional-planning study, a transportation engineer collects data on traffic. The data of Table 13-17 give measurements of nonwork, home-based, trip-production rates for suburban, medium-density households (Y) as a function of the number of persons per household (X_1) and the number of vehicles per household (X_2). In this case, the predictor variables are integers.

The correlation matrix for the sample of 15 is given by:

	X_1	X_2	Y
X_1	1.000	0.000	0.931
X_2		1.000	0.264
Y			1.000

The intercorrelation between the two predictor variables is zero because the data were collected on a grid. One advantage of having no intercorrelation is that the potential for irrational regression coefficients is minimized. The predictor–criterion correlation coefficient for the number of persons per household is large ($R^2 = 0.868$), while that for the number of vehicles per household is small ($R^2 = 0.070$). This may suggest that the bivariate equation $\hat{Y} = f(X_1)$ may be as accurate as the multiple regression model $\hat{Y} = f(X_1, X_2)$.

A multiple regression analysis produced the following equation:

$$\hat{Y} = -1.023 + 1.4467X_1 + 0.7100X_2 \tag{13-59}$$

The correlation coefficient is 0.968, which gives $R^2 = 0.937$. Because the intercorrelation between X_1 and X_2 is zero, the sum of the squares of the two predictor–criterion correlation coefficients equals the coefficient of multiple determination:

$$R^2 = R_{1Y}^2 + R_{2Y}^2 \tag{13-60a}$$

$$0.937 = (0.931)^2 + (0.264)^2 \tag{13-60b}$$

This equality holds only when there is no intercorrelation. It also shows that the multiple regression model explains 7% more variation than the bivariate model based solely on X_1.

The standard error of estimate is 0.6153, which gives $S_e/S_y = 0.271$. The total F statistic is 89.6. For $k = \nu_1 = 2$ and $u = \nu_2 = 12$, this is statistically significant at even the smallest level of significance. Both of these statistics support the conclusion suggested by the correlation coefficient that the model is highly accurate.

The standardized partial regression coefficients are $t_1 = 0.931$ and $t_2 = 0.264$. Because the intercorrelation coefficient is 0, the t values equal the predictor–criterion correlation coefficients. Again, the number of persons per household is shown to be the more important predictor.

The intercept of Equation 13-59 is –1.023, which is irrational, as a value of 0 would be expected. If both X_1 and X_2 are 0, then Y should be 0. Furthermore, the magnitude of the intercept is large; that is, it is approximately 25% of the mean of Y and larger then the value of Y for $X_1 = 1$ and $X_2 = 0$. This suggests that another model structure may be needed to realistically represent the data.

The residuals are given in Table 13-17, and they also suggest that the linear model of Equation 13-59 is an incorrect structure. There are very noticeable local biases. For each value of X_2 the residuals for $X_1 = 1$ and $X_1 = 5$ are negative; additionally, for each value of X_2 the residuals for $X_1 = 3$ and $X_1 = 4$ are positive. The trends for each value of X_2 suggest that a nonlinear structure would be more appropriate.

13.6.6. Estimation of the Reaeration Coefficient

To estimate the biochemical oxygen demand of a stream, the environmental engineer needs an estimate of the reaeration coefficient. The reaeration coefficient varies with the stream velocity (X_1, in meters per second), the water depth (X_2, in meters), and the water temperature (X_3, in °C). Table 13-18 gives a set of 21 measurements of the three predictor variables and estimated reaeration coefficients (Y). The correlation matrix is:

	X_1	X_2	X_3	Y
X_1	1.000	0.839	0.139	0.436
X_2		1.000	0.240	0.456
X_3			1.000	0.621
Y				1.000

TABLE 13-18 Multiple Regression Analysis of the Stream Reaeration Coefficient (Y) on Stream Velocity (X_1, m/sec), Water Depth (X_2, m), and Water Temperature (X_3, °C)

X_1	X_2	X_3	Y	$\hat{Y}$	e
1.4	1.24	10	2.16	2.69	0.80
0.6	0.87	9	2.50	2.57	0.07
2.2	1.87	13	2.54	3.55	1.01
1.5	0.99	11	2.60	3.06	0.46
0.5	0.85	14	2.71	2.97	0.26
1.6	0.97	8	3.00	2.83	–0.17
1.3	1.12	19	3.02	3.70	0.68
0.8	1.21	11	3.03	2.85	–0.18
1.6	1.64	14	3.27	3.41	0.14
2.3	1.91	6	3.42	2.98	–0.44
1.0	0.70	15	3.42	3.21	–0.21
1.9	1.62	12	3.45	3.34	–0.11
1.3	1.01	17	3.49	3.51	0.02
1.8	1.43	15	3.58	3.54	–0.04
1.7	1.05	15	3.61	3.47	–0.14
0.7	0.62	18	3.69	3.36	–0.33
2.2	1.58	12	3.85	3.43	–0.42
1.4	1.21	15	3.88	3.39	–0.49
1.8	2.07	20	3.99	4.04	0.05
1.5	1.59	16	4.51	3.55	–0.96
2.8	2.38	24	4.74	4.75	0.01

The intercorrelation is high ($|\boldsymbol{R}_{11}| = 0.276$) because of the high correlation between X_1 and X_2. The water temperature is not highly correlated with either velocity or depth. The regression analysis of the 21 sample values yields:

$$\hat{Y} = 1.508 + 0.3315X_1 + 0.09731X_2 + 0.08672X_3 \tag{13-61}$$

All the regression coefficients are rational in sign. Equation 13-61 results in a correlation coefficient of 0.715, which means that the equation explains 51.2% of the total variation. The standard error of estimate is 0.497 ($S_e/S_y = 0.758$), which suggests that the equation provides estimates of Y that are not highly accurate. For the ANOVA test on the null hypothesis H_0: $\beta_1 = \beta_2 = \beta_3 = 0$, and the computed F is 5.935 for $k = \nu_1 = 3$ and $u = \nu_2 = 17$. Critical F values for selected levels of significance (α) and the decisions are as follows:

α	F_α	Decision
0.001	8.73	Accept H_0
0.005	6.16	Accept H_0
0.010	5.18	Reject H_0
0.050	3.20	Reject H_0

Thus, for the commonly used levels of 1 and 5%, the null hypothesis would be rejected; however, for more restrictive levels of less than 1%, the null hypothesis would be accepted. The limiting probability for acceptance is about 0.6%.

Of the three predictor variables, the water temperature (X_3) would provide the highest R^2 for a bivariate model ($R^2 = 0.386$). The multiple regression model of Equation 13-61 explains 51.2%. Thus, the addition of X_1 and X_2 increases the explained variance by 12.6%. This means that more than one predictor variable should be used, but it does not necessarily mean that all three predictor variables are needed. It may be worthwhile developing two-predictor models: $\hat{Y} = f(X_1, X_3)$ and $\hat{Y} = f(X_2, X_3)$. Because X_1 and X_2 are highly correlated, both are probably not needed, so a two-predictor model may provide estimates that are just as accurate as the three-predictor model.

The standardized partial regression coefficients are $t_1 = 0.299$, $t_2 = 0.070$, and $t_3 = 0.563$. These suggest that X_3 is most important, followed by X_1, and finally X_2, which appears to be unimportant. This differs somewhat from the predictor–criterion correlation coefficients, which suggest that X_3 is most important, but that X_1 and X_2 are of equal importance. Because velocity is probably physically more important than depth, the t values probably give a more realistic assessment of the relative importance of the predictor variables than are given by the predictor–criterion correlation coefficients.

13.6.7. Estimating Slope Stability

Slope stability is very important to engineers who design with unrestrained slopes. The stability number is an index used by geotechnical engineers to assess the safety of a slope. Among others, the slope stability is a function of the angle of friction of the soil (X_1, in degrees) and the slope angle (X_2, in degrees). Twelve laboratory measurements were made of the stability number for selected values of the angle of friction and the slope angle (see Table 13-19). The correlation matrix for the twelve tests is:

	X_1	X_2	Y
X_1	1.000	0.520	0.415
X_2		1.000	–0.543
Y			1.000

The intercorrelation is moderate; the predictor–criterion correlation coefficients are not high, with the two values only representing 17.2 and 29.5% of the total variation in the stability number.

The least-squares regression equation is:

$$\hat{Y} = 0.04234 + 0.002842X_1 - 0.005093X_2 \tag{13-62}$$

TABLE 13-19 Multiple Regression Analysis of Slope Stability Values

X_1	X_2	Y	$\hat{Y}$	e
10	5	0.053	0.045	–0.008
10	10	0.005	0.020	0.015
20	5	0.081	0.074	–0.007
20	12	0.036	0.038	0.002
20	18	0.012	0.008	–0.004
30	6	0.102	0.097	–0.005
30	13	0.058	0.061	0.003
30	21	0.021	0.021	0.000
40	9	0.105	0.110	0.005
40	14	0.079	0.085	0.006
40	21	0.052	0.049	–0.003
40	26	0.027	0.024	–0.003

Note: X_1 = friction angle (degrees), X_2 = slope angle (degrees), and Y = stability number.

The multiple correlation coefficient for Equation 13-62 is 0.981, which means that the equation explains 96.2% of the variation in the measured values of the stability number. The standard error of estimate is 0.0072, which gives $S_e/S_y = 0.214$. The goodness-of-fit statistics suggest that Equation 13-62 provides accurate estimates of the stability number. The total F statistic is 115.1, with $k = \nu_1 = 2$ and $u = \nu_2 = 9$; because the critical value for a level of significance of 0.1% is 16.4, the ANOVA would suggest that the two slope coefficients of Equation 13-62 are significantly different from 0.

Note that in spite of the relatively low predictor–criterion correlation coefficients (i.e., 0.415 and –0.543), the total R^2 is very high. The intercorrelation actually enhances the reliability of the equation.

The goodness-of-fit statistics are revealing, but they do not give an entirely accurate assessment of the model. The standardized partial regression coefficients for X_1 and X_2 are $t_1 = 0.96$ and $t_2 = -1.04$. Because the latter is greater than 1, it is considered irrational and suggests that b_2 gives an unrealistic indication of the rate of change of Y with respect to X_2. The high accuracy indicated by the goodness-of-fit statistics suggests that the model is acceptable for estimation; however, the t_2 value suggests that the model may not be reliable upon extrapolation or for assessing model sensitivity.

The residuals (see Table 13-19) should be evaluated. Figure 13-2 shows a plot of the residuals as a function of the angle of friction (ordinate) and the slope angle (abscissa). While the sample size is probably too small to be very conclusive, there appears to be a region of positive residuals between friction angles of 9° and 15° and two regions of negative residuals, friction angles less than 6° and greater than 18°. This would suggest that the structure of the linear model may be incorrect.

The intercept also appears to be a problem. If both X_1 and X_2 are zero, then the stability number equals 0.042, which does not seem reasonable.

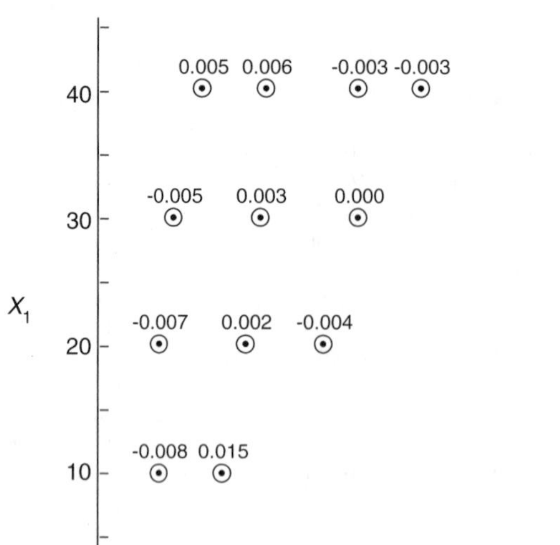

FIGURE 13-2 Residuals of a multiple regression model vs. predictor variables X_1 and X_2.

13.7. SIMULATION IN CURVILINEAR MODELING

The confidence intervals and hypothesis tests presented in Chapter 12 apply to the linear bivariate regression model with an intercept coefficient. They can be derived analytically because the model is quite simple. As the complexity of the model structure increases, it is less likely that statistics such as the sampling distribution of parameters can be derived analytically. Simulation provides a means of identifying a model of the sampling distributions of statistics related to complex models. In some cases, exact relationships may not be possible, but simulation can provide approximations.

Example 13-4: Simulation of the Power Model Intercept

Equation 12-56 applies to the linear model of Equation 12-28. When fitting the coefficients of Equation 13-37, Equation 12-56 does not apply. While Equation 12-56 could be applied to the logarithmic model of Equation 13-40, it would be based on the logarithms and it would reflect the variation of c and not b_0. The distribution of b_0 in Equation 13-37 in the y-space can be simulated. Equation 12-56 suggests that the distribution of b_0 would depend on the population standard error of estimate σ_e in the y-space, the distribution of the independent variable X, and the sample size, n. To develop a general relationship for computing S_{e, b_0} would require many simulation runs for various population values.

To demonstrate the simulation approach, the following procedure is used:

1. Assume that the random variable X is normally distributed with mean μ_X and variance σ_X^2.
2. Assume population standard error, ε.
3. Assume population values for the intercept β_0 and slope β_1 coefficients, where the population value of c in $\alpha = \ln_e \beta_0$.
4. Specify the seed for the random-number generator, if required, and add the number of simulation runs, Ns.
5. Iterate over the number of simulation runs.

   a. Generate n values of x_i using the model:

$$x_i = \mu_X + z_i \sigma_X \tag{13-63}$$

   in which z_i is a standard normal deviate.

   b. Make a log transform of the x_i values:

$$w_i = \ln_e x_i \tag{13-64}$$

   c. Generate logarithmic values of Z_i:

$$Z_i = \alpha + \beta_1 w_i + z_i \varepsilon \tag{13.65}$$

   d. Regress the sample Z_i values on the w_i values and obtain sample estimates c and b_1 for Equation 13-40.
   e. Transform the sample estimate of c to a sample estimate of b_0 using Equation 13-41.
   f. Use the value of b_0 to create a histogram that will show the distribution of b_0.
6. For the histogram of step 5f, identify the distribution of b_0.
7. Repeat the procedure for other values of the standard error in the y-space, sample sizes, and μ_X and σ_X to identify how each variable influences the distribution of b_0. Develop a general model that corresponds to Equation 12-56 for the linear model.

The above procedure was applied for one set of parameters, specifically $Ns = 10^5$, $\beta_0 = 3$, $\beta_1 = 0.5$, $\mu_X = 5$, $\sigma_X = 1$, $\sigma_e = 2$. The procedure was executed for sample sizes of 20, 25, and 30. The results are given in Table 13-20. Column 1 of Table 13-20 shows the cell bounds of the histogram on the sample coefficient b_0. An increment of 0.15 was used to provide 21 frequency cells. The values in column 2 are the proportion of the sample values of b_0 in each cell. Because a population value of 3 was used to generate the samples, the sample values should scatter around a central value of 3, which is the case, with 50.16% of the values above 3 and 49.84% below 3. This reflects sampling variation in the simulation process. While 50% of the values lie above and below the population, the sample probabilities shown in column 2 indicate a slightly nonsymmetric distribution with a greater proportion of the values in the upper tail of the distribution. While the confidence interval on b_0 for the linear model (Equation 12-58) uses the symmetric t distribution, it could not be applied for the power model unless interest was only in a very approximate confidence interval.

The probability histograms for samples sizes of 25 and 30 are given in columns 3 and 4 of Table 13-20. They center about the population means of 3, but the histograms show less scatter than that for sample sizes of 20. This is expected because

Table 13-20 Simulation of Sampling Distribution of the Intercept Coefficient β_0 of the Power Model

	$Ns = 10^5$	$Ns = 10^5$	$Ns = 10^5$	$Ns = 5 \times 10^4$	$Ns = 10^4$	$Ns = 10^5$
Sample b_0	$n = 20$	$n = 25$	$n = 30$	$n = 20$	$n = 20$	$n = 20$
	0.0000	0.0000	0.0000	0.0000	0.0000	0.0000
1.50						
	0.0000	0.0000	0.0000	0.0000	0.0000	0.0000
1.65						
	0.0000	0.0000	0.0000	0.0000	0.0000	0.0000
1.80						
	0.0000	0.0000	0.0000	0.0000	0.0000	0.0000
1.95						
	0.0000	0.0000	0.0000	0.0000	0.0001	0.0000
2.10						
	0.0003	0.0002	0.0000	0.0004	0.0004	0.0004
2.25						
	0.0021	0.0008	0.0004	0.0023	0.0022	0.0024
2.40						
	0.0132	0.0061	0.0035	0.0132	0.0140	0.0138
2.55						
	0.0564	0.0418	0.0303	0.0569	0.0577	0.0579
2.70						
	0.1600	0.1560	0.1491	0.1608	0.1664	0.1589
2.85						
	0.2696	0.2950	0.3180	0.2682	0.2572	0.2662
3.00						
	0.2566	0.2856	0.3054	0.2569	0.2593	0.2568
3.15						
	0.1500	0.1474	0.1433	0.1497	0.1515	0.1511
3.30						
	0.0620	0.0503	0.0404	0.0622	0.0633	0.0627
3.45						
	0.0207	0.0131	0.0079	0.0203	0.0192	0.0210
3.60						
	0.0066	0.0029	0.0014	0.0067	0.0066	0.0065
3.75						
	0.0018	0.0007	0.0003	0.0016	0.0015	0.0017
3.90						
	0.0005	0.0001	0.0000	0.0005	0.0004	0.0005
4.05						
	0.0001	0.0000	0.0000	0.0002	0.0002	0.0001
4.20						
	0.0001	0.0000	0.0000	0.0001	0.0000	0.0000
4.35						
	0.0000	0.0000	0.0000	0.0000	0.0000	0.0000
	3.0078	3.0060	3.0050	3.0071	3.0058	3.0074

larger sample sizes produce more accurate parameter estimates. Both distributions show the same tendency for slight nonsymmetry as shown with the histogram for samples sizes of 20.

To show the effect of the simulation size, analyses were made for *Ns* of 5×10^4 and 10^4, which are summarized in columns 5 and 6, respectively. Because these are for sample sizes (n) of 20, the histograms can be compared with column 2. The largest absolute difference between columns 2 and 5 is 0.0014, which is relatively small (0.5%). This reflects sampling variations. For the smaller simulation size of column 6, the largest absolute difference is 0.0124, which is 4.6%. This is too large to be considered sampling variation of b_0. Instead, it reflects the need to have a simulation length greater than 10^4 for this problem.

The effect of the seed value for the random-number generator was studied. Column 7 contains the histogram for the case of $n = 20$ and $Ns = 10^5$, which are the same parameters as for column 2, but for a different seed. The largest absolute difference is 0.0034, which is a 1.3% difference. This may be beyond the bounds of expected sampling variation, but it does suggest that, if the procedure was actually being applied, analyses should be made for several seed values and the results averaged or weighted.

13.8. PROBLEMS

13-1. A set of data for predicting a student's score on a test (Y) in a probability and statistics course includes the following variables: X_1 is the student's grade point average (GPA); X_2 is the student's age; X_3 is the number of hours the student studied for the test; and X_4 is the number of credit hours for which the student is currently enrolled. The data set is based on a sample of 24 students, which results in the following correlation matrix:

	X_1	X_2	X_3	X_4	Y
X_1	1.00	0.53	0.20	–0.36	0.62
X_2		1.00	–0.12	0.24	0.25
X_3			1.00	–0.41	0.47
X_4				1.00	–0.18
Y					1.00

Discuss the potential accuracy of a relationship between Y and the X variables. Also, discuss the rationality of the correlation coefficient and the level of interconnection between the predictor variables.

13-2. A data set of 32 stream sections includes the following variables: X_1 is the depth of flow; X_2 is the roughness of the channel bottom; X_3 is the slope of the stream bed; X_4 is the mean soil particle diameter; and Y is the erosion rate of the channel bed. The data yield the following correlation matrix:

	X_1	X_2	X_3	X_4	Y
X_1	1.00	–0.26	–0.66	–0.08	0.76
X_2		1.00	–0.31	0.87	–0.27
X_3			1.00	0.06	0.61
X_4				1.00	–0.43
Y					1.00

Discuss: (a) the potential accuracy of a relationship between Y and all of the X variables; (b) the predictor variable that would yield the most accurate bivariate model; (c) the rationality of the correlation coefficients; (d) the potential effect of intercorrelation on the rationality of the regression coefficients.

13-3. Using the principle of least squares, derive the normal equations for the following model:

$$\hat{Y} = b_1 X_1 + b_2 X_2 + b_3 X_3$$

13-4. The cost of producing power (P) in mills per kilowatt hour is a function of the load factor (L) in percent and the cost of coal (C) in cents per million Btu. Perform a regression analysis to develop a regression equation for predicting P using the following data:

L	C	P
85	15	4.1
80	17	4.5
70	27	5.6
74	23	5.1
67	20	5.0
87	29	5.2
78	25	5.3
73	14	4.3
72	26	5.8
69	29	5.7
82	24	4.9
89	23	4.8

13-5. Visitation to lakes for recreation purposes varies directly with the population of the nearest city or town and inversely with the distance between the lake and city. Use the data below to develop a regression equation to predict

D	P	V
10	10,500	22,000
22	27,500	98,000
31	18,000	24,000
18	9,000	33,000
28	31,000	41,000
12	34,000	140,000
21	22,500	78,000
13	19,000	110,000
33	12,000	13,000

visitation (V) in person-days per year, using as predictors the population (P) of the nearby town and the distance (D) in miles between the lake and the city.

13-6. Starting with Equation 13-8, derive Equation 13-2.

13-7. Find the standardized partial regression coefficient (t) for the following R matrix:

$$R = \begin{array}{c} \\ X_1 \\ X_2 \\ Y \end{array} \begin{array}{c} \begin{array}{ccc} X_1 & X_2 & Y \end{array} \\ \begin{bmatrix} 1.0 & 0.3 & 0.37 \\ & 1.0 & 0.71 \\ & & 1.00 \end{bmatrix} \end{array}$$

13-8. Determine the determinant for the following matrix of intercorrelations. Discuss the implications for model rationality.

$$R_{11} = \begin{bmatrix} 1.0 & -0.62 \\ -0.62 & 1.0 \end{bmatrix}$$

13-9. Determine the determinant for the following matrix of intercorrelations. Discuss the implications for model rationality.

$$R_{11} = \begin{bmatrix} 1 & 0.3 & 0.7 \\ & 1 & 0.5 \\ & & 1 \end{bmatrix}$$

13-10. Determine the determinant for the following matrix of intercorrelations. Discuss the implications for model rationality.

$$R_{11} = \begin{bmatrix} 1 & -0.7 & 0.25 \\ & 1 & -0.20 \\ & & 1.0 \end{bmatrix}$$

13-11. Discuss the advantages and disadvantages of the multiple correlation coefficient as a goodness-of-fit statistic.

13.12. Discuss the advantages and disadvantages of the standard error of estimate as a goodness-of-fit statistic.

13-13. Using the method of least squares, derive the normal equations for the model $\hat{Y} = b_1X + b_2X^2$. Evaluate the coefficients using the data of Problem 12-10.

13-14. Using the data of Problem 12-47, fit a linear model and a quadratic polynomial and make a complete assessment of each. Which model is best for prediction? Support your choice.

13-15. Using the data of Problem 12-48, fit a linear model and a quadratic polynomial and make a complete assessment of each. Which model is best for prediction? Support your choice.

13-16. If a fifth-order polynomial is fitted to a data set with a sample size of six, what is the coefficient of determination? What is the standard error of estimate? What general implication arises from this?

13-17. For a set of data based on a sample size of 24, the coefficients of determination (R) for linear, quadratic, and cubic polynomials are 0.53, 0.64, and 0.75. Perform an analysis of variance. At a 5% level of significance, which model would be recommended?

13-18. For a set of data based on a sample size of 11, the coefficients of determination (R) for linear, quadratic, and cubic polynomials are 0.41, 0.46, and 0.59. Perform an analysis of variance. At a 5% level of significance, which model would be recommended?

13-19. For a set of data based on a sample size of 18, the coefficients of determination (R) for linear, quadratic, cubic, and quartic polynomials are 0.47, 0.58, 0.61, and 0.74, respectively. Perform an ANOVA. At a 5% level of significance, which model would be recommended?

13-20. Using the method of least squares, derive the normal equations for the model $\hat{Y} = ax^b$. Evaluate the coefficients using the data of Problem 12-10. Solve the problem (a) directly (without transformation) and (b) by converting the equation to linear form by making a logarithmic transformation. Compare and discuss the two cases.

13-21. Derive the normal equations for the following model:

$$\hat{y} = a x_1^b x_2^c$$

13-22. Using the data of Problem 12-47, fit a linear model and a power model. Make a complete assessment of each. Which model is best for prediction? Support your choice.

13-23. Using the data of Problem 12-48, fit a linear model and a power model. Make a complete assessment of each. Which model is best for prediction? Support your choice.

13-24. Using the data of Problem 13-4, fit a multiple power model. Make a complete assessment of the model.

13-25. Using the method of least squares, derive the normal equations for the model $\hat{Y} = b_0 + b_1\sqrt{X}$. Evaluate the coefficients using the data of Problem 12-10.

13.9. SIMULATION PROJECTS

Using the data of Table C-1, C-2, or C-3, calibrate a multiple linear regression model for Y vs. the three predictor (X) variables and assess the accuracy of the model. Compare the multiple regression model to each of the three bivariate models of Project 12-1 and decide if the multiple regression model provides significantly greater accuracy than the bivariate models.

Chapter 14
Reliability Analysis of Components

TABLE OF CONTENTS

14.1. INTRODUCTION

Uncertainties in civil engineering systems can be mainly attributed to ambiguity and vagueness in defining the variables and parameters of the systems and their relations. The ambiguity component is generally due to noncognitive sources, which include (1) physical randomness, (2) statistical uncertainty due to the use of limited information to estimate the characteristics of the variables and parameters, and (3) model uncertainties due to simplifying assumptions in analytical and prediction models, simplified methods, and idealized representations of real performances. The vagueness-related uncertainty is due to cognitive sources, including (1) definition of certain variables or parameters, such as structural performance (failure or survival), quality, deterioration, skill and experience of construction workers and engineers, environmental impact of projects, and conditions of existing structures; (2) other human factors; and (3) definition of the interrelationships among the parameters of the problems, especially for complex systems. These uncertainty types were discussed in Chapter 1.

Reliability considerations are important in the analysis and design of engineering systems. The reliability of a system is commonly stated in reference to some performance criteria. The need for the reliability analysis stems from acknowledging the presence of uncertainty in the definition, understanding, modeling, and behavior prediction of the system that models a project. The objective of the analysis is the assurance of some level of reliability. An engineered system has numerous sources of uncertainties. The absolute safety of the system cannot be guaranteed due to the unpredictability of the demand conditions for the system, our inability to obtain and express the properties and performances of the system accurately, the use of simplified assumptions in predicting the behavior of the system under consideration, limitations in the numerical methods used, and human factors (e.g., errors or omissions). However, the likelihood of unacceptable performances can be limited to a reasonable level. Estimation of this likelihood, even when used to compare various design alternatives, is an important task for an engineer. The objective of this chapter is to provide an introduction to the area of reliability assessment.

14.2. TIME TO FAILURE

The early developments in reliability were based on assessing the time to failure of an engineered component (e.g., a mechanical or electrical product). In these applications, the reliability of a component was defined as the probability that the component would perform according to a specified performance criterion for at least a specified period under specified conditions. The performance period T can be defined as the time to failure. Distance, number of cycles, or any other convenient and suitable measure can be used instead of time. The performance period is a random variable that has a probability density function $f_T(t)$ and a cumulative distribution function $F_T(t)$. Therefore, the reliability function $R_T(t)$ can be defined as:

$$R_T(t) = \int_t^{\infty} f_T(t)dt = 1 - F_T(t) \tag{14-1}$$

The expected value (mean) for the time to failure, μ_T, can be computed as:

$$\mu_T = E(T) = \int_0^{\infty} t f_T(t)dt \tag{14-2}$$

The result from Equation 14-2 can be viewed as the expected life. It can be shown that the expected life can be computed as:

$$\mu_T = E(T) = \int_0^\infty R_T(t)dt \tag{14-3}$$

The variance of the time to failure (or life), σ_T^2, can be computed as:

$$\sigma_T^2 = \text{Var}(T) = \int_0^\infty (t-\mu_T)^2 f_T(t)dt \tag{14-4}$$

It can be shown that the variance of life can be computed as:

$$\sigma_T^2 = \text{Var}(T) = \int_0^\infty 2tR_T(t)dt - \left(\int_0^\infty R_T(t)dt\right)^2 \tag{14-5}$$

Another aspect of the reliability problem in the context of the time to failure is the assessment of failure rate. In general, the failure rate of a component can be time variant as the component ages. In reliability analysis, it is common to measure this failure rate with the hazard function. The hazard function $h_T(t)$ can be viewed as the probability that the component will fail in the next time increment Δt, given that it survived up to some time t. This conditional probability can be expressed as:

$$P(t < T \le t + \Delta T | T > t) = \frac{P(t < T \le t + \Delta T)}{P(T > t)} \tag{14-6}$$

Dividing both sides of Equation 14-6 with ΔT and taking the limiting case as $\Delta T \to 0$, the hazard function becomes:

$$h_T(t) = \frac{f_T(t)}{R_T(t)} \tag{14-7}$$

Therefore, based on the definition of the hazard function, as the failure rate increases, the hazard function increases.

Experimentally developed hazard functions for many components can be modeled as a bathtub distribution, as shown schematically in Figure 14-1. In this typical bathtub distribution, three regions can be identified: the warranty-failures region, the chance-failures region, and the aging-failures region. The high failure rate (hazard function) in the first region is generally attributed to several factors related to manufacturing errors, imperfections, deviations from standards, poor quality, and human factors. This region represents the break-in period and is commonly covered by a warranty by the manufacturer. The hazard function in this region can be reduced

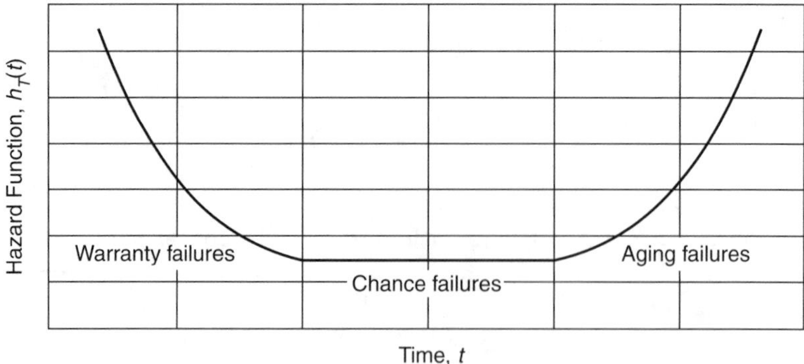

FIGURE 14-1 The hazard function.

by increasing the quality control at the production level. The hazard level in the second region represents the hazard during the service life of the component. In the third region, the hazard level starts to increase again due to aging. During this region, a decision needs to be made regarding replacement, life extension, or repair of the component.

Example 14-1: Time to Failure for a Construction Crane

A major construction company has 100 construction cranes that are within the same size and capacity class and have the same level of use. The records of the company indicate that the average time between failures for these cranes is 2 years. Assuming that the failure rate is constant, the hazard function is given by:

$$h_T(t) = \lambda = \frac{1}{2} = 0.5 \tag{14-8}$$

where λ is the failure rate. Using Equation 14-2, with $\mu_T = 2$ years, the following $f_T(t)$ results:

$$f_T(t) = 0.5\exp(-0.5t) \tag{14-9}$$

Therefore, the cumulative distribution function is:

$$F_T(t) = 1 - \exp(-0.5t) \tag{14-10}$$

Equations 14-9 and 14-10 are the density and the cumulative functions for the exponential distribution as described in Section 5.5. The reliability function is given by:

$$R_T(t) = \exp(-0.5t) \tag{14-11}$$

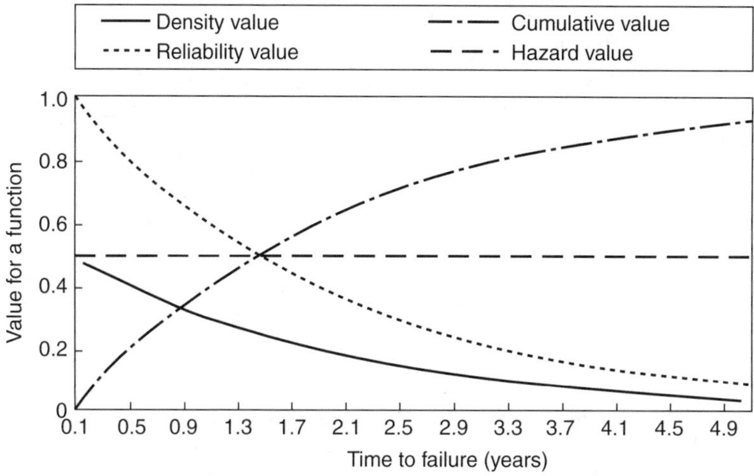

FIGURE 14-2 Density, cumulative, reliability, and hazard functions.

The density, cumulative, reliability, and hazard functions are shown in Figure 14-2. In this example, the reliability of a crane decreases exponentially as a function of time, with a constant hazard level.

14.3. RELIABILITY OF COMPONENTS

The reliability of a component can be defined as the probability that the component meets some specified demands under specified environmental conditions. For example, the reliability of a structural component (e.g., a beam) can be defined as the probability that the structural strength (e.g., ultimate moment capacity) exceeds the applied loads (e.g., the moment due to the total combined loads). This definition of the reliability R_e can be expressed mathematically as:

$$R_e = P(R > L) \tag{14-12}$$

where R is the structural resistance or strength of the component, and L is the load effect expressed in the same units as the resistance R. In general, R can be the resistance, supply, capacity, or input for the component, and L can be the load effect, demand, or needs. Although most of the methods presented in this section were used for structural reliability assessment, they can be used for nonstructural problems.

In general, R and L in Equation 14-12 are functions of several random variables that are called the *basic random variables*. For example, the moment capacity of a beam is a function of the stress–strain characteristics of the materials that make up the beam, the dimensions of the cross section, and the definition of capacity for the beam. In this case, the variables, such as yield stress, width of the cross section of the beam, and height of the cross section, are basic random variables. In reliability analysis, it is common to define a performance function that describes the performance of the component in meeting the demand. The performance function Z is expressed as:

$$Z = R - L \tag{14-13a}$$

or

$$Z = g(X_1, X_2, \ldots, X_n) \tag{14-13b}$$

where $X_1, X_2, \ldots, X_n$ are n basic random variables for R and L, and $g(.)$ is a function that describes the relationship between R–L and the basic random variables. The performance function in Equation 14-13 is expressed such that failure of the component results in a negative sign for the function (i.e., $Z < 0$), survival of the component results in a positive sign for the function (i.e., $Z > 0$), and the limit state for the beam results in $Z = 0$.

The probability of failure P_f for the component is defined as:

$$P_f = P(R < L) \tag{14-14a}$$

$$= 1 - R_e \tag{14-14b}$$

In terms of the performance function, the probability of failure is expressed as:

$$P_f = P(Z < 0) \tag{14-15a}$$

$$= P[g((X_1, X_2, \ldots, X_n) < 0)] \tag{14-15b}$$

For the general case, where the basic random variables can be correlated, the probability of failure can be determined by solving the following integral:

$$P_f = \underset{\text{over } Z \leq 0}{\int \cdots \int} f_X(x_1, x_2, \ldots, x_n)\, dx_1 dx_2 \ldots dx_n \tag{14-16}$$

where f_X is the joint probability density function (PDF) of the random vector $\boldsymbol{X} = \{X_1, X_2, \ldots, X_n\}$, and the integration is performed over the region where $Z = g(.) < 0$. Determining the probability of failure by evaluating the integral of Equation 14-16 can be a difficult task.

In this section, methods for assessing the reliability of components are presented and discussed. These methods include the first-order reliability method (FORM), the advanced second-moment (ASM) method, the direct Monte Carlo simulation (MCS) method, and variance-reduction techniques (VRT) in simulation.

14.4. FIRST-ORDER RELIABILITY METHOD

In order to avoid some computational difficulties of Equation 14-16 in the evaluation of the probability of failure, the FORM is used in structural reliability analysis. The approximate mean and standard deviation of Z in Equation 14-13 (μ_Z and σ_Z, respectively) are estimated using the information on means, μ_{X_i}, standard deviations, σ_{X_i}, or coefficient of variations (δ_{X_i}) of the basic random variables, X_i, and the correlation coefficients among the basic variables. Expanding $g(.)$ in Equation 14-13 using a Taylor series expansion about the mean values of the basic random variables and truncating the series at the first-order terms, the first-order approximate mean, and variance of Z, respectively, are given by:

$$\mu_Z \approx g(\mu_{X_1}, \mu_{X_2}, \ldots, \mu_{X_n}) \tag{14-17}$$

and

$$\sigma_Z^2 \approx \sum_{i=1}^{n} \sum_{j=1}^{n} \left(\frac{\partial Z}{\partial X_i}\right)_\mu \left(\frac{\partial Z}{\partial X_j}\right)_\mu \mathrm{COV}(X_i, X_j) \tag{14-18}$$

where the partial derivatives of Z are evaluated at the mean values of the basic random variables, and $\mathrm{COV}(X_i, X_j)$ is the covariance of X_i and X_j. For statistically uncorrelated random variables, Equation 14-18 can be expressed as:

$$\sigma_Z^2 \approx \sum_{i=1}^{n} \left(\frac{\partial Z}{\partial X_i}\right)_\mu^2 \sigma_{X_i}^2 \tag{14-19}$$

The second-order mean (considering the square term in the Taylor series expansion) can be used to improve the accuracy of the estimation of the mean as follows:

$$\mu_Z \approx g(\mu_{X_1}, \mu_{X_2}, \ldots, \mu_{X_n}) + \frac{1}{2}\sum_{i=1}^{n} \sum_{j=1}^{n} \left(\frac{\partial^2 Z}{\partial X_i \partial X_j}\right)_\mu \mathrm{COV}(X_i, X_j) \tag{14-20}$$

Again the partial derivatives are evaluated at the mean values of the basic random variables. The second-order variance can be similarly computed. For practical purposes, the use of the second-order mean and the first-order variance can be adequate for some engineering applications.

A measure of reliability can be defined by introducing a parameter β, called the *reliability index*, as:

$$\beta = \frac{\mu_Z}{\sigma_Z} \tag{14-21}$$

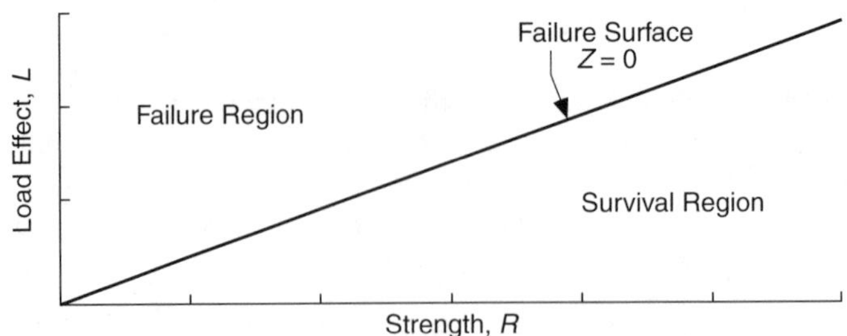

FIGURE 14-3 Performance space for special case.

If Z is assumed to have a normal probability distribution, the probability of failure P_f is given by:

$$P_f = 1 - \Phi(\beta) \tag{14-22}$$

in which Φ is the cumulative probability distribution function of the standard normal variate (Table A-1) and β is the reliability index.

An illustrative special case for the structural reliability problem is given for the following performance function:

$$Z = R - L \tag{14-23}$$

This two-random-variable case is shown in Figure 14-3. The performance region (sample space) in this figure is divided into two mutually exclusive and collectively exhaustive events: failure and survival. Assuming normal probability distributions for R and L results in Z being normally distributed with the following mean and variance, respectively:

$$\mu_Z = \mu_R - \mu_L \tag{14-24}$$

and

$$\sigma_Z^2 = \sigma_R^2 + \sigma_L^2 \tag{14-25}$$

where μ is the mean value, and σ is the standard deviation. The random variables R, L, and Z are shown in Figure 14-4. The probability of failure according to this figure can be computed as the area under $f_Z(z)$ for $z < 0$. This area equals the cumulative distribution function of Z evaluated at $z = 0$; that is,

$$P_f = P(Z < 0) \tag{14-26}$$

or

$$P_f = F_Z(0) \tag{14-27}$$

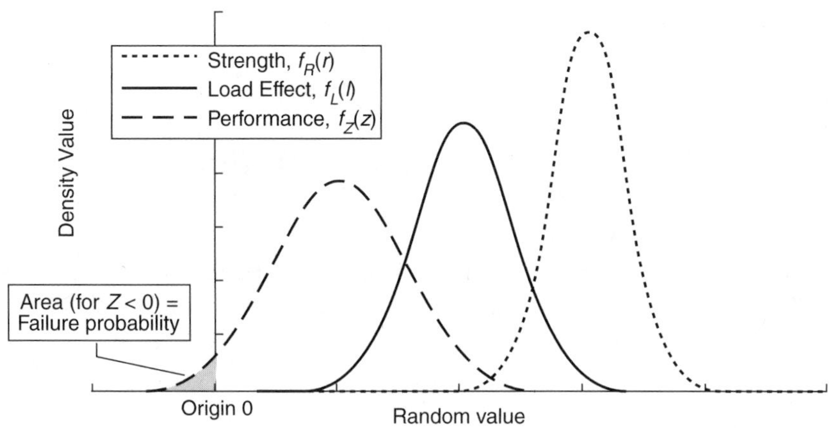

FIGURE 14-4 Random variables for special case.

Because Z is normally distributed, the probability of failure can be computed as:

$$P_f = \Phi\left(\frac{0-\mu_Z}{\sigma_Z}\right) \tag{14-28a}$$

$$= \Phi(-\beta) = 1-\Phi(\beta) = \Phi\left(-\frac{\mu_R-\mu_L}{\sqrt{\sigma_R^2+\sigma_L^2}}\right) \tag{14-28b}$$

The reliability index according to Equation 14-28b is given by:

$$\beta = \frac{\mu_R-\mu_L}{\sqrt{\sigma_R^2+\sigma_L^2}} \tag{14-29}$$

Another illustrative special case for the structural reliability problem is given for the case where R and L have lognormal probability distributions. In this case, the performance function should be expressed as:

$$Z = \frac{R}{L} \tag{14-30}$$

where failure is defined as $Z < 1$, survival is $Z > 1$, and the limit state is $Z = 1$. In this case, Z is lognormally distributed. It can be shown that the probability of failure can be computed as:

$$P_f = P(Z < 1) \tag{14-31a}$$

resulting in the following expression:

$$P_f = \Phi(-\beta) = \Phi\left[\frac{-\ln\left(\frac{\mu_R}{\mu_L}\sqrt{\frac{\delta_L^2+1}{\delta_R^2+1}}\right)}{\sqrt{\ln[(\delta_R^2+1)(\delta_L^2+1)]}}\right] \tag{14-31b}$$

where μ is the mean value, and δ is the coefficient of variation. The reliability index in this case is given by:

$$\beta = \frac{\ln\left(\frac{\mu_R}{\mu_L}\sqrt{\frac{\delta_L^2+1}{\delta_R^2+1}}\right)}{\sqrt{\ln[(\delta_R^2+1)(\delta_L^2+1)]}} \tag{14-32}$$

The first-order reliability method has some shortcomings. According to this method, the function $g(.)$ in Equation 14-13b is linearized at the mean values of the basic random variables. Where $g(.)$ is nonlinear, significant error may be introduced by neglecting higher-order terms. This approximate method also may yield different results for different mechanically equivalent formulations of the same problem (e.g., stress formulation vs. strength formulation). Moreover, the method completely ignores the information on distributions of the basic random variables. In other words, the method results in accurate reliability assessments only where the basic random variables are normally distributed and the function $g(.)$ is linear.

Example 14-2: Tensile Reliability of a Truss Member

A tension member in a truss has an ultimate tensile strength T with a mean value of 120 kips and a standard deviation of 10 kips. The tension load L in this member has a mean value of 80 kips and standard deviation of 20 kips. Normal probability distributions are assumed for T and L. The reliability index β for the member can be computed using Equation 14-29 as:

$$\beta = \frac{120-80}{\sqrt{10^2+20^2}} = 1.79 \tag{14-33}$$

Therefore, the probability of failure according to Equation 14-28b is:

$$P_f = \Phi(-1.79) = 1-\Phi(1.79) = 0.03673 \tag{14-34}$$

Example 14-3: Flexural Reliability of a Beam

The following performance function describes the flexural behavior of a simply supported beam of a span length $\sqrt{2}\,L$ that carries a uniform load W:

$$Z = YS - \frac{WL^2}{4} \tag{14-35}$$

where Y is the yield stress of the material of the beam, and S is the elastic section modulus. In this example, failure is defined as yielding at the extreme material fibers of the cross section of the beam. The mean values and standard deviations of the variables are given in Table 14-1.

The following first-order mean of Z can be computed by substituting the mean values of the basic random variables in Equation 14-35:

$$\mu_Z \approx (38)(100) - \frac{(0.3)(180)^2}{4} = 1370 \text{ kip-in.} \tag{14-36}$$

Assuming the random variables to be statistically uncorrelated, the first-order variance is computed using Equation 14-19 as:

$$\sigma_Z^2 \approx (100)^2(1.9)^2 + (38)^2(5)^2 + \left(\frac{180^2}{4}\right)^2(0.075)^2 + \left(\frac{0.3(180)}{2}\right)^2(9)^2 = 500305.25 \tag{14-37}$$

Therefore, $\sigma_Z \approx 707.32$ kip-in.

The coefficient of variation is 707.32/1370 = 0.516. A measure of reliability can be computed using Equation 14-21 as:

$$\beta = \frac{\mu_Z}{\sigma_Z} = 1.9369 \tag{14-38}$$

If Z is assumed to have a normal probability distribution, the probability of failure, P_f, is given by:

$$P_f = 1 - \Phi(1.9369) = 0.02638 \tag{14-39}$$

TABLE 14-1 Mean Values and Standard Deviations for Random Variables Used in Example 14-3

Random variable	Mean value	Standard deviation	Coefficient of variation
Y	38 ksi	1.9 ksi	0.05
S	100 in.^{3}	5 in.^{3}	0.05
W	0.3 kips/in.	0.075 kips/in.	0.25
L	180 in.	9 in.	0.05

14.5. ADVANCED SECOND-MOMENT METHOD

14.5.1. Uncorrelated, Normally Distributed Random Variables

According to this method, the basic normal random variables, which are uncorrelated, are transformed into a reduced (or normalized) coordinate system $\boldsymbol{X}' = \{X_1',\ X_2',\ \ldots,\ X_n'\}$ according to the following transformation:

$$X_i' = \frac{X_i - \mu_{X_i}}{\sigma_{X_i}} \quad \text{for } i = 1, 2, \ldots, n \tag{14-40}$$

It should be noted that the resulting reduced variables have the following mean and standard deviation:

$$\mu_{X_i'} = 0$$

and

$$\sigma_{X_i'} = 1$$

For the following fundamental performance function:

$$Z = R - L \tag{14-41}$$

The transformation results in the following expression for the performance function:

$$Z = R'\sigma_R + \mu_R - (L'\sigma_L + \mu_L) \tag{14-42}$$

The limit state (i.e., $Z = 0$) can be expressed as:

$$L' = R'\frac{\sigma_R}{\sigma_L} + \frac{\mu_R - \mu_L}{\sigma_L} \tag{14-43}$$

The resulting performance function in the reduced coordinate system is shown in Figure 14-5. By comparing Figures 14-3 and 14-5, it is evident that the limit state in the reduced coordinates does not pass through the origin. It can be shown that in the reduced coordinates, the shortest distance from the origin to the limit state is given by:

$$\beta = \frac{\mu_R - \mu_L}{\sqrt{\sigma_R^2 + \sigma_L^2}} \tag{14-44}$$

The resulting distance is the same as the reliability index given by Equation 14-29. Therefore, it can be stated that a measure of safety in the form of the reliability index

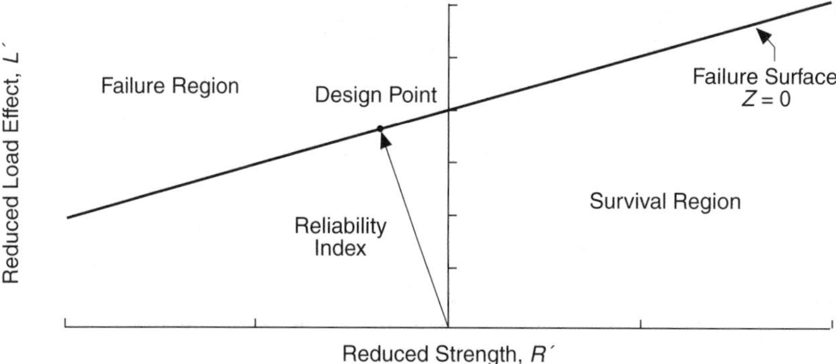

FIGURE 14-5 Performance space in reduced coordinates.

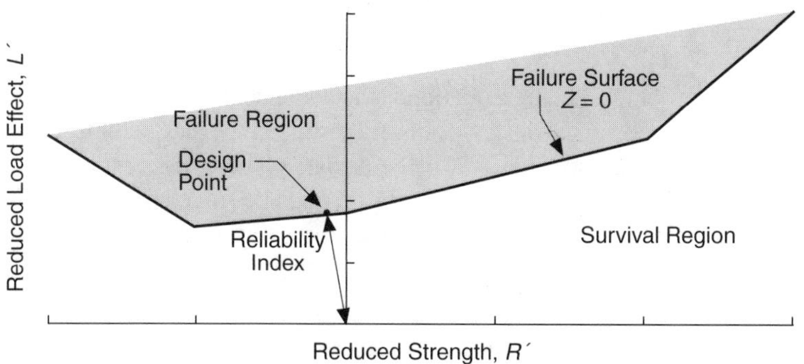

FIGURE 14-6 Performance space in reduced coordinates: nonlinear case.

β is defined by the advanced second-moment method as the shortest distance from the origin to the failure surface (limit state) in the reduced coordinates. The reliability index according to this definition is commonly referred to as the *Hasofer–and–Lind index*. This definition of β is valid for a nonlinear performance function of the type:

$$Z = g(X_1, X_2) \tag{14-45}$$

such that the failure region is $Z < 0$, the survival region is $Z > 0$, and the failure surface (limit state) is $Z = 0$. The nonlinear case of two random variables is shown in Figure 14-6. The point on the failure surface that corresponds to the shortest distance is called the *design* (or *failure*) *point*. It can be shown that it is the most likely failure point. This point has some significance in defining partial safety factors in reliability-based design codes. For the nonlinear case, it is possible to obtain the β value by approximating the failure surface by a linear tangent at the design point. Numerically, the approximation can be achieved by a first-order Taylor series approximation of the failure surface at the design point. The procedure is similar to the Taylor series expansion performed in Equations 14-17 and 14-18, with the expansion point being the design point instead of the mean values. The resulting reliability index based on this approximation of the failure surface should be exactly the same as the result from the case without approximation (i.e., nonlinear failure surface).

For the general case where the nonlinear performance function is:

$$Z = g(X_1, X_2, \ldots, X_n) \tag{14-46}$$

the reliability index has the same definition as the shortest distance from the origin to the failure surface in the reduced coordinates. Mathematically, this definition can be expressed as:

$$\beta = \min_{x'} \sqrt{\sum_{i=1}^{n} x'^2} \quad \text{such that } g(\boldsymbol{x}) = 0 \tag{14-47}$$

where the square root term in this equation is the distance in the reduced coordinates; the point $\boldsymbol{x}'$ corresponds to the design point and results in the least (shortest) distance. It is common to use the notation $\boldsymbol{x}^*$ for the design point in the regular coordinate and $\boldsymbol{x}'^*$ in the reduced coordinates. Equation 14-47 can be solved using one of several nonlinear optimization methods. For example, the solution can be achieved iteratively by simultaneously solving the following set of $(n + 1)$ equations:

$$\alpha_i^* = \frac{\left(\dfrac{\partial g}{\partial X_i'}\right)_*}{\sqrt{\displaystyle\sum_{i=1}^{n}\left(\dfrac{\partial g}{\partial X_i'}\right)_*^2}} \quad \text{for } i = 1, 2, \ldots, n \tag{14-48a}$$

$$x_i'^* = -\alpha_i^* \beta \quad \text{for } i = 1, 2, \ldots, n \tag{14-48b}$$

$$Z = g(x_1^*, x_2^*, \ldots, x_n^*) = 0 \tag{14-48c}$$

Using the following relation between a random variable and its reduced form:

$$x_i^* = \mu_{X_i} + \sigma_{X_i} x_i'^* = \mu_{X_i} - \alpha_i^* \sigma_{X_i} \beta \tag{14-49}$$

Equation 14-48c can be expressed as:

$$Z = g[(\mu_{X_1} - \alpha_1^* \sigma_{X_1} \beta), (\mu_{X_2} - \alpha_2^* \sigma_{X_2} \beta), \ldots, (\mu_{X_n} - \alpha_n^* \sigma_{X_n} \beta)] = 0 \tag{14-50}$$

A subscript or superscript of * indicates the evaluation of the noted quantity (or term) at the design point. The set of equations represents a first-order (linear) approximation of the failure surface. The α_i' values are the directional cosines of the tangent hyperplane in the reduced coordinates. The directional cosines are indicators of the sensitivity of the performance function to variations in the corresponding basic random variables. The solution of Equations 14-48 and 14-50 can be achieved by performing the following computational steps:

1. Assume an initial value for the design point. It is common to start with the mean values of the basic random variables. The design point in the reduced coordinates should then be computed using:

$$x_i' = \frac{x_i - \mu_{X_i}}{\sigma_{X_i}} \quad \text{for } i = 1, 2, \ldots, n \tag{14-51}$$

2. Evaluate the directional cosines at the failure point. The partial derivatives that are needed for computing the directional cosines (Equation 14-48a) can be obtained as:

$$\left(\frac{\partial g}{\partial X_i'}\right)_* = \left(\frac{\partial g}{\partial X_i}\frac{\partial X_i}{\partial X_i'}\right)_* = \left(\frac{\partial g}{\partial X_i}\right)_* \sigma_{X_i} \quad \text{for } i = 1, 2, \ldots, n \tag{14-52}$$

3. Solve the following equation for the root β:

$$g[(\mu_{X_1} - \alpha_1^* \sigma_{X_1}\beta), (\mu_{X_2} - \alpha_2^* \sigma_{X_2}\beta), \ldots, (\mu_{X_n} - \alpha_n^* \sigma_{X_n}\beta)] = 0 \tag{14-53}$$

4. Using the β obtained from step 3, evaluate a new design point using the following equation:

$$x_i^* = \mu_{X_i} - \alpha_i^* \sigma_{X_i}\beta \quad \text{for } i = 1, 2, \ldots, n \tag{14-54}$$

5. Repeat steps 1 to 4 until convergence of β is obtained.

The important relation between the probability of failure and the reliability index can be maintained as:

$$P_f = 1 - \Phi(\beta) \tag{14-55}$$

Equation 14-55 results in the exact probability if the performance function is linear and the basic random variables are normally distributed. For the case of a nonlinear performance function, the error depends on the level of nonlinearity of the performance function. This error is due to approximating the performance function by a tangent hyperplane at the design point. Although the reliability index does not have any error, P_f is approximate. In reliability problems, it is common to deal with nonlinear performance functions with a relatively small level of nonlinearity, resulting in very small errors in the estimated P_f. Therefore, in these cases, for all practical purposes, P_f is of sufficient accuracy.

Example 14-4: Flexural Reliability of a Beam: Nonlinear Case

The performance function describing the flexural behavior of a simply supported beam of a span length $\sqrt{2}L$ that carries a uniform load W was discussed in Example 14-3. It is given by:

$$Z = YS - \frac{WL^2}{4}$$

where Y is the yield stress of the material of the beam, and S is the elastic section modulus. In this example, failure is defined as yielding at the extreme material fibers of the cross section of the beam. The probabilistic characteristics of the variables are given in Table 14-2.

In this case, the random variables are assumed to have normal probability distributions. The partial derivatives of the performance function with respect to the reduced variables are:

$$\left(\frac{\partial g}{\partial Y'}\right)_* = s^* \sigma_Y \tag{14-56a}$$

$$\left(\frac{\partial g}{\partial S'}\right)_* = y^* \sigma_S \tag{14-56b}$$

$$\left(\frac{\partial g}{\partial w'}\right)_* = \frac{-l^{*2}}{4} \sigma_w \tag{14-56c}$$

$$\left(\frac{\partial g}{\partial L'}\right)_* = \frac{-w^* l^*}{2} \sigma_L \tag{14-56d}$$

Now, the previously described iterative steps can be performed as shown in Table 14-3. The solution in this case converges in two iterations; therefore, the probability of failure is:

$$P_f = 1 - \Phi(1.83653) = 0.0341 \tag{14-56e}$$

The FORM solution in Example 14-3 produced a probability of failure of 0.02638; therefore, the relative error of the FORM solution is –23%.

TABLE 14-2 Probabilistic Characteristics of Random Variables in Example 14-4

Random variable	Mean value	Standard deviation	Coefficient of variation	Distribution type
Y	38 ksi	1.9 ksi	0.05	Normal
S	100 in.^{3}	5 in.^{3}	0.05	Normal
W	0.3 kips/in.	0.075 kips/in.	0.25	Normal
L	180 in.	9 in.	0.05	Normal

TABLE 14-3 Iterations Using the Advanced Second-Moment Method for Example 14-4

Iteration number	Random variable	Design point	$\left(\frac{\partial g}{\partial X'}\right)_*$	Directional cosine, α_i^*	New design point that satisfies $g(.) = 0$, according to Equation 14-53
1	Y	38	190	0.26862	37.05645020
	S	100	190	0.26862	97.51697430
	W	0.3	–607.5	–0.85890	0.41908722
	L	180	–243	–0.34350	185.71618600
					0.00078733 = sum
				Therefore,	$\beta = 1.84874$
2	Y	37.056450	185.28	0.23732	37.17189830
	S	97.516970	185.28	0.23732	97.82078500
	W	0.419087	–646.70	–0.82830	0.41409282
	L	185.716200	–350.24	–0.44860	187.41490300
					–0.00056764 = sum
				Therefore,	$\beta = 1.83653$

14.5.2. Uncorrelated, Nonnormally Distributed Random Variables

The advanced second-moment method deals with nonnormal probability distributions for the basic random variables by determining an equivalent normal distribution at the design point. In the iterative solution, this is performed at each iteration. The equivalent normal distribution at the design point x^* for a nonnormal basic random variable X with a density function $f_X(x)$ and cumulative distribution function $F_X(x)$ is defined by the following mean, μ_X^N, and standard deviation, σ_X^N:

$$\mu_X^N = x^* - \Phi^{-1}[F_X(x^*)]\sigma_X^N \tag{14-57}$$

and

$$\sigma_X^N = \frac{\phi[\Phi^{-1}(F_X(x^*))]}{f_X(x^*)} \tag{14-58}$$

where $\Phi(z)$ is the cumulative distribution function for the standard normal evaluated at z, and $\phi(z)$ is the density function of the standard normal. The computational steps in this case are revised as follows:

1. Assume an initial value for the design point. It is common to start with the mean values of the basic random variables. The design point in the reduced coordinates should then be computed using:

$$x_i' = \frac{x_i - \mu_{X_i}}{\sigma_{X_i}} \quad \text{for } i = 1, 2, \ldots, n \tag{14-59}$$

2. Evaluate the equivalent normal distributions for the nonnormal basic random variables at the design point using Equations 14-57 and 14-58.
3. Evaluate the directional cosines at the design point. The partial derivatives that are needed for computing the directional cosines (Equation 14-48a) can be obtained in this case as:

$$\left(\frac{\partial g}{\partial X_i'}\right)_* = \left(\frac{\partial g}{\partial X_i}\frac{\partial X_i}{\partial X_i'}\right)_* = \left(\frac{\partial g}{\partial X_i}\right)_* \sigma_{X_i}^N \quad \text{for } i = 1, 2, \ldots, n \tag{14-60}$$

4. Solve the following equation for the root β:

$$Z = g[(\mu_{X_1}^N - \alpha_1^* \sigma_{X_1}^N \beta), (\mu_{X_2}^N - \alpha_2^* \sigma_{X_2}^N \beta), \ldots, (\mu_{X_n}^N - \alpha_n^* \sigma_{X_n}^N \beta)] = 0 \tag{14-61}$$

5. Using the β obtained from step 4, evaluate a new design point using the following equation:

$$x_i^* = \mu_{X_i}^N - \alpha_i^* \sigma_{X_i}^N \beta \quad \text{for } i = 1, 2, \ldots, n \tag{14-62}$$

6. Repeat steps 1 to 5 until convergence of β is obtained.

The important relation between the probability of failure and the reliability index can be maintained as:

$$P_f = 1 - \Phi(\beta) \tag{14-63}$$

Example 14-5: Flexural Reliability of a Beam, Nonlinear Nonnormal Case

The performance function for the flexural behavior of a simply supported beam of a span length $\sqrt{2}\,L$ that carries a uniform load W was discussed in Examples 14-3 and 14-4. It is given by:

$$Z = YS - \frac{WL^2}{4}$$

where Y is the yield stress of the material of the beam, and S is the elastic section modulus. Failure was defined as yielding at the extreme material fibers of the cross section of the beam. The mean values, standard deviations, and distribution types of the variables are given in Table 14-4.

Now, the previously described iterative steps can be performed as shown in Table 14-5 to obtain the reliability index β. The parameters of the lognormal distributions for Y, S, W, and L are given by:

Random variable	Mean value	Standard deviation	First parameter (μ_Y)	Second parameter (σ_Y)
Y	38 ksi	1.9 ksi	3.634	0.04997
S	100 in.^{3}	5 in.^{3}	4.604	0.04997
W	0.3 kips/in.	0.075 kips/in.	−1.234	0.2462
L	180 in.	9 in.	5.192	0.04997

Now the equivalent normal can be computed for these nonnormal random variables as shown in Table 14-5. The starting design point is assumed to be the mean values of the random variables. Two iterations are needed in this case to obtain the reliability index; therefore, the probability of failure is:

$$P_f = 1 - \Phi(\beta) = 1 - \Phi(1.7526) = 0.0398$$

The assumption of normal probability distributions for all random variables in Example 14-4 produced a probability of failure of 0.0341; therefore, in this case, the relative difference due to this assumption is –14%.

TABLE 14-4 Probabilistic Characteristics of Random Variables Used in Example 14-5

Random variable	Mean value	Standard deviation	Coefficient of variation	Distribution type
Y	38 ksi	1.9 ksi	0.05	Lognormal
S	100 in.^{3}	5 in.^{3}	0.05	Lognormal
W	0.3 kips/in.	0.075 kips/in.	0.25	Lognormal
L	180 in.	9 in.	0.05	Lognormal

TABLE 14-5 Iterations Using the Advanced Second-Moment Method for Example 14-5

Iteration number	Random variable	Mean for equivalent normal	Standard deviation for equivalent normal	Directional cosine, α_i^*	New design point
1	Y	37.9526	1.8988	0.27275	36.930
	S	99.8752	4.9969	0.27275	97.190
	W	0.2909	0.0739	−0.85840	0.416
	L	179.7753	8.9944	−0.33830	185.780
				Therefore,	β = 1.97252
2	Y	37.9390	1.8450	0.20587	37.27290
	S	99.8390	4.8560	0.20587	98.08650
	W	0.2674	0.1024	−0.92370	0.43321
	L	179.6760	9.2830	−0.24920	183.73050
				Therefore,	β = 1.7526

Example 14-6: Flexural Reliability of a Beam, Nonlinear Nonnormal Case

The performance function described in Example 14-5 is used herein. In this example, the mean values, standard deviations, and distribution types of the variables are given in Table 14-6. The difference between this example and Example 14-5 is in the distribution types. In this example, a mixture of distribution types is used. The extreme-value (largest) type I distribution is a continuous probability distribution with the following density and cumulative functions:

$$f_X(x) = \alpha e^{-\alpha(x-u)} \exp[-e^{-\alpha(x-u)}] \tag{14-64a}$$

$$F_X(x) = \exp[-e^{-\alpha(x-u)}] \tag{14-64b}$$

where α and u are the parameters of the distribution. The mean and variance of X in terms of these parameters are given, respectively, by:

$$\mu_X = u + \frac{0.577216}{\alpha} \tag{14-65a}$$

$$\sigma_X^2 = \frac{\pi^2}{6\alpha^2} \tag{14-65b}$$

The parameters as functions of the mean and variance are:

$$u = \mu_X - \frac{0.577216}{\alpha} = 0.2662 \tag{14-65c}$$

$$\alpha = \frac{\pi}{\sqrt{6}\sigma_X} = 17.10 \tag{14-65d}$$

The iterative steps can now be performed as shown in Table 14-7 to obtain the reliability index β; therefore, the probability of failure is:

$$P_f = 1 - \Phi(\beta) = 1 - \Phi(1.7155) = 0.043128$$

TABLE 14-6 Probabilistic Characteristics of Random Variables Used in Example 14-6

Random variable	Mean value	Standard deviation	Coefficient of variation	Distribution type
Y	38 ksi	1.9 ksi	0.05	Normal
S	100 in.^{2}	5 in.^{2}	0.05	Lognormal
W	0.3 kips/in.	0.075 kips/in.	0.25	Extreme value type I, largest
L	180 in.	9 in.	0.05	Normal

TABLE 14-7 Iterations Using the Advanced Second-Moment Method for Example 14-6

Iteration number	Random variable	Mean for equivalent normal	Standard deviation for equivalent normal	Directional cosine, α_i^*	New design point
1	Y	38.	1.9	0.2776	36.92
	S	99.8752	4.9969	0.2774	97.04
	W	0.2873	0.0717	–0.8485	0.4118
	L	180.	9.	–0.3550	186.5
				Therefore,	$\beta = 2.04694$
2	Y	38.	1.9	0.1712	37.44
	S	99.8340	4.849	0.1663	98.45
	W	0.2521	0.1134	–0.9165	0.4309
	L	180.	9.	–0.3210	185.0
				Therefore,	$\beta = 1.7198$
3	Y	38.	1.9	0.1667	37.46
	S	99.86	4.919	0.1641	98.48
	W	0.2416	0.1205	–0.9182	0.4314
	L	180.0	9.	–0.3196	184.9
				Therefore,	$\beta = 1.7155$

14.5.3. Correlated Random Variables

The previous sections dealt with uncorrelated random variables that can be normally or nonnormally distributed. In engineering and the sciences, some of the basic random variables in Equation 14-13 can be correlated. For correlated basic random variables, X_i for $i = 1,2, \ldots, n$, that have respective means (μ_{X_i}) and standard deviations (σ_{X_i}), a pairwise covariance matrix (Cov) is needed. This covariance matrix is denoted as $\boldsymbol{C}$ and can be constructed as:

$$\boldsymbol{C} = \text{Cov} = \begin{bmatrix} \sigma_{X_1}^2 & \text{Cov}(X_1, X_2) & \ldots & \text{Cov}(X_1, X_n) \\ \text{Cov}(X_2, X_1) & \sigma_{X_2}^2 & \ldots & \text{Cov}(X_2, X_n) \\ \ldots & \ldots & \ldots & \ldots \\ \text{Cov}(X_n, X_1) & \text{Cov}(X_n, X_2) & \ldots & \sigma_{X_n}^2 \end{bmatrix} \quad (14\text{-}66)$$

The random variables in the reduced coordinates according to Equation 14-40 are defined as:

$$X_i' = \frac{X_i - \mu_{X_i}}{\sigma_{X_i}} \quad \text{for } i = 1, 2, \ldots, n$$

The covariance matrix of these reduced random variables (**C'**) is their correlation matrix as given by:

$$C' = \begin{bmatrix} 1 & \rho_{12} & \cdots & \rho_{1n} \\ \rho_{21} & 1 & \cdots & \rho_{2n} \\ \vdots & \vdots & \vdots & \vdots \\ \rho_{n1} & \rho_{n2} & \cdots & 1 \end{bmatrix} \tag{14-67}$$

where ρ_{ij} is the correlation coefficient between X_i and X_j. In order to use the first-order, second-moment method or the advanced second-moment method, the vector of reduced correlated random variables ($\boldsymbol{X}'$) needs to be transformed to a vector of noncorrelated random variables ($\boldsymbol{Y}$) using:

$$\boldsymbol{X}' = \boldsymbol{TY} \tag{14-68a}$$

or

$$\boldsymbol{Y} = \boldsymbol{T}^{\mathrm{T}}\boldsymbol{X}' \tag{14-68b}$$

where the superscript T indicates the transpose of a matrix, and $\boldsymbol{T}$ is a transformation matrix computed from $\boldsymbol{C}'$ as the orthonormal eigenvectors corresponding to the eigenvalues of $\boldsymbol{C}'$. The eigenvalues of $\boldsymbol{C}'$ can be computed by solving for the roots of the following determinant that is based on $\boldsymbol{C}'$:

$$|\boldsymbol{C}' - E\boldsymbol{I}| = 0 \tag{14-69}$$

where $\boldsymbol{I}$ is an $(n \times n)$ identity matrix with ones on the diagonal and zeros for all off-diagonal elements, and E is an eigenvalue. The result of solving Equation 14-69 is a vector of eigenvalues $\boldsymbol{E}$. The transformation matrix $\boldsymbol{T}$ can be constructed column by column, where the ith column of $\boldsymbol{T}$ is the orthonormal eigenvector corresponding to E_i in the vector $\boldsymbol{E}$. The ith eigenvector ($\boldsymbol{V}_i$) results from solving the following equation:

$$(\boldsymbol{C}' - E_i\boldsymbol{I})\boldsymbol{V}_i = 0 \tag{14-70}$$

The normalized (i.e., orthonormal) vector can be obtained by dividing each element of the vector ($\boldsymbol{V}_i$) by the sum of squares of all the elements. The transformation matrix $\boldsymbol{T}$ is obtained by assembling the n orthonormal vectors in the same order as the eigenvalues. These matrices are related as follows:

$$\boldsymbol{D} = \boldsymbol{T}^{\mathrm{T}}\boldsymbol{C}'\boldsymbol{T} \tag{14-71a}$$

$$\boldsymbol{C}' = \boldsymbol{TDT}^{\mathrm{T}} \tag{14-71b}$$

where the superscript T indicates the transpose of a matrix, and $\boldsymbol{D}$ is a diagonal matrix that contains the eigenvalues of $\boldsymbol{C}'$. The transpose is computed by changing the position of each element in the matrix from the *i*th column and *j*th row to the *j*th column and *i*th row. For orthogonal and orthonormal matrices, their transposes are the same as their inverses. It can be shown that:

$$\boldsymbol{C}' = \boldsymbol{T}^{\mathrm{T}}\boldsymbol{C}_y\boldsymbol{T} \tag{14-72a}$$

$$\boldsymbol{C}_y = \boldsymbol{T}\boldsymbol{C}'\boldsymbol{T}^{\mathrm{T}} \tag{14-72b}$$

where $\boldsymbol{C}_y$ is the covariance of $\boldsymbol{Y}$. The transformation objective can also be achieved using the numercial method of Cholesky factorization of the correlation matrix ($\boldsymbol{C}'$) as follows:

$$\boldsymbol{Y} = \boldsymbol{L}^{-1}(\boldsymbol{X}')^{\mathrm{T}} \tag{14-73}$$

where $\boldsymbol{L}$ is the lower triangular matrix obtained by Cholesky factorization of the correlation matrix ($\boldsymbol{C}'$).

Once the transformation matrix $\boldsymbol{T}$ is obtained, the limit-state function of Equation 14-13 should be expressed in terms of the $\boldsymbol{Y}$ random variables instead of the $\boldsymbol{X}$ random variables. The $\boldsymbol{Y}$ random variables expressed in terms of the $\boldsymbol{X}$ random variables need to be substituted in Equation 14-13 as follows:

$$Z = g(X_1, X_2, \ldots, X_n) = g(Y_1, Y_2, \ldots, Y_n) \tag{14-74}$$

where the $\boldsymbol{Y}$ random variables are related to the $\boldsymbol{X}$ random variables as follows:

$$\boldsymbol{X} = [\sigma_X^N]\boldsymbol{T}\boldsymbol{Y} + \{\mu_X^N\} \tag{14-75}$$

where $[\sigma]$ is a diagonal matrix of standard deviations, in the case of nonnormal distributions of standard deviations of equivalent normal distributions, and $\{\mu\}$ is a vector of mean values, in the case of nonnormal distributions of mean values of equivalent normal distributions.

Example 14-7: Correlated Random Variables

In this case, two correlated random variables that define a limit state are used to demonstrate the reliability analysis. The limit state is given by:

$$Z = g(X_1, X_2) = g(Y_1, Y_2) \tag{14-76}$$

The correlation matrix for the two random variables is:

$$\boldsymbol{C}' = \begin{bmatrix} 1 & \rho \\ \rho & 1 \end{bmatrix} \tag{14-77}$$

where ρ is the correlation coefficient for the two random variables. The eigenvalues can be obtained by solving the following equation:

$$\begin{vmatrix} 1-E & \rho \\ \rho & 1-E \end{vmatrix} = 0 \tag{14-78a}$$

or

$$(1-E)^2 - \rho^2 = 0 \tag{14-78b}$$

The two roots are:

$$E = 1 \pm \rho \tag{14-79}$$

The eigenvectors can be determined by solving the following two equations that correspond to the eigenvalues of (1 – ρ) and (1 + ρ), respectively:

$$\begin{bmatrix} 1-(1-\rho) & \rho \\ \rho & 1-(1-\rho) \end{bmatrix} \begin{bmatrix} V_1 \\ V_2 \end{bmatrix} = \begin{bmatrix} 0 \\ 0 \end{bmatrix} \tag{14-80a}$$

$$\begin{bmatrix} 1-(1+\rho) & \rho \\ \rho & 1-(1+\rho) \end{bmatrix} \begin{bmatrix} V_1 \\ V_2 \end{bmatrix} = \begin{bmatrix} 0 \\ 0 \end{bmatrix} \tag{14-80b}$$

These two equations result in the following eigenvectors, respectively:

$$\begin{bmatrix} V_1 \\ V_2 \end{bmatrix} = \begin{bmatrix} 1 \\ -1 \end{bmatrix} \tag{14-81a}$$

$$\begin{bmatrix} V_1 \\ V_2 \end{bmatrix} = \begin{bmatrix} 1 \\ 1 \end{bmatrix} \tag{14-81b}$$

The respective normalized eigenvectors (i.e., orthonormal vectors) are given by:

$$\begin{bmatrix} V_1 \\ V_2 \end{bmatrix} = \begin{bmatrix} 0.707 \\ -0.707 \end{bmatrix} \tag{14-82a}$$

$$\begin{bmatrix} V_1 \\ V_2 \end{bmatrix} = \begin{bmatrix} 0.707 \\ 0.707 \end{bmatrix} \tag{14-82b}$$

Therefore, the transformation matrix $\boldsymbol{T}$ is given by:

$$\boldsymbol{T} = \begin{bmatrix} 0.707 & 0.707 \\ -0.707 & 0.707 \end{bmatrix} \tag{14-83}$$

Using Equation 14-75, the $\boldsymbol{X}$ random variables can be expressed in terms of the $\boldsymbol{Y}$ random variables as follows:

$$\begin{bmatrix} X_1 \\ X_2 \end{bmatrix} = \begin{bmatrix} \sigma_{X_1} & 0 \\ 0 & \sigma_{X_2} \end{bmatrix} \begin{bmatrix} 0.707 & 0.707 \\ -0.707 & 0.707 \end{bmatrix} \begin{bmatrix} Y_1 \\ Y_2 \end{bmatrix} + \begin{bmatrix} \mu_{X_1} \\ \mu_{X_2} \end{bmatrix} \tag{14-84}$$

The resulting expressions for the $\boldsymbol{X}$ random variables can now be substituted in the limit-state function of Equation 14-76 to obtain a limit-state function in terms of $\boldsymbol{Y}$. Now, the steps of the advanced second moment can be followed using the $\boldsymbol{Y}$ variables, not the $\boldsymbol{X}$ variables. In the case of nonnormal random variables, Equation 14-84 needs to be reevaluated in every iteration, as the equivalent mean and standard deviation that appear in this equation might change from one iteration to another.

14.6. SIMULATION METHODS

14.6.1. Direct Monte Carlo Simulation

Monte Carlo simulation techniques can be used to estimate the probabilistic characteristics of the functional relationship Z in Equation 14-13. Monte Carlo, or direct simulation, consists of drawing samples of the basic variables according to their probabilistic characteristics and then feeding them into the performance function. If it is known that failure occurs when $g(.) < 0$, then an estimate of the probability of failure, P_f, can be found by:

$$P_f = \frac{N_f}{N} \tag{14-85}$$

where N_f is the number of simulation cycles when $g(.) < 0$, and N is the total number of simulation cycles. As N approaches infinity, P_f approaches the true probability of failure. The accuracy of Equation 14-85 can be evaluated in terms of its bias and error variance, herein called *variance*. The estimator of the failure probability according to Equation 14-85 should be unbiased. For a small probability of failure and/or a small number of simulation cycles, the variance of P_f can be quite large. Consequently, it may take a large number of simulation cycles to achieve a specified accuracy, especially as the probability of failure is unknown. The variance of the estimated probability of failure can be computed by assuming each simulation cycle constitutes a Bernoulli trial. Therefore, the number of failures in N trials can be considered to follow a binomial distribution. Then, the variance of the estimated probability of failure can be approximately computed as:

$$\text{Var}\left((P_f) \approx \frac{(1-P_f)P_f}{N}\right) \tag{14-86}$$

It is recommended to measure the statistical accuracy of the estimated probability of failure by computing its coefficient of variation as:

$$\text{COV}(P_f) \approx \frac{\sqrt{\frac{P_f(1-P_f)}{N}}}{P_f} \tag{14-87}$$

It is evident from Equations 14-86 and 14-87 that as N approaches infinity, Var(P_f) and COV(P_f) approach zero.

Example 14-8: Flexural Reliability of a Beam, Nonlinear Normal Case

The performance function describing the flexural behavior of a simply supported beam of a span length $\sqrt{2}\,L$ that carries a uniform load W was discussed in Examples 14-3 and 14-4. It is given by:

$$Z = YS - \frac{WL^2}{4}$$

where Y is the yield stress of the material of the beam, and S is the elastic section modulus. In this example, failure is defined as yielding at the extreme material fibers of the cross section of the beam. The probabilistic characteristics of the random variables are given in Table 14-2.

Using direct Monte Carlo simulation, the random variables Y, S, W, and L were randomly generated and substituted in the performance function. Failures were then counted by monitoring the sign of the resulting evaluations of the performance function (negative means failure). Then, the probability of failure was estimated as N_f/N. Also, the COV(P_f) was computed using Equation 14-87. The number of simulation cycles N was varied to illustrate convergence of the simulation process. The results are shown in Figure 14-7 using a nonarithmetic scale for the number of simulation cycles (N) in order to show the effect of N on P_f and COV(P_f) at small as well as large N values.

In the classical use of simulation-based methods, all of the basic random variables are randomly generated, and Equation 14-13 is evaluated. Failures are then counted depending on the resulting sign of Equation 14-13. The probability of failure is estimated as the ratio of the number of failures to the total number of

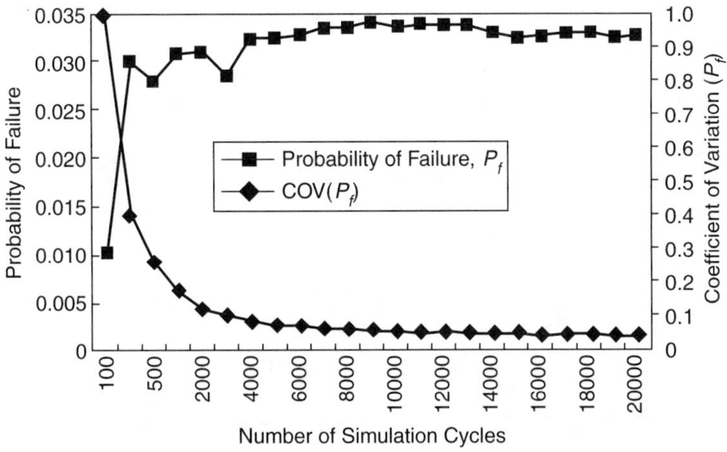

FIGURE 14-7 Results for Example 14-8.

simulation cycles. Therefore, for smaller probabilities of failure, larger numbers of simulation cycles are needed to estimate the probability of failure within an acceptable level of statistical error. The amount of computer time necessary for this method could be relatively large for small failure probabilities, but the analytical effort is relatively small.

The efficiency of simulation can be improved considerably by using variance-reduction techniques. However, the level of computational difficulty increases. Importance sampling and conditional expectation combined with antithetic variates variance-reduction techniques are commonly used for structural reliability assessment. These methods were determined to be highly efficient and converge to the correct probability of failure in a relatively small number of simulation cycles.

14.6.2. Importance Sampling Method

The probability of failure of a structure according to the performance function of Equation 14-13 is provided by the integral of Equation 14-16. In evaluating this integral with direct simulation, the efficiency of the simulation process depends on the magnitude of the probability of failure (i.e., the location of the most likely failure point or design point as defined in Section 14.5). The deeper the location of the design point in the failure domain, the larger the necessary simulation effort to obtain failures; in other words, smaller failure probabilities require larger numbers of simulation cycles. This deficiency can be addressed by using importance sampling (IS). In this method, the basic random variables are generated according to some carefully selected probability distributions with mean values that are closer to the design point than their original (actual) probability distributions. It should be noted that the design point is not known in advance; the analyst can only guess. Therefore, failures are obtained more frequently and the simulation efficiency is increased. To compensate for the change in the probability distributions, the results of the simulation cycles should be corrected. The fundamental equation for this method is given by:

$$P_f = \frac{1}{N}\sum_{I=1}^{N} I_f \frac{f_X(x_{1i}, x_{2i}, \ldots, x_{ni})}{h_X(x_{1i}, x_{2i}, \ldots, x_{ni})} \tag{14-88}$$

where N is the number of simulation cycles, $f_X(x_{1i}, x_{2i}, \ldots, x_{ni})$ is the original joint density function of the basic random variables evaluated at the ith generated values of the basic random variables, $h_X(x_{1i}, x_{2i}, \ldots, x_{ni})$ is the selected joint density function of the basic random variables evaluated at the ith generated values of the basic random variables, and I_f is the failure indicator function that takes values of either 1 for failure or 0 for survival. In Equation 14-88, $h_X(\mathbf{x})$ is called the *sampling density function* or the *importance function.*

Efficiency (and thus the required number of simulation cycles) depends on the choice of the sampling density function. A number of procedures for selecting the sampling density functions have been suggested and are beyond the scope of this chapter.

14.6.3. Conditional Expectation Method

The performance function for a fundamental structural reliability assessment case is given by:

$$Z = R - L \tag{14-89a}$$

where R is the function of structural strength or resistance, and L is the function of the corresponding load effect. Therefore, the probability of failure, P_f, is given by:

$$P_f = P(Z < 0) = P(R < L) \tag{14-89b}$$

For a randomly generated value of L (or R), say l_i (or r_i), the probabilities of failure are given by:

$$P_{f_i} = \text{Prob}(R < l_i) = F_R(l_i) \tag{14-90a}$$

or

$$P_{f_i} = \text{Prob}(L > r_i) = 1 - F_L(r_i) \tag{14-90b}$$

where F_R and F_L are cumulative distribution functions of R and L, respectively. In this formulation, R and L are assumed to be statistically uncorrelated random variables. Thus, for N simulation cycles, the mean value of the probability of failure is given by the following equation:

$$\bar{P}_f = \frac{\sum_{i=1}^{N} P_{f_i}}{N} \tag{14-91}$$

The variance (Var) and the coefficient of variation (COV) of the estimated (i.e., sample) probability of failure, respectively, are given by:

$$\text{Var}(\bar{P}_f) = \frac{\sum_{i=1}^{N} (P_{f_i} - \bar{P}_f)^2}{N(N-1)} \tag{14-92a}$$

and

$$\text{COV}(\bar{P}_f) = \frac{\sqrt{\text{Var}(\bar{P}_f)}}{\bar{P}_f} \tag{14-92b}$$

For simplicity of notation, COV($\bar{P}_f$) is sometimes referred to as COV(P_f).

For the general performance function given by Equation 14-13, the conditional expectation (CE) method can be utilized by randomly generating all the basic random variables except one variable, called the *control variable* X_k. The randomly generated variables should be selected as the ones of least variability, and the resulting conditional expectation can be evaluated by some known expression (e.g., the cumulative distribution function of the control random variable that was not randomly generated). This method can be used for any performance function with any probability distributions for the random variables. The only limitation is that the control random variable, X_k, must be statistically uncorrelated to the $g_k(.)$ as shown in the following equation:

$$P_f = E_{X_i:\, i=1,2,\ldots,n\ \&\ i \neq k}[F_{X_i}(g_k(X_i\text{: } i = 1, 2, \ldots, n\ \&\ i \neq k))] \tag{14-93a}$$

where $g_k(X_i\text{: } i = 1, 2, \ldots, n\ \&\ i \neq k)$ is defined as follows:

$$P_f = \text{Prob}\,[X_k < g_k(X_i\text{: } i = 1, 2, \ldots, n\ \&\ i \neq k)] \tag{14-93b}$$

According to this method, the variance of the estimated quantity is reduced by removing the variability of the control variable from the simulation process. In addition, the method converges to the correct probability of failure in a relatively small number of simulation cycles.

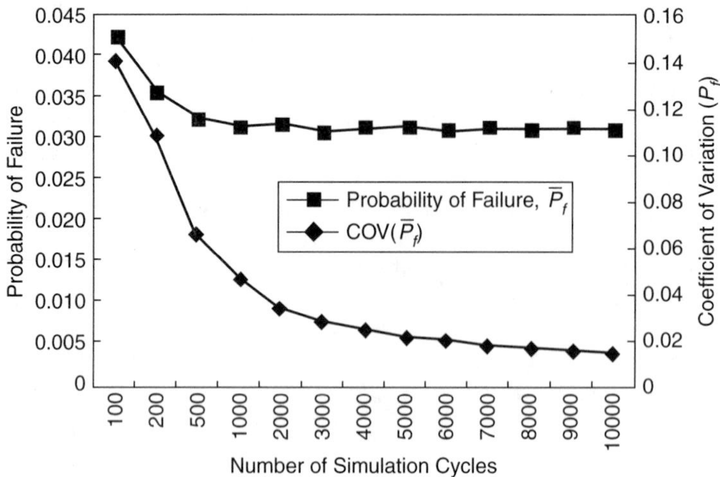

FIGURE 14-8 Results for Example 14-9.

Example 14-9: Flexural Reliability of a Beam, Nonlinear Normal Case

The performance function for the flexural behavior of a simply supported beam of a span length $\sqrt{2}\,L$ that carries a uniform load W was discussed in Examples 14-3, 14-4, and 14-8. It is given by:

$$Z = YS - \frac{WL^2}{4}$$

where W is the uniform load, L is the span length, Y is the yield stress of the material of the beam, and S is the elastic section modulus. In this example, failure is defined as yielding at the extreme material fibers of the cross section of the beam. The probabilistic characteristics of the random variables are given in Table 14-2.

Using conditional expectation with W as the control variable, Y, S, and L were randomly generated, and the cumulative distribution function of W was used to compute the probability of failure at each simulation cycle. Then, the probability of failure was estimated as the average probability of failure based on all the simulation cycles. Also, the COV($\bar{P}_f$) was computed. The number of simulation cycles N was varied to illustrate convergence of the simulation process. The results are shown in Figure 14-8 using a nonarithmetic scale for N.

14.6.4. Generalized Conditional Expectation Method

The conditional expectation method can be generalized to produce the generalized conditional expectation (GCE) method by allowing the number of the control variables to be larger than one. Equation 14-93 can be generalized as follows:

$$P_f = E_{X_i:\, i = 1, 2, \ldots, n \,\&\, i \notin k}[P_f(\mathbf{X_k})] \tag{14-94}$$

where $\mathbf{X_k}$ is a vector of control random variables, $X_{k1}, X_{k2}, \ldots, X_{km}$, and $P_f(\mathbf{X_k})$ is the probability of failure evaluated in the dimensions of $X_{k1}, X_{k2}, \ldots, X_{km}$. This probability can be evaluated using any method, such as moment methods, importance sampling, or conditional expectation.

The computational steps according to this generalized approach are summarized in the following:

1. The performance function should be defined according to Equation 14-13.
2. The control random variables, $\mathbf{X_k} = (X_{k1}, X_{k2}, \ldots, X_{km})$, are selected on the basis of reducing the dimensionality of the problem to the space of the control random variables. All other random variables, X_i: $i \notin [k_1, k_2, \ldots, k_m]$, are considered the conditional random variables, and they are generated randomly according to their respective probability distributions.
3. Therefore, the reliability assessment problem is reduced to N evaluations of the probability term (Prob) in the following expression:

$$P_f = E_{X_i:\, i = 1, 2, \ldots, n \,\&\, i \notin k}\{\text{Prob}[g(X_1, X_2, \ldots, X_{k1}, X_{k2}, \ldots, X_{km}, \ldots, X_n) < 0]\} \tag{14-95}$$

   for N number of simulation cycles.
4. In the nongeneralized conditional expectation method, the expression "Prob[$g(x_1, x_2, \ldots, X_{k1}, X_{k2}, \ldots, X_{km}, \ldots, x_n) < 0$]" with $m = 1$ is evaluated using the cumulative distribution function (CDF) of X_k as given in Equation 14-93. It should be noted that the lower cases for $x_1, x_2, \ldots, x_n$ in this expression indicate generated values of the random variables $X_1, X_2, \ldots, X_n$. In the generalized approach, the value of m is larger than one; therefore, the probability expression can be evaluated using any suitable (or convenient) method. For example, the FORM, advanced second-moment method, importance sampling, conditional expectation (CE), or other structural reliability method can be used for this purpose. The choice of the m random variables of $\mathbf{X_k}$ should be based on the intended method for the evaluation of the probability expression. Care should be exercised in selecting the m variables that result in simplifying the reliability evaluation, in addition to reducing the dimensionality of the problem. The simplification can be, for example, in the form of (1) reducing a nonlinear performance function into a linear function that is used in the probability expression, (2) using a closed-form expression for evaluating the probability expression, and (3) removing random variables with nonnormal probability distributions from the expression. These concepts are best explained using Examples 14-10 and 14-11.

14.6.5. Antithetic Variates Method

According to the antithetic variates (AV) variance-reduction technique, negative correlation between different cycles of simulation is induced in order to decrease the variance of the estimated mean value. If U is a random number uniformly distributed in the range [0,1] and is used in a computer run to determine the probability of failure $P_{f_i}^{(1)}$, the 1 – U can be used in another run to determine the probability of failure $P_{f_i}^{(2)}$. Therefore, the probability of failure in the ith simulation cycle is given by:

$$P_{f_i} = \frac{P_{f_i}^{(1)} + P_{f_i}^{(2)}}{2} \tag{14-96}$$

Then, the mean value of probability of failure can be calculated by Equation 14-91. The AV variance-reduction technique can be used in combination with the nongeneralized or generalized conditional expectation VRT. The negative correlation can be achieved by using, for example, the inverse transformation method for generating the random variables as discussed in Chapter 7. In the random-number-generation process for each simulation cycle, say the ith cycle, a set of uncorrelated random numbers based on the uniform random variable U is used in the first stage of the ith cycle to determine the probability of failure $P_{f_i}^{(1)}$. In the second stage of the ith cycle, a complementary set of uncorrelated random numbers based on the random variable (1 – U) is used to determine the probability of failure $P_{f_i}^{(2)}$. Therefore, the probability of failure in the ith simulation cycle is given by Equation 14-96. The result is simply additional reduction in the variance of the estimated quantity and expedited convergence.

Example 14-10: Flexural Reliability of a Beam, Nonlinear Normal and Nonnormal Cases

Consider the first-yield failure mode of a structural steel section subjected to a bending moment loading. The performance function is given by:

$$Z = YS - M \tag{14-97}$$

where Y is the yield stress of material, S is the elastic section modulus, and M is the moment effect due to applied loading. The statistical characteristics of the variables are shown in Table 14-8. These variables are assumed to be statistically uncorrelated.

The probability of failure of the structural component according to the first-yield failure mode can be expressed as:

TABLE 14-8 Statistical Characteristics of Variables for Example 14-10

Random variable	Mean value	COV	Case 1	Case 2
Y	275.52 MPa	0.125	Normal	Lognormal
S	8.19×10^{-4} m^3	0.050	Normal	Lognormal
M	1.13×10^5 N-m	0.200	Normal	Type I, largest

TABLE 14-9 Results of Conditional Expectation Method in Example 14-10

Simulation method	Number of cycles	Estimated failure probability $\overline{P}_f$	COV($\overline{P}_f$)
Case 1: normal			
Direct Monte Carlo	200,000	0.00128	0.0625
Conditional expectation (CE)	40,000	0.00118	0.0460
Generalized CE	500	0.00118	0.0380
Case 2: nonnormal			
Direct Monte Carlo	100,000	0.00325	0.0560
Conditional expectation (CE)	2000	0.00319	0.0460
Generalized CE	500	0.00300	0.0240

$$P_f = \text{Prob}(Z < 0) = \text{Prob}(M > YS) \tag{14-98}$$

The control random variable, in this case, is selected as the random variable M, because it has the largest coefficient of variation (COV). Therefore, conditioning on Y and S, the cumulative distribution function of M was used to evaluate P_f in each simulation cycle. The variables Y and S were randomly generated using the inverse transformation method, then the CE variance-reduction technique was used to estimate the probability of failure. To improve the performance of the method, the AV variance-reduction technique was combined with CE to estimate the probability of failure.

In this example, two cases were considered, normal random variables (case 1) and nonnormal random variables (case 2), as shown in Table 14-8. For the nonnormal case, the probability of failure according to the specified performance function and the distribution types for the ith simulation cycle is given by:

$$P_{f_i} = 1 - F_M(y_i s_i) = 1 - \exp\{-\exp[-\alpha(y_i s_i - u)]\} \tag{14-99}$$

in which α and u are the parameters of type I, largest extreme value distribution for the assumed probabilistic characteristics of M, and F_M is the cumulative distribution function of M. The sample mean and COV of the failure probability were then determined using Equations 14-91 and 14-92b, respectively, and are shown in Table 14-9 for the normal and nonnormal probability distributions.

For the purpose of comparison, P_f was recalculated using the ASM method. The results are 1.1×10^{-3} and 3×10^{-3} for the normal and nonnormal probability distributions, respectively. For the ASM method, P_f was determined as:

$$P_f = 1 - \Phi(\beta) \tag{14-100}$$

where Φ is the CDF of the standard normal probability distribution, and β is the reliability or reliability index.

Using the generalized conditional expectation (GCE) method, M and Y were selected as the control variables, and S was randomly generated. For the ith simulation cycle, the probability expression as described in Equation 14-95 is given by:

$$P_{f_i} = \text{Prob}[Ys_i - M < 0] \tag{14-101}$$

where s_i is a randomly generated value of S. The probability expression was then evaluated for the normal probability distributions (case 1), as follows:

$$P_{f_i} = 1 - \Phi\left[\frac{\mu_Y s_i - \mu_M}{\sqrt{s_i^2\sigma_Y^2 + \sigma_M^2}}\right] \tag{14-102}$$

where μ is the mean value and σ is the standard deviation. For the nonnormal probability distributions (case 2), the ASM method was used to determine P_{f_i}. Then the mean value and COV of the failure probability were determined using Equations 14-91 and 14-92b, respectively, for N simulation cycles. The resulting statistical characteristics of P_f are shown in Table 14-9 for the normal and nonnormal probability distributions.

Example 14-11: Flexural Reliability of a Portal Frame, Linear Nonnormal Case

The probability of occurrence of the most critical mode of a one-bay, one-story portal frame subjected to horizontal and vertical loads is of interest in this example. The performance function for the most critical failure mode is given by:

$$Z = 4M_1 + 2M_2 - 1500k - 1000 \tag{14-103}$$

where M_1 and M_2 are the plastic capacities for the two columns, and k is the wind load factor. The columns are assumed to be weaker than the beam. The assumed probabilistic characteristics of the basic random variables are summarized in Table 14-10. Because the columns have different plastic moment capacities from the beam, the plastic hinge at the right joint in the most critical failure mechanism is on the column. In this example, combinations of the ASM method, the CE method, the GCE, and the AV method of the variance-reduction techniques were used to estimate the probability of failure of the frame. The control variable was selected as k for the CE method, and M_1 and k for the GCE method. The results are shown in Table 14-11. The ASM method resulted in a probability of failure of 4.8×10^{-3}.

It is evident from Examples 14-10 and 14-11 that the assessment of failure probability based on nonlinear performance function using the GCE method can be reduced to averaging N evaluations of the probability of structural failure according

TABLE 14-10 Probabilistic Characteristics of Random Variables for Example 14-11

Variable	Mean value	Coefficient of variation	Distribution type
M_1	4.068×10^5 N-m (300 kip-ft)	0.15	Normal
M_2	6.102×10^5 N-m (450 kip-ft)	0.10	Normal
k	0.3	0.33	Normal

TABLE 14-11 Simulation Results of Example 14-11

Simulation method	Number of cycles	Estimated failure probability $\overline{P}_f$	COV($\overline{P}_f$)
Direct Monte Carlo	10,000	0.00475	0.046
Conditional expectation (CE)	5000	0.00467	0.078
Generalized CE	500	0.00481	0.057

to a linear expression of the performance function. This transformation can be achieved by carefully selecting the control random variables. The probabilistic evaluation of the linear expression was performed, in the two examples, using the ASM method. Other methods could have been used to achieve this objective. The choice of the ASM method was for the purpose of illustrating merging moment reliability methods with conditional expectation in Monte Carlo simulation. This concept can be greatly utilized in complex performance functions to transform them into computationally manageable formats.

Another unique aspect of the GCE method is the reduction of the structural reliability assessment problem that has a performance function with nonnormal probability distributions to N probabilistic evaluations based on a performance expression with only normal probability distributions. These evaluations can be determined under certain conditions using closed-form equations. A combination of the above two aspects of the GCE method can be utilized in solving structural reliability problems.

14.7. RELIABILITY-BASED DESIGN

Reliability-based design of structures requires the consideration of three components: (1) loads, (2) structural strength, and (3) methods of reliability analysis. The two primary approaches for reliability-based design are (1) direct reliability-based design, and (2) load and resistance factor design (LRFD). The direct reliability-based design approach can include both level 2 and/or level 3 reliability methods. Level 2 reliability methods are based on the moments (mean and variance) of random variables, sometimes with a linear approximation of nonlinear limit states; level 3 reliability methods, however, use the complete probabilistic characteristics of the random variables. In some cases, level 3 reliability analysis is not possible because of a lack of complete information on the full probabilistic characteristics of the random variables. Also, the computational difficulty of level 3 methods sometimes discourages their use. The LRFD approach is considered a level 1 reliability method. Level 1 reliability methods utilize partial safety factors (PSFs) that are reliability based, but the methods do not require explicit use of the probabilistic description of the variables.

The design of any structural system or element must provide for adequate safety and proper functioning of that system or element regardless of what philosophy of design is used. The structural systems or elements must have adequate strength to permit proper functioning during their intended service life. The performance of a structural element is defined by a set of requirements stated in terms of tests and measurements of how well the structure serves various or intended functions over its service life. Reliability and risk measures can be considered as performance measures, specified as target reliability levels (or as a target reliability index, β_0). The selected reliability levels of a particular structural element reflect the probability of failure of that element. These levels can be set based on implied levels in the currently used design practice with some calibration, or they can be based on cost–benefit analysis. Three methods can be used to select a target reliability value: (1) agree upon a reasonable value in cases of novel structure without history, (2) calibrate reliability levels implied in currently used design codes, or (3) choose a target reliability level that minimizes total expected costs over the service life of, for example, a marine structure when dealing with design for which failures result in only economic losses and consequences.

The reliability-based design approaches for a system start with the definition of a mission and an environment for a system. Then, the general dimensions and arrangements, structural member sizes, scantlings, and details must be assumed. The weight of the structure can then be estimated to ensure its conformance to a specified limit. Using an assumed operational or usage profile, analysis of the system produces stochastic responses. The resulting responses can be adjusted using modeling uncertainty estimates that are based on any available results of full-scale or large-scale testing.

The reliability-based design procedure also requires defining performance functions that correspond to limit states for significant failure modes. In general, the problem can be considered as one of supply and demand. Failure of a structural element occurs when the supply (i.e., strength of the element) is less than the demand (i.e., loading on the element). On the other hand, reliability of this element is achieved when the supply is greater than the demand.

Reliability-based design methods offer many advantages and benefits; for example, they:

1. Provide the means for management of uncertainty in loading, strength, and degradation mechanisms.
2. Provide consistency in reliability.
3. Result in efficient and possibly economical use of materials.
4. Provide compatibility and reliability consistency across materials, such as steel grades, aluminum, and composites.

5. Allow for future changes as a result of information gained in prediction models, and material and load characterization.
6. Provide directional cosines and sensitivity factors that can be used for defining future research and development needs.
7. Allow for performing time-dependent reliability analysis that can form the bases for life-expectancy assessment, life extension, and development of inspection and maintenance strategies.
8. Are consistent with many steel construction, bridge, building, concrete, mechanical system, and marine industries, among others.
9. Allow for performing system reliability analysis.

14.7.1. Direct Reliability-Based Design

The direct reliability-based design method uses all available information about the basic variables (including correlation) and does not simplify the limit state in any manner. Using reliability assessment methods, each reliability index, β, for all modes at all levels must be computed and compared with target reliability indices. The design can be iteratively performed using reliability checking according to the following condition:

$$\beta \geq \beta_0 \tag{14-104}$$

14.7.2. Load and Resistance Factor Design

The second approach (LRFD) of reliability-based design consists of the requirement that a factored (reduced) strength of a structural component is larger than a linear combination of factored (magnified) load effects as given by the following general format:

$$\phi R_n \geq \sum_{i=1}^{m} \gamma_i L_{ni} \tag{14-105}$$

where ϕ is the strength factor, R_n is the nominal (or design) strength, γ_i is the load factor for the ith load component out of n components, and L_{ni} is the nominal (or design) value for the ith load component out of m components.

In this approach, load effects are increased, and strength is reduced, by multiplying the corresponding characteristic (nominal) values with factors, which are called *strength* (resistance) and *load factors*, respectively, or *partial safety factors*. The characteristic value of some quantity is the value used in current design practice and is usually equal to a certain percentile of the probability distribution of that quantity. The load and strength factors are different for each type of load and strength. Generally, the higher the uncertainty associated with a load, the higher the corresponding load factor; also, the higher the uncertainty associated with strength, the lower the corresponding strength factor. These factors are determined probabilistically to correspond to a prescribed level of reliability or safety. It is also common to consider two classes of performance function that correspond to strength and serviceability requirements.

The advanced second-moment method (ASM) can be used to estimate partial safety factors such as those found in the design format of Equation 14-105. At the failure point $(R^*, L_1^*, \ldots, L_m^*)$, the limit state of Equation 4-105 is given by:

$$g = R^* - L_1^* - \ldots - L_m^* = 0 \tag{14-106}$$

or, in a general form:

$$g(X) = g(x_1^*, x_2^*, \ldots, x_m^*) = 0 \tag{14-107}$$

For the given target reliability index β_0, probability distributions and statistics (means and standard deviations) of the load effects and the coefficient of variation of the strength, the mean value of the resistance, and the partial safety factors can be determined by iterative solution of the ASM. The mean value of the resistance and the design point can be used to compute the required mean partial design safety factors as follows:

$$\phi = \frac{R^*}{\mu_R} \tag{14-108}$$

$$\gamma_i = \frac{L_i^*}{\mu_{L_i}} \tag{14-109}$$

The strength factors are generally less than one, whereas the load factors are greater than one.

In developing design-code provisions for structures, it is sometimes necessary to follow the current design practice to ensure consistent levels of reliability over various types of structures. Calibrations of existing design codes are necessary to make the new design formats as simple as possible and to put them in a form that is familiar to the users or designers. Moreover, the partial safety factors for the new codes should provide consistent levels of reliability. For a given reliability index β and probability characteristics for the resistance and the load effects, the partial safety factors determined by the ASM approach might be different for various failure modes for the same structural component. Therefore, the calculated partial safety factors should be adjusted to maintain the same values for all loads at different failure modes by the strength factor ϕ for a given set of load factors. The following algorithm can be used to accomplish this objective:

1. For a given value of the reliability index β, probability distributions and statistics of the load variables, and the coefficient of variation for the strength, compute the mean strength needed to achieve the target reliability using the first-order reliability method as outlined in the previous sections.
2. With the mean value for R computed in step 1, the partial safety factor can be revised for a given set of load factors as follows:

$$\phi' = \frac{\sum_{i=1}^{m} \gamma_i' \mu_{L_i}}{\mu_R} \tag{14-110}$$

where ϕ' is the revised strength factor; μ_{L_i} and μ_R are the mean values of the loads and strength variables, respectively; and γ_i' , $i = 1, 2, \ldots, m$, are the given set of load factors.

Example 14-12: Unstiffened Plate Panel Under Uniaxial Compression

Plates are important components in structures; therefore, they should be designed for a set of failure modes such as yielding, buckling, and fatigue of critical connecting components. This example considers only a simply supported rectangular plate of size a by b under uniaxial compressive stress due to stillwater and wave bending of a ship structure. The limit state for this case is given by:

$$g = F_u - f_{SW} - f_W \tag{14-111}$$

where F_u is the strength of the plate (stress), f_{SW} is external stress due to stillwater bending, and f_W is external stress due to wave bending.

The partial safety factors were computed using a target reliability index β of 3.0. The advanced second-moment method requires the probabilistic characteristics of F_u, f_{SW}, and f_W. The plate strength, F_u, is assumed to have a lognormal distribution with a coefficient of variation of 0.04 to 0.08. The mean value of F_u is determined to meet a target reliability level β. The stillwater load effect, f_{SW}, is due to stillwater bending that can be assumed to follow a normal distribution with a coefficient of variation of 0.2. The wave load effect, f_W, is due to waves that can be assumed to follow an extreme value distribution (type I, largest) with a coefficient of variation of 0.1. The mean values of stillwater and waves are considered in the study in the form of a ratio of wave/stillwater loads that ranges from 1.5 to 1.7. F_u, f_{SW}, and f_W are assumed to have mean-to-nominal ratios of 1.0, 1.0, and 1.03, respectively.

The ratios of means for strength/stillwater load and the partial safety factors for a target reliability of 3.0 are computed as summarized in Tables 14-12 and 14-13, respectively, and in Figure 14-9. Based on these results, the following preliminary values for partial safety factors are recommended for demonstration purposes:

TABLE 14-12 Computed Ratios of Means for Strength/Stillwater Load

	Ratios of means for wave/stillwater load		
COV(F_u)	**1.5**	**1.6**	**1.7**
0.04	3.43035	3.5695	3.70977
0.08	3.6375	3.7817	3.9271

TABLE 14-13 Computed Partial Safety Factors

Partial safety factors	Ratios of means for wave/stillwater load 1.5	1.6	1.7
Strength reduction factor (ϕ)			
COV(F_u) = 0.04	0.960338	0.961079	0.961747
COV(F_u) = 0.08	0.863684	0.86526	0.86679
Stillwater load factor (γ_S)			
COV(F_u) = 0.04	1.301221	1.283616	1.267817
COV(F_u) = 0.08	1.28566	1.270806	1.257081
Wave load factor (γ_W)			
COV(F_u) = 0.04	1.328696	1.341832	1.352955
COV(F_u) = 0.08	1.237262	1.250783	1.262827

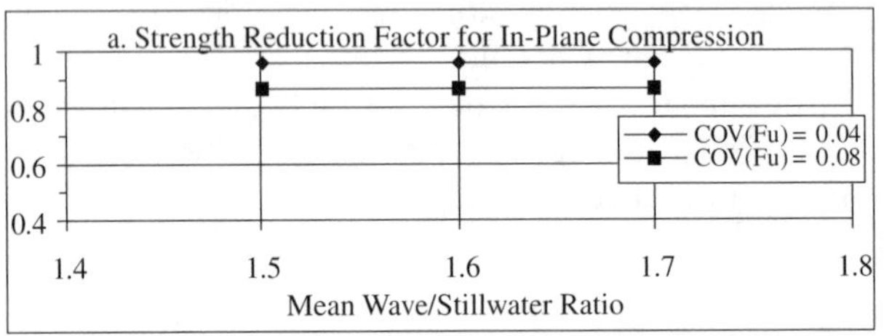

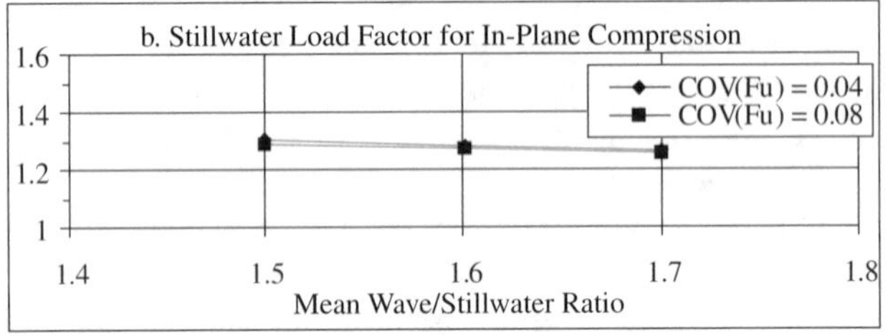

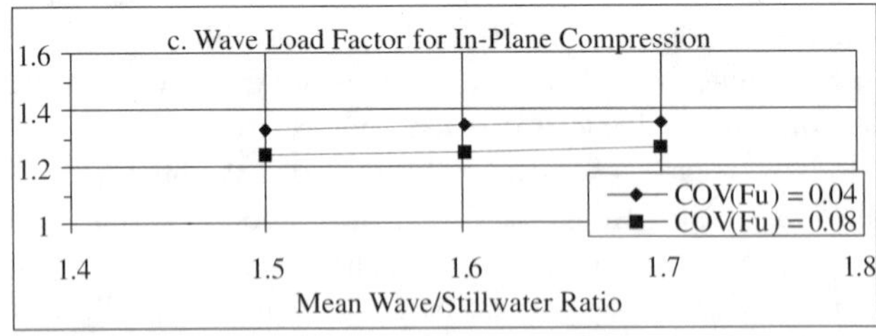

FIGURE 14-9 Partial safety factors for plates under uniaxial compressive stress.

Strength reduction factor (ϕ) = 0.85(1.03) = 0.88
Stillwater load factor (γ_S) = 1.3
Wave load factor (γ_W) = 1.25

where 1.03 is the mean-to-nominal ratio for F_u. As indicated previously for a given β and probabilistic characteristics for the strength and the load effects, the partial

TABLE 14-14 Probabilistic Characteristics of Random Variables

Random variable	Mean	COV	Distribution type
F_u	Not provided	0.06	Lognormal
f_S	1	0.2	Normal
f_W	1.6	0.1	Type I, largest

safety factors determined by the ASM approach might be different for different failure modes. For this reason, calibration of the strength factor ϕ is often necessary to maintain the same values for all γ load factors. The following numerical example illustrates revising the strength factor for a given set of load factors. For instance, given $\gamma_S = 1.3$, $\gamma_W = 1.2$, and the probabilistic characteristics of the random variables as shown in Table 14-14, the corresponding strength factor ϕ was calculated for a target reliability level of $\beta = 3.0$. Using the ASM, the mean of F_u was found to be 3.66. With the mean value known, Equation 14-111 gives:

$$\phi = \frac{\gamma_S\mu_S + \gamma_W\mu_W}{\mu_{F_u}}(1.03) = \frac{1.3(1) + 1.2(1.6)}{3.66}(1.03) = 0.91$$

Because the strength F_u has a mean-to-nominal ratio of 1.03, this ratio had to be revised by multiplying ϕ by 1.03.

14.8. APPLICATION: STRUCTURAL RELIABILITY OF A PRESSURE VESSEL

Pressure vessels are structures that are designed to support internal pressure, in addition to dead load, wind load, earthquake load, temperature, piping load, impact, and cyclic load. In this example, the reliability of a pressure vessel subjected to wind load is assessed. The example vessel is 120 ft high and 6.0 ft in diameter. The vessel is shown in Figure 14-10 with the assumed wind load distribution. The vessel is assumed to have a uniform shell thickness. The bending moment due to the wind load at the base of the vessel is considered to define the performance function. The first yield of the cross section was assumed to be the failure mode of interest.

The wind load (W_1, W_2, and W_3) as shown in Figure 14-10 is given by the following empirical equation according to the American National Standard Institute (1972):

$$W_i = 0.00256CK_iGDV^2 \tag{14-112}$$

where, in this example, W_i is the total design wind load per unit length of the vessel in pounds per foot; C is the net pressure coefficient; K_i is the velocity pressure coefficient, which depends on the exposure type and height above ground level; G is the gust factor, which depends on the exposure type and dynamic response

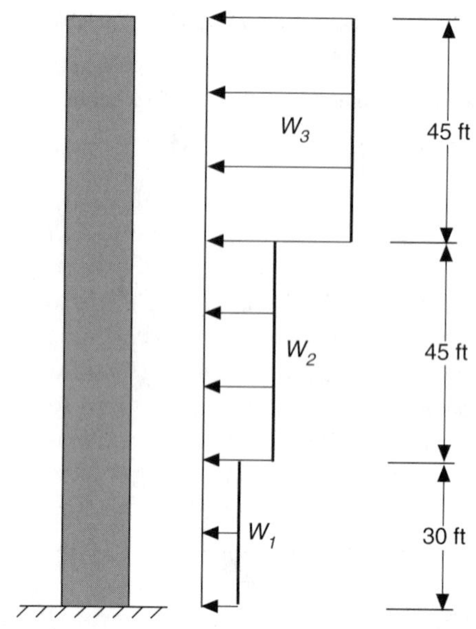

FIGURE 14-10 Wind load distribution on a pressure vessel.

characteristics of the structure; D is the effective diameter in feet, which considers the effect of ladders, insulation, and pipes attached to the vessel; and V is the basic wind speed in miles per hour recorded at 30 ft above ground level. Therefore, the load effect (L) in the form of a bending moment at the base is:

$$L = W_1 \frac{30^2}{2} + W_2(45)\left(30 + \frac{45}{2}\right) + W_3(45)\left(30 + 45 + \frac{45}{2}\right) \tag{14-113}$$

The moment capacity (R) of the circular hollow cross section at the base of the vessel is given by:

$$R = \pi Y R_d^2 T \tag{14-114}$$

where Y is the yield strength of the material, which is assumed to be steel; R_d is the radius of the vessel; and T is the wall thickness.

Using Equations 14-113 and 14-114, the performance function (Z) can be expressed as:

$$Z = R - L \tag{14-115}$$

or

$$Z = \pi Y R_d^2 T - W_1 \frac{30^2}{2} + W_2(45)\left(30 + \frac{45}{2}\right) + W_3(45)\left(30 + 45 + \frac{45}{2}\right) \tag{14-116a}$$

TABLE 14-15 Probabilistic Characteristics of Random Variables for Pressure Vessel

Random variable	Mean value	Coefficient of variation	Probability distribution
C	0.672	0.10	Normal
G	1.283	0.12	Lognormal
V	63 mph	0.16	Extreme value type II
D	8.3 ft	0.10	Lognormal
K	1.0	0.10	Lognormal
T	1/8, 3/16, 1/4, 5/16 in.	0.05	Lognormal
R_d	3.08 ft	0.05	Lognormal
F	38 ksi	0.10	Normal
$X_1 = TR^2$	0.0988, 0.1482, 0.1976, 0.2470 in.^{3}	0.08666	Lognormal
$X_2 = KGD$	10.6489 ft	0.18652	Lognormal
$X = X_1/X_2$	0.0096, 0.0144, 0.0192, 0.0240	0.20632	Lognormal

Substituting Equation 14-112 into Equation 14-116a gives:

$$Z = \pi Y R_d^2 T - 0.00256 CGDV^2[450K_1 + 2362.5K_2 + 4387.5K_3] \quad \text{(14-116b)}$$

Because K_1, K_2, and K_3 are fully correlated with the assumed mean values of 1, 1.2, and 1.4, respectively, they can be normalized with respect to K_1. Therefore, Equation 14-116b can be expressed as:

$$Z = \pi Y R_d^2 T - 24.1344 CGDKV^2 \quad \text{(14-117)}$$

where $K = K_1$. Table 14-15 provides the assumed probabilistic characteristics of the random variables. Several thicknesses are examined in this example as shown in Table 14-15. The random variables X_1, X_2, and X are used in the example to reduce the number of the random variables in the problem. The reduction is based on the property that the product or division of random variables with lognormal distribution is a random variable that has a lognormal distribution. The wind speed was assumed to have a probability distribution of extreme value (largest) type II. This distribution has the following probability density and cumulative function, respectively:

$$f_v(v) = \frac{k}{v}\left(\frac{u}{v}\right)^k \exp[-(u/v)^{-k}] \quad \text{(14-118)}$$

and

$$F_v(v) = \exp[-(u/v)^{-k}] \quad \text{(14-119)}$$

where u and k are the parameters of the distribution. These parameters are related to the mean and variance, respectively, as follows:

TABLE 14-16 Results of the Advanced Second-Moment Method for Pressure Vessel

Thickness (in.)	Safety index (β)	Failure probability
1/8	2.18	0.014629
3/16	2.80	0.002555
1/4	3.19	0.000711
5/16	3.47	0.000260

$$E(V) = u\Gamma\left(1 - \frac{1}{k}\right) \tag{14-120}$$

and

$$\mathrm{Var}(V) = u^2\left[\Gamma\left(1 - \frac{2}{k}\right) - \Gamma^2\left(1 - \frac{1}{k}\right)\right] \tag{14-121}$$

where Γ(.) is the gamma function as described in Chapter 5 and tabulated in Appendix A. The mean and coefficient of variation for the wind speed V provided in Table 14-15 give the following parameters:

$$u = 58.369 \tag{14-122a}$$

and

$$k = 8.85 \tag{14-122b}$$

Using the probabilistic characteristics of the random variables as shown in Table 14-15, the advanced second-moment method (Table 14-16) was used to assess the reliability of the vessel according to the performance function of Equation 14-117. Nonnormal random variables were transformed to normal random variables using Equations 14-57 and 14-58. The random variables were assumed to be uncorrelated. The resulting safety indices and failure probabilities for increasing numbers of simulation cycles and the four thicknesses are shown in Tables 14-17 and 14-18.

Direct Monte Carlo simulation was also used to estimate the failure probability. Utilizing the property that the product or division of random variables with lognormal distribution is a random variable that has a lognormal distribution, the performance function was simplified as follows:

$$Z = \pi X_1 Y - 24.1344 C X_2 V^2 \tag{14-123}$$

Generating the random variables and substituting in Equation 14-117, failure can be counted based on the sign of Z, then the probability of failure can be estimated using Equation 14-85. The results of direct simulation for a ¼-in.-thick vessel are shown in Table 14-17. The number of simulation cycles was increased from 500 to

TABLE 14-17 Simulation Results for Pressure Vessel with 1/4-in. Thickness

Number of simulation cycles (N)	Direct simulation (P_f)	Conditional expectation $\bar{P}_f$	Conditional expectation $COV(\bar{P}_f)$	Conditional expectation with antithetic variates $\bar{P}_f$	Conditional expectation with antithetic variates $COV(\bar{P}_f)$
500	0.000000 [a]	0.0008440	0.0642	0.0007427	0.0370
1000	0.000000 [a]	0.0008029	0.05085	0.0007472	0.0288
5000	0.000000 [a]	0.0007415	0.0208	0.0007362	0.0118
20,000	0.0004500	0.0007451	0.0107	0.0007389	0.0063

[a] Number of failures = 0 for these cases.

TABLE 14-18 Simulation Results for Pressure Vessel for All Thicknesses Using 5000 Simulation Cycles

Mean thickness (in.) ($\bar{T}$)	Direct simulation (P_f)	Conditional expectation ($\bar{P}_f$)	Conditional expectation $COV(\bar{P}_f)$	Conditional expectation with antithetic variates $\bar{P}_f$	Conditional expectation with antithetic variates $COV(\bar{P}_f)$
1/8	0.015000	0.015561	0.0199	0.015459	0.0111
3/16	0.003000	0.002640	0.0206	0.002622	0.0117
1/4	0.000000 [a]	0.000742	0.0208	0.000736	0.0118
5/16	0.000000 [a]	0.000276	0.0208	0.000274	0.0118

[a] Number of failures = 0 for these cases.

20,000 in increments. Also, the reliability of the vessel was estimated using conditional expectation without and with antithetic variates variance-reduction techniques. In these methods, the wind speed V was selected as the control variable. The cumulative distribution function of V (Equation 14-119) was used to compute the probability of failure as follows:

$$P_f = P(Z < 0) \tag{14-124a}$$

$$= P(\pi X_1 Y - 24.1344 C X_2 V^2 < 0) \tag{14-124b}$$

$$= P\left(V > \sqrt{0.1301707\frac{XY}{C}}\right) \tag{14-124c}$$

$$= 1 - F_V\left(\sqrt{0.1301707\frac{XY}{C}}\right) \tag{14-124d}$$

Equations 14-91 and 14-92 were then used to estimate the failure probability and its coefficient of variation. The results are also shown in Table 14-18.

Table 14-17 indicates that as the number of simulation cycles is increased, the accuracy of the estimated failure probability increases. Also, the table shows the inability of direct simulation to estimate the failure probability using small numbers of simulation cycles. Based on this table, a value for N of 5000 simulation cycles was selected to study the other thicknesses in Table 14-18. The results are shown in Table 14-18. The table shows that the vessel reliability improves as the thickness is increased. Also, the table indicates that as the failure probability becomes small, the direct simulation method becomes incapable of estimating it.

14.9. PROBLEMS

14-1. For the following performance function, determine the safety index (β) using (a) the first-order reliability method, and (b) the advanced second-moment method:

$$Z = X_1 X_2 - \sqrt{X_3}$$

The noncorrelated random variables are assumed to have the following probabilistic characteristics:

Random variable	Mean value	Coefficient of variation	Distribution type
X_1	1	0.25	Normal
X_2	5	0.05	Normal
X_3	4	0.20	Normal

14-2. For the following performance function, determine the safety index (β) using (a) the first-order reliability method, and (b) the advanced second-moment method:

$$Z = X_1 X_2 - \sqrt{X_3}$$

The noncorrelated random variables are assumed to have the following probabilistic characteristics:

Random variable	Mean value	Coefficient of variation	Distribution type
X_1	1	0.25	Lognormal
X_2	5	0.05	Lognormal
X_3	4	0.20	Lognormal

14-3. Redo Problem 14-1 using (a) the direct Monte Carlo simulation method, and (b) the conditional expectation method. Use 100 and 2000 simulation cycles to estimate the failure probability and its coefficient of variation.

14-4. Redo Problem 14-2 using (a) the direct Monte Carlo simulation method, and (b) the conditional expectation method. Use 100 and 2000 simulation cycles to estimate the failure probability and its coefficient of variation.

14-5. For the following performance function, determine the safety index (β) using (a) the first-order reliability method, and (b) the advanced second-moment method:

$$Z = X_1 - X_2 - X_3$$

The uncorrelated random variables are assumed to have the following probabilistic characteristics:

Random variable	Mean value	Coefficient of variation	Distribution type
X_1	10	0.25	Normal
X_2	5	0.05	Normal
X_3	3	0.20	Normal

14-6. For the following performance function, determine the safety index (β) using (a) the first-order reliability method, and (b) the advanced second-moment method:

$$Z = X_1 - X_2 - X_3$$

The noncorrelated random variables are assumed to have the following probabilistic characteristics:

Random variable	Mean value	Coefficient of variation	Distribution type
X_1	10	0.25	Normal
X_2	5	0.05	Normal
X_3	3	0.20	Type I, largest

14-7. For the following performance function, determine an estimated failure probability and its coefficient of variation using (a) the direct simulation method, and (b) the conditional expectation method with 100 and 2000 cycles:

$$Z = X_1 - X_2 - X_3 + I_1 \sin\left(\frac{I_2}{10} W\right)$$

The last term in this equation introduces noise to the performance function. Assume that the constants $I_1 = 1$ and $I_2 = 1$. The noncorrelated random variables are assumed to have the following probabilistic characteristics, with W having a standard deviation of 1:

Random variable	Mean value	Coefficient of variation	Distribution type
X_1	10	0.25	Normal
X_2	5	0.05	Normal
X_3	3	0.20	Normal
W	0	Not applicable (COV = 1)	Normal

14-8. In Problem 14-7, study the effect of changing the level of noise in the performance function. Investigate the effect of using $I_1 = 1$, 5, and 10 given that $I_2 = 1$ or 10 using the conditional expectation method, with $N = 10$, 100, 1000, and 2000 cycles. Discuss your results.

14-9. Redo Problem 14-7 using (a) importance sampling, and (b) conditional expectation with antithetic variates.

14-10. Redo Problem 14-8 using (a) importance sampling, and (b) conditional expectation with antithetic variates.

14-11. The stability of a vehicle, such as a truck, can be measured using a simplified static model as follows:

$$\text{Stability} = \frac{T}{2H}$$

where T is the nonrandom truck width (i.e., center of the right front tire to the center of the left front tire), and H is the random location of the center of gravity of the vehicle from street level. As the ratio approaches a mean limit $L = 1.0$, the tendency for the vehicle to roll is increased. Assuming that H and L are random variables with coefficients of variations of 0.15 and 0.1, respectively, perform parametric analysis by assessing failure (rollover) probabilities as functions of typical T (5 to 7 ft) and mean H (2 to 4 ft) values with mean $L = 1.0$. Assume (a) normal and (b) lognormal probability distributions for all the random variables.

14-12. Use the stability model for a truck as provided in Problem 14-11 to compute the mean H that you would permit for loading a truck by stacking loads vertically so that the failure (rollover) probability does not exceed 0.01 for mean values of T of 5 to 10 ft. Use lognormal probability distributions.

14-13. Catchment basins are used to protect residential areas and highways from mud slides. A mudslide basin has a capacity of C, which is a random variable with a coefficient of variation of 0.2. The expected 25-year maximum inflow is denoted F with a coefficient of variation of 1.0. Develop an expression for computing the reliability index and the failure probability assuming lognormal probability distributions for both C and F. Compute the reliability index and the failure probability assuming that the mean ratio of C/F is 1.0, 1.2, 1.4, 1.6, 1.8, 2.0, 2.2, 2.4, 2.6, 2.8, and 3.0. What mean ratio would you recommend to protect a highway section with average traffic volume of 5000 vehicles per day at an average speed of 50 mph? Discuss and justify your selection.

14-14. Use the reliability model for a catchment basin as provided in Problem 14-13 to compute the mean C that you would permit for a flow F so that the failure probability does not exceed 0.01. Provide your answer in the form of a mean C/F ratio. Perform parametric analysis by varying the target failure probability as follows: 0.01, 0.005, 0.001, 0.0005, and 0.0001.

14-15. The change in the length of a rod due to axial force P is given by:

$$\Delta L = \frac{PL}{AE}$$

where L is the length of the rod, P is the applied axial force, A is the cross-sectional area of the rod, and E is the modulus of elasticity. Assume that the change in length of the rod must not exceed 0.15 in. Using 2000 simulation cycles in (a) the direct Monte Carlo simulation method, and (b) the conditional expectation method, determine the failure probability of the rod in meeting this limit. Assume that the noncorrelated random variables have the following probabilistic characteristics:

Random variable	Mean value	Coefficient of variation	Distribution type
P	100 kips	0.35	Lognormal
L	20 in.	0.05	Normal
E	30,000 ksi	0.10	Lognormal
A	1 in.^{2}	0.05	Normal

14-16. For the rod in Problem 14-15, study the effect of increasing the number of simulation cycles on the estimated failure probability. Use the following numbers of simulation cycles: 20, 100, 500, 1000, and 2000. Provide your results in the form of plots of estimated failure probability and its coefficient of variation as a function of the number of simulation cycles.

14-17. The ultimate moment capacity, M, of an under-reinforced concrete rectangular section is given by:

$$M = A_s f_y\left(d - \frac{a}{2}\right)$$

where

$$a = \frac{A_s f_y}{0.85 b f_c'}$$

in which the following are random variables: A_s is the cross-sectional area of the reinforcing steel, f_y is the yield stress (strength) of the steel, d is the distance from the reinforcing steel to the top fibers of the beam, b is the width of the beam, and f_c' is the ultimate stress (strength) of the concrete. Assume that the random variables are statistically noncorrelated with the following characteristics:

Random variable	Mean value	Coefficient of variation	Distribution type
$F = f_y A_s$	40,000 lb	0.15	Lognormal
d	15 in.	0.05	Lognormal
b	12 in.	0.05	Normal
f_c'	3000 psi	0.20	Lognormal

The applied random moment has a coefficient of variation of 0.2 and lognormal distribution type. Using simulation methods, determine the mean value of the applied moment that results in a failure probability of about 0.001 and a coefficient of variation of the estimated failure probability of, at the most, 0.05.

14-18. Use the performance function and probabilistic characteristics of the random variables of Problem 14-5, except for the mean value of X_l, to compute the partial safety factors for a target reliability level of 3.0.

14-19. Use the performance function and probabilistic characteristics of the random variables of Problem 14-6, except for the mean value of X_l, to compute the partial safety factors for a target reliability level of 3.0.

14-20. Use the performance function and probabilistic characteristics of the random variables of Problem 14-1, except for the mean value of X_l, to compute the partial safety factors for a target reliability level of 3.0.

14-21. Use the performance function and probabilistic characteristics of the random variables of Problem 14-2, except for the mean value of X_l, to compute the partial safety factors for a target reliability level of 3.0.

14.10. SIMULATION PROJECTS

This section provides additional work items for the problem statements of the four simulation projects introduced at the end of previous chapters.

14.10.1. Structural Beam Study

Using the project information provided in Chapter 5, investigate the use of variance-reduction methods for assessing the reliability of the beam. Assess the increase in simulation efficiency as a result of using variance-reduction methods. Perform parametric analysis on the results produced for this project. Investigate the effects of individually changing mean values and standard deviations of input random variables. Investigate the effects of individually changing distribution types of input random variables.

14.10.2. Stream Erosion Study

Using the project information provided in Chapters 1 to 5, investigate the use of variance-reduction methods for studying stream erosion.

14.10.3. Traffic Estimation Study

Using the project information provided in Chapters 1 to 5, investigate the use of variance-reduction methods for estimating traffic.

14.10.4. Water Evaporation Study

Using the project information provided in Chapters 1 to 5, investigate the use of variance-reduction methods for studying water evaporation.

Chapter 15
Reliability and Risk Analysis of Systems

TABLE OF CONTENTS

15.1. INTRODUCTION

The process of engineering analysis and design can be systematically performed within a systems framework. Generally, an engineering project can be modeled to include a segment of the environment that interacts significantly with it, defining an engineering system. The limits of the system are drawn based on the nature of the project, class of performances (including failures) under consideration, and the objectives of the analysis.

The first step in solving any engineering problem is to define the architecture of the system. The definition of a system can be based on observations about its source and data elements, interaction among the elements, and behavior of the system. Each level of knowledge obtained about an engineering problem can be said to define a system of the problem. As additional levels of knowledge are added to previous ones, higher epistemological levels of system definition and description are generated which, taken together, form a hierarchy of such system descriptions.

In the previous chapter, the reliability assessment discussion was limited to the failure of one component according to one failure mode that is expressed by the performance function of Equation 14-13. In general, a component can fail in one of several failure modes. The treatment of multiple failure modes requires modeling the component behavior as a system. Also, a system can be defined as an assemblage of several components that serves some function or purpose. Again, a multicomponent system can fail according to several failure modes. In this case, one or more component failures are necessary for system failure. Modeling multimode components or multicomponent systems can be based on similar concepts. The objective of this chapter is to introduce the fundamental concepts of system analysis for the purpose of reliability assessment at the system level.

System analysis requires the recognition and modeling of some system characteristics that include (1) post-failure behavior of a component, (2) the contribution of a component failure or failure-mode occurrence to failure of the system, (3) the statistical correlation among failure modes and failure of the component, and (4) the definition of failure at the system level. The post-failure behavior of a component is needed in order to determine the remaining effect or contribution of the component, after its failure, to subsequent system response or failure. For example, in structural engineering, components can be ideally classified into brittle or ductile components according to their potential failure modes. Components lose their strength completely after failure according to a brittle failure mode; therefore, these components can be removed from the structural analysis of a system upon failure. On the other hand, components that fail according to a ductile failure mode maintain complete or substantial partial force resistance at increasing levels of deformation after failure; therefore, they continue to contribute to the behavior of the system after failure, and their contribution must be considered in the analysis of the system.

The contribution to failure of the system by a component failure or failure-mode occurrence depends on the level of redundancy in the system. The failure of some components can lead to the failure of the system, whereas the failure of other components does not result in system failure but can weaken the system.

The statistical correlation among failure modes and failure of components can have a large effect on the reliability of the system. The difficulty in assessing this correlation can result in approximate assessment of system reliability or interval assessment of the reliability by considering the extreme conditions of statistically noncorrelated and fully correlated failure modes and failure of the components.

Finally, the reliability assessment of a system requires defining failure at the system level. For example, in a structural system, the failure definition of a system can be that the remaining (or surviving) elements of a system become structurally unstable. The results of system reliability can be used in risk studies and risk-based decision making.

15.2. RELIABILITY OF SYSTEMS

15.2.1. Systems in Series

A system in series can be represented as shown in Figure 15-1. The failure of one of the n components shown in the figure can result in failure of the system. This type of system is also referred to as a *weakest-link system*. The connectivity of the components as shown in Figure 15-1 is a logical connectivity in terms of the contribution of component failure to system failure. The logical connectivity can be different than the physical connectivity of the components in the real system. For example, the statistically determinate truss shown in Figure 15-2 is a weakest-link system, because the failure of a component in the system results in the failure of the system. The system model for the truss is shown in Figure 15-3. The difference between the physical connectivity (Figure 15-2) and the logical connectivity (Figure 15-3) can be observed by comparing the two figures.

The reliability of a system in series depends on the level of dependence among the underlying failure events for the components (called E_i, $i = 1, 2, \ldots, n$). The survival of the system ($\overline{E}$) can be modeled as:

$$\overline{E} = \overline{E}_1 \cap \overline{E}_2 \cap \ldots \cap \overline{E}_{n-1} \cap \overline{E}_n \tag{15-1}$$

where $\overline{E}_i$ is the survival event of component i. For independent failure events of the components, the probability of failure of the system can be computed as:

$$P(E) = 1 - P(\overline{E}) \tag{15-2a}$$

or

$$P(E) = 1 - [1 - P(E_1)][1 - P(E_2)]\ldots[1 - P(E_{n-1})][1 - P(E_n)] \tag{15-2b}$$

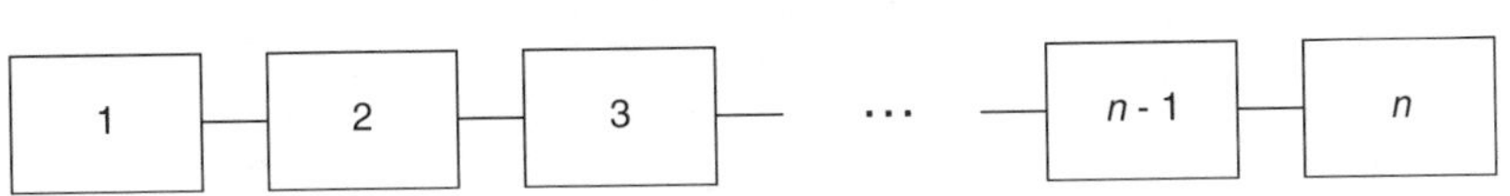

FIGURE 15-1 A system of n components in series.

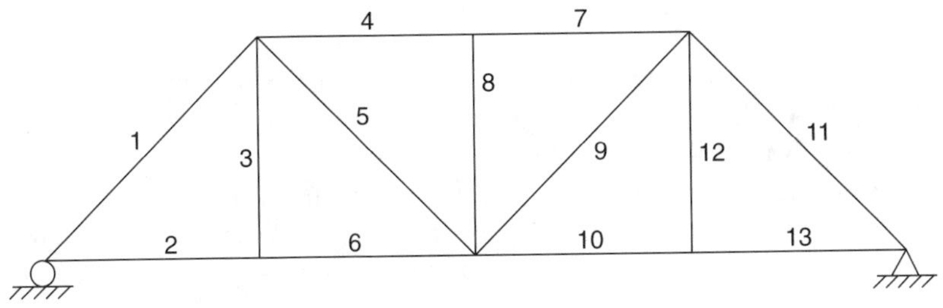

FIGURE 15-2 A truss structural system.

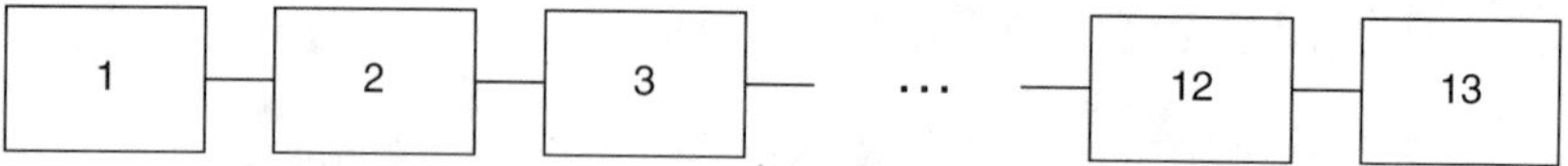

FIGURE 15-3 A system of 13 components in series for the truss.

For fully dependent events, the following relations hold:

$$E_i \subseteq (E_i \cap E_j) \tag{15-3a}$$

and

$$E_j \subseteq (E_i \cap E_j) \tag{15-3b}$$

Therefore, the following relationship holds:

$$P(E_i) \leq P(E_i \cap E_j) \tag{15-4a}$$

and

$$P(E_j) \leq P(E_i \cap E_j) \tag{15-4b}$$

For a system of n components with fully dependent failure events, the probability of survival of the system can be determined from Equation 15-1 as:

$$P(\bar{E}) = P(\bar{E}_1 \cap \bar{E}_2 \cap \ldots \cap \bar{E}_{n-1} \cap \bar{E}_n) \tag{15-5}$$

Therefore, extending Equations 15-4a and 15-4b with Equation 15-5 gives:

$$P(E) = \min_{i=1}^{n} [P(E_i)] \tag{15-6}$$

or

$$P(\bar{E}) = \max_{i=1}^{n} [P(\bar{E}_i)] \tag{15-7}$$

Equations 15-5 to 15-7 are limited to cases where the underlying performance functions for the failure events have positive statistical correlation. The results from Equations 15-2b and 15-7 provide upper and lower limits on the probability of failure of the system. The results are useful in cases where the positive dependence levels between failure events are unknown. Using Equations 15-2b and 15-7, the probability of failure of the system (P_f) can, therefore, be expressed as:

$$\max_{i=1}^{n} (P_{f_i}) \leq P_f \leq 1 - (1 - P_{f_1})(1 - P_{f_2})\ldots(1 - P_{f_{n-1}})(1 - P_{f_n}) \tag{15-8a}$$

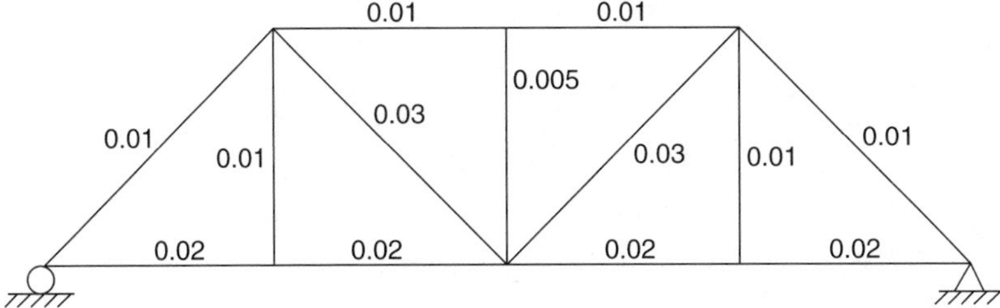

FIGURE 15-4 Example truss structural system with failure probabilities of truss members.

or

$$\max_{i=1}^{n}(P_{f_i}) \le P_f \le 1 - \prod_{i=1}^{n}(1 - P_{f_i}) \tag{15-8b}$$

where P_{f_i} is the failure probability of the ith component, for $i = 1, 2, \ldots, n$.

Example 15-1: Reliability of a Truss

For the truss shown in Figure 15-4, the truss members can be either in compression or tension depending on the applied loads and the weight of the truss. Given some loading that is random and using statics, the member forces of the truss can be determined. For tension members, the yield strength of the members can be used to determine their failure probabilities. The compression members of the truss can fail either by compressive yielding or buckling. Using the smaller strength, the failure probabilities can be computed for these members. Assuming the failure probabilities of the truss members as shown in Figure 15-4, the failure probability of the truss system can be computed using Equation 15-8 as:

$$\max(0.01, 0.02, 0.03, 0.005) \le P_f \le 1 - (1 - 0.01)^6(1 - 0.02)^4(1 - 0.03)^2(1 - 0.005) \tag{15-9}$$

or

$$0.03 \le P_f \le 0.1870 \tag{15-10}$$

The lower bound in Equation 15-10 is the failure probability of the truss system for failure events that are dependent, whereas the upper bound is the failure probability for the independent case. For real trusses, a level of dependence exists that can be related to both the strength and loading aspects of the reliability assessment problem at the member level. The computation of the dependence level can be difficult due to the various ill-defined factors that affect it, such as manufacturing and construction factors for the truss members and the contribution of dead load to the load effect in the members. Therefore, a failure probability at the system level expressed according to Equation 15-10 is commonly used.

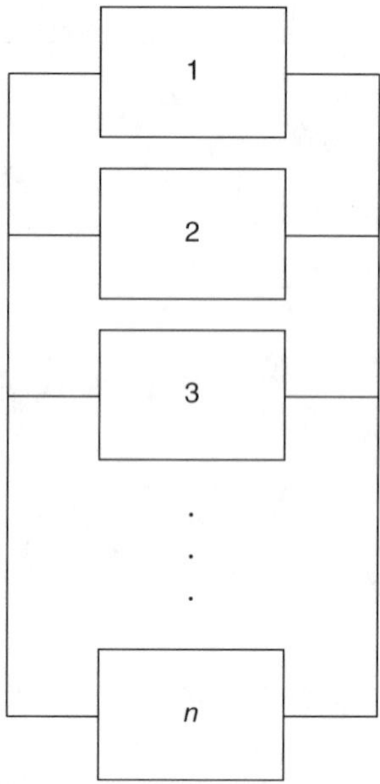

FIGURE 15-5 A system of n components in parallel.

15.2.2. Systems in Parallel

A system in parallel can be represented as shown in Figure 15-5. The n components that are shown in the figure must fail in order to result in system failure. This type of system is also referred to as a *redundant system*. The connectivity of the components as shown in Figure 15-5 is again a logical connectivity in terms of the contribution of failure of the components to the system failure.

The reliability of a system in parallel depends on the level of dependence among the underlying failure events for the components (called E_i, $i = 1, 2, \ldots, n$). The failure of the system (E) can be modeled as:

$$E = E_1 \cap E_2 \cap \ldots \cap E_{n-1} \cap E_n \tag{15-11}$$

or the survival of the system ($\overline{E}$) can be modeled as:

$$\overline{E} = \overline{E}_1 \cup \overline{E}_2 \cup \ldots \cup \overline{E}_{n-1} \cup \overline{E}_n \tag{15-12}$$

where $\overline{E}_i$ is the survival event of component i. Equation 15-11 can be evaluated using conditional probabilities as was discussed in Chapter 3.

The probability of failure of the system (P_f) for a system in parallel can be expressed similar to Equation 15-8 as:

$$\prod_{i=1}^{n} P_{f_i} \le P_f \le \min_{i=1}^{n}(P_{f_i}) \tag{15-13}$$

where P_{f_i} is the failure probability of the ith component, for $i = 1, 2, \ldots, n$. These bounds can be wide. Narrower bounds require the knowledge of correlation levels between the underlying failure events or conditional probabilities.

Example 15-2: Reliability of a Piping System in Parallel

The piping system shown in Figure 15-6 has four components with the indicated failure probabilities. The components are in parallel as shown in the figure. The failure of the system is defined as the flow failure from point A to B. Therefore, for the system to fail, all the components must fail. Assuming the survival events of the pipes to be independent, the survival of the system ($\overline{E}$) can be modeled as:

$$\overline{E} = \overline{E}_1 \cup \overline{E}_2 \cup \overline{E}_3 \cup \overline{E}_4 \tag{15-14a}$$

where $\overline{E}_i$ is the survival event of component i. The failure event (E) for the system can be expressed as:

$$E = E_1 \cap E_2 \cap E_3 \cap E_4 \tag{15-14b}$$

where E_i is the failure event for component i. Therefore, the failure probability of the system can be computed as:

$$P_f = (0.01)(0.03)(0.001)(0.02) = 6 \times 10^{-9} \tag{15-14c}$$

The resulting small failure probability reflects the high reliability due to redundancy in the piping system.

In this example, the failure of a pipe was assumed not to affect the performance of other pipes that did not fail. If such an effect exists, then fault or event tree analyses as discussed in Sections 15.2.4 and 15.2.5 should be used with conditional probabilities. In this case, the performance of nonfailing pipes, given the failure of one or more pipes, must be determined for the purpose of evaluating the system reliability.

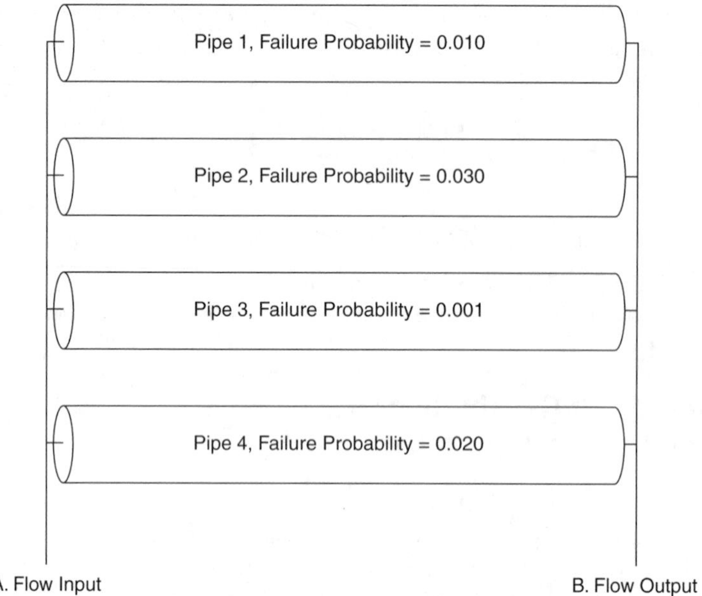

FIGURE 15-6 A piping system in parallel.

15.2.3. Mixed Systems in Series and Parallel

In general, many engineering problems can be modeled using systems with components that are mixed in parallel and series. The representation in this case can be in the form of combined plots using Figure 15-1 for the portions of a system that are in series and Figure 15-5 for the portions of the system that are in parallel. The computations can also be performed using the mathematical formulations presented for the systems in series and parallel. The example that follows illustrates how to model mixed systems and how to compute their probabilities.

Example 15-3: Reliability of a Mixed Piping System

The piping system shown in Figure 15-7 has nine components with the indicated failure probabilities. The components are in series and parallel as shown in the figure. The failure of the system is defined as the flow failure from point A to B.

The pipes 1, 2, 3, and 4 are similar to the pipes discussed in Example 15-2; therefore, they can be represented by a single pipe with a failure probability of 6×10^{-9}. Similarly, pipes 6, 7, 8, and 9 are similar to the pipes discussed in Example 15-2; therefore, they can be represented by a single pipe with a failure probability of 6×10^{-9}. As a result of these two representations, the system can be modeled as shown in Figure 15-8. The resulting system in series can now be evaluated using Equation 15-2b as follows:

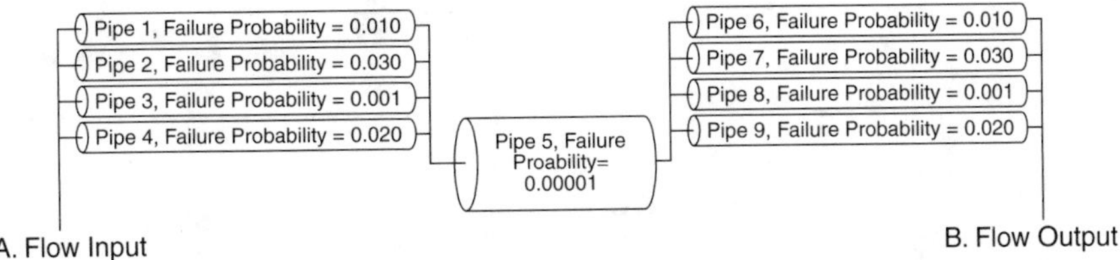

FIGURE 15-7 A piping system.

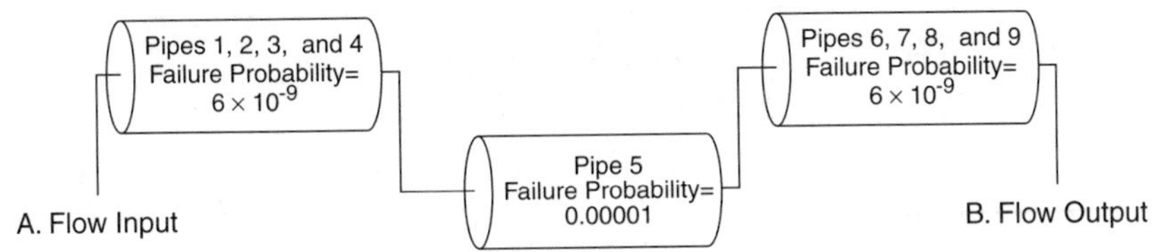

FIGURE 15-8 A revised piping system.

$$P_f = 1-[1-(6\times 10^{-9})][1-(6\times 10^{-9})](1-0.00001) = 0.0000100 \quad (15\text{-}15)$$

The resulting small failure probability reflects the high reliability due to redundancy in the parallel segments of the piping system.

15.2.4. Fault Tree Analysis

The reliability evaluation of complex engineering systems requires analytical methods that are systematic and methodical. Fault tree analysis (FTA) is a method that can be used for this purpose. The method starts by defining a top event, which is commonly selected as an adverse event. An engineering system can have more than one top event. For example, a building might have the following top events for the purpose of reliability assessment: power failure, sewer failure, water interruption, fire, or structural failure. Then, each top event should be examined using the following logic: In order for the top event to occur, other events must occur. As a result, a set of lower level events is defined. Also, the form in which these lower level events are logically connected (i.e., in parallel or in series) needs to be defined. The connectivity of these events is expressed using gates. The "AND" or "OR" gates should be used as shown in Figure 15-9. Lower-level events are classified into the following types:

1. *Basic events*, which cannot be decomposed further into lower-level events. They are the lowest events that can be obtained. For these events, failure probabilities must be obtained.
2. *Events that can be decomposed further*, which can be decomposed further to lower levels; therefore, they should be decomposed until the basic events are obtained.
3. *Undeveloped events*, which are not basic and can be decomposed further; however, because they are not important, they are not developed further. Usually, the

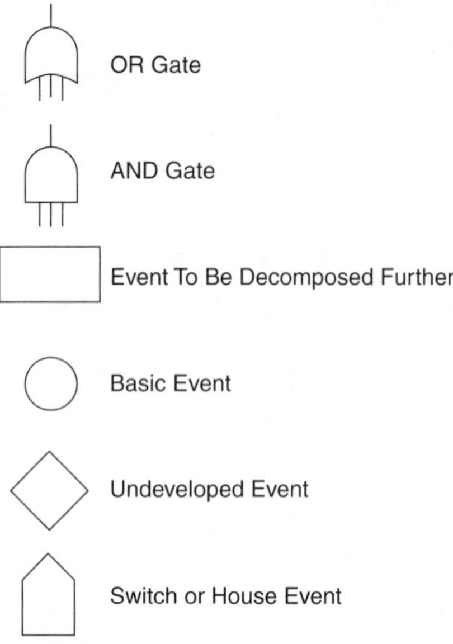

FIGURE 15-9 Symbols used in fault tree analysis.

probabilities of these events are very small or the effect of their occurrence on the system is negligible or can be controlled or mediated.

4. *Switch (or house) events*, which are not random and can be turned on or off with full control.

The symbols shown in Figure 15-9 are used for these events. Also, a continuation symbol is shown, which is used to break up a fault tree into several parts for the purpose of fitting it on several pages.

Fault tree analysis requires the development of a tree-looking diagram for the system; therefore, the tree shows failure paths and scenarios that can result in the occurrence of a top event. The construction of the tree should be based on the building blocks and the logic gates in Figure 15-9.

The FTA outcome of interest is the occurrence probability of the top event. Because the top event was decomposed into basic events, its occurrence can be stated using the AND and OR of the basic events. The resulting statement can be restated by replacing the AND with the intersection of the corresponding basic events, and the OR with the union of the corresponding basic events. The occurrence probability of the top event can then be computed by evaluating the probabilities of the unions and intersections of the basic events. The dependence between these events, sometimes called *common cause*, affects the resulting probability based on the concepts covered in Chapter 3. Computation of fault trees is illustrated using an example at the end of this section.

For a large fault tree, computation of the occurrence probability of the top event can be difficult because of its size. In this case, more efficient approaches are necessary to assess the reliability of a system, such as the minimal cut set approach. According to this approach, each cut set is defined as a set of basic events where the joint occurrence of these basic events results in the occurrence of the top event. A minimal cut set is a cut set with the condition that the nonoccurrence of any one basic event from this set results in the nonoccurrence of the top event. Therefore, a minimal cut set can be viewed as a subsystem in parallel. In general, systems have more than one minimal cut sets. The occurrence of the top event of the system can, therefore, be due to any one of these minimal cut sets. As a result, the system can be viewed as the union of all the minimal cut sets for the system. Computation of fault trees using minimal cut sets is illustrated using an example at the end of this section.

Alternatively, for large fault trees, computation of the occurrence probability of the top event can be based on the minimal link set approach. According to this approach, each link set is defined as a set of basic events where the joint nonoccurrence of these basic events results in the nonoccurrence of the top event. Minimal link sets are then defined similar to the minimal cut sets. In general, this approach is more difficult to deal with than the minimal cut set approach.

Efficient computational algorithms are available to evaluate minimal cut sets. A recommended algorithm is based on a "top–down" method for generating minimal cut sets. This algorithm was used to develop the main source code for the FTA program. The following algorithm summarizes the analytical procedure for computing the minimal cut sets of a fault tree:

1. Provide a unique label for each gate.
2. Label each basic event.
3. Set up a two-cell array:
   [.] [.]
4. Place the top event gate label in the first row, first column:
   [Top] [.]
5. Scan each row from left to right replacing:
   a. Each OR gate by a vertical arrangement defining the input event to the gate, and
   b. Each AND gate by a horizontal arrangement defining the input events to the gate. For example, the following table sequence can be generated for an AND top gate with two gates below (Gate 1 of OR type, and Gate 2 of AND type):

      [Top(AND)] [.]
      or
      [Gate1(OR)] [Gate2(AND)]

      Assuming Gate 1 has two events (1 and 2):
      [Event 1] [Gate2]
      [Event 2] [Gate2]

      Assuming Gate 2 has two events (3 and 4):
      [Event 1] [Event 3] [Event 4]
      [Event 2] [Event 3] [Event 4]
6. When no gate events remain, each row is a cut set.
7. Remove all nonminimal combinations of events such that only minimal cut sets remain.

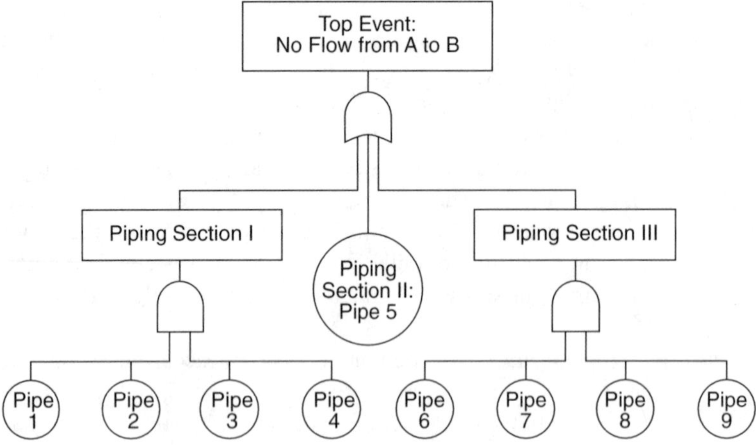

FIGURE 15-10 Fault tree model for a piping system.

8. Compute the occurrence probability for each minimal cut set as the products of the probabilities of its underlying events.
9. Compute the system (top event) occurrence probabilities as the sum of the occurrence probabilities of all the minimal cut sets.

This procedure can be used to assess the minimal cut sets for a fault tree.

Example 15-4: Reliability of a Piping System Using Fault Tree Analysis

The piping system that was discussed in Example 15-3 is reexamined here. The piping system, as shown in Figure 15-7, consists of nine components with the indicated failure probabilities. The components are in series and parallel as shown in the figure. The top event is defined as the flow failure from point A to B. The system is broken down into three piping sections as follows:

1. Section I: pipes 1, 2, 3, and 4
2. Section II: pipe 5
3. Section III: pipes 6, 7, 8, and 9

The fault tree shown in Figure 15-10 was constructed using AND and OR gates. The events that are shown in the figure are failure events. The occurrence probability of the top event (P_f) can be computed as:

$$\begin{aligned} P_f = P[&(P1 \text{ AND } P2 \text{ AND } P3 \text{ AND } P4) \\ &\text{OR } (P5) \text{ OR } (P6 \text{ AND } P7 \text{ AND } P8 \text{ AND } P9)] \end{aligned} \tag{15-16}$$

where Pi is the failure event of the ith pipe. Replacing AND and OR with intersection and union operators produces the following:

$$P_f = P[(P1 \cap P2 \cap P3 \cap P4) \cup (P5) \cup (P6 \cap P7 \cap P8 \cap P9)] \tag{15-17}$$

In Equation 15-17, the failure probabilities of pipe section I ($P[S1]$) can be computed based on the assumption of independent pipe failures as:

$$\begin{aligned} P(S1) &= P[(P1 \cap P2 \cap P3 \cap P4)] \\ &= P(P1)P(P2)P(P3)P(P4) \\ &= (0.01)(0.03)(0.001)(0.02) \\ &= 6 \times 10^{-9} \end{aligned} \tag{15-18a}$$

Similarly, the failure probabilities of sections II and III, respectively, are given by:

$$P(S2) = 0.00001 \tag{15-18b}$$

$$P(S3) = P(S1) = 6 \times 10^{-9} \tag{15-18c}$$

Now, the occurrence probability of the top event can be computed as:

$$\begin{aligned} P_f = {} & P[S1 \cup S2 \cup S3] \\ = {} & P(S1) + P(S2) + P(S3) \\ & - P(S1)P(S2) - P(S1)P(S3) - P(S2)P(S3) \\ & + P(S1)P(S2)P(S3) \end{aligned} \tag{15-19}$$

Substituting the values of the failure probabilities gives:

$$\begin{aligned} P_f &= (6 \times 10^{-9}) + 0.00001 + (6 \times 10^{-9}) \\ &\quad - (6 \times 10^{-9})(0.00001) - (6 \times 10^{-9})(6 \times 10^{-9}) - (0.00001)(6 \times 10^{-9}) \\ &\quad + (6 \times 10^{-9})(0.00001)(6 \times 10^{-9}) \\ &= 0.0000100 \end{aligned} \tag{15-20}$$

Example 15-5: Reliability of a Piping System Using Minimal Cut Sets

The piping system that was discussed in Examples 15-3 and 15-4 is reexamined here. The piping system, as shown in Figure 15-7, consists of nine components with the indicated failure probabilities. The components are in series and parallel as shown in the figure. The top event is defined as the flow failure from point A to B. In this case, the following are example cut sets:

$$\begin{aligned} \text{Example cut sets} = \{ & (P1, P5, P7), (P1, P2, P5, P8), (P3, P4, P5, P7, P9) \\ & (P2, P3, P5, P7), (P1, P2, P3, P4) \} \end{aligned} \tag{15-21}$$

where Pi is the failure event of the ith pipe. The minimal cut sets are:

$$\text{Minimal cut sets} = \{(P1, P2, P3, P4), (P6, P7, P8, P9), (P5)\} \tag{15-22}$$

The occurrence of the top event is, therefore, considered as the union of the minimal cut sets. Each minimal cut set is considered to be the intersection of its basic events. The occurrence probabilities of the minimal cut sets can be computed as the product of the occurrence probabilities of its events. For example, the first minimal cut set $(P1,P2,P3,P4)$ has a probability of:

$$P(P1 \cap P2 \cap P3 \cap P4) = (0.01)(0.03)(0.001)(0.02) = 6 \times 10^{-9} \tag{15-23}$$

Similarly, the probabilities of all the minimal cut sets can be computed. The occurrence probability of the top event can then be evaluated as a system in series of the minimal cut sets. The resulting occurrence probability is the same as the previous example (i.e., 0.0000100). It should be noted that minimal cut sets that include some common basic events are dependent sets; therefore, the occurrence probability of their union should be based on conditional probabilities as given by Equations 3-10, 3-27, and 3-28.

Using the top–down algorithm produces the following minimal cut sets:

Top event
$S1$ replaced by $(P1)$ and $(P2)$ and $(P3)$ and $(P4)$
$S2$ replaced by $(P5)$
$S3$ replaced by $(P6)$ and $(P7)$ and $(P8)$ and $(P9)$

15.2.5. Event Tree Analysis

Event tree analysis is a method to study the effect of an initiating event on a system. The initiating event can be the failure of a component in the system or the top event as was defined in fault tree analysis. The effect of an initiating event on a system depends on what might happen next and the sequence of occurrence. As a result, several possible scenarios can be developed with possibly very severe impact on the system. In reliability analysis of systems, it is important to identify these scenarios, their occurrence probabilities, and effects on the system. In some systems, an initiating event might appear to have limited consequence on the system, but as its occurrence is combined with the occurrence of other events, the result can have significant, possibly devastating, effects on the system.

Event tree analysis was used in the nuclear industry to investigate the effect of accidents or other initiating events on the integrity of nuclear reactors. The event tree shown in Figure 15-11 was adapted from a reactor safety study performed for the Nuclear Regulatory Commission, called the WASH 1400 study. The initiating event was considered for the purpose of illustration to be a pipe break with a probability of $P(E)$. The first subsequent event was selected to be electric power failure with a failure probability $P(E1)$; therefore, the electric power survival probability is $[1 - P(E1)]$. The second subsequent event selected was failure of the emergency core cooling system with a probability of $P(E2)$. The third subsequent event selected was failure of the fission product removal system with a probability of $P(E3)$. The fourth subsequent event selected was failure of the containment (reactor) integrity with a probability of $P(E4)$. The selection of these subsequent events and their order are for the purpose of illustration. In real systems, such selections should be based on knowledge of the system, its components, and their interactions, functions, and performance. The resulting event tree has 16 branches as shown in Figure 15-11. The subsequent events at any level are not necessarily limited to two. In this example, the treatment was kept simple by considering two subsequent events at each level. In general, it is possible to consider several levels of failure for each subsequent event; however, the tree can grow to a large size. Sometimes, it is possible to identify some unimportant tree branches because of their relatively very small probabilities and small consequences. These branches can be pruned and ignored.

After completing construction of the tree, the branch probabilities can be computed as the product of all the probabilities of all the events along a branch as shown in Figure 15-11. The assumption behind this mathematical operation is that the underlying events are independent, an assumption that must be examined carefully for its validity. Also, we need to define the consequences for each branch. The consequences can be expressed in different units depending on their types. It is possible, for example, to define the consequences as property damage with estimates in dollars, or potential human injury with estimates in numbers. Also, it is possible to keep the assessed consequences in linguistic terms as shown in Figure 15-11.

Because all the tree branches are possible failure scenarios, they can be considered as failure modes for the system with varying levels of consequence. Branches that produce high probabilities with significant consequences are referred to as *high-risk branches* and should be targeted for risk improvement. The meaning of risk in this case is the combination of both occurrence probability and consequences. The combination can be expressed in the form of a product as follows:

$$\text{Risk} = \text{occurrence probability} \times \text{occurrence consequence} \tag{15-24}$$

The risk improvement can be achieved either by reducing the occurrence probability through system or component changes or by reducing the potential consequences.

Example 15-6: Reliability of a Nuclear Reactor Using Event Tree Analysis

The event tree shown in Figure 15-11 is used in this example to illustrate the computations of branch probabilities. The assumed annual occurrence probabilities for the events E, $E1$, $E2$, $E3$, and $E4$ are for the purpose of illustration and are given by:

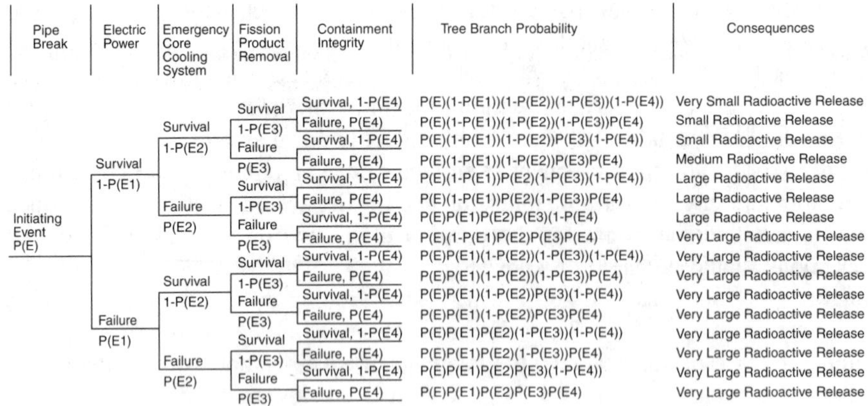

FIGURE 15-11 Event tree model for reactor safety.

$$P(E) = 1\text{E-04} \tag{15-25a}$$

$$P(E1) = 1\text{E-07} \tag{15-25b}$$

$$P(E2) = 1\text{E-05} \tag{15-25c}$$

$$P(E3) = 1\text{E-06} \tag{15-25d}$$

$$P(E4) = 1\text{E-05} \tag{15-25e}$$

where 1E-04 = 1×10^{-4}. Table 15-1 shows the computation of the branch probabilities and approximate expressions for computing the branch probabilities. Because the event probabilities as given by Equations 15-25a to 15-25e are small, the product of a small number by one minus another small number is approximately equal to the small number; that is,

$$P(Ei)[1 - P(Ej)] \approx P(Ei) \tag{15-26}$$

where $P(Ei)$ and $P(Ej)$ are small numbers.

Table 15-1 shows that branch 1 has the largest occurrence probability, but it has a small consequence according to Figure 15-11. According to Figure 15-11, branches 8 to 16 have the highest level of consequence. From these branches, branch 9 has the highest occurrence probability; therefore, branch 9 can be considered as the highest risk branch. The occurrence probability of this branch can be reduced by improving the failure probability of the pipe breakage ($P[E]$) or improving the electric power failure probability ($P[E1]$). When options for system improvement of these two events have been identified, then cost–benefit analysis using decision trees can be utilized to select the best option. Decision analysis is discussed in Section 15.4.

TABLE 15-1 Computation of Branch Probabilities in Event Tree Analysis

Branch no.	Branch probability expression from Figure 15-11	Approximate expression for branch probability	Branch probability
1	$P(E)[1 - P(E1)][1 - P(E2)][1 - P(E3)][1 - P(E4)]$	$P(E)$	1E-04
2	$P(E)[1 - P(E1)][1 - P(E2)][1 - P(E3)]P(E4)$	$P(E)P(E4)$	1E-09
3	$P(E)[1 - P(E1)][1 - P(E2)]P(E3)[1 - P(E4)]$	$P(E)P(E3)$	1E-10
4	$P(E)[1 - P(E1)][1 - P(E2)]P(E3)P(E4)$	$P(E)P(E3)P(E4)$	1E-15
5	$P(E)[1 - P(E1)]P(E2)[1 - P(E3)][1 - P(E4)]$	$P(E)P(E2)$	1E-09
6	$P(E)[1 - P(E1)]P(E2)[1 - P(E3)]P(E4)$	$P(E)P(E2)P(E4)$	1E-14
7	$P(E)[1 - P(E1)]P(E2)P(E3)[1 - P(E4)]$	$P(E)P(E2)P(E3)$	1E-15
8	$P(E)[1 - P(E1)]P(E2)P(E3)P(E4)$	$P(E)P(E2)P(E3)P(E4)$	1E-20
9	$P(E)P(E1)[1 - P(E2)][1 - P(E3)][1 - P(E4)]$	$P(E)P(E1)$	1E-11
10	$P(E)P(E1)[1 - P(E2)][1 - P(E3)]P(E4)$	$P(E)P(E1)P(E4)$	1E-16
11	$P(E)P(E1)[1 - P(E2)]P(E3)[1 - P(E4)]$	$P(E)P(E1)P(E3)$	1E-17
12	$P(E)P(E1)[1 - P(E2)]P(E3)P(E4)$	$P(E)P(E1)P(E3)P(E4)$	1E-22
13	$P(E)P(E1)P(E2)[1 - P(E3)][1 - P(E4)]$	$P(E)P(E1)P(E2)$	1E-16
14	$P(E)P(E1)P(E2)[1 - P(E3)]P(E4)$	$P(E)P(E1)P(E2)P(E4)$	1E-21
15	$P(E)P(E1)P(E2)P(E3)[1 - P(E4)]$	$P(E)P(E1)P(E2)P(E3)$	1E-22
16	$P(E)P(E1)P(E2)P(E3)P(E4)$	None	1E-27

15.3. RISK ANALYSIS

The concept of risk is used to assess and evaluate uncertainties associated with an event. Risk can be defined as the potential of losses as a result of a system failure and can be measured by pairing the probability of occurrence of an event or a failure scenario and the outcomes or consequences associated with occurrence of that event. This pairing can be represented by the following equation or graphically, as shown in Figure 15-12:

$$\text{Risk} = \{(p_1, c_1),(p_2, c_2),(p_3, c_3), \ldots, (p_x, c_x)\} \tag{15-27}$$

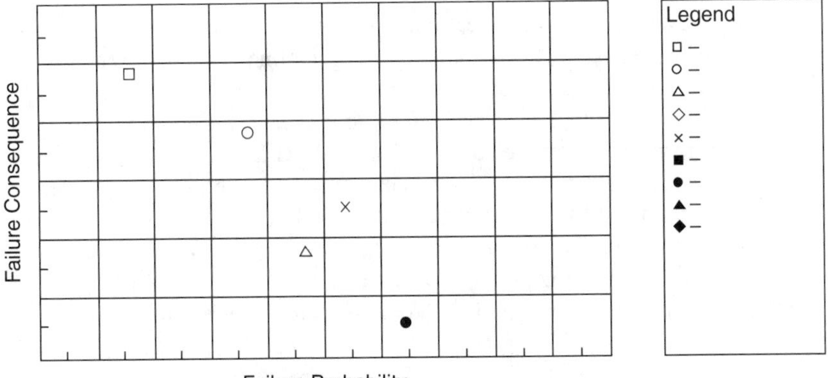

FIGURE 15-12 Example risk plot.

In this equation p_x is the occurrence probability of event x, and c_x is the occurrence consequence or outcome of the event. Risk is commonly evaluated as the product of likelihood of occurrence and the impact of an accident as provided in Equation 15-24.

The risk for a system results from the interaction of natural hazards with a system, aging and degradation of the systems, and human and organizational factors. Consequently, risk can be classified into voluntary and involuntary depending on whether or not, respectively, the events leading to the risk are under the control of the persons at risk. Society, in general, accepts a higher level of voluntary risk than involuntary risk. The losses associated with events can be classified into reversible and irreversible (e.g., property and human losses, respectively).

The population-size effect should be considered in risk studies because society responds differently for risks associated with a large population in comparison to a small population. For example, a fatality rate of 1 in 100,000 per event for an affected population of 10 results in an expected fatality of 10E-4 per event, whereas the same fatality rate per event for an affected population of 10,000,000 results in an expected fatality of 100 per event. The impact of the two scenarios is the same on the society. The size of the population at risk should be considered as a factor in setting the acceptable risk level.

The risk assessment process answers three questions: (a) What can go wrong? (b) What is the likelihood that it will go wrong? (c) What are the consequences if it does go wrong? Several methods have been developed to perform risk assessment, including preliminary hazard analysis (PrHA); HAZOP; failure modes and effects analysis (FMEA); failure modes, effects, and criticality analysis (FMECA); fault tree analysis (FTA); and event tree analysis (ETA). Each of these methods of risk assessment is suitable at certain stages of the system life cycle.

The reliability of a system can be improved or hindered by the combination of individual elements in a system; therefore, the occurrence probability and consequence are used to determine the risk associated with the system. When applying risk-based technology (RBT) methods to system safety analysis, the following interdependent primary activities are to be considered: (1) risk assessment, (2) risk management, and (3) risk communication. These activities when applied consistently provide a useful means for developing safety guidelines and requirements to the point where hazards are controlled at predetermined levels.

Risk assessment is a technical and scientific process by which the risk of given situations for a system are modeled and quantified. Risk assessment provides qualitative and quantitative data to decision makers for later use in risk management.

Commonly used risk assessment methods can be divided into how the risk is determined by quantitative or qualitative analysis. Qualitative risk analysis uses expert opinion to evaluate the probability and consequence of a hazard. Quantitative analysis relies on statistical methods and databases that identify the probability and consequence of a hazard. The safety review/audit, checklist, what-if, preliminary hazard analysis, and HAZOP are normally considered qualitative techniques.

Failure modes and effects analysis, fault tree analysis, and event tree analysis are generally considered quantitative risk assessment techniques. The selection of a quantitative or qualitative method depends upon the availability of data for evaluating the hazard and the level of comfort of those performing the risk assessments. Risk management addresses the processes by which system operators, managers, and owners make safety decisions, initiate regulatory changes, and choose different system configurations based on the data generated in the risk assessment. Risk management involves using information from the previously described risk assessment stage to make educated decisions about different configurations and operational parameters of a system. Therefore, the safety of the system can be maintained, and the involved risks in operating the system can be controlled.

Risk management allows making decisions based on risk assessment and other considerations, including economical, political, environmental, legal, reliability, producibility, and safety, among others. Despite society's attempts to prevent accidents, governmental agencies are often reactive in the development of regulations.

Answering the question of "how safe is safe enough?" is difficult and varied due to different perceptions and understandings of risk. The point at which risk is considered acceptable constitutes a definition of the term "safety." Risk acceptance is a complex subject and is often controversial. Determining acceptable levels of risk is important to establish the required risk performance of a system to be considered safe. If a system has a risk value higher than the risk acceptance level, risk reduction or mitigation measures should be taken to address safety concerns and improve the system. One difficulty with this process is defining acceptable safety levels for activities, industries, structures, and systems. Because the acceptance of risk depends upon societal perceptions and priorities, acceptance criteria do not depend on risk values alone. Acceptable levels of risk are commonly implicit values defined by decisions that guide the design and management of the life cycles of systems. Several methods assist in determining acceptable levels of risk and are summarized as follows:

1. *Risk conversion factors*: Address the attitudes of the public about risk by comparing risk categories and provide an estimate for converting risk acceptance values into different risk categories.
2. *Farmer's curve*: Provides an estimated curve for a cumulative probability risk profile for certain consequences (e.g., death) and demonstrates graphical regions of risk acceptance/nonacceptance.
3. *Revealed preferences*: Categorize societal preferences for voluntary and involuntary exposure to risk by comparing risks and benefits for various activities.
4. *Evaluating magnitude of consequences*: Compares the probabilities of risks to the magnitudes of the consequences for various industries to determine acceptable risk levels based on consequences.
5. *Risk effectiveness*: Provides a ratio for comparing cost to the magnitude of risk reduction. For a cost–benefit decision criterion, a risk reduction effort should not be pursued if costs outweigh benefits. This may not coincide with societal values about safety.
6. *Risk comparison*: Provides a comparison of various activities, industries, and procedures and is best suited to comparing risks of the same types.

Risk managers make decisions based on risk assessment and other considerations, including economics, politics, environment, law, reliability, producibility, safety, and other factors. To determine acceptable risk, managers must analyze alternatives before deciding on the best choice. In some industries, an acceptable risk has been defined by consensus. For example, the U.S. Nuclear Regulatory Commission requires that reactors be designed such that the probability of a large radioactive release to the environment from a reactor is less than 10^{-6} per year. Risk levels for certain carcinogens and pollutants have also been given acceptable concentration levels based on some assessment of acceptable risk.

Risk acceptance for many other activities is not stated explicitly. Often the level of risk acceptance with various activities is implied. Society has responded to risk by developing ways to balance risk against potential benefit. Measuring the safety levels accepted for various risks provides a means of assessing societal values. These threshold values of acceptable risk depend on a variety of issues, including the activity type, the industry, the users, and the society as a whole.

Because risk can be defined minimally as the combination of the probability of failure of an event and its consequences, target reliability levels constitute a definition of acceptable risk, on the failure probability dimension, that does not explicitly consider failure consequences. Target reliability levels are commonly used in developing structural design codes and rules based on calibrating new codes using existing ones. According to the code calibration process, an assumption is made that society has determined an implicit acceptable risk level in current design practices. Hence, future design codes can be based on these implicit levels by determining target reliability levels using reliability methods of designs resulting from the current practices adjusted for achieving reliability consistency in future designs.

Figure 15-13 shows implicit risk levels in current practices of designing engineering systems. Target reliability levels can be used for risk-based design methods. These methods should be developed so that they are compatible with the target risk levels determined for this purpose. Risk-based design methods at the system and component levels developed in this example are based on uncertainty modeling and analysis.

Unfortunately, it often takes a disaster to stimulate action in regard to safety issues. Although communication is necessary, it is important that risk management is kept separate from risk assessment in order to lend credibility to the assessment of risk without biasing the evaluation. Especially in a qualitative assessment of risk where "expert judgment" plays a role in decisions, it is important to allow the risk assessors to be free of the political pressures that managers encounter. However, there must be communication linking the risk assessors and risk managers together. The risk assessors need to assist the risk managers in making a decision. While the managers should not be involved in making any risk assessment, they should be involved in presenting to the assessors the questions that need to be answered.

To determine "acceptable risk" several steps should be taken: (1) define alternatives, (2) specify the objectives and measures for effectiveness, (3) identify consequences of the alternatives, (4) quantify values for the consequences, and (5) analyze the alternatives to select the best choice. Risk managers must consider various other

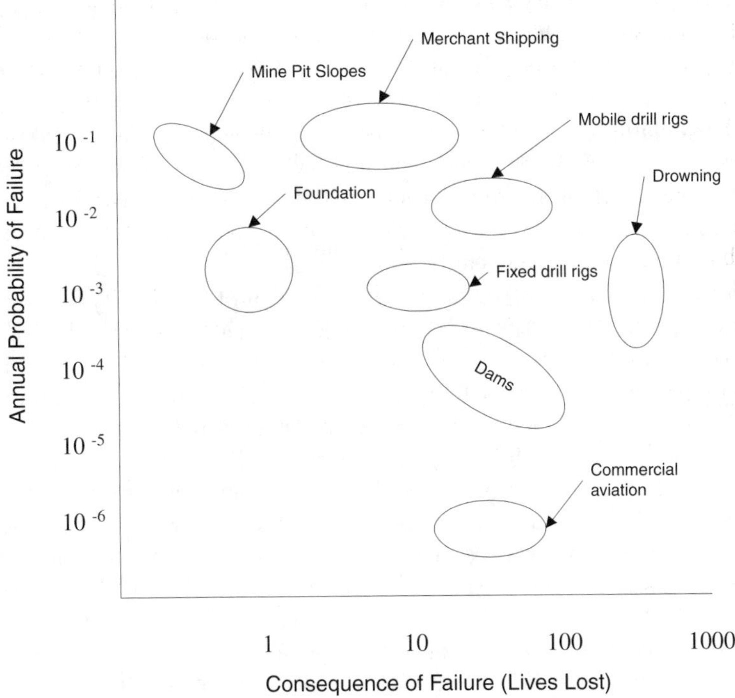

FIGURE 15-13 Implicit risk levels in current practices.

factors; for example, a manager might make a decision based on cost and risk using decision trees.

Risk communication can be defined as an interactive process of exchange of information and opinion among individuals, groups, and institutions. This definition of risk communication delineates it from risk-message transmittal from experts to nonexperts. Risk communication should be an interactive (i.e., two-way) process. However, this relatively simple definition does not imply that risk communication is easy; technical information about controversial issues must be skillfully delivered by risk managers and communicators who might be viewed as adversaries to the public. Risk communication between risk assessors and risk managers is necessary to effectively apply risk assessments in decision making. Risk managers must participate in determining the criteria for determining what risk is acceptable and unacceptable. This communication between the risk managers and risk assessors is necessary for a better understanding of risk analysis in making decisions.

Risk communication provides the vital link between the risk assessors, risk managers, and the public to help understand risk. However, there is a common misconception that risk communication can lead to harmony among the involved parties, which is not necessarily true all the time. Risk communication is a complex dynamic process that needs to be handled with extreme care by experts, especially after disasters. Risk managers need to establish contingency plans for risk communication of disasters. The added pressure by the media and public in a disaster situation can create miscommunication that might be difficult to undo or remedy.

Reliability of a system can be defined as its ability to fulfill its design functions for a specified time period under specified environmental conditions. This ability is commonly measured using probabilities. Reliability is, therefore, the occurrence probability of the complementary event to failure, or:

$$\text{Reliability} = 1 - \text{failure probability} \tag{15-28}$$

Based on this definition, reliability is one component of risk. Safety can be defined as evaluating risk acceptability for the system, thus making it a component of risk management.

After performing risk and safety analysis, system improvement in terms of risk can be achieved in one or more ways: (1) consequence reduction in magnitude or uncertainty, (2) failure-probability reduction in magnitude or uncertainty, and (3) reexamination of acceptable risk. It is common in engineering to give attention to reducing both the magnitude and uncertainty of failure probability because doing so offers more system variables that can be controlled by analysts than the other two cases. As a result, it is common to perform reliability-based design of systems. However, the other two cases should be examined for possible solution as they might offer some innovative system improvement options.

A complete treatment of risk methods is beyond the scope of this book, and only introductory concepts and definitions are provided in this section.

15.4. RISK-BASED DECISION ANALYSIS

The risk of an event was defined in Section 15.3 as a combination of both its occurrence probability and its occurrence consequence. The combination can be in the form of a product as given in Equation 15-24. Alternatively, it can be considered to be an ordered pair of occurrence probability and its occurrence consequence (i.e., probability, consequences). For several events of interest, risk plots can be produced using these ordered pairs for the events and a coordinate system of occurrence probability and consequence as shown in Figure 15-12. The risk plot can be used to easily identify high-risk events, high-probability events, or high-consequence events.

Performing risk analysis requires making various decisions to reduce the risk of a system. Assuming the system consists of equipment, components, and details, the decisions can include, for example, what and when to inspect components or equipment, which inspection methods to use, assessing the significance of detected damage, and repair/replace actions. These decisions are important in operating, using, and maintaining the system. The risk aspect of the analysis requires obtaining and utilizing information about failure likelihood and consequences. Engineering decisions of these types need to be made using a systematic framework that considers all facets of a decision problem. The decision framework is called the *decision model.*

The objective of this section is to introduce a decision model (a systematic framework) for decision making in the risk analysis. In order to construct a decision model, the following elements of the decision model must be defined:

1. Objectives of decision analysis
2. Decision variables
3. Decision outcomes
4. Associated probabilities and consequences

The components of the decision model are described in the following sections.

15.4.1. Objectives of Decision Analysis

Engineering decision problems can be classified into single- and multiple-objective problems. Example objectives are minimizing the total expected cost, maximizing safety, maximizing the total expected utility value, and maximizing the total expected profit. Decision analysis requires the definition of these objectives. For cases of multiple objectives, the objectives should be stated in the same units, and weight factors that can be used to combine the objectives should be assigned.

15.4.2. Decision Variables

The decision variables for the decision model also must be defined. The decision variables are the feasible options or alternatives available to the decision maker at any stage of the decision-making process. Also, ranges of values that can be taken by the decision variables should be defined. Decision variables can include, for example, what and when to inspect components or equipment, which inspection methods to use, assessing the significance of detected damage, and repair/replace decisions. Therefore, assigning a value to a decision variable means making a decision at that point of a decision-making process. These points within the decision-making process are called *decision nodes*. The decision nodes are identified in the model using a rectangle or square symbol (■).

15.4.3. Decision Outcomes

The decision outcomes for the decision model also must be defined. The decision outcomes are the events that can happen as a result of a decision. They are random in nature, and their occurrence cannot be fully controlled by the decision maker. Example decision outcomes can include, for example, the outcomes of an inspection (detection or nondetection of a damage) and the outcomes of a repair (satisfactory or nonsatisfactory repair). Therefore, the decision outcomes with the associated occurrence probabilities must be defined. The decision outcomes can occur after making a decision at points within the decision-making process, called *chance nodes*. The chance nodes are identified in the model using the circle symbol (●).

15.4.4. Associated Probabilities and Consequences

The decision variables take values that can have associated costs. These costs can be considered as the direct consequences of making these decisions. The decision outcomes have both consequences and occurrence probabilities. The probabilities are necessary due to the random (chance) nature of these outcomes. The consequences can include, for example, the cost of failure due to damage that was not detected by an inspection method.

15.4.5. Decision Trees

The elements of a decision model should be considered in a systematic form in order to make decisions that meet the objectives of the decision-making process. Decision trees are commonly used to examine the available information for the purpose of decision making. Figure 15-14 shows a general layout for an example decision tree. The decision tree includes the decision and chance nodes. The decision nodes are followed by possible actions (or alternatives, Ai) that can be selected by a decision maker. The chance nodes are followed by outcomes (or chances, Oj) that can happen without the complete control of the decision maker. The actions have costs associated with them (CAi), whereas the outcomes have both probabilities (P[Oj]) and consequences (COj). Each line followed from the beginning (left end) of the tree to the end (right end) of the tree is called a *tree branch*. Each branch represents a possible scenario of decisions and possible outcomes that can happen. The total expected cost for each branch can be computed as shown in Figure 15-14. The most suitable decisions can then be selected such that the minimum total expected cost is obtained.

In general, utility values can be used instead of cost values. Decision analysis using utility values is not discussed in here. Also, the decision tree shown in Figure 15-14 was developed for the purpose of its use as a computational form, where values for probability and cost can be recorded and computed. It should be noted that, in computing the total expected cost of an action, the cost of an action i (CAi) should not be multiply counted from all the chance-node branches. It should be included once. The figure shows this cost in each branch in order to facilitate the computation of the branch cost.

Example 15-7: Decision Analysis for Selection of an Inspection Strategy

The objective here is to develop an inspection strategy for a selected component of a limited section of a refinery plant. This study is for illustration purposes and is

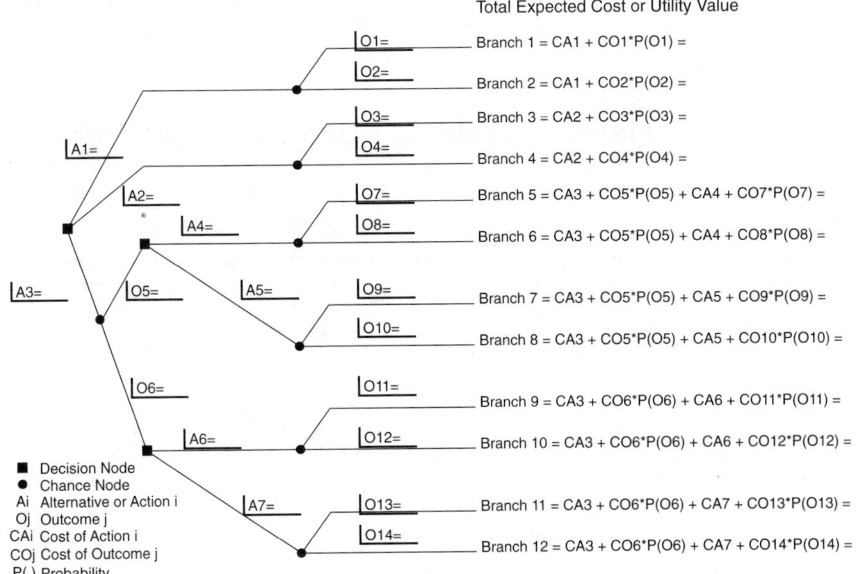

FIGURE 15-14 Example decision tree.

based on hypothetical inspection strategies, probabilities, costs, and consequences. The inspection strategy must be selected using decision analysis.

The first step in decision analysis is to select a component for inspection, possibly based on risk analysis, fault trees, and event trees. High-risk components can be selected for this purpose. The shell of a regenerator in a refinery was chosen to illustrate how an inspection strategy could be selected. We need to select and define candidate inspection strategies based on previous experience and knowledge of the system and logistics of inspection. For the purpose of illustration, only three candidate inspection strategies are considered: internal visual inspection, external visual inspection, and internal and external ultrasonic testing. The probabilities and costs in this example were chosen to provide a range of values to show the scope of the selection procedure.

Figure 15-15 shows the decision tree with the three decision strategies. Each branch provides a portion of the tree and is shown in a separate figure. Figures 15-16A, B, and C show the three portions of the decision tree for the three strategies.

In Figure 15-16A to C, the outcome of an inspection strategy is either detection or nondetection of a defect. The detection probability is denoted $P1$, and the nodetection probability is $P2$. These outcomes originate from a chance node. The cost of these outcomes for detection is $C2 = 0$, while $C4$ is the cost of the consequences of nondetection of a defect. The cost of an inspection strategy is denoted $C1$. At the end of one of these branches is a chance node that can lead to damage to the component due to inspection or no damage to the component due to inspection. The probabilities $P3$ and $P4$ are used to denote the damage probability and the no-damage

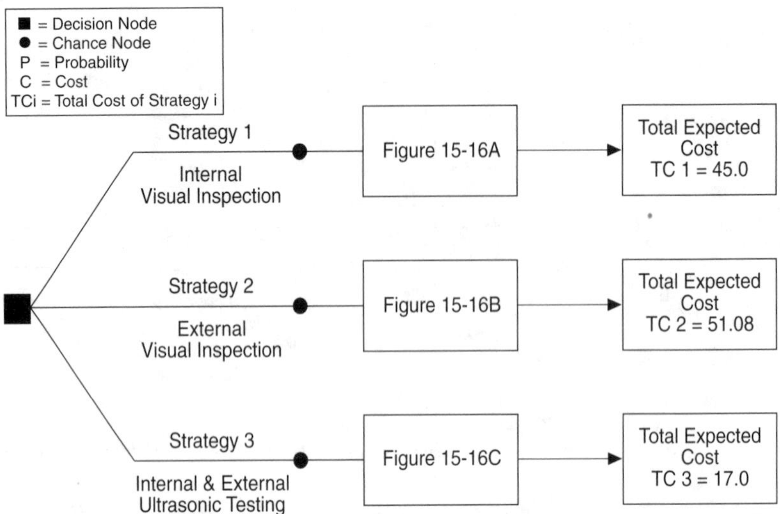

FIGURE 15-15 Decision tree for inspection.

probability, respectively. The associated cost for damage is $C3$; no damage, zero. The probability and cost estimates were assumed for each inspection strategy according to its portion of the decision tree as shown in Figure 15-16A to C.

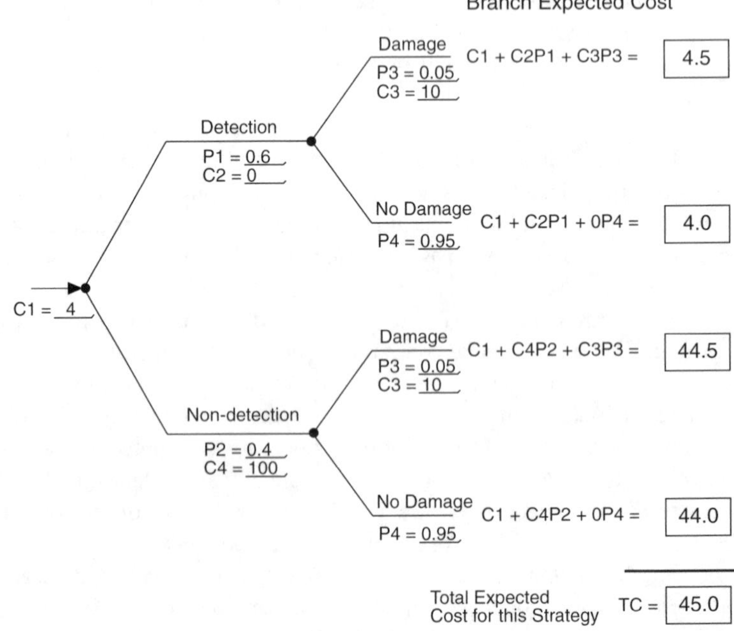

FIGURE 15-16A Portion of the decision tree for strategy 1.

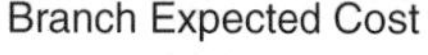

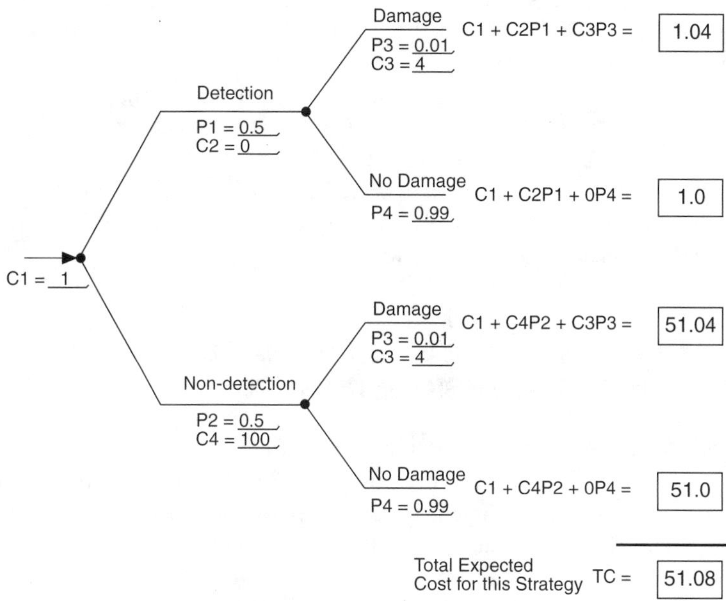

FIGURE 15-16B Portion of the decision tree for strategy 2.

Branch Expected Cost

Damage
P3 = 0.05
C3 = 10
C1 + C2P1 + C3P3 =
10.5
Detection
P1 = 0.4
C2 = 0
No Damage
P4 = 0.95
C1 + C2P1 + 0P4 =
10.0
C1 = 10
Damage
P3 = 0.05
C3 = 10
C1 + C4P2 + C3P3 =
16.5
Non-detection
P2 = 0.4
C4 = 15
No Damage
P4 = 0.95
C1 + C4P2 + 0P4 =
16.0
Total Expected
Cost for this Strategy
TC =
17.0

FIGURE 15-16C Portion of the decision tree for strategy 3.

The total expected cost for each branch was computed by summing up the product of the pairs of cost and probability along the branch. Then total expected cost for the inspection strategy was obtained by adding up the total expected costs of the branches of its portion of the decision tree.

The total expected costs of the strategies are summarized in Figure 15-15. Assuming that the decision objective is to minimize the total expected cost, then strategy 3 should be selected as the optimal strategy. Although strategy 3 is the most expensive (its $C1$ value is the highest), its total expected cost ($TC3$) is the smallest. This result can be attributed to the ability of strategy 3 to detect smaller defects and therefore reduce the consequences of nondetection.

15.5. APPLICATION: SYSTEM RELIABILITY OF A POST-TENSIONED TRUSS

Trusses are structural systems commonly used for roofs or bridges. An example truss is shown in Figure 15-17. This truss has 13 members with loads applied at truss joints L_1, L_2, and L_3 as shown. The loads are of two types: dead load (DL) and live load (LL). The loads were given the mean values shown in the figure. The coefficient of variation of the load effect (i.e., member forces) was assumed to be 0.25, and its distribution type was assumed to be normal. In order to strengthen this truss, a post-tensioning cable was added as shown in Figure 15-17 in the form of a dashed line along the bottom chord of the truss. The cable was anchored at truss joints L_0 and L_4.

The truss members are assumed to have the mean tensile capacity and mean compressive capacity shown in Table 15-2. The coefficient of variation for the tensile and compressive capacities was assumed to be 0.1. A normal distribution was assumed for the capacities. The post-tensioning cable was assumed to have mean

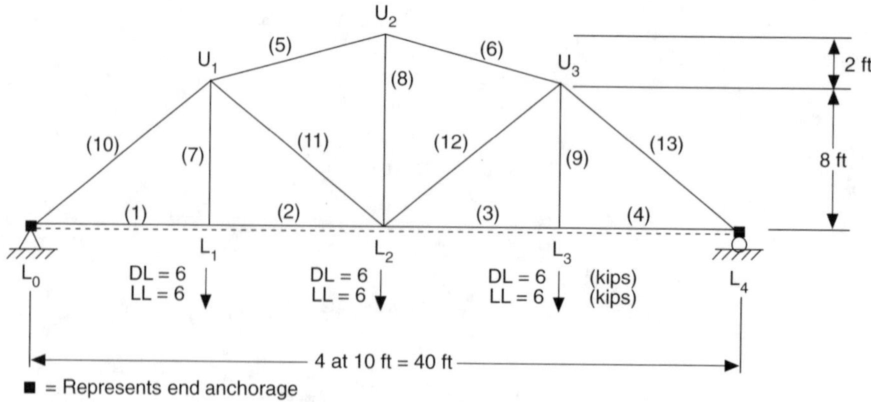

FIGURE 15-17 Post-tensioned truss.

TABLE 15-2 Mean Strength Values for Truss Members

Truss member	Mean tensile capacity (kips)	Mean compressive capacity (kips)
$L_0\ L_1$	44	40
$L_1\ L_2$	44	40
$U_1\ U_2$	37	52
$L_1\ U_1$	25	12
$L_2\ U_2$	26	13
$L_0\ U_1$	27	64
$L_2\ U_1$	5	7

TABLE 15-3 Mean Member Forces for the Truss

Truss member	Mean tensile force (mean compressive force) (kips)
$L_0\ L_1$	10.8
$L_1\ L_2$	10.8
$U_1\ U_2$	(24.5)
$L_1\ U_1$	12
$L_2\ U_2$	9.6
$L_0\ U_1$	(28.8)
$L_2\ U_1$	1.9
Cable	11.7

tensile capacity of 33.75 kips, a coefficient of variation of 0.1, and normal probability distribution. The post-tensioning force was assumed to be deterministic with a value of 10 kips.

Stiffness structural analysis of the truss subjected to the mean dead and live loads and post-tensioning force resulted in the member forces shown in Table 15-3. Using Equation 14-28b, the failure probabilities of the truss members were computed. The member forces and their strength values are as shown in Tables 15-3 and 15-2, respectively. The resulting failure probabilities are shown in Table 15-4.

TABLE 15-4 Failure Probabilities for the Truss Members

Truss member	Failure probability
$L_0\ L_1$	1.28×10^{-12}
$L_1\ L_2$	1.28×10^{-12}
$U_1\ U_2$	3.13×10^{-4}
$L_1\ U_1$	4.34×10^{-4}
$L_2\ U_2$	1.88×10^{-4}
$L_0\ U_1$	1.48×10^{-4}
$L_2\ U_1$	4.32×10^{-6}

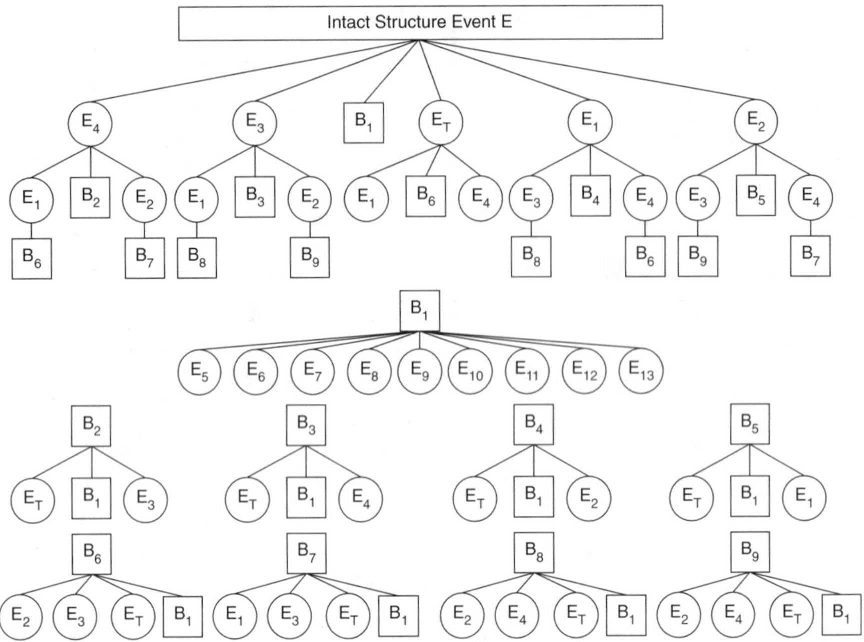

FIGURE 15-18 Event tree model for post-tensioned truss.

Event tree analysis was used to evaluate the reliability of the truss system. Starting with an intact post-tensioned truss, the truss elements were failed one by one to develop the branches of the event tree, then subsequent member failures were introduced until a system failure was obtained. For each partially failed system, the failure probabilities of the remaining truss members were evaluated. These failure probabilities can be viewed as conditional probabilities for the condition of partial failure. The failure of the truss members was assumed to be brittle; that is, members were without post-failure (or residual) strength. The resulting event tree is shown in Figure 15-18. In the figure, E_i means failure of the ith truss member, E_T means cable failure, and B_j is a continuation reference code for the tree. The occurrence probability of each branch of the tree was computed by multiplying the conditional failure probabilities of all the truss members along the branch. The failure probability of the system (P_f) was then computed as a system in series with the branches as the elements of the system, using Equation 15-8 as follows:

$$\max_{i=1}^{n} (P_{f_i}) \le P_f \le 1-(1-P_{f_1})(1-P_{f_2})\ldots(1-P_{f_{n-1}})(1-P_{f_n}) \qquad (15\text{-}29)$$

where P_{f_i} is the failure probability of the ith branch, for $i = 1, 2, \ldots, n$ branches of the event tree. The lower and upper limits correspond to fully dependent and independent branch occurrence (i.e., failure scenarios). The result of the system failure-probability evaluation is:

$$0.434 \times 10^{-3} \leq P_f \leq 1.990 \times 10^{-3} \tag{15-30}$$

The failure probability of the original truss without post-tensioning is:

$$1.350 \times 10^{-3} \leq P_f \leq 7.390 \times 10^{-3} \tag{15-31}$$

From Equations 15-30 and 15-31, it can be observed that the failure probability was improved approximately by a factor of 3.5 due to post-tensioning.

15.6. PROBLEMS

15-1. The following system consists of components with the indicated nonfailure probabilities:

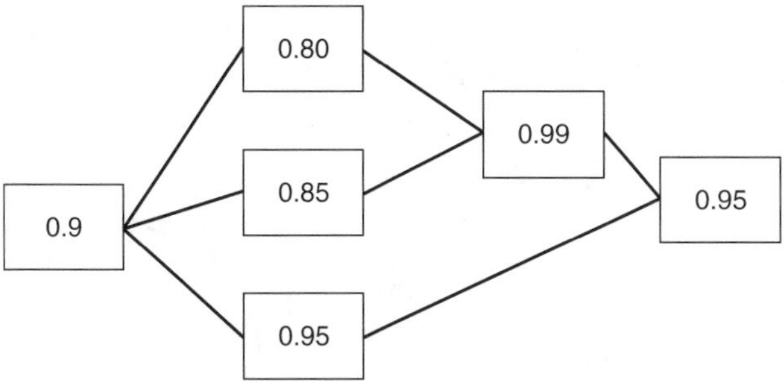

Compute the reliability of the system assuming independent failure events for the components.

15-2. For the system described in Problem 15-1, develop a fault tree model and evaluate the minimal cut set. Determine the reliability of the system.

15-3. The following system consists of components with the indicated nonfailure probabilities:

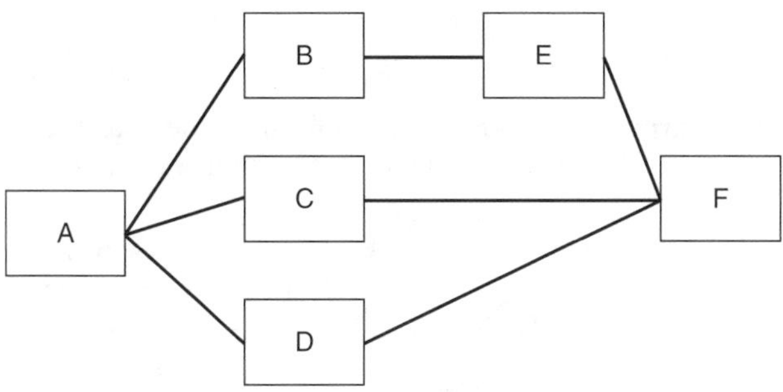

Compute the reliability of the system assuming independent failure events for the components.

15-4. For the system described in Problem 15-3, develop a fault tree model and evaluate the minimal cut set. Determine the reliability of the system.

15-5. For the system described in Problem 15-3, develop a fault tree model and evaluate the minimal cut set using the algorithm at the end of Section 15.2.4. Determine the reliability of the system.

15-6. Develop a fault tree analysis program that evaluates the minimal cut set using the algorithm at the end of Section 15.2.4. Determine the reliability of several systems to demonstrate your program.

15-7. Select an engineering system that interests you, define the system, define a system failure criterion, and develop (a) a fault tree model for the system, and (b) an event tree model.

15-8. The following system consists of six components with the following reliability values and associated costs:

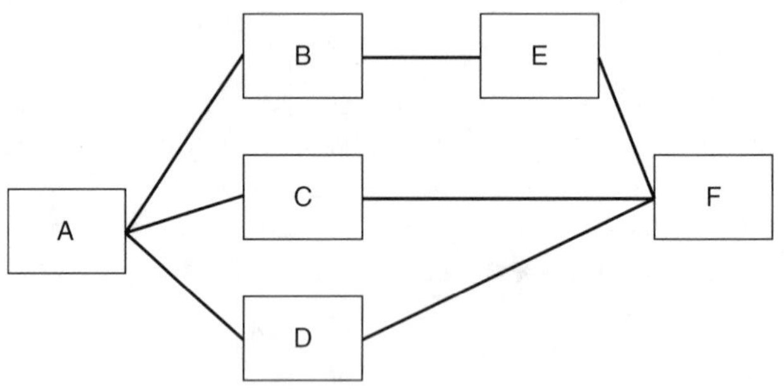

Components	Reliability	Initial (or component) cost	Failure cost
A	0.99	100	1000
B	0.95	150	2000
C	0.85	150	4000
D	0.90	100	2000
E	0.85	120	1000
F	0.99	500	5000

(a) Develop an event tree model for the system, and (b) using risk analysis, select one of the following two improvement options:

Option	New reliability	New initial (or component) cost	New failure cost
Replace component B with a new design, B'	0.99	200	800
Replace component C with a new design, C'	0.9	250	3500

15-9. A system consists of n identical components in series with each component having a reliability of p. The failure events of the components can be assumed to be independent. Compute the reliability of the system as functions of p and n. Plot system reliability as functions of p and n for the ranges of p = [0.9, 0.99, 0.999] and n = [3, 5, 10, 100, 1000].

15-10. A system consists of n identical components in parallel with each component having a reliability of p. The failure events of the components can be assumed to be independent. Compute the reliability of the system as functions of p and n. Plot system reliability as functions of p and n for the ranges of p = [0.9, 0.99, 0.999] and n = [3, 5, 10, 100, 1000].

15-11. A system consists of n_1 sets of n_2 identical components in series with each component having a reliability of p. The n_1 sets are in parallel. The failure events of the components can be assumed to be independent. Compute the reliability of the system as functions of p, n_1, and n_2. Plot system reliability as functions of p, n_1, and n_2 for the ranges of p = [0.9, 0.99, 0.999], n_1 = [3, 5, 10, 100], and n_2 = [3, 5, 10, 100].

15-12. A system consists of n_1 sets of n_2 identical components in parallel, with each component having a reliability of p. The n_1 sets are in series. The failure events of the components can be assumed to be independent. Compute the reliability of the system as functions of p, n_1, and n_2. Plot system reliability as functions of p, n_1, and n_2 for the ranges of p = [0.9, 0.99, 0.999], n_1 = [3, 5, 10, 100], and n_2 = [3, 5, 10, 100].

15-13. A system consists of N identical components, with each component having a reliability of p. The failure of the system is defined as the failure of any n out of N components. The failure events of the components can be assumed to be independent. Compute the reliability of the system as functions of p and n for $N = 20$. Plot system reliability as functions of p and n for the ranges of p = [0.9, 0.99, 0.999] and n = [1, 3, 5, 10].

15-14. For the system described in Problem 15-13, develop a fault tree model, and evaluate the minimal cut set for $N = 20$, $n = 3$, and $p = 0.9$. Determine the reliability of the system.

15-15. For the system described in Problem 15-13, develop a fault tree model and evaluate the minimal cut set for $N = 20$, $n = 5$, and $p = 0.9$. Determine the reliability of the system.

15-16. For the system described in Problem 15-13, develop a fault tree model and evaluate the minimal cut set for $N = 20$, $n = 10$, and $p = 0.9$. Determine the reliability of the system.

15-17. A system consists of N components of types 1 and 2: N_1 is the number of components of type 1, N_2 is the number of components of type 2, and $N = N_1 + N_2$. The components have a reliability each of p_1 and p_2, respectively. The failure of the system is defined as the failure of any n_1 out of N_1 components of type 1, and any n_2 out of N_2 components of type 2. The failure events of the components can be assumed to be independent. Compute the reliability of the system as functions of p_1, n_1, p_2, n_2, $N_1 = 10$, and $N_2 = 20$.

Plot system reliability as functions of p_1, n_1, p_2, and n_2 for the ranges of $p_1 = p_2 = [0.9, 0.99, 0.999]$, and $n_1 = n_2 = [1, 2, 3]$.

15-18. For the system described in Problem 15-17, develop a fault tree model and evaluate the minimal cut set for $N_1 = 10$, $N_2 = 20$, $n_1 = 2$, $n_2 = 3$, $p_2 = 0.99$, and $p_2 = 0.9$. Determine the reliability of the system.

15-19. For the system described in Problem 15-17, develop a fault tree model and evaluate the minimal cut set for $N_1 = 10$, $N_2 = 20$, $n_1 = 5$, $n_2 = 10$, $p_2 = 0.99$, and $p_2 = 0.9$. Determine the reliability of the system.

15-20. For the system described in Problem 15-17, develop a fault tree model and evaluate the minimal cut set for $N_1 = 10$, $N_2 = 20$, $n_1 = 8$, $n_2 = 15$, $p_2 = 0.99$, and $p_2 = 0.9$. Determine the reliability of the system.

15.7. SIMULATION PROJECTS

This section provides additional work items for the problem statement of the first simulation project introduced at the end of previous chapters. The projects deals with the failure and reliability analysis of a structural beam. Using the project information provided in Chapter 5, investigate the effects of redundancy in structural systems, such as a multigirder bridge, on the reliability of systems. Suggest a measure of redundancy based on system reliability assessments.

Chapter 16
Bayesian Methods

TABLE OF CONTENTS

16.1. INTRODUCTION

Engineers and scientists commonly need to solve a problem and make decisions based on limited information about one or more of the parameters of the problem. The types of information available to them can be:

1. Objective information based on experimental results or observations
2. Subjective information based on experience, intuition, other previous problems similar to the one under consideration, or the physics of the problem

The first type of information can be dealt with using the theories of probability and statistics as described in the previous chapters. In this type, probability is interpreted as the frequency of occurrence assuming sufficient repetitions of the problem, its outcomes, and parameters, as a basis of the information. The second type of information is subjective and can depend on the engineer or analyst studying the problem. In this type, uncertainty that exists needs to be dealt with using probabilities. However, the definition of probability is not the same as the first type because it is viewed herein as a subjective probability that reflects the state of knowledge of the engineer or the analyst.

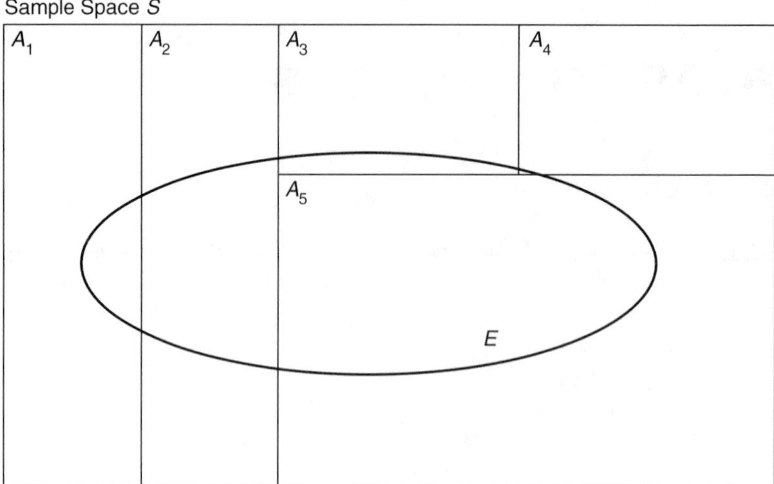

FIGURE 16-1 Bayes' theorem.

16.2. BAYESIAN PROBABILITIES

It is common in engineering to encounter problems with both objective and subjective types of information. In these cases, it is desirable to utilize both types of information to obtain solutions or make decisions. The subjective probabilities are assumed to constitute a prior knowledge about a parameter, with gained objective information (or probabilities). Combining the two types produces posterior knowledge. The combination is performed based on Bayes' theorem as described in Chapter 3.

If A_1, A_2, ..., A_n represent the prior (subjective) information, or a partition of a sample space S, and $E \subset S$ represents the objective information (or arbitrary event) as shown in Figure 16-1, the theorem of total probability states that:

$$P(E) = P(A_1)P(E|A_1) + P(A_2)P(E|A_2) + \ldots + P(A_n)P(E|A_n) \tag{16-1}$$

This theorem is very important in computing the probability of the event E, especially in practical cases where the probability cannot be computed directly but the probabilities of the partitioning events and the conditional probabilities can be computed.

Bayes' theorem is based on the same conditions of partitioning and events as the theorem of total probability and is very useful in computing the posterior (or reverse) probability of the type $P(A_i|E)$, for $i = 1, 2, \ldots, n$. The posterior probability can be computed as follows:

$$P(A_i|E) = \frac{P(A_i)P(E|A_i)}{P(A_1)P(E|A_1) + P(A_2)P(E|A_2) + \ldots + P(A_n)P(E|A_n)} \tag{16-2}$$

The denominator of this equation is $P(E)$, which is based on the theorem of total probability. According to Equation 16-2, the prior knowledge, $P(A_i)$, is updated using the objective information, $P(E)$, to obtain the posterior knowledge, $P(A_i|E)$.

Example 16-1: Defective Products in Manufacturing Lines

Consider line 3 of the three manufacturing lines of Example 3-6. The three lines manufacture 20, 30, and 50% of the components, respectively. The quality assurance department of the producing factory determined that the probability of having defective products from lines 1, 2, and 3 are 0.1, 0.1, and 0.2, respectively. The following events were defined:

$$L_1 = \text{Component produced by line 1} \tag{16-3a}$$

$$L_2 = \text{Component produced by line 2} \tag{16-3b}$$

$$L_3 = \text{Component produced by line 3} \tag{16-3c}$$

$$D = \text{Defective component} \tag{16-3d}$$

and the following probabilities are given:

$$P(D|L_1) = 0.1 \tag{16-4a}$$

$$P(D|L_2) = 0.1 \tag{16-4b}$$

$$P(D|L_3) = 0.2 \tag{16-4c}$$

Because these events are not independent, the joint probabilities can be determined as follows:

$$P(D \cap L_1) = P(D|L_1)P(L_1) = 0.1(0.2) = 0.02 \tag{16-5a}$$

$$P(D \cap L_2) = P(D|L_2)P(L_2) = 0.1(0.3) = 0.03 \tag{16-5b}$$

$$P(D \cap L_3) = P(D|L_3)P(L_3) = 0.2(0.5) = 0.10 \tag{16-5c}$$

The theorem of total probability can be used to determine the probability of a defective component as follows:

$$\begin{aligned} P(D) &= P(D|L_1)P(L_1) + P(D|L_2)P(L_2) + P(D|L_3)P(L_3) \\ &= 0.1(0.2) + 0.1(0.3) + 0.2(0.5) = 0.02 + 0.03 + 0.10 \\ &= 0.15 \end{aligned} \tag{16-6}$$

Therefore, on the average, 15% of the components produced by the factory are defective.

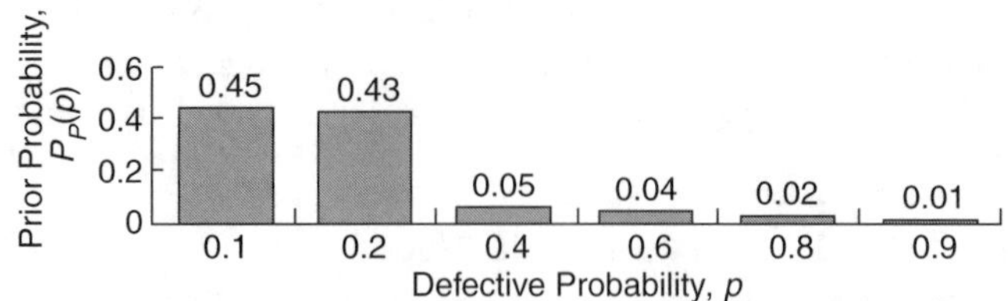

FIGURE 16-2 Prior probability distribution for defective probability of line 3.

Because of the high contribution of line 3 to the defective probability, a quality assurance engineer subjected the line to further analysis. The defective probability for line 3 was assumed to be 0.2. An examination of the source of this probability revealed that it is subjective and is also uncertain. A better description of this probability can be as shown in Figure 16-2 in the form of a prior discrete distribution for the probability. The distribution is denoted as $P_P(p)$. The mean defective component probability $\bar{p}(D)$ based on this distribution is:

$$\begin{aligned}\bar{p}(D) &= 0.1(0.45) + 0.2(0.43) + 0.4(0.05) + 0.6(0.04) + 0.8(0.02) + 0.9(0.01) \\ &= 0.200\end{aligned} \tag{16-7}$$

Now assume that a component from line 3 was tested and found to be defective; the subjective prior distribution of Figure 16-2 should be revised to reflect the new (objective) information. The revised distribution is called the *posterior distribution* $[P'_P(p)]$ and can be computed using Equation 16-2 as follows:

$$P'_P(0.1) = \frac{0.45(0.1)}{0.2} = 0.225 \tag{16-8a}$$

Similarly, the following posterior probabilities can be computed:

$$P'_P(0.2) = \frac{0.43(0.2)}{0.2} = 0.430 \tag{16-8b}$$

$$P'_P(0.4) = \frac{0.05(0.4)}{0.2} = 0.100 \tag{16-8c}$$

$$P'_P(0.6) = \frac{0.04(0.6)}{0.2} = 0.120 \tag{16-8d}$$

$$P'_P(0.8) = \frac{0.02(0.8)}{0.2} = 0.80 \tag{16-8e}$$

$$P'_P(0.9) = \frac{0.01(0.9)}{0.2} = 0.045 \tag{16-8f}$$

The resulting probabilities in Equations 16-8a to 16-8f add up to 1. Also, the average probability of 0.2 can be viewed as a normalizing factor for computing these probabilities. The mean defective component probability $\bar{p}(D)$ based on the posterior distribution is:

$$\begin{aligned}\bar{p}(D) &= 0.1(0.225) + 0.2(0.430) + 0.4(0.100) + 0.6(0.120) + 0.8(0.080) + 0.9(0.045) \\ &= 0.325\end{aligned} \tag{16-9a}$$

The posterior mean probability (0.325) is larger than the prior mean probability (0.200). The increase is due to the detected failure in the test. Now assume that a second component from line 3 was tested and found to be defective; the posterior distribution of Equations 16-8a to 16-8f needs to be revised to reflect the new (objective) information. The revised posterior distribution builds on the posterior distribution of Equations 16-8a to 16-8f, treating it as a prior distribution. Performing similar computations as in Equations 16-8 and 16-9 results in the posterior distribution shown in Table 16-1 in the column "Post. 2 *D*". The average defective component probability $\bar{p}(D)$ is also given in the table. The last row in the table is the average nondefective component probability $\bar{p}(ND)$ for cases where a nondefective component results from a test. This value $\bar{p}(ND)$ can be computed in a manner similar to Equation 16-7 or 16-9. For example, the $\bar{p}(ND)$ in the case of a nondefective test according to the prior distribution is:

$$\begin{aligned}\bar{p}(ND) &= (1-0.1)(0.225)+(1-0.2)(0.430)+(1-0.4)(0.100) \\ &\quad +(1-0.6)(0.120)+(1-0.8)(0.080)+(1-0.9)(0.045) \\ &= 0.800\end{aligned} \tag{16-9b}$$

The computations for other cases are similarly performed as shown in Table 16-1. It should be noted that

$$\bar{p}(D)+\bar{p}(ND) = 1.0 \tag{16-10}$$

Now assume that a third component from line 3 was tested and found to be nondefective; the posterior distribution in column "Post. 2 *D*" of Table 16-1 should be revised to reflect the new (objective) information. The revised distribution is the posterior distribution $[P_P'(p)]$ and can be computed using Equation 16-2 as follows:

$$P_P'(0.1) = \frac{0.0692(1-0.1)}{0.4883} = 0.1275 \tag{16-11a}$$

Similarly, the following posterior probabilities can be computed:

$$P_P'(0.2) = \frac{0.2646(1-0.2)}{0.4883} = 0.4335 \tag{16-11b}$$

TABLE 16-1 Prior and Posterior Distributions for Line 3

Probability, p	$P(p)$	Post. 1 D	Post. 2 D	Post. 3 ND	Post. 4 D	Post. 5 D	Post. 6 D	Post. 7 D	Post. 8 D	Post. 9 D	Post. 10 D
0.1	0.45	0.225	0.0692308	0.127599244	0.0358090	0.0070718	0.0011355	0.0001638	2.22693E-05	2.912E-06	3.703E-07
0.2	0.43	0.430	0.2646154	0.433522369	0.2433245	0.0961062	0.0308633	0.0089068	0.002421135	0.0006332	0.0001611
0.4	0.05	0.100	0.1230769	0.151228733	0.1697613	0.1341016	0.0861300	0.0497125	0.027026626	0.0141359	0.0071914
0.6	0.04	0.120	0.2215385	0.181474480	0.3055703	0.3620744	0.3488266	0.3020033	0.246280127	0.1932203	0.1474458
0.8	0.02	0.080	0.1969231	0.080655325	0.1810787	0.2860835	0.3674882	0.4242131	0.461254413	0.4825060	0.4909318
0.9	0.01	0.045	0.1246154	0.025519849	0.0644562	0.1145626	0.1655564	0.2150004	0.262995429	0.3095016	0.3542696
$\bar{p}$ (D)	0.20	0.325	0.5116923	0.356332703	0.5063660	0.6227868	0.6930255	0.7357556	0.764764597	0.7862698	0.8029643
$\bar{p}$ (ND)	0.80	0.675	0.4883077	0.643667297	0.4936340	0.3772132	0.3069745	0.2642444	0.235235403	0.2137302	0.1970357

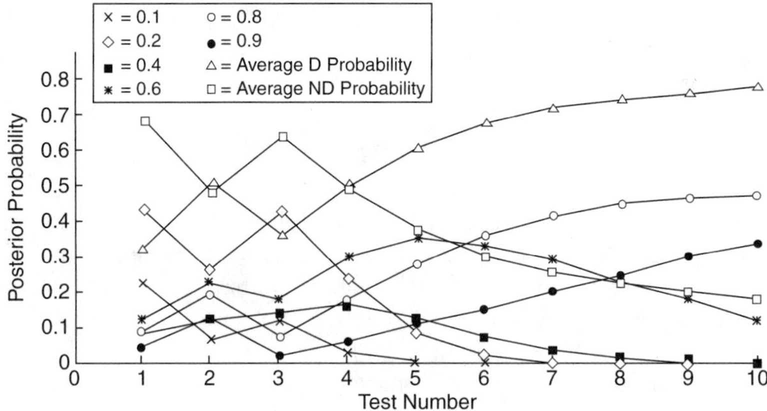

FIGURE 16-3 Posterior distributions for line 3.

$$P_P'(0.4) = \frac{0.1231(1-0.4)}{0.4883} = 0.1512 \tag{16-11c}$$

$$P_P'(0.6) = \frac{0.2215(1-0.6)}{0.4883} = 0.1815 \tag{16-11d}$$

$$P_P'(0.8) = \frac{0.1969(1-0.8)}{0.4883} = 0.0807 \tag{16-11e}$$

$$P_P'(0.9) = \frac{0.1246(1-0.9)}{0.4883} = 0.0255 \tag{16-11f}$$

The resulting probabilities in Equations 16-11a to 16-11f add up to 1. The probability $\bar{p}\,(ND)$ of 0.4883 was used in these calculations. The results of these calculations and the mean probability $\bar{p}\,(D)$ are shown in Table 16-1. It can be noted from the table that the mean defective component probability decreases as nondefective components are obtained through testing.

If the next seven tests result in defective components, the resulting posterior distributions are shown in Table 16-1. The results are also shown in Figure 16-3. It can be observed from the figure that the average probability is approaching one as more and more defective tests are obtained. Also, the effect of a nondefective component on the posterior probabilities can be seen in this figure.

16.3. BAYESIAN ESTIMATION OF PARAMETERS

16.3.1. Discrete Parameters

For an unknown parameter Θ, a prior distribution for the parameters can be subjectively determined and expressed using a probability mass function as:

$$P_{\Theta}(\theta_i) = P(\Theta = \theta_i) \quad \text{for } i = 1, 2, \ldots n \tag{16-12a}$$

or, in an abbreviated form, as:

$$P_{\Theta}(\theta_i) = P(\theta_i) \quad \text{for } i = 1, 2, \ldots, n \tag{16-12b}$$

Therefore, the parameter Θ is assumed to take n discrete values with probabilities given by Equation 16-12. The distribution of Θ reflects the uncertainty in this parameter, including its randomness. It is possible to have a parameter that is not random, but uncertain, thus requiring description through a distribution as given by Equation 16-12.

Now assume that new (objective) information ε was obtained. Using Equation 16-2, the posterior distribution of the parameter can be obtained as:

$$P(\theta_i|\varepsilon) = \frac{P(\theta_i)P(\varepsilon|\theta_i)}{P(\theta_1)P(\varepsilon|\theta_1) + P(\theta_2)P(\varepsilon|\theta_2) + \ldots + P(\theta_n)P(\varepsilon|\theta_n)} \tag{16-13a}$$

where $P(\theta_i|\varepsilon)$ is the conditional probability of θ_i given ε, or the posterior probability for θ_i; $P(\theta_i)$ is the prior probability as given by Equation 16-12; and $P(\varepsilon|\theta_i)$ is the probability of obtaining the new information (ε) given a certain value (θ_i) for the parameter. The following notation for the posterior distribution is also common:

$$P'(\theta_i) = \frac{P(\theta_i)P(\varepsilon|\theta_i)}{P(\theta_1)P(\varepsilon|\theta_1) + P(\theta_2)P(\varepsilon|\theta_2) + \ldots + P(\theta_n)P(\varepsilon|\theta_n)} \tag{16-13b}$$

where $P'(\theta_i)$ is the conditional probability of θ_i given ε, or the posterior probability for θ_i.

Using the prior distribution of the parameter Θ given by Equation 16-12, the expected value of the parameter can be computed as:

$$E(\Theta) = \sum_{i=1}^{n} \theta_i P(\theta_i) \tag{16-14}$$

Based on the posterior distribution, the expected value of Θ can be computed as:

$$E(\Theta|\varepsilon) = \sum_{i=1}^{n} \theta_i P'(\theta_i) \tag{16-15}$$

In many engineering problems, the parameter Θ can be used to define a probability distribution of a random variable X. The probability distribution of X can be either for a discrete random variable in the form of a probability mass function,

$P_X(x)$, or for a continuous random variable in the form of a density function, $f_X(x)$. The Bayesian estimation of the parameter can be used to compute Bayesian probabilities that are obtained using the information gained about the parameters. For example, the probability that X is less than some value x_0 can be computed using the prior distribution as:

$$P(X < x_0) = \sum_{i=1}^{n} P(X < x_0 | \theta_i) P(\theta_i) \tag{16-16}$$

or

$$F_X(x_0) = \sum_{i=1}^{n} F_X(x_0 | \theta_i) P(\theta_i) \tag{16-17}$$

where $F_X(x)$ is the cumulative distribution function of X evaluated at x_0. Using the posterior distribution results in the following expressions:

$$P(X < x_0) = \sum_{i=1}^{n} P(X < x_0 | \theta_i) P'(\theta_i) \tag{16-18}$$

or

$$F_X(x_0) = \sum_{i=1}^{n} F_X(x_0 | \theta_i) P'(\theta_i) \tag{16-19}$$

Example 16-2: The Duration of a Construction Activity, Discrete Case

A construction engineer is interested in modeling the duration of a construction activity for the purpose of project scheduling and control. Based on previous experiences with similar activities, the average duration of the activity is uncertain and can be from 5 to 10 days. The engineer assumed a prior probability mass function for the average duration (D) as shown in Figure 16-4. The expected average duration based on this distribution according to Equation 16-14 is given by:

$$\begin{aligned} E(D) &= 5(0.1) + 6(0.1) + 7(0.2) + 8(0.3) + 9(0.2) + 10(0.1) \\ &= 7.7 \text{ days} \end{aligned} \tag{16-20}$$

Assuming that the duration of activity X has an exponential probability distribution, the probability of completing the activity within 8 days can be computed using Equation 16-17 as:

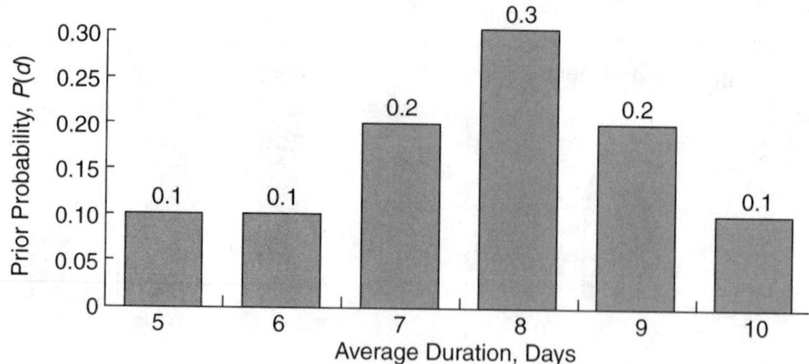

FIGURE 16-4 Prior probability distribution for average duration of a construction activity.

$$P(X \le 8) = \sum_{i=1}^{n} F_X(8 \text{ days} | D_i) P(D_i) \tag{16-21a}$$

or

$$\begin{aligned} P(X \le 8) &= \sum_{i=1}^{n} \left(1 - \exp\left(-\frac{8}{D_i}\right)\right) P(D_i) \\ &= 0.1\left(1 - \exp\left(-\frac{8}{5}\right)\right) + 0.1\left(1 - \exp\left(-\frac{8}{6}\right)\right) + 0.2\left(1 - \exp\left(-\frac{8}{7}\right)\right) \\ &\quad + 0.3\left(1 - \exp\left(-\frac{8}{8}\right)\right) + 0.2\left(1 - \exp\left(-\frac{8}{9}\right)\right) + 0.1\left(1 - \exp\left(-\frac{8}{10}\right)\right) \\ &= 0.080 + 0.074 + 0.136 + 0.190 + 0.118 + 0.055 = 0.652 \end{aligned} \tag{16-21b}$$

Now assume that an activity of this type was completed in another similar construction site in less than 8 days; the prior distribution in Figure 16-4 can be revised to reflect the new (objective) information. The revised (or posterior) distribution $[P'(d)]$ can be computed using Equation 16-13 as follows:

$$P'(5) = \frac{0.080}{0.652} = 0.122 \tag{16-22a}$$

Similarly, the following posterior probabilities can be computed:

$$P'(6) = \frac{0.074}{0.652} = 0.113 \tag{16-22b}$$

$$P'(7) = \frac{0.136}{0.652} = 0.209 \tag{16-22c}$$

$$P'(8) = \frac{0.190}{0.652} = 0.291 \tag{16-22d}$$

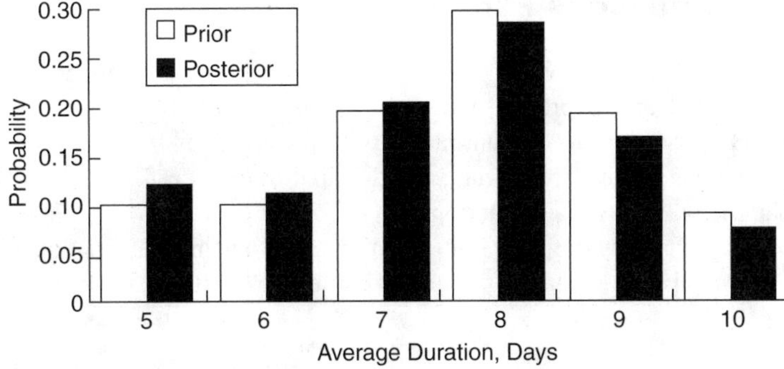

FIGURE 16-5 Prior and posterior probability distributions for average duration of a construction activity.

$$P'(9) = \frac{0.118}{0.652} = 0.181 \tag{16-22e}$$

$$P'(10) = \frac{0.055}{0.652} = 0.084 \tag{16-22f}$$

The resulting probabilities in Equations 16-22a to 16-22f add up to 1 and are shown in Figure 16-5 with the prior probabilities. Also, the average probability of 0.652 can be viewed as a normalizing factor for computing these probabilities. The mean duration $E(D)$ based on the posterior distribution is:

$$\begin{aligned} E(D) &= 5(0.122) + 6(0.113) + 7(0.209) + 8(0.291) + 9(0.181) + 10(0.084) \\ &= 7.55 \text{ days} \end{aligned} \tag{16-23}$$

The resulting posterior average duration is smaller than the prior average duration. The probability of completing the activity within 8 days can be now be revised using the posterior distribution as:

$$P(X \leq 8) = \sum_{i=1}^{n} F_X(8 \text{ days}|D_i)P'(D_i) \tag{16-24a}$$

or

$$\begin{aligned} P(X \leq 8) &= \sum_{i=1}^{n}\left(1 - \exp\left(-\frac{8}{D_i}\right)\right)P'(D_i) \\ &= 0.122\left(1 - \exp\left(-\frac{8}{5}\right)\right) + 0.113\left(1 - \exp\left(-\frac{8}{6}\right)\right) + 0.209\left(1 - \exp\left(-\frac{8}{7}\right)\right) \\ &\quad + 0.291\left(1 - \exp\left(-\frac{8}{8}\right)\right) + 0.181\left(1 - \exp\left(-\frac{8}{9}\right)\right) + 0.084\left(1 - \exp\left(-\frac{8}{10}\right)\right) \\ &= 0.098 + 0.083 + 0.142 + 0.184 + 0.106 + 0.046 = 0.660 \end{aligned} \tag{16-24b}$$

16.3.2. Continuous Parameters

For an unknown parameter Θ, a prior distribution for the parameters can be subjectively determined and expressed using a probability density function $f_\Theta(\theta)$. The parameter Θ is assumed to be continuous with probabilities that can be computed based on its density function. Again, the distribution of Θ reflects the uncertainty in this parameter, including its randomness.

Now assume that new (objective) information ε was obtained. Using Equation 16-2, the posterior distribution for the parameter can be obtained as:

$$f_\Theta(\theta|\varepsilon) = \frac{f_\Theta(\theta)P(\varepsilon|\theta)}{\int_{-\infty}^{\infty} P(\varepsilon|\theta)f_\Theta(\theta)d\theta} \tag{16-25}$$

where $f_\Theta(\theta)$ is the prior density function of Θ; $f_\Theta(\theta|\varepsilon)$ is the posterior density function of Θ; and $P(\varepsilon|\theta)$ is the probability of obtaining the new information (ε) given a certain value for the parameter θ. The probability $P(\varepsilon|\theta)$ is called the *likelihood function* $L(\theta)$. The following notations for the posterior distribution are also common:

$$f'_\Theta(\theta) = f'_\Theta(\theta|\varepsilon) = \frac{f_\Theta(\theta)L(\theta)}{\int_{-\infty}^{\infty} L(\theta)f_\Theta(\theta)d\theta} \tag{16-26}$$

where $f'_\Theta(\theta|\varepsilon)$ is the conditional density function of θ given ε, or the posterior density function of Θ.

Using the prior density function of the parameter Θ, the expected value of the parameter can be computed as:

$$E(\Theta) = \int_{-\infty}^{\infty} \theta f_\Theta(\theta)d\theta \tag{16-27}$$

Based on the posterior distribution, the expected value of Θ can be computed as:

$$E(\Theta|\varepsilon) = \int_{-\infty}^{\infty} \theta f'_\Theta(\theta|\varepsilon)d\theta \tag{16-28}$$

In many engineering problems, the parameter Θ can be used to define a probability distribution of a random variable X. The probability distribution of X can be either for a discrete random variable in the form of a probability mass function, $P_X(x)$, or for a continuous random variable in the form of a density function, $f_X(x)$. The Bayesian estimation of the parameter can be used to compute Bayesian probabilities that are obtained with the information gained about the parameters. For example, the probability that X is less than some value x_0 can be computed using the prior distribution as:

$$P(X < x_0) = \int_{-\infty}^{\infty} P(X < x_0|\theta) f_\Theta(\theta) d\theta \tag{16-29}$$

or

$$F_X(x_0) = \int_{-\infty}^{\infty} F_X(x_0|\theta) f_\Theta(\theta) d\theta \tag{16-30}$$

where $F_X(x_0)$ is the cumulative distribution function of X evaluated at x_0. Using the posterior distribution results in the following expression:

$$P(X < x_0) = \int_{-\infty}^{\infty} P(X < x_0|\theta) f'_\Theta(\theta) d\theta \tag{16-31}$$

or

$$F_X(x_0) = \int_{-\infty}^{\infty} F_X(x_0|\theta) f'_\Theta(\theta) d\theta \tag{16-32}$$

Example 16-3: The Duration of a Construction Activity, Continuous Case

The construction engineer in Example 16-2 is interested in modeling the duration of a construction activity for the purpose of project scheduling and control. Based on previous experiences with similar activities, the average duration of the activity is uncertain and is assumed again to be from 5 to 10 days. In this example, the engineer assumed a prior probability density function for the average duration (D) to be uniform, as shown in Figure 16-6. The expected average duration based on this distribution according to Equation 16-27 is given by:

$$\begin{aligned} E(D) &= \int_{-\infty}^{\infty} \theta f_\Theta(\theta) d\theta \\ &= \int_{5}^{10} \theta(0.2) d\theta = 7.5 \text{ days} \end{aligned} \tag{16-33}$$

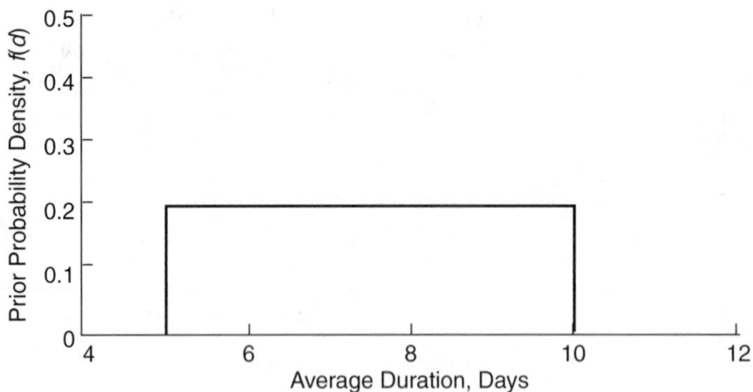

FIGURE 16-6 Prior probability distribution for average duration of a construction activity.

Assuming the duration of the activity (X) to have an exponential probability distribution, the probability of completing the activity within 8 days can be computed using Equation 16-30 as:

$$P(X \le 8) = \int_{-\infty}^{\infty} F_X(x_0|\theta) f_\Theta(\theta) d\theta \tag{16-34a}$$

or

$$P(X \le 8) = \int_{5}^{10} \left(1 - \exp\left(-\frac{8}{\theta}\right)\right)(0.2)\, d\theta = 0.6626 \tag{16-34b}$$

Now assume that an activity of this type was completed in another similar construction site in less than 8 days; the prior distribution in Figure 16-6 can be revised to reflect the new (objective) information. The revised (or posterior) distribution $[f'_D(d)]$ can be computed using Equation 16-26. The likelihood function $L(\theta)$ in Equation 16-26 is defined as the conditional probability of obtaining the outcome of the experiment given a value for the average activity duration D (i.e., the parameter θ). This conditional probability can be computed using the cumulative distribution function of the exponential distribution as:

$$L(d) = 1 - \exp\left(-\frac{8}{\theta}\right) \tag{16-35}$$

Therefore, using Equation 16-26 results in the following:

$$f'_D(d) = \frac{f_D(d)\left[1 - \exp\left(-\frac{8}{d}\right)\right]}{\int_{-\infty}^{\infty}\left[1 - \exp\left(-\frac{8}{d}\right)\right] f_D(\theta) d\theta} \tag{16-36}$$

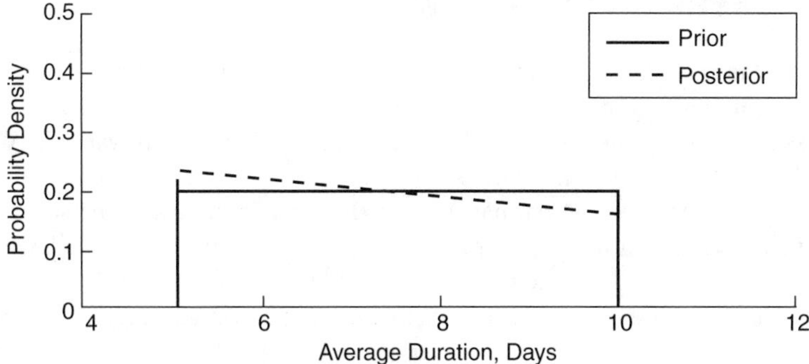

FIGURE 16-7 Prior and posterior probability density functions for average duration of a construction activity.

Substituting Equation 16-34b into Equation 16-36 produces:

$$\begin{aligned} f'_D(d) &= \frac{0.2}{0.6626}\left[1-\exp\left(-\frac{8}{d}\right)\right] \quad \text{for } 5 \le d \le 10 \text{ days} \\ &= 0.302\left[1-\exp\left(-\frac{8}{d}\right)\right] \end{aligned} \tag{16-37}$$

The area under the density function given by Equation 16-37 is 1. The prior and posterior density functions are shown in Figure 16-7. The mean duration $E(D)$ based on the posterior distribution is:

$$\begin{aligned} E(D) &= \int_{-\infty}^{\infty} \theta f'_\Theta(\theta)d\theta \\ &= \int_{5}^{10} 0.302\theta\left(1-\exp\left(-\frac{8}{\theta}\right)\right)d\theta = 7.33 \text{ days} \end{aligned} \tag{16-38}$$

The resulting posterior average duration is smaller than the prior average duration. The probability of completing the activity within 8 days can now be revised using the posterior distribution as:

$$P(X \le 8) = \int_{-\infty}^{\infty} F_X(x_0|\theta)f'_\Theta(\theta)d\theta \tag{16-39a}$$

or

$$P(X \le 8) = \int_{5}^{10} 0.302\left(1-\exp\left(-\frac{8}{\theta}\right)\right)^2 d\theta = 0.671 \text{ days} \tag{16-39b}$$

16.4. BAYESIAN STATISTICS

The Bayesian methods that were developed in the previous sections can be used in the statistical analysis of data. In this section, two cases are considered to illustrate their use in statistical analysis. The first case deals with a random variable X that is normally distributed with a known standard deviation. The mean value of the random variable is of interest and is estimated using Bayesian methods. In the second case, the random variable X is also normally distributed, but its standard deviation is unknown. In this case, both the mean value and the variance of the random variable are of interest and are estimated using Bayesian methods.

16.4.1. Mean Value with Known Variance

A random variable X is considered to be normally distributed with a known variance σ^2. The mean value of the random variable is of interest and is unknown. The prior distribution of the unknown mean (μ) is normal, with a mean value and variance of μ_0 and σ_0^2, respectively. New (objective) information was obtained by a sample of size n. The mean value based on the sample is $\bar{X}$. We are interested in determining the posterior distribution of the mean. Using Equation 16-26, the following expression can be established:

$$f'(\mu) = \frac{f(\mu)L(\mu)}{\int_{-\infty}^{\infty} L(\mu)f(\mu)d\mu} \tag{16-40}$$

where $f(\mu)$ is the prior density function of μ, which is normal with mean and variance of μ_0 and σ_0^2, respectively [i.e., $N(\mu_0, \sigma_0^2)$]; $f'(\mu)$ is the posterior density function of the unknown mean μ; and $L(\mu)$ is the likelihood function for the sample of size n. The likelihood function can be computed as the product of n values of the density function of the normal distribution with a mean μ and standard deviation σ, each evaluated at a sampled value x_i. The product can be expressed as:

$$L(\mu) = \frac{1}{(2\pi)^{n/2}\sigma^n}\exp\left[-\frac{1}{2}\sum_{i=1}^{n}\left(\frac{x_i-\mu}{\sigma}\right)^2\right] \tag{16-41}$$

It can be shown that by substituting Equation 16-41 in Equation 16-40, the resulting $f'(\mu)$ is normally distributed with the following mean value and variance, respectively:

$$\mu' = \frac{n\bar{X}\sigma_0^2 + \mu_0\sigma^2}{n\sigma_0^2 + \sigma^2} \tag{16-42a}$$

and

$$\sigma'^2 = \frac{\sigma_0^2\sigma^2}{n\sigma_0^2 + \sigma^2} \tag{16-42b}$$

The resulting μ' and σ' are the posterior mean and standard deviation of the unknown mean value μ. Using the normal posterior distribution, any Bayesian probabilities of interest for the random variable X can be computed.

The prior and posterior mean values and variances can also be used in other aspects of statistical analysis such as confidence intervals and hypothesis testing. For example, they can be used to establish the following prior confidence interval on the mean:

$$\mu_0 - z_{\alpha/2}\sigma \leq \mu \leq \mu_0 + z_{\alpha/2}\sigma \tag{16-43a}$$

Also, they can be used to establish the following posterior confidence interval:

$$\mu' - z_{\alpha/2}\sigma' \leq \mu \leq \mu' + z_{\alpha/2}\sigma' \tag{16-43b}$$

The confidence level is $(1 - \alpha)$. In a similar approach, prior hypothesis testing and posterior hypothesis testing can be performed.

Example 16-4: Mean Concrete Strength

An existing reinforced concrete structure must be evaluated for its strength. The mean concrete strength is an important quantity for this evaluation. A review of the design and construction documents, which are about 15 years old, indicated that the mean concrete strength has a mean and variance as follows:

$$\mu_0 = 3000 \text{ psi}$$

and

$$\sigma_0^2 = 90{,}000 \text{ (psi)}^2$$

These values can be considered to constitute prior information about the mean strength. Based on the construction practice during the period of construction of the building, it was common to produce concrete with a standard deviation (σ) of 400 psi.

Due to aging and environmental factors, μ_0 and σ_0 are uncertain. Therefore, it was decided to perform nondestructive testing of the concrete strength. A certified testing company produced a sample size (n) of 10, with a sample mean of 3500 psi. This information can be used to update the prior mean and variance of the average concrete strength using Equations 16-42a and 16-42b as follows:

$$\mu' = \frac{10(3500)(90{,}000) + (3000)(400)^2}{10(90{,}000) + (400)^2} = 3425 \text{ psi} \tag{16-44a}$$

and

$$\sigma'^2 = \frac{90{,}000(400)^2}{10(90{,}000) + (400)^2} = 13{,}585 \text{ (psi)}^2 \tag{16-44b}$$

Therefore, the posterior standard deviation is 117 psi.

The prior mean value and variance can be used to establish the following prior 95% confidence interval on the mean (using Equation 16-43a):

$$3000 - 1.96(300) \le \mu \le 3000 + 1.96(300) \tag{16-45a}$$

or

$$2412 \text{ psi } \le \mu \le 3588 \text{ psi} \tag{16-45b}$$

The posterior mean and variance can be used to establish the following posterior 95% confidence interval (using Equation 16-43b):

$$3425 - 1.96(117) \le \mu \le 3415 + 1.96(117) \tag{16-46a}$$

or

$$3197 \text{ psi } \le \mu \le 3653 \text{ psi} \tag{16-46b}$$

By comparing Equations 16-45 and 16-46, it can be observed that the nondestructive testing resulted in reducing the width of the confidence interval (i.e., reducing the uncertainty in the mean concrete strength).

16.4.2. Mean Value with Unknown Variance

The random variable X in this case is considered to be normally distributed with a unknown mean value (μ) and unknown variance (σ^2). Both the mean value and variance of the random variable are of interest but are unknown. The prior joint distribution of the unknown mean (μ) and unknown variance (σ^2) is assumed to be normal-gamma, which is defined as the product of a normal distribution for the mean (μ) and a gamma distribution for the variance (σ^2). The prior information about the mean and variance is based on a sample of size N with sample mean and variance of $\overline{X}_0$ and S_0^2, respectively. New (objective) information was obtained by a sample of size n. The mean value and variance based on the sample are $\overline{X}$ and S^2, respectively. We are interested in determining the posterior distribution of the mean and variance. Using Equation 16-26, it can be shown that the posterior distribution is also a normal-gamma.

The posterior mean ($\overline{X}'$) and posterior variance (S'^2) can be shown to be:

$$\overline{X}' = \frac{N\overline{X}_0 + n\overline{X}}{n'} \qquad \text{(16-47a)}$$

and

$$S'^2 = \frac{(N-1)S_0^2 + N\overline{X}_0^2 + (n-1)S^2 + n\overline{X}^2 - n'\overline{X}'^2}{n'-1} \qquad \text{(16-47b)}$$

where

$$n' = N + n \qquad \text{(16-47c)}$$

The resulting values from Equations 16-47a to 16-47b are the posterior mean and standard deviation of the unknown mean and variance.

Example 16-5: Mean and Variance of Concrete Strength

The existing reinforced concrete structure that was discussed in Example 16-4 is reexamined here. In this example, both the mean and variance of the concrete strength are of interest. Assume the following prior information about the concrete strength: mean strength $\overline{X}_0$ = 3000 psi; variance of strength S_0^2 = 90,000 (psi)^2; and sample size N = 10. These values are considered to constitute prior information about the concrete strength. Due to aging and environmental factors, $\overline{X}_0$ and S_0^2 are uncertain. Therefore, it was decided to perform nondestructive testing of the concrete strength. A certified testing company produced a sample of size (n) of 20, with a sample mean of 3500 psi and a sample standard deviation of 400 psi. This information can be used to update the prior mean and variance of the average concrete strength using Equations 16-47a to 16-47c as follows:

$$\overline{X}' = \frac{10(3000) + 20(3500)}{10 + 20} = 3333 \text{ psi} \qquad \text{(16-48a)}$$

and

$$\begin{aligned} S'^2 &= \frac{(10-1)(90{,}000) + 10(3000)^2 + (20-1)(400)^2 + 20(3500)^2 - (10+20)(3333)^2}{10+20-1} \\ &= 190{,}230 \text{ (psi)}^2 \end{aligned} \qquad \text{(16-48b)}$$

where

$$n' = 10 + 20 = 30 \qquad \text{(16-48c)}$$

Therefore, the posterior standard deviation is 436 psi.

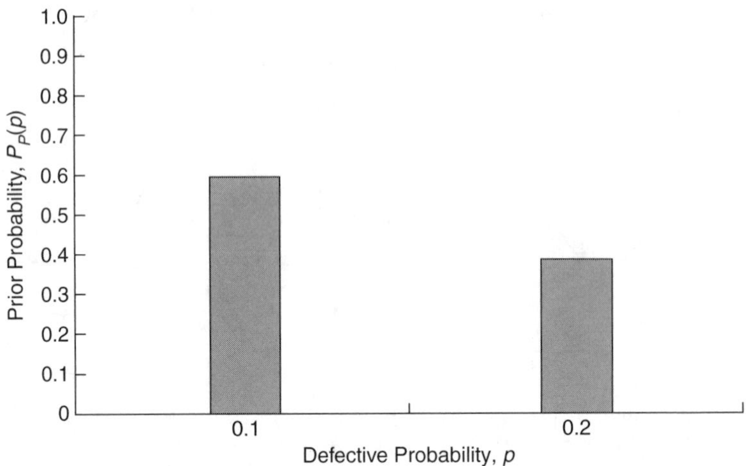

FIGURE 16-8 Prior probability distribution for defective probability of an assembly line.

16.5. APPLICATIONS

16.5.1. Sampling from an Assembly Line

An assembly line produces components of automobiles that have a defective probability distribution as shown in Figure 16-8. This prior distribution is called $P_P(p)$. The mean probability $\overline{P}$ based on this distribution is:

$$\overline{P} = 0.1(0.6) + 0.2(0.4) = 0.140 \tag{16-49}$$

For a sample of size n, the probability $P(D)$ of obtaining one or more defective components can be computed based on the binomial distribution as:

$$\begin{aligned} P(D) &= 0.6\left[1 - \binom{n}{0}(0.1)^0(1-0.1)^n\right] + 0.4\left[1 - \binom{n}{0}(0.2)^0(1-0.2)^n\right] \\ &= 0.6[1-(0.9)^n] + 0.4[1-(0.8)^n] \end{aligned} \tag{16-50}$$

The probability according to Equation 16-50 is shown in Figure 16-9 for different values of n.

Now assume that a sample of size $n = 4$ was tested and found to contain defective components; the subjective prior distribution of Figure 16-8 now must be revised to reflect the new (objective) information. From Figure 16-9, the probability of the sample of size $n = 4$ containing defective components is 0.4425. The posterior distribution $[P'_P(p)]$ can then be computed using Equation 16-13 as follows:

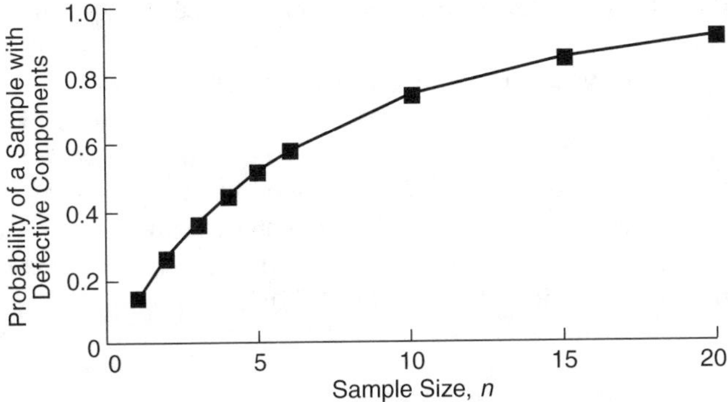

FIGURE 16-9 Probability of a sample with defective components.

$$P'_P\,(0.1) = \frac{0.6[1-(1-0.1)^4]}{0.4425} = 0.4663 \tag{16-51a}$$

Similarly, the second posterior probability can be computed as follows:

$$P'_P\,(0.2) = \frac{0.4[1-(1-0.2)^4]}{0.4425} = 0.5337 \tag{16-51b}$$

The resulting probabilities in Equation 16-51 add up to 1. Also, the average probability of 0.4425 can be viewed as a normalizing factor for computing these probabilities. The mean probability $\overline{P}$ based on the posterior distribution is:

$$\begin{aligned}\overline{P} &= 0.1(0.4663) + 0.2(0.5337) \\ &= 0.1534\end{aligned} \tag{16-52}$$

The result shows an increase from the average probability of 0.140 as given by Equation 16-49 due to the detected failure by testing.

16.5.2. Scour around Bridge Piers

The scour around a bridge pier develops due to the infrequent occurrence of extreme flow in a river and is of interest to engineers. The mean scour depth is an important quantity for the evaluation of bridge piers in streams. A review of several bridge piers in streams indicated that the mean annual scour has a mean and variance as follows:

$$\mu_0 = 3 \text{ in.}$$

and

$$\sigma_0^2 = 1 \text{ in.}^2$$

These values can be considered to constitute prior information about the mean scour. Based on the physics of scour, it can be considered to be random variable with a high variability level with an assumed coefficient of variation of 0.5. Therefore, scour can be assumed to have, in this case, a known or assumed standard deviation (σ) of $0.5 \times 3 = 1.5$ in.

In order to improve estimates of the annual scour depth, a sample of five bridges was analyzed for estimating the annual scour, and a sample mean of 2 in. was obtained. This information can be used to update the prior mean and variance of the average scour using Equations 16-42a and 16-42b as follows:

$$\mu' = \frac{5(2)(1) + 3(1.5)^2}{5(1) + (1.5)^2} = 2.31 \text{ in.} \tag{16-53}$$

and

$$\sigma'^2 = \frac{1(1.5)^2}{5(1) + (1.5)^2} = 0.31 \text{ (in.)}^2 \tag{16-54}$$

Therefore, the posterior standard deviation is 0.56 in.

The prior mean value and variance can be used to establish the following prior 95% confidence interval on the mean (using Equation 16-43a):

$$3 - 1.96(1) \le \mu \le 3 + 1.96(1) \tag{16-55a}$$

or

$$1.04 \text{ in.} \le \mu \le 4.96 \text{ in.} \tag{16-55b}$$

The posterior mean and variance can be used to establish the following posterior 95% confidence interval (using Equation 16-43b):

$$2.31 - 1.96(0.56) \le \mu \le 2.31 + 1.96(0.56) \tag{16-56a}$$

or

$$1.21 \text{ in.} \le \mu \le 3.41 \text{ in.} \tag{16-56b}$$

By comparing Equations 16-55 and 16-56, it can be observed that the information gained by the sample has resulted in reducing the width of the confidence interval (i.e., reducing the uncertainty in the mean concrete strength). However, the range might still be considered large, requiring the collection of more information.

16.6. PROBLEMS

16-1. The accident probability at a new intersection is of interest to a traffic engineer. The engineer subjectively estimated the weekly accident probability as follows:

Weekly accident probability	Subjective probability of accident probability
0.1	0.30
0.2	0.40
0.4	0.20
0.6	0.05
0.8	0.04
0.9	0.01

Solve the following:

a. What is the average accident probability based on the prior information?

b. Given an accident in the first week of traffic, update the distribution of the accident probability.

c. What is the new average accident probability based on the posterior information?

d. Given an accident in the first and second weeks and no accidents in the third week of traffic, update the distribution of the accident probability.

e. What is the average accident probability after the second week?

f. Given no additional accidents for the weeks 4, 5, 6, 7, 8, 9, and 10, update the distribution and the average accident probability. Plot your results. Discuss the results.

16-2. The probability of back injuries of construction workers at XYZ Corporation is of interest to a safety engineer. The engineer subjectively estimated the annual probability as follows:

Annual probability	Subjective probability of annual probability
0.05	0.70
0.10	0.20
0.20	0.04
0.60	0.02
0.80	0.02
0.90	0.02

Solve the following:

a. What is the average annual probability based on the prior information?

b. Given no back injuries in a year, update the distribution of the annual probability.

c. What is the new average annual probability based on the posterior information?

d. Given injuries in the first and second years and no injuries in the following year, update the distribution of the annual probability.

e. What is the average annual probability after the second year?

f. Given no additional injuries for the years 4, 5, 6, 7, 8, 9, and 10, update the distribution and the average annual probability. Plot your results, and discuss the results.

16-3. In Problem 16-1, assume that a weekly rate is used with a uniform prior distribution over the range 0.1 to 0.5 per week. Use an exponential likelihood function. Determine the posterior distribution based on additional information that an accident occurred in less than 2 weeks.

16-4. In Problem 16-2, assume that an annual rate is used with a prior uniform distribution over the range 10 to 18 per year. Use an exponential likelihood function. Determine the posterior distribution based on additional information that an injury occurred in less than 2 weeks (i.e., 0.04 year).

16-5. Redo Problem 16-3 assuming the weekly rate has a prior distribution that is exponential with a mean of 0.2 per week.

16-6. Redo Problem 16-4 assuming the annual rate has a prior distribution that is exponential with a mean of 15 per year.

16-7. The mean thickness of a plate girder used in a suspension bridge needs to be estimated using ultrasonic testing. The mean thickness can be assumed to have a normal probability distribution with a mean (μ) and standard deviation (σ). The standard deviation (σ) is known to be 1.8 mm. The construction drawings indicate a mean thickness of 15 mm and a standard deviation of 2.0 mm. A sample of $n = 20$ was collected. The sample resulted in a mean of 16 mm. Determine: (a) the posterior mean thickness, (b) the posterior standard deviation, and (c) the 95% confidence intervals for thickness based on prior and posterior information.

16-8. The mean thickness of the shell of a pressure vessel needs to be estimated using ultrasonic testing. The mean thickness can be assumed to have a normal probability distribution with a mean (μ) and standard deviation (σ). The standard deviation (σ) is known to be 0.5 mm. The construction drawings indicate a mean thickness of 635 mm and a standard deviation of 1.0 mm. A sample of $n = 20$ was collected. The sample resulted in a mean of 625 mm. Determine: (a) the posterior mean thickness, (b) the posterior standard deviation, and (c) the 95% confidence intervals for the mean thickness based on prior and posterior information. Does the pressure vessel meet the specification of a mean thickness of at least 628 mm using two-sided hypothesis testing with a significance level of 5%?

16-9. The air quality in a building is of interest to an environmental engineer. The mean air quality is measured in the form of the amount of pollutant in the air in ppm. The mean amount of pollutant can be assumed to have a normal probability distribution with a mean (μ) and standard deviation (σ). The standard deviation of pollutant (σ) is known to be 0.2 ppm. The air quality based on the records of the building has a mean of 3 ppm and a standard deviation of 0.2 ppm. A sample of $n = 10$ was collected. The sample resulted in a mean of 2.7 ppm. Determine: (a) the posterior mean pollutant concentration, (b) the posterior standard deviation, and (c) the 95% confidence intervals for the mean concentration based on prior and posterior information.

16-10. Redo Problem 16-9 using a sample size $n = 100$ with the same mean and standard deviation. Compare and discuss your results.

Appendix A
Probability and Statistics Tables

TABLE OF CONTENTS

A-1. Cumulative Distribution Function of Standard Normal ($\Phi(z)$)

z	$\Phi(z)$	z	$\Phi(z)$	z	$\Phi(z)$	z	$\Phi(z)$	z	$\Phi(z)$	z	$\Phi(z)$	z	$\Phi(z)$	z	$\Phi(z)$
0.00	0.500000	0.50	0.691463	1.00	0.841345	1.50	0.933193	2.00	0.977250	2.50	0.993790	3.00	0.998650	3.50	0.999767
0.01	0.503989	0.51	0.694975	1.01	0.843752	1.51	0.934478	2.01	0.977784	2.51	0.993963	3.01	0.998694	3.51	0.999776
0.02	0.507978	0.52	0.698468	1.02	0.846136	1.52	0.935744	2.02	0.978308	2.52	0.994132	3.02	0.998736	3.52	0.999784
0.03	0.511966	0.53	0.701944	1.03	0.848495	1.53	0.936992	2.03	0.978822	2.53	0.994297	3.03	0.998777	3.53	0.999792
0.04	0.515954	0.54	0.705401	1.04	0.850830	1.54	0.938220	2.04	0.979325	2.54	0.994457	3.04	0.998817	3.54	0.999800
0.05	0.519939	0.55	0.708840	1.05	0.853141	1.55	0.939429	2.05	0.979818	2.55	0.994614	3.05	0.998856	3.55	0.999807
0.06	0.523922	0.56	0.712260	1.06	0.855428	1.56	0.940620	2.06	0.980301	2.56	0.994766	3.06	0.998893	3.56	0.999815
0.07	0.527904	0.57	0.715661	1.07	0.857690	1.57	0.941792	2.07	0.980774	2.57	0.994915	3.07	0.998930	3.57	0.999821
0.08	0.531882	0.58	0.719043	1.08	0.859920	1.58	0.942947	2.08	0.981237	2.58	0.995060	3.08	0.998965	3.58	0.999828
0.09	0.535857	0.59	0.722405	1.09	0.862143	1.59	0.944083	2.09	0.981691	2.59	0.995201	3.09	0.998999	3.59	0.999835
0.10	0.539829	0.60	0.725747	1.10	0.864334	1.60	0.945201	2.10	0.982136	2.60	0.995339	3.10	0.999032	3.60	0.999841
0.11	0.543796	0.61	0.729069	1.11	0.866500	1.61	0.946301	2.11	0.982571	2.61	0.995473	3.11	0.999065	3.61	0.999847
0.12	0.547759	0.62	0.732371	1.12	0.868643	1.62	0.947384	2.12	0.982997	2.62	0.995604	3.12	0.999096	3.62	0.999853
0.13	0.551717	0.63	0.735653	1.13	0.870862	1.63	0.948449	2.13	0.983414	2.63	0.995731	3.13	0.999126	3.63	0.999858
0.14	0.555671	0.64	0.738914	1.14	0.872857	1.64	0.949497	2.14	0.983823	2.64	0.995855	3.14	0.999155	3.64	0.999864
0.15	0.559618	0.65	0.742154	1.15	0.874928	1.65	0.950529	2.15	0.984223	2.65	0.995975	3.15	0.999184	3.65	0.999869
0.16	0.563560	0.66	0.745374	1.16	0.876976	1.66	0.951543	2.16	0.984614	2.66	0.996093	3.16	0.999211	3.66	0.999874
0.17	0.567494	0.67	0.748572	1.17	0.878999	1.67	0.952540	2.17	0.984997	2.67	0.996207	3.17	0.999238	3.67	0.999879
0.18	0.571423	0.68	0.751748	1.18	0.881000	1.68	0.953521	2.18	0.985371	2.68	0.996319	3.18	0.999264	3.68	0.999883
0.19	0.575345	0.69	0.754903	1.19	0.882977	1.69	0.954486	2.19	0.985738	2.69	0.996427	3.19	0.999289	3.69	0.999888

0.20	0.579260	0.70	0.758036	1.20	0.884930	1.70	0.955435	2.20	0.986097	2.70	0.996533	3.20	0.999313	3.70	0.999892
0.21	0.583166	0.71	0.761148	1.21	0.886860	1.71	0.956367	2.21	0.986447	2.71	0.996636	3.21	0.999336	3.71	0.999896
0.22	0.587064	0.72	0.764238	1.22	0.888767	1.72	0.957284	2.22	0.986791	2.72	0.996736	3.22	0.999359	3.72	0.999900
0.23	0.590954	0.73	0.767305	1.23	0.890651	1.73	0.958185	2.23	0.987126	2.73	0.996833	3.23	0.999381	3.73	0.999904
0.24	0.594835	0.74	0.770350	1.24	0.892512	1.74	0.959071	2.24	0.987455	2.74	0.996928	3.24	0.999402	3.74	0.999908
0.25	0.598706	0.75	0.773373	1.25	0.894350	1.75	0.959941	2.25	0.987776	2.75	0.997020	3.25	0.999423	3.75	0.999912
0.26	0.602568	0.76	0.776373	1.26	0.896165	1.76	0.960796	2.26	0.988089	2.76	0.997110	3.26	0.999443	3.76	0.999915
0.27	0.606420	0.77	0.779350	1.27	0.897958	1.77	0.961636	2.27	0.988396	2.77	0.997197	3.27	0.999462	3.77	0.999918
0.28	0.610262	0.78	0.782305	1.28	0.899727	1.78	0.962462	2.28	0.988696	2.78	0.997282	3.28	0.999481	3.78	0.999922
0.29	0.614092	0.79	0.785236	1.29	0.901475	1.79	0.963273	2.29	0.988989	2.79	0.997365	3.29	0.999499	3.79	0.999925
0.30	0.617912	0.8	0.788145	1.30	0.903199	1.80	0.964070	2.30	0.989276	2.80	0.997445	3.30	0.999516	3.80	0.999928
0.31	0.621720	0.81	0.791030	1.31	0.904902	1.81	0.964852	2.31	0.989556	2.81	0.997523	3.31	0.999533	3.81	0.999931
0.32	0.625517	0.82	0.793892	1.32	0.906583	1.82	0.965621	2.32	0.989830	2.82	0.997599	3.32	0.99955	3.82	0.999933
0.33	0.629301	0.83	0.796731	1.33	0.908241	1.83	0.966375	2.33	0.990097	2.83	0.997673	3.33	0.999566	3.83	0.999936
0.34	0.633072	0.84	0.799546	1.34	0.909877	1.84	0.967116	2.34	0.990358	2.84	0.997744	3.34	0.999581	3.84	0.999938
0.35	0.636831	0.85	0.802337	1.35	0.911492	1.85	0.967843	2.35	0.990613	2.85	0.997814	3.35	0.999596	3.85	0.999941
0.36	0.640576	0.86	0.805105	1.36	0.913085	1.86	0.968557	2.36	0.990863	2.86	0.997882	3.36	0.999610	3.86	0.999943
0.37	0.644309	0.87	0.807850	1.37	0.914656	1.87	0.969258	2.37	0.991106	2.87	0.997948	3.37	0.999624	3.87	0.999946
0.38	0.648027	0.88	0.810570	1.38	0.916207	1.88	0.969946	2.38	0.991344	2.88	0.998012	3.38	0.999637	3.88	0.999948
0.39	0.651732	0.89	0.813267	1.39	0.917735	1.89	0.970621	2.39	0.991576	2.89	0.998074	3.39	0.99965	3.89	0.999950
0.40	0.655422	0.9	0.815940	1.40	0.919243	1.90	0.971284	2.40	0.991802	2.90	0.998134	3.40	0.999663	3.90	0.999952
0.41	0.659097	0.91	0.818589	1.41	0.920730	1.91	0.971933	2.41	0.992024	2.91	0.998193	3.41	0.999675	3.91	0.999954
0.42	0.662757	0.92	0.821214	1.42	0.922196	1.92	0.972571	2.42	0.992240	2.92	0.998250	3.42	0.999687	3.92	0.999956
0.43	0.666402	0.93	0.823815	1.43	0.923641	1.93	0.973197	2.43	0.992451	2.93	0.998305	3.43	0.999698	3.93	0.999958
0.44	0.670032	0.94	0.826391	1.44	0.925066	1.94	0.973810	2.44	0.992656	2.94	0.998359	3.44	0.999709	3.94	0.999959
0.45	0.673645	0.95	0.828944	1.45	0.926471	1.95	0.974412	2.45	0.992857	2.95	0.998411	3.45	0.999720	3.95	0.999961
0.46	0.677242	0.96	0.831473	1.46	0.927855	1.96	0.975002	2.46	0.993053	2.96	0.998462	3.46	0.999730	3.96	0.999963
0.47	0.680823	0.97	0.833977	1.47	0.929219	1.97	0.975581	2.47	0.993244	2.97	0.998511	3.47	0.999740	3.97	0.999964
0.48	0.684387	0.98	0.836457	1.48	0.930563	1.98	0.976148	2.48	0.993431	2.98	0.998559	3.48	0.999749	3.98	0.999966
0.49	0.687933	0.99	0.838913	1.49	0.931888	1.99	0.976705	2.49	0.993613	2.99	0.998605	3.49	0.999758	3.99	0.999967

A-1. Cumulative Distribution Function of Standard Normal ($\Phi(z)$) (cont.)

z	$1 - \Phi(z)$	z	$1 - \Phi(z)$	z	$1 - \Phi(z)$	z	$1 - \Phi(z)$	z	$1 - \Phi(z)$	z	$1 - \Phi(z)$
4.00	3.17E-05	4.50	3.40E-06	5.00	2.87E-07	6.00	9.87E-10	7.00	1.28E-12	8.00	6.66E-16
4.05	2.56E-05	4.55	2.68E-06	5.10	1.70E-07	6.10	5.30E-10	7.10	6.24E-13	8.10	2.22E-16
4.10	2.07E-05	4.60	2.11E-06	5.20	9.96E-08	6.20	2.82E-10	7.20	3.01E-13	8.20	1.11E-16
4.15	1.66E-05	4.65	1.66E-06	5.30	5.79E-08	6.30	1.49E-10	7.30	1.44E-13		
4.20	1.33E-05	4.70	1.30E-06	5.40	3.33E-08	6.40	7.77E-11	7.40	6.80E-14		
4.25	1.07E-05	4.75	1.02E-06	5.50	1.90E-08	6.50	4.02E-11	7.50	3.20E-14		
4.30	8.54E-06	4.80	7.93E-07	5.60	1.07E-08	6.60	2.06E-11	7.60	1.50E-14		
4.35	6.81E-06	4.85	6.17E-07	5.70	5.99E-09	6.70	1.04E-11	7.70	7.00E-15		
4.40	5.41E-06	4.90	4.79E-07	5.80	3.32E-09	6.80	5.23E-12	7.80	3.00E-15		
4.45	4.29E-06	4.95	3.71E-07	5.90	1.82E-09	6.90	2.60E-12	7.90	1.50E-15		

A-2. Critical Values for the Student's *t* Distribution ($t_{\alpha,k}$)

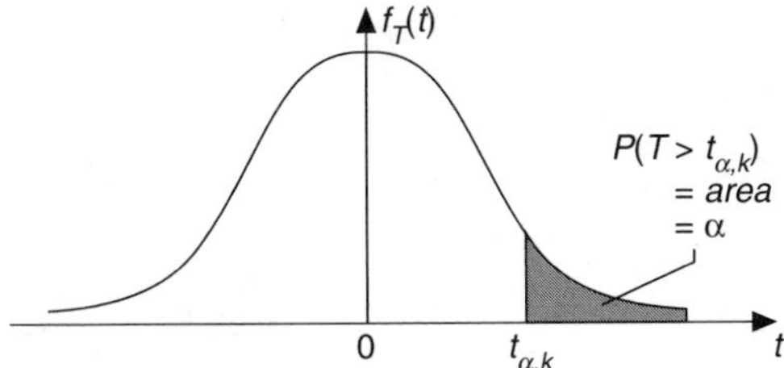

For Upper Tail Values

Degrees of freedom, k	Level of significance, α							
	0.2500	**0.1000**	**0.0500**	**0.0250**	**0.0100**	**0.0050**	**0.0025**	**0.0005**
1	1.000	3.078	6.314	12.706	31.821	63.657	127.321	536.627
2	0.816	1.886	2.920	4.303	6.965	9.925	14.089	31.599
3	0.765	1.638	2.353	3.182	4.541	5.841	7.453	12.924
4	0.741	1.533	2.132	2.776	3.747	4.604	5.598	8.610
5	0.727	1.476	2.015	2.571	3.365	4.032	4.773	6.869
6	0.718	1.440	1.943	2.447	3.143	3.707	4.317	5.959
7	0.711	1.415	1.895	2.365	2.998	3.499	4.029	5.408
8	0.706	1.397	1.860	2.306	2.896	3.355	3.833	5.041
9	0.703	1.383	1.833	2.262	2.821	3.250	3.690	4.781
10	0.700	1.372	1.812	2.228	2.764	3.169	3.581	4.587
11	0.697	1.363	1.796	2.201	2.718	3.106	3.497	4.437
12	0.695	1.356	1.782	2.179	2.681	3.055	3.428	4.318
13	0.694	1.350	1.771	2.160	2.650	3.012	3.372	4.221
14	0.692	1.345	1.761	2.145	2.624	2.977	3.326	4.140
15	0.691	1.341	1.753	2.131	2.602	2.947	3.286	4.073
16	0.690	1.337	1.746	2.120	2.583	2.921	3.252	4.015
17	0.689	1.333	1.740	2.110	2.567	2.898	3.222	3.965
18	0.688	1.330	1.734	2.101	2.552	2.878	3.197	3.922
19	0.688	1.328	1.729	2.093	2.539	2.861	3.174	3.883
20	0.687	1.325	1.725	2.086	2.528	2.845	3.153	3.850
21	0.686	1.323	1.721	2.080	2.518	2.831	3.135	3.819
22	0.686	1.321	1.717	2.074	2.508	2.819	3.119	3.792
23	0.685	1.319	1.714	2.069	2.500	2.807	3.104	3.768
24	0.685	1.318	1.711	2.064	2.492	2.797	3.091	3.745
25	0.684	1.316	1.708	2.060	2.485	2.787	3.078	3.725
26	0.684	1.315	1.706	2.056	2.479	2.779	3.067	3.707
27	0.684	1.314	1.703	2.052	2.473	2.771	3.057	3.690
28	0.683	1.313	1.701	2.048	2.467	2.763	3.047	3.674
29	0.683	1.311	1.699	2.045	2.462	2.756	3.038	3.659
30	0.683	1.310	1.697	2.042	2.457	2.750	3.030	3.646
35	0.682	1.306	1.690	2.030	2.438	2.724	2.996	3.591
40	0.681	1.303	1.684	2.021	2.423	2.704	2.971	3.551
45	0.680	1.301	1.679	2.014	2.412	2.690	2.952	3.520
50	0.679	1.299	1.676	2.009	2.403	2.678	2.937	3.496

A-2. Critical Values for the Student's t Distribution ($t_{\alpha,k}$) (cont.)

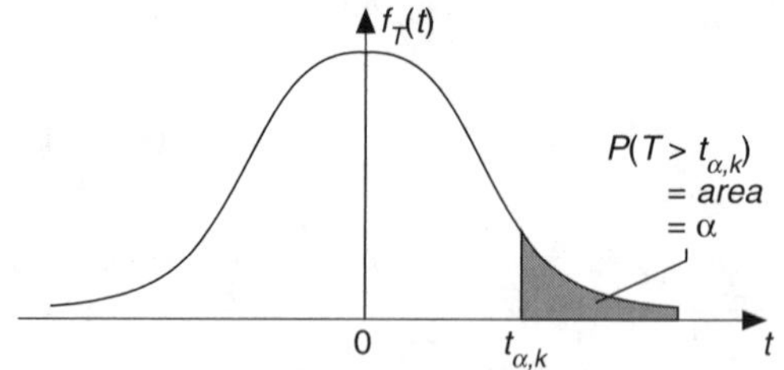

For Upper Tail Values

Degrees of freedom, k	Level of significance, α							
	0.2500	**0.1000**	**0.0500**	**0.0250**	**0.0100**	**0.0050**	**0.0025**	**0.0005**
55	0.679	1.297	1.673	2.004	2.396	2.668	2.925	3.476
60	0.679	1.296	1.671	2.000	2.390	2.660	2.915	3.460
65	0.678	1.295	1.669	1.997	2.385	2.654	2.906	3.447
70	0.678	1.294	1.667	1.994	2.381	2.648	2.899	3.435
80	0.678	1.292	1.664	1.990	2.374	2.639	2.887	3.416
90	0.677	1.291	1.662	1.987	2.368	2.632	2.878	3.402
100	0.677	1.290	1.660	1.984	2.364	2.626	2.871	3.390
125	0.676	1.288	1.657	1.979	2.357	2.616	2.858	3.370
150	0.676	1.287	1.655	1.976	2.351	2.609	2.849	3.357
200	0.676	1.286	1.653	1.972	2.345	2.601	2.839	3.340
∞	0.6745	1.2816	1.6448	1.9600	2.3267	2.5758	2.8070	3.2905

A-3. Critical Values for the Chi-Square Distribution ($c_{\alpha,k} = \chi^2_{\alpha,k}$)

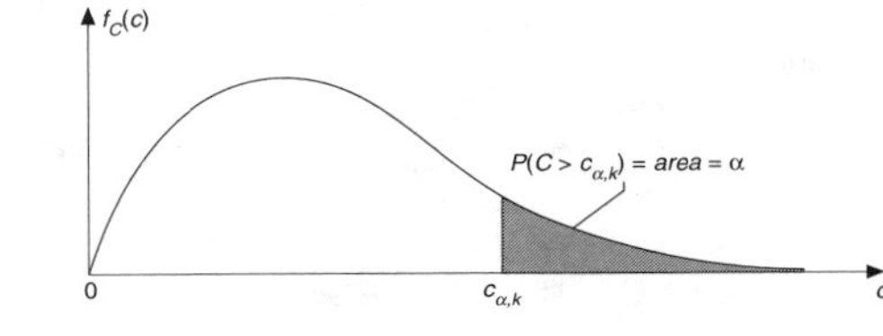

For upper tail values

Degrees of freedom, k	Level of significance, α														
	0.9990	**0.9950**	**0.9900**	**0.9750**	**0.9500**	**0.9000**	**0.7000**	**0.5000**	**0.2000**	**0.1000**	**0.0500**	**0.0200**	**0.0100**	**0.0050**	**0.0010**
1	0.000	0.000	0.000	0.001	0.004	0.016	0.148	0.455	1.642	2.706	3.842	5.405	6.637	7.905	10.809
2	0.002	0.010	0.020	0.051	0.102	0.211	0.713	1.386	3.219	4.604	5.995	7.822	9.221	10.589	13.691
3	0.024	0.071	0.115	0.216	0.352	0.584	1.424	2.366	4.642	6.252	7.817	9.841	11.325	12.819	16.292
4	0.090	0.205	0.297	0.484	0.711	1.064	2.195	3.357	5.989	7.782	9.492	11.660	13.280	14.824	18.432
5	0.209	0.411	0.553	0.831	1.145	1.610	3.000	4.352	7.291	9.237	11.073	13.385	15.088	16.762	20.751
6	0.377	0.673	0.871	1.236	1.635	2.204	3.828	5.349	8.559	10.646	12.596	15.033	16.810	18.550	22.677
7	0.597	0.988	1.237	1.688	2.167	2.833	4.671	6.346	9.804	12.070	14.070	16.624	18.471	20.270	24.527
8	0.850	1.341	1.642	2.179	2.732	3.489	5.527	7.344	11.031	13.363	15.512	18.171	20.082	21.938	26.318
9	1.135	1.728	2.086	2.699	3.324	4.168	6.393	8.343	12.243	14.686	16.925	19.683	21.654	23.563	28.061
10	1.446	2.152	2.555	3.244	3.938	4.864	7.266	9.342	13.443	15.990	18.311	21.165	23.194	25.154	29.763
11	1.819	2.597	3.047	3.815	4.574	5.576	8.148	10.341	14.633	17.278	19.681	22.623	24.755	26.714	31.431
12	2.188	3.064	3.568	4.402	5.225	6.303	9.034	11.340	15.813	18.551	21.030	24.059	26.246	28.249	33.070
13	2.577	3.560	4.102	5.006	5.890	7.041	9.926	12.340	16.986	19.814	22.367	25.477	27.717	29.878	34.683
14	3.018	4.066	4.653	5.624	6.568	7.789	10.821	13.339	18.152	21.067	23.691	26.879	29.169	31.376	36.272
15	3.449	4.588	5.226	6.260	7.260	8.546	11.721	14.339	19.312	22.310	25.000	28.266	30.605	32.857	37.842
16	3.894	5.135	5.807	6.905	7.960	9.311	12.624	15.339	20.466	23.546	26.301	29.640	32.027	34.321	39.392
17	4.350	5.687	6.400	7.560	8.670	10.083	13.530	16.338	21.616	24.771	27.593	31.002	33.435	35.771	40.926
18	4.864	6.251	7.004	8.225	9.388	10.864	14.440	17.337	22.761	25.992	28.877	32.353	34.831	37.208	42.444
19	5.351	6.825	7.627	8.904	10.114	11.650	15.351	18.338	23.902	27.206	30.148	33.694	36.216	38.633	43.949
20	5.848	7.422	8.252	9.587	10.849	12.442	16.625	19.337	25.039	28.415	31.416	35.026	37.591	40.046	45.440
21	6.398	8.018	8.886	10.278	11.590	13.238	17.182	20.337	26.173	29.619	32.678	36.350	38.957	41.449	46.919
22	6.919	8.622	9.528	10.976	12.336	14.040	18.100	21.337	27.304	30.817	33.933	37.666	40.314	42.843	48.387

A-3. Critical Values for the Chi-Square Distribution ($c_{\alpha,k} = \chi^2_{\alpha,k}$) (cont.)

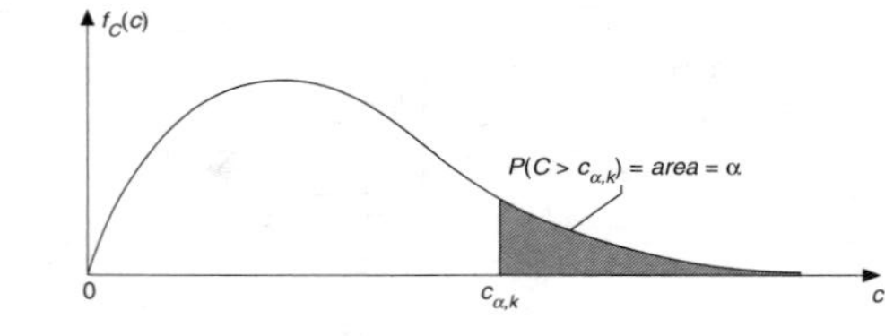

For upper tail values

Degrees of freedom, k	Level of significance, α														
	0.9990	**0.9950**	**0.9900**	**0.9750**	**0.9500**	**0.9000**	**0.7000**	**0.5000**	**0.2000**	**0.1000**	**0.0500**	**0.0200**	**0.0100**	**0.0050**	**0.0010**
23	7.447	9.247	10.187	11.685	13.088	14.846	19.020	22.337	28.431	32.012	35.178	38.975	41.662	44.228	49.845
24	8.027	9.869	10.846	12.397	13.845	15.657	19.943	23.337	29.556	33.199	36.421	40.277	43.004	45.604	51.293
25	8.576	10.498	11.510	13.115	14.607	16.471	20.866	24.337	30.678	34.384	37.660	41.573	44.338	46.973	52.732
26	9.130	11.132	12.190	13.837	15.377	17.291	21.792	25.337	31.796	35.566	38.894	42.863	45.665	48.334	54.162
27	9.735	11.789	12.868	14.565	16.149	18.113	22.718	26.336	32.913	36.745	40.119	44.147	46.986	49.688	55.584
28	10.306	12.348	13.551	15.304	16.925	18.938	23.646	27.336	34.028	37.920	41.344	45.426	48.301	51.036	56.998
29	10.882	13.092	14.240	16.042	17.705	19.766	24.576	28.336	35.140	39.092	42.565	46.699	49.610	52.378	58.405
30	11.509	13.767	14.943	16.784	18.488	20.598	25.507	29.336	36.251	40.261	43.782	47.968	50.914	53.713	59.805
35	14.554	17.160	18.494	20.562	22.461	24.793	30.176	34.336	41.780	46.063	49.811	54.250	57.363	60.313	66.714
40	17.846	20.669	22.139	24.423	26.508	29.055	34.879	39.337	47.261	51.796	55.753	60.443	63.710	66.802	73.490
45	21.183	24.275	25.878	28.357	30.611	33.355	39.591	44.337	52.721	57.497	61.652	66.562	69.976	73.201	80.160
50	24.609	27.957	29.685	32.349	34.763	37.693	44.319	49.336	58.157	63.159	67.501	72.619	76.172	79.523	86.740
55	28.111	31.702	33.549	36.390	38.957	42.064	49.061	54.336	63.571	68.789	73.308	78.625	82.309	85.780	93.243
60	31.678	35.503	37.465	40.474	43.187	46.463	53.814	59.336	68.966	74.390	79.078	84.586	88.396	91.982	99.679
65	35.303	39.353	41.424	44.596	47.449	50.887	58.578	64.336	74.344	79.966	84.817	90.507	94.438	98.134	106.057
70	38.980	43.246	45.423	48.750	51.739	55.333	63.351	69.335	79.709	85.521	90.528	96.393	100.441	104.243	112.383
80	46.466	51.145	53.523	57.147	60.391	64.282	72.920	79.335	90.400	96.572	101.876	108.075	112.344	116.348	124.901
90	54.104	59.171	61.738	65.641	69.126	73.295	82.515	89.335	101.048	107.559	113.143	119.654	124.130	128.324	137.267
100	61.869	67.303	70.049	74.216	77.929	82.362	92.133	99.335	111.662	118.493	124.340	131.147	135.820	140.193	149.505
125	81.726	88.007	91.166	95.941	100.178	105.217	116.254	12.335	138.071	145.638	152.092	159.580	164.706	169.493	179.653
150	102.073	109.122	112.655	117.980	122.692	128.278	140.460	149.334	164.345	172.577	179.579	187.683	193.219	198.380	209.310
200	143.807	152.224	156.421	162.724	168.279	174.838	189.051	199.334	216.605	226.017	233.993	243.191	249.455	255.281	267.579

A-4. Critical Values for the *F* Distribution ($f_{\alpha,k,u} = f_{\alpha,\nu_1,\nu_2}$)

Upper values for 5% (in first row) and 1% (in second row) significance levels, α

Second degrees of freedom, *u*	First degrees of freedom,[a] *k*														
	1	**2**	**3**	**4**	**5**	**6**	**7**	**8**	**9**	**10**	**11**	**12**	**13**	**14**	**15**
1	161.	200.	216.	225.	230.	234.	237.	239.	241.	242.	243.	244.	245.	245.	246.
	4052.	5000.	5403.	5625.	5764.	5859.	5928.	5981.	6022.	6056.	6083.	6106.	6126.	6143.	6157.
2	18.51	19.00	19.16	19.25	19.30	19.33	19.35	19.37	19.38	19.40	19.40	19.41	19.42	19.42	19.43
	98.50	99.00	99.17	99.25	99.30	99.33	99.36	99.37	99.39	99.40	19.41	99.42	19.42	19.43	19.43
3	10.13	9.55	9.28	9.12	9.01	8.94	8.89	8.85	8.81	8.79	8.76	8.74	8.73	8.71	8.70
	34.12	30.82	29.46	28.71	28.24	27.91	27.67	27.49	27.34	27.23	27.13	27.05	26.99	26.93	26.87
4	7.71	6.94	6.59	6.39	6.26	6.16	6.09	6.04	6.00	5.96	5.94	5.91	5.89	5.87	5.86
	21.20	18.00	16.69	15.98	15.52	15.21	14.98	14.80	14.66	14.55	14.45	14.37	14.31	14.25	14.20
5	6.61	5.79	5.41	5.19	5.05	4.95	4.88	4.82	4.77	4.74	4.70	4.68	4.66	4.64	4.62
	16.26	13.27	12.06	11.39	10.97	10.67	10.46	10.29	10.16	10.05	9.96	9.89	9.82	9.77	9.72
6	5.99	5.14	4.76	4.53	4.39	4.28	4.21	4.15	4.10	4.06	4.03	4.00	3.98	3.96	3.94
	13.75	10.92	9.78	9.15	8.75	8.47	8.26	8.10	7.98	7.87	7.79	7.72	7.66	7.60	7.56
7	5.59	4.74	4.35	4.12	3.97	3.87	3.79	3.73	3.68	3.64	3.60	3.57	3.55	3.53	3.51
	12.25	9.55	8.45	7.85	7.46	7.19	6.99	6.84	6.72	3.62	6.54	6.47	6.41	6.36	6.31
8	5.32	4.46	4.07	3.84	3.69	3.58	3.50	3.44	3.39	3.35	3.31	3.28	3.26	3.24	3.22
	11.26	8.65	9.59	7.01	6.63	6.37	6.18	6.03	5.91	5.81	5.73	5.67	5.61	5.56	5.52
9	5.12	4.26	3.86	3.63	3.48	3.37	3.29	3.23	3.18	3.14	3.10	3.07	3.05	3.03	3.01
	10.56	8.02	6.99	6.42	6.06	5.30	5.61	5.47	5.35	5.26	5.18	5.11	5.05	5.01	4.96
10	4.96	4.10	3.71	3.48	3.33	3.22	3.14	3.07	3.02	2.98	2.94	2.91	2.89	2.86	2.85
	10.04	7.56	6.55	5.99	5.64	5.39	5.20	5.06	4.94	4.85	4.77	4.71	4.65	4.60	4.56

[a] For *k* values larger than 15, refer to pages 605 to 607.

A-4. Critical Values for the F Distribution ($f_{\alpha,k,u} = f_{\alpha,\nu_1,\nu_2}$) (cont.)

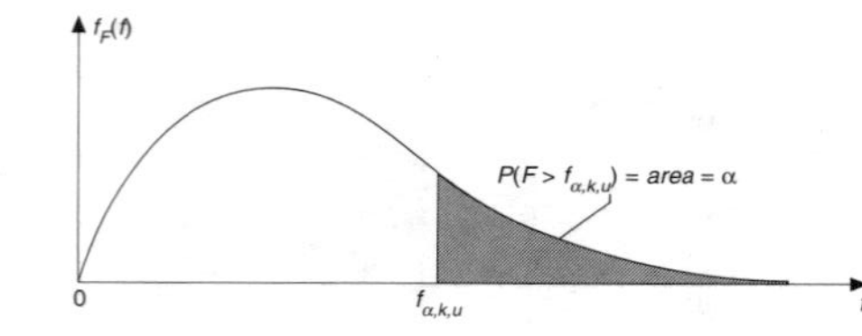

Upper values for 5% (in first row) and 1% (in second row) significance levels, α

Second degrees of freedom, u	First degrees of freedom, k														
	1	2	3	4	5	6	7	8	9	10	11	12	13	14	15
11	4.84	3.98	3.59	3.36	3.20	3.09	3.01	2.95	2.90	2.85	2.82	2.79	2.76	2.74	2.72
	9.65	7.21	6.22	5.67	5.32	5.07	4.89	4.74	4.63	4.54	4.46	4.40	4.34	4.29	4.25
12	4.75	3.89	3.49	3.26	3.11	3.00	2.91	2.85	2.80	2.75	2.72	2.69	2.66	2.64	2.62
	9.33	6.93	5.95	5.41	5.06	4.82	4.64	4.50	4.39	4.30	4.22	4.16	4.10	4.05	4.01
13	4.67	3.81	3.41	3.18	3.03	2.92	2.83	2.77	2.71	2.67	2.63	2.60	2.58	2.55	2.53
	9.07	6.70	5.74	5.21	4.86	4.62	4.44	4.30	4.19	4.10	4.02	3.96	3.91	3.86	3.82
14	4.60	3.74	3.34	3.11	2.96	2.85	2.76	2.70	2.65	2.60	2.57	2.53	2.51	2.48	2.46
	8.86	6.51	5.56	5.04	4.69	4.46	4.28	4.14	4.03	3.94	3.86	3.80	3.75	3.70	3.66
15	4.54	3.68	3.29	3.06	2.90	2.79	2.71	2.64	2.59	2.54	2.51	2.48	2.45	2.42	2.40
	8.68	6.36	5.42	4.89	4.56	4.32	4.14	4.00	3.89	3.80	3.37	3.67	3.61	3.56	3.52
17	4.45	3.59	3.20	2.96	2.81	2.70	2.61	2.55	2.49	2.45	2.41	2.38	2.35	2.33	2.31
	8.40	6.11	5.19	4.67	4.34	4.10	3.93	3.79	3.68	3.59	3.52	3.46	3.40	3.35	3.31
20	4.35	3.49	3.10	2.87	2.71	2.60	2.51	2.45	2.39	2.35	2.31	2.28	2.25	2.22	2.20
	8.10	5.85	4.94	4.43	4.10	3.87	3.70	3.56	3.46	3.37	3.29	3.23	3.18	3.13	3.09
25	4.24	3.39	2.99	2.76	2.60	2.49	2.40	2.34	2.28	2.24	2.20	2.16	2.14	2.11	2.09
	7.77	5.57	4.68	4.18	3.85	3.63	3.46	3.32	3.22	3.13	3.06	2.99	2.94	2.89	2.85
30	4.17	3.32	2.92	2.69	2.53	2.42	2.33	2.27	2.21	2.16	2.13	2.09	2.06	2.04	2.01
	7.56	5.39	4.51	4.02	3.70	3.47	3.30	3.17	3.07	2.98	2.91	2.84	2.79	2.74	2.70
40	4.08	3.23	2.84	2.61	2.45	2.34	2.25	2.18	2.12	2.08	2.04	2.00	1.97	1.95	1.92
	7.31	5.18	4.31	3.83	3.51	3.29	5.12	2.99	2.89	2.80	2.73	2.66	2.61	2.56	2.52

50	4.03	3.18	2.79	2.56	2.40	2.29	2.20	2.13	2.07	2.03	1.99	1.95	1.92	1.89	1.87
	7.17	5.03	4.20	3.72	3.41	3.19	3.02	2.89	2.78	2.76	2.63	2.56	2.51	2.46	2.42
75	3.97	3.12	2.73	2.49	2.34	2.22	2.13	2.06	2.01	1.96	1.92	1.88	1.85	1.83	1.80
	6.99	4.90	4.05	3.58	3.27	3.05	2.89	2.76	2.65	2.57	2.49	2.43	2.38	2.33	2.29
100	3.94	3.09	2.70	2.48	2.31	2.19	2.10	2.03	1.97	1.93	1.89	1.85	1.82	1.79	1.77
	6.90	4.82	3.98	3.51	3.21	2.99	2.82	2.69	2.59	2.50	2.43	2.37	2.31	2.27	2.22
200	3.89	3.04	2.65	2.42	2.26	2.14	2.06	1.98	1.93	1.88	1.84	1.80	1.77	1.74	1.72
	6.76	4.71	3.88	3.41	3.11	2.89	2.73	2.60	2.50	2.41	2.34	2.27	2.22	2.17	2.13
500	3.86	3.01	2.62	2.39	2.23	2.12	2.03	1.96	1.90	1.85	1.81	1.77	1.74	1.71	1.69
	6.69	4.65	3.82	3.36	3.05	2.84	2.68	2.55	2.44	2.36	2.28	2.22	2.17	2.12	2.07
1000	3.85	3.00	2.61	2.38	2.22	2.11	2.02	1.98	1.89	1.84	1.80	1.76	1.73	1.70	1.68
	6.66	4.63	3.80	3.34	3.04	2.82	2.66	2.53	2.43	2.34	2.27	2.20	2.15	2.10	2.06
10,000	3.84	3.00	2.61	2.37	2.21	2.10	2.01	1.94	1.88	1.83	1.79	1.75	1.72	1.69	1.67
	6.64	4.61	3.78	3.32	3.02	2.80	2.64	2.51	2.41	2.32	2.25	2.19	2.13	2.08	2.04

Second degrees of freedom, *u*	First degrees of freedom, *k*											
	17	**20**	**25**	**30**	**40**	**50**	**75**	**100**	**200**	**500**	**1000**	**10,000**
1	247.	248.	249.	250.	251.	252.	253.	253.	254.	254.	254.	254.
	6181.	6209.	6240.	6261.	6287.	6303.	6324.	6334.	6350.	6360.	6363.	6364.
2	19.44	19.45	19.46	19.46	18.47	19.48	19.48	19.49	19.49	19.49	19.49	19.49
	99.44	99.45	99.46	99.47	99.47	99.48	99.49	99.49	99.49	99.50	99.50	99.49
3	8.68	8.66	8.63	8.62	8.59	8.58	8.56	8.55	8.54	8.53	8.53	8.52
	26.79	26.69	26.58	26.50	26.41	26.35	26.28	26.24	26.18	26.15	26.14	26.12
4	5.83	5.80	5.77	5.75	5.72	5.70	5.68	5.66	5.65	5.64	5.63	5.63
	14.11	14.02	13.91	13.84	13.75	13.69	13.61	13.58	13.52	13.49	13.47	13.46
5	4.59	4.56	4.52	4.50	4.46	4.44	4.42	4.41	4.39	4.37	4.37	4.37
	9.64	9.55	9.45	9.38	9.29	9.24	9.17	9.13	9.08	9.04	9.03	9.02
6	3.91	3.87	3.83	3.81	3.77	3.75	3.73	3.71	3.69	3.68	3.67	3.67
	7.48	7.40	7.30	7.23	7.14	7.09	7.02	6.99	6.93	6.90	6.89	6.88
7	3.48	3.44	3.40	3.38	3.34	3.32	3.29	3.27	3.25	3.24	3.23	3.23
	6.24	6.16	6.06	5.99	5.91	5.86	5.79	5.75	5.70	5.67	5.66	5.65

A-4. Critical Values for the *F* Distribution ($f_{\alpha,k,u} = f_{\alpha,\nu_1,\nu_2}$) (cont.)

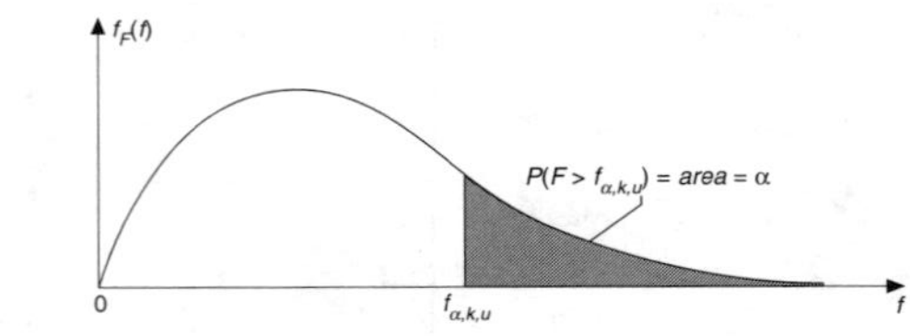

Upper values for 5% (in first row) and 1% (in second row) significance levels, α

Second degrees of freedom, *u*	First degrees of freedom, *k*											
	17	**20**	**25**	**30**	**40**	**50**	**75**	**100**	**200**	**500**	**1000**	**10,000**
8	3.19	3.15	3.11	3.08	3.04	3.02	2.99	2.97	2.95	2.94	2.93	2.93
	5.44	5.36	5.26	5.20	5.12	5.07	5.00	4.96	4.91	4.88	4.87	4.86
9	2.97	2.94	2.89	2.86	2.83	2.80	2.77	2.76	2.73	2.72	2.71	2.71
	4.89	4.81	4.71	4.65	4.57	4.52	4.45	4.41	4.36	4.33	4.32	4.31
10	2.81	2.77	2.73	2.70	2.66	2.64	2.60	2.59	2.56	2.55	2.54	2.54
	4.49	4.41	3.31	4.25	4.17	4.12	4.05	4.01	3.96	3.93	3.92	3.91
11	2.69	2.65	2.60	2.57	2.53	2.51	2.47	2.46	2.43	2.42	2.41	2.41
	4.18	4.10	4.01	3.94	3.86	3.81	3.74	3.71	3.66	3.62	3.61	3.60
12	2.58	2.54	2.50	2.47	2.43	2.40	2.37	2.35	2.32	2.31	2.30	2.30
	3.94	3.86	3.76	3.70	3.62	3.57	3.50	3.47	3.41	3.38	3.37	3.36
13	2.50	2.46	2.41	2.38	2.34	2.31	2.28	2.26	2.23	2.22	2.21	2.21
	3.75	3.66	3.57	3.51	3.43	3.38	3.31	3.27	3.22	3.19	3.18	3.17
14	2.43	2.39	2.34	2.31	2.27	2.24	2.21	2.19	2.16	2.14	2.14	2.13
	3.59	3.51	3.41	3.35	3.27	3.22	3.15	3.11	3.06	3.03	3.02	3.00
15	2.37	2.33	2.28	2.25	2.20	2.18	2.14	2.12	2.10	2.08	2.07	2.07
	3.45	3.37	3.28	3.21	3.13	3.08	3.01	2.98	2.92	2.89	2.88	2.87
17	2.27	2.23	2.18	2.15	2.10	2.08	2.04	2.02	1.99	1.97	1.97	1.96
	3.24	3.16	3.07	3.00	2.92	2.87	2.80	2.76	2.71	2.68	2.66	2.65
20	2.17	2.12	2.07	2.04	1.99	1.97	1.93	1.91	1.88	1.86	1.85	1.85
	3.02	2.94	2.84	2.78	2.69	2.64	2.57	2.54	2.48	2.44	2.43	2.42

25	2.05	2.01	1.96	1.92	1.87	1.84	1.80	1.78	1.75	1.73	1.72	1.72
	2.78	2.70	2.60	2.54	2.45	2.40	2.33	2.29	2.23	2.19	2.18	2.17
30	1.98	1.93	1.88	1.84	1.79	1.76	1.74	1.70	1.66	1.64	1.63	1.62
	2.63	2.55	2.45	2.39	2.30	2.25	2.17	2.13	2.07	2.03	2.02	2.01
40	1.89	1.84	1.78	1.74	1.69	1.66	1.61	1.59	1.55	1.53	1.52	1.51
	2.45	2.37	2.27	2.20	2.11	2.06	1.98	1.94	1.87	1.83	1.82	0.18
50	1.83	1.78	1.73	1.69	1.63	1.60	1.55	1.52	1.48	1.46	1.45	1.44
	2.35	2.27	2.17	2.10	2.01	1.95	1.87	1.82	1.76	1.71	1.70	1.68
75	1.76	1.71	1.65	1.61	1.55	1.52	1.47	1.44	1.39	1.36	1.35	1.34
	2.22	2.13	2.03	1.96	1.87	1.81	1.72	1.67	1.60	1.55	1.53	1.52
100	1.73	1.68	1.62	1.57	1.52	1.48	1.42	1.39	1.34	1.31	0.30	1.28
	2.15	2.07	1.97	1.89	1.80	1.74	1.65	1.60	1.52	1.47	1.45	1.43
200	1.67	1.62	1.56	1.52	1.46	1.41	1.35	1.32	1.26	1.22	1.21	1.19
	2.06	1.97	1.87	1.79	1.69	1.63	1.53	1.48	1.39	1.33	1.30	1.28
500	1.64	1.59	1.53	1.48	1.42	1.38	1.31	1.28	1.21	1.16	1.14	1.12
	2.00	1.92	1.81	1.74	1.63	1.57	1.47	1.41	1.31	1.23	1.20	1.17
1000	1.63	1.58	1.52	1.47	1.41	1.36	1.30	1.26	1.19	1.13	1.11	1.08
	1.98	1.90	1.79	1.72	1.61	1.54	1.44	1.38	1.28	1.19	1.16	1.12
10,000	1.62	1.57	1.51	1.46	1.40	1.35	1.28	1.25	1.17	1.11	1.08	1.03
	1.97	1.88	1.77	1.70	1.59	1.53	1.42	1.36	1.25	1.16	1.11	1.05

A-5. Critical Values for the Pearson Correlation Coefficient for the Null Hypothesis H_0: $\rho = 0$ and Both the One-Tailed Alternative H_A: $|\rho| > 0$ and the Two-Tailed Alternative H_A: $\rho \neq 0$

	Level of significance for a one-tailed test					
	0.1000	**0.0500**	**0.0250**	**0.0100**	**0.0050**	**0.0005**
	Level of significance for a two-tailed test					
Degrees of freedom	**0.2000**	**0.1000**	**0.0500**	**0.0200**	**0.0100**	**0.0010**
1	0.9511	0.9877	0.9969	0.9995	0.9999	1.0000
2	0.8001	0.9000	0.9500	0.9800	0.9900	0.9990
3	0.6871	0.8053	0.8783	0.9343	0.9587	0.9912
4	0.6083	0.7293	0.8114	0.8822	0.9172	0.9741
5	0.5509	0.6694	0.7545	0.8329	0.8745	0.9508
6	0.5068	0.6215	0.7067	0.7888	0.8343	0.9249
7	0.4716	0.5823	0.6664	0.7498	0.7976	0.8982
8	0.4428	0.5495	0.6319	0.7154	0.7646	0.8721
9	0.4187	0.5214	0.6020	0.6850	0.7348	0.8471
10	0.3980	0.4972	0.5760	0.6581	0.7079	0.8233
11	0.3801	0.4762	0.5529	0.6339	0.6836	0.8010
12	0.3645	0.4574	0.5324	0.6120	0.6614	0.7800
13	0.3506	0.4409	0.5139	0.5922	0.6411	0.7604
14	0.3383	0.4258	0.4973	0.5742	0.6226	0.7419
15	0.3272	0.4124	0.4821	0.5577	0.6055	0.7247
16	0.3170	0.4000	0.4683	0.5425	0.5897	0.7084
17	0.3076	0.3888	0.4556	0.5285	0.5750	0.6932
18	0.2991	0.3783	0.4438	0.5154	0.5614	0.6853
19	0.2914	0.3687	0.4329	0.5033	0.5487	0.6652
20	0.2841	0.3599	0.4227	0.4921	0.5368	0.6524
21	0.2774	0.3516	0.4133	0.4816	0.5256	0.6402
22	0.2711	0.3438	0.4044	0.4715	0.5151	0.6287
23	0.2652	0.3365	0.3961	0.4623	0.5051	0.6177
24	0.2598	0.3297	0.3883	0.4534	0.4958	0.6073
25	0.2545	0.3233	0.3809	0.4451	0.4869	0.5974
26	0.2497	0.3173	0.3740	0.4372	0.4785	0.5880
27	0.2452	0.3114	0.3673	0.4297	0.4706	0.5790
28	0.2408	0.3060	0.3609	0.4226	0.4629	0.5703
29	0.2365	0.3009	0.3550	0.4158	0.4556	0.5620
30	0.2326	0.2959	0.3493	0.4093	0.4487	0.5541
40	0.2018	0.2573	0.3044	0.3578	0.3931	0.4896
60	0.1650	0.2109	0.2500	0.2948	0.3248	0.4078
120	0.1169	0.1496	0.1779	0.2104	0.2324	0.2943
250	0.0808	0.1035	0.1230	0.1455	0.1608	0.2038

A-6. Uniformly Distributed Random Numbers

0.538246	0.181648	0.172614	0.450166	0.293027	0.030195	0.757836	0.915061
0.663357	0.368934	0.516388	0.656254	0.284258	0.906335	0.329788	0.054487
0.035771	0.053784	0.424573	0.942479	0.293872	0.326815	0.862351	0.358055
0.513560	0.165508	0.667312	0.878444	0.414203	0.100839	0.555287	0.685601
0.880006	0.069305	0.854410	0.371911	0.751341	0.128446	0.678679	0.514995
0.748794	0.902497	0.629615	0.662531	0.932879	0.018376	0.683876	0.554810
0.115441	0.207278	0.887853	0.812124	0.082143	0.939258	0.666874	0.582525
0.953369	0.543997	0.806486	0.707493	0.503949	0.489926	0.774467	0.248617
0.243600	0.537111	0.181388	0.619277	0.131852	0.131876	0.361814	0.582682
0.610186	0.411580	0.339972	0.080869	0.429448	0.822770	0.632690	0.863227
0.848375	0.043973	0.071429	0.713405	0.562010	0.716050	0.536620	0.357681
0.102922	0.201752	0.617270	0.416471	0.371492	0.633301	0.857578	0.483474
0.009326	0.912932	0.113850	0.331600	0.852807	0.626191	0.035676	0.581386
0.801494	0.365068	0.548750	0.480788	0.032959	0.906331	0.291263	0.706212
0.682049	0.946008	0.960047	0.830463	0.186225	0.123762	0.674147	0.012839
0.610019	0.495159	0.165360	0.207562	0.676065	0.843231	0.581861	0.669062
0.203327	0.400473	0.257740	0.474822	0.439519	0.722845	0.274275	0.862333
0.081627	0.889037	0.446990	0.102859	0.384948	0.785505	0.655849	0.305504
0.866440	0.209892	0.558315	0.422571	0.277399	0.548600	0.009868	0.121337
0.514767	0.666866	0.498019	0.253118	0.081162	0.306135	0.762659	0.284422
0.739733	0.030429	0.054100	0.524432	0.654229	0.393821	0.928477	0.784502
0.131999	0.540034	0.884054	0.508152	0.767278	0.646574	0.211862	0.912805
0.569512	0.983024	0.488274	0.554083	0.860977	0.868964	0.304604	0.729149
0.388688	0.592203	0.241637	0.331246	0.380662	0.696245	0.029850	0.363413
0.518582	0.634681	0.409353	0.468436	0.724978	0.222090	0.357835	0.507892
0.204704	0.364712	0.046972	0.526575	0.705034	0.352802	0.077436	0.715174
0.613146	0.281786	0.635090	0.433333	0.976778	0.148786	0.442014	0.232516
0.914132	0.089797	0.111000	0.346466	0.855181	0.945228	0.293689	0.531055
0.898415	0.959250	0.005809	0.260049	0.150843	0.637507	0.171769	0.151870
0.543517	0.275269	0.629999	0.428829	0.744025	0.279269	0.913490	0.597542
0.091718	0.646875	0.173754	0.647365	0.987107	0.588587	0.721938	0.361474
0.995853	0.625377	0.043191	0.390483	0.148443	0.070920	0.716346	0.441175
0.926216	0.579495	0.436100	0.147287	0.713980	0.209226	0.016912	0.305757
0.055266	0.975932	0.836327	0.782469	0.843648	0.679524	0.812039	0.250478
0.159596	0.599053	0.513586	0.140141	0.537635	0.326994	0.619062	0.017493
0.014355	0.195823	0.393806	0.776234	0.603296	0.184318	0.394712	0.678653
0.264856	0.362599	0.297700	0.918237	0.213690	0.865163	0.976820	0.562911
0.555641	0.166171	0.179553	0.601542	0.953832	0.795291	0.764679	0.127128
0.739608	0.903922	0.626750	0.518962	0.933695	0.027938	0.587163	0.737179
0.043369	0.136388	0.674675	0.193588	0.440263	0.035475	0.610216	0.143710
0.587513	0.174713	0.072123	0.535015	0.597938	0.555897	0.680197	0.424194
0.222986	0.153264	0.420138	0.383530	0.859170	0.120550	0.131152	0.254565
0.292917	0.952745	0.120721	0.812756	0.292032	0.261342	0.850707	0.002310
0.901735	0.147683	0.610115	0.151896	0.980433	0.046558	0.462072	0.222302
0.435405	0.327907	0.588422	0.100345	0.704154	0.708261	0.043857	0.929596
0.775609	0.463232	0.617348	0.189444	0.742318	0.519705	0.233550	0.903274
0.266295	0.492168	0.796914	0.705457	0.509377	0.802423	0.808290	0.427691

A-7. Critical Values for the Kolmogorov–Smirnov One-Sample Test

Sample size (n)	Level of significance (α)				
	.20	**.15**	**.10**	**.05**	**.01**
1	.900	.925	.950	.975	.995
2	.684	.726	.776	.842	.929
3	.565	.597	.642	.708	.828
4	.494	.525	.564	.624	.733
5	.446	.474	.510	.565	.669
6	.410	.436	.470	.521	.618
7	.381	.405	.438	.486	.577
8	.358	.381	.411	.457	.543
9	.339	.360	.388	.432	.514
10	.322	.342	.368	.410	.490
11	.307	.326	.352	.391	.468
12	.295	.313	.338	.375	.450
13	.284	.302	.325	.361	.433
14	.274	.292	.314	.349	.418
15	.266	.283	.304	.338	.404
16	.258	.274	.295	.328	.392
17	.250	.266	.286	.318	.381
18	.244	.259	.278	.309	.371
19	.237	.252	.272	.301	.363
20	.231	.246	.264	.294	.356
25	.21	.22	.24	.27	.32
30	.19	.20	.22	.24	.29
35	.18	.19	.21	.23	.27
Over 35	$\frac{1.07}{\sqrt{n}}$	$\frac{1.14}{\sqrt{n}}$	$\frac{1.22}{\sqrt{n}}$	$\frac{1.36}{\sqrt{n}}$	$\frac{1.63}{\sqrt{n}}$

A-8. Values of the Gamma Function

$$\Gamma(t) = \int_0^\infty r^{t-1} \exp(-r)dr$$

t	$\Gamma(t)$	t	$\Gamma(t)$	t	$\Gamma(t)$	t	$\Gamma(t)$	t	$\Gamma(t)$
1.000	1.00000	1.200	0.91817	1.400	0.88726	1.600	0.89352	1.800	0.93138
1.005	0.99714	1.205	0.91686	1.405	0.88700	1.605	0.89409	1.805	0.93272
1.010	0.99433	1.210	0.91558	1.410	0.88676	1.610	0.89468	1.810	0.93408
1.015	0.99156	1.215	0.91433	1.415	0.88655	1.615	0.89529	1.815	0.93545
1.020	0.98884	1.220	0.91311	1.420	0.88636	1.620	0.89592	1.820	0.93685
1.025	0.98617	1.225	0.91192	1.425	0.88618	1.625	0.89657	1.825	0.93826
1.030	0.98355	1.230	0.91075	1.430	0.88604	1.630	0.89724	1.830	0.93969
1.035	0.98097	1.235	0.90962	1.435	0.88591	1.635	0.89793	1.835	0.94114
1.040	0.97844	1.240	0.90852	1.440	0.88581	1.640	0.89864	1.840	0.94261
1.045	0.97595	1.245	0.90745	1.445	0.88572	1.645	0.89937	1.845	0.94410
1.050	0.97350	1.250	0.90640	1.450	0.88566	1.650	0.90012	1.850	0.94561
1.055	0.97110	1.255	0.90539	1.455	0.88562	1.655	0.90088	1.855	0.94714
1.060	0.96874	1.260	0.90440	1.460	0.88560	1.660	0.90167	1.860	0.94869
1.065	0.96643	1.265	0.90344	1.465	0.88561	1.665	0.90247	1.865	0.95025
1.070	0.96415	1.270	0.90250	1.470	0.88563	1.670	0.90330	1.870	0.95184
1.075	0.96192	1.275	0.90160	1.475	0.88568	1.675	0.90414	1.875	0.95345
1.080	0.95973	1.280	0.90072	1.480	0.88575	1.680	0.90500	1.880	0.95507
1.085	0.95757	1.285	0.89987	1.485	0.88584	1.685	0.90588	1.885	0.95672
1.090	0.95546	1.290	0.89904	1.490	0.88595	1.690	0.90678	1.890	0.95838
1.095	0.95339	1.295	0.89824	1.495	0.88608	1.695	0.90770	1.895	0.96006
1.100	0.95135	1.300	0.89747	1.500	0.88623	1.700	0.90864	1.900	0.96177
1.105	0.94935	1.305	0.89672	1.505	0.88640	1.705	0.90960	1.905	0.96349
1.110	0.94740	1.310	0.89600	1.510	0.88659	1.710	0.91057	1.910	0.96523
1.115	0.94547	1.315	0.89531	1.515	0.88680	1.715	0.91157	1.915	0.96699
1.120	0.94359	1.320	0.89464	1.520	0.88704	1.720	0.91258	1.920	0.96877
1.125	0.94174	1.325	0.89400	1.525	0.88729	1.725	0.91361	1.925	0.97058
1.130	0.93993	1.330	0.89338	1.530	0.88757	1.730	0.91467	1.930	0.97240
1.135	0.93816	1.335	0.89278	1.535	0.88786	1.735	0.91574	1.935	0.97424
1.140	0.93642	1.340	0.89222	1.540	0.88818	1.740	0.91683	1.940	0.97610
1.145	0.93471	1.345	0.89167	1.545	0.88851	1.745	0.91793	1.945	0.97798
1.150	0.93304	1.350	0.89115	1.550	0.88887	1.750	0.91606	1.950	0.97988
1.155	0.93141	1.355	0.89066	1.555	0.88924	1.755	0.92021	1.955	0.98180
1.160	0.92980	1.360	0.89018	1.560	0.88964	1.760	0.92137	1.960	0.98374
1.165	0.92823	1.365	0.88974	1.565	0.89005	1.765	0.92256	1.965	0.98570
1.170	0.92670	1.370	0.88931	1.570	0.89049	1.770	0.92376	1.970	0.98768
1.175	0.92520	1.375	0.88891	1.575	0.89094	1.775	0.92499	1.975	0.98969
1.180	0.92373	1.380	0.88854	1.580	0.89142	1.780	0.92623	1.980	0.99171
1.185	0.92229	1.385	0.88818	1.585	0.89191	1.785	0.92749	1.985	0.99375
1.190	0.92089	1.390	0.88785	1.590	0.89243	1.790	0.92877	1.990	0.99581
1.195	0.91951	1.395	0.88755	1.595	0.89296	1.795	0.93007	1.995	0.99790
								2.000	1.00000

A-9. Critical Values for the Duncan Multiple Range Test for a 5% Level of Significance and Selected Degrees of Freedom (*df*) and *p* Groups

Degrees of freedom	*p* = 2	3	4	5	6	7	8	9	10
1	17.97								
2	6.09	6.09							
3	4.50	4.52	4.52						
4	3.93	4.01	4.03	4.03					
5	3.64	3.75	3.80	3.81	3.81				
6	3.46	3.59	3.65	3.68	3.69	3.70			
7	3.34	3.48	3.55	3.59	3.61	3.62	3.63		
8	3.26	3.40	3.48	3.52	3.55	3.57	3.57	3.58	
9	3.20	3.34	3.42	3.47	3.50	3.52	3.54	3.54	3.55
10	3.15	3.29	3.38	3.43	3.47	3.49	3.51	3.52	3.52
11	3.11	3.26	3.34	3.40	3.44	3.46	3.48	3.49	3.50
12	3.08	3.23	3.31	3.37	3.41	3.44	3.46	3.47	3.48
13	3.06	3.20	3.29	3.35	3.39	3.42	3.44	3.46	3.47
14	3.03	3.18	3.27	3.33	3.37	3.40	3.43	3.44	3.46
15	3.01	3.16	3.25	3.31	3.36	3.39	3.41	3.43	3.45
16	3.00	3.14	3.23	3.30	3.34	3.38	3.40	3.42	3.44
17	2.98	3.13	3.22	3.28	3.33	3.37	3.39	3.41	3.43
18	2.97	3.12	3.21	3.27	3.32	3.36	3.38	3.40	3.42
19	2.96	3.11	3.20	3.26	3.31	3.35	3.38	3.40	3.41
20	2.95	3.10	3.19	3.25	3.30	3.34	3.37	3.39	3.41
25	2.92	3.07	3.16	3.23	3.28	3.31	3.35	3.37	3.39
30	2.89	3.03	3.13	3.20	3.25	3.29	3.32	3.35	3.37
40	2.86	3.01	3.10	3.17	3.22	3.27	3.30	3.33	3.35
60	2.83	2.98	3.07	3.14	3.20	3.24	3.28	3.31	3.33
120	2.80	2.95	3.04	3.12	3.17	3.22	3.25	3.29	3.31
∞	2.77	2.92	3.02	3.09	3.15	3.19	3.23	3.27	3.29

Appendix B
Taylor Series Expansion

TABLE OF CONTENTS

B-1. TAYLOR SERIES

A Taylor series is commonly used as a basis of approximation in engineering analysis. The objective of this section is to provide a review of the Taylor series.

A Taylor series is the sum of functions based on continually increasing derivatives. For a function $f(x)$ that depends on only one independent variable x, the value of the function at point $x_0 + h$ can be approximated by the following Taylor series:

$$f(x_0+h) = f(x_0) + hf^{(1)}(x_0) + \frac{h^2}{2!}f^{(2)}(x_0) + \frac{h^3}{3!}f^{(3)}(x_0) + \dots + \frac{h^n}{n!}f^{(n)}(x_0) + R_{n+1} \tag{B-1a}$$

in which x_0 is some base value (or starting value) of the independent variable x; h is the distance between x_0 and the point x at which the value of the function is needed — that is, $h = x - x_0$; $n!$ is the factorial of n and $n! = n(n-1)(n-2)\cdots 1$; R_{n+1} is the remainder of the Taylor series expansion, as the expression can have an infinite number of terms; and the superscript (n) on the function $f^{(n)}$ indicates the nth derivative of the function $f(x)$. The derivatives of $f(x)$ in Equation B-1a are evaluated at x_0. Equation B-1a can also be written in the following form:

$$f(x_0+h) = \sum_{k=0}^{\infty}\frac{h^k}{k!}f^{(k)}(x_0) \tag{B-1b}$$

where $0! = 1$ by convention. Equations B-1a and B-1b are based on the assumption that continuous derivatives exist in an interval that includes the points x and x_0.

Equations B-1a and B-1b can be used with a finite number of terms to produce approximations for the evaluated function at $f(x_0 + h)$. The order of the approximation is defined by the order of the highest derivative that is included in the approximation. For example, the first-order approximation is defined by the terms of the Taylor series expansion up to the first derivative as follows:

$$f(x_0 + h) \approx f(x_0) + hf^{(1)}(x_0) \tag{B-2a}$$

Similarly, the second- and third-order approximations, respectively, are given by:

$$f(x_0 + h) \approx f(x_0) + hf^{(1)}(x_0) + \frac{h^2}{2!}f^{(2)}(x_0) \tag{B-2b}$$

and

$$f(x_0 + h) \approx f(x_0) + hf^{(1)}(x_0) + \frac{h^2}{2!}f^{(2)}(x_0) + \frac{h^3}{3!}f^{(3)}(x_0) \tag{B-2c}$$

The accuracy of the approximation improves as the order of the approximation is increased.

The importance of the individual terms of the Taylor series depends on the nature of the function and the distance h. The higher-order terms become more important as the nonlinearity of the function increases and the difference $x - x_0$ increases. For linear functions, only the term with the first derivative is necessary, because terms with higher-order derivatives are zero. As the separation distance h increases, the importance of including the nonlinear terms usually increases. Thus, the error in an estimate decreases as more terms of the Taylor series are included.

A real advantage of using a Taylor series to evaluate a function is that it provides the means to approximately evaluate complex functions that cannot be easily evaluated otherwise. Even if it is only possible to compute an approximation of the first derivative, an estimated value of the function can be made using the linear term of the Taylor series expansion and a small distance h from a known boundary value or another approximately evaluated point. The benefits become evident in cases where the function f cannot be expressed in a closed form.

Example B-1: Nonlinear Polynomials

Taylor series are used to evaluate a function that is of practical interest but with an unknown expression. If the expression of the function is known, then the Taylor series approximation is not necessary. In this example, the true function will be assumed to be known in order to demonstrate the computational aspects of Taylor series and the sizes of the remainder terms in the series. For this purpose, the following fourth-order polynomial is assumed to be the true function:

$$f(x) = \frac{1}{8}(x^4) - \frac{1}{6}(x^3) + \frac{1}{2}(x^2) - \frac{1}{2}(x) + 2 \tag{B-3}$$

The function has the following derivatives:

$$f^{(1)}(x) = 0.5(x^3) - 0.5(x^2) + x - 0.5 \tag{B-4}$$

$$f^{(2)}(x) = 1.5(x^2) - x + 1 \tag{B-5}$$

$$f^{(3)}(x) = 3x - 1 \tag{B-6}$$

$$f^{(4)}(x) = 3 \tag{B-7}$$

$$f^{(5)}(x) = 0 \tag{B-8}$$

Assume a boundary condition to a define a base point (x_0) of 1.0. Then, the value of the function and its derivatives are:

$$f(x) = 1.95833 \tag{B-9a}$$

$$f^{(1)}(x) = 0.5 \tag{B-9b}$$

$$f^{(2)}(x) = 1.5 \tag{B-9c}$$

$$f^{(3)}(x) = 2 \tag{B-9d}$$

$$f^{(4)}(x) = 3 \tag{B-9e}$$

$$f^{(5)}(x) = 0 \tag{B-9f}$$

Assume that a value of the function is needed for a value of x of 1.5. The separation distance h is 0.5. The values of the individual terms of the fourth-order Taylor series are shown in row 1 of Table B-1(A). For $x = 1.5$, the linear term is dominant, with the magnitude of the higher-order terms being less significant. Because Equation B-3 is a fourth-order polynomial, the fourth-order Taylor series gives the exact solution. As the separation distance increases, the contribution made by each term of the Taylor series expansion changes. For x from 2 to 3, the second-order term is the most important, with less emphasis given to the linear term and more emphasis to the cubic term. For $x = 3.5$, the cubic term is the most important, while for $x = 4$, the fourth-order term is most important. These results illustrate the interaction between the importance of the terms and both the nonlinearity of the function and the separation distance. The cumulative values for the terms of the Taylor series are given in Table B-1(B), and the differences between the cumulative values and the true values are given in Table B-1(C). The differences show the need to consider the higher-order terms as the separation distance increases. This is clearly evident from Figure B-1, which shows the cumulative values from Table B-1(B) as a function of the separation distance h.

TABLE B-1 (A) Values of the Individual Terms of the Taylor Series Expansion of Equation B-1 for the Quadratic Function of Equation B-3; (B) Cumulative Function of the Taylor Series Expansion; and (C) Difference between the Cumulative Function and the True Value of the Function

Part	x	h	$hf^{(1)}(x_0)$	$\frac{h^2}{2!}f^{(2)}(x_0)$	$\frac{h^3}{3!}f^{(3)}(x_0)$	$\frac{h^4}{4!}f^{(4)}(x_0)$
(A) Individual terms	1.5	0.5	0.25	0.1875	0.04167	0.0078125
	2.0	1.0	0.50	0.7500	0.33333	0.1250000
	2.5	1.5	0.75	1.6875	1.12500	0.6328125
	3.0	2.0	1.00	3.0000	2.66667	2.0000000
	3.5	2.5	1.25	4.6875	5.20833	4.8828125
	4.0	3.0	1.50	6.7500	9.00000	10.1250000
(B) Cumulative function	1.5		2.20833	2.39583	2.43750	2.44531
	2.0		2.45833	3.20833	3.54167	3.66667
	2.5		2.70833	4.39538	5.52083	6.15365
	3.0		2.95833	5.95833	8.62500	10.62500
	3.5		3.20833	7.89583	13.10417	17.98698
	4.0		3.45833	10.20833	19.20833	29.33333
(C) Errors	1.5		–0.23698	–0.04948	–0.00781	0.00000
	2.0		–1.20833	–0.45833	–0.12500	0.00000
	2.5		–3.44531	–1.75781	–0.63281	0.00000
	3.0		–7.66667	–4.66667	–2.00000	0.00000
	3.5		–14.77865	–10.09115	–4.88281	0.00000
	4.0		–25.87500	–19.12500	–10.12500	0.00000

B-2. COMMON TAYLOR SERIES

In this section, example Taylor series expressions are provided for some commonly used functions. These series might be familiar to the readers from previous courses in mathematics. For example, the exponential evaluation to the base e of x can be expressed by the following series:

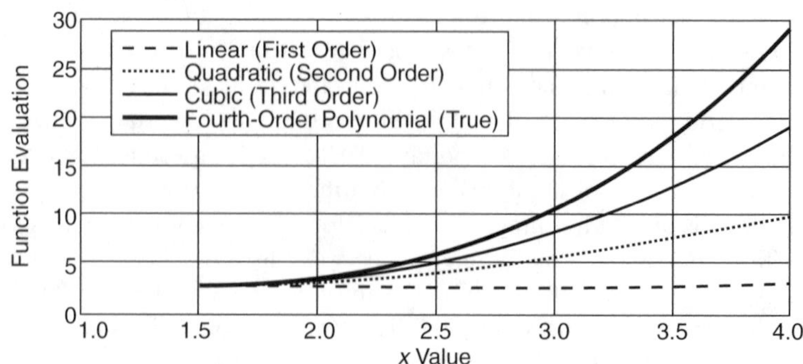

FIGURE B-1 Taylor series approximation of a fourth-order polynomial.

$$e^x = 1 + x + \frac{x^2}{2!} + \frac{x^3}{3!} + \ldots = \sum_{k=0}^{\infty} \frac{x^k}{k!} \tag{B-10}$$

This equation is valid only for x that is finite. Similarly, the natural logarithm of x can be expressed using a Taylor series as:

$$\ln(x) = (x-1) - \frac{(x-1)^2}{2!} + \frac{(x-1)^3}{3!} - \ldots = \sum_{k=0}^{\infty} (-1)^{k-1} \frac{(x-k)^k}{k!} \quad \text{for } 0 < x \le 2 \tag{B-11}$$

The sine and cosine functions can also be expressed using Taylor series as follows, respectively:

$$\sin(x) = x - \frac{x^3}{3!} + \frac{x^5}{5!} - \ldots = \sum_{k=0}^{\infty} (-1)^k \frac{x^{2k+1}}{(2k+1)!} \tag{B-12}$$

$$\cos(x) = 1 - \frac{x^2}{2!} + \frac{x^4}{4!} - \ldots = \sum_{k=0}^{\infty} (-1)^k \frac{x^{2k}}{(2k)!} \tag{B-13}$$

Equations B-12 and B-13 are valid only for x values that are finite. The reciprocal of $(1 - x)$, where $|x| < 1$, can be expressed by the following Taylor series:

$$\frac{1}{1-x} = 1 + x + x^2 + x^3 + \ldots = \sum_{k=0}^{\infty} x^k \tag{B-14}$$

These Taylor series can be used to evaluate their corresponding functions at any point x using a base value of $x_0 = 0$ and increment h.

B-3. APPLICATIONS: TAYLOR SERIES EXPANSION OF THE SQUARE ROOT

The square-root function is used in this section to illustrate the Taylor series expansion. The square-root function can be expressed as:

$$f(x) = \sqrt{x} \tag{B-15}$$

To evaluate the Taylor series, the derivatives of the function are computed as follows:

$$f^{(1)}(x) = \frac{1}{2}x^{-0.5} \tag{B-16a}$$

$$f^{(2)}(x) = -\frac{1}{4}x^{-1.5} \tag{B-16b}$$

$$f^{(3)}(x) = \frac{3}{8}x^{-2.5} \tag{B-16c}$$

Higher-order derivatives can be developed similarly. For a base point $x_0 = 1$ and $h = 0.001$, the four terms of the Taylor series produce the following estimate for the square root of 1.001:

$$\begin{aligned} f(1.001) &= \sqrt{1.001} \approx \sqrt{1} + 0.5(0.001)(1)^{-0.5} \\ &\quad - \frac{1}{4(2!)}(0.001)^2(1)^{-1.5} + \frac{3}{8(3!)}(0.001)^3(1)^{-2.5} \end{aligned} \tag{B-17a}$$

or

$$f(1.001) \approx 1 + 0.5 \times 10^{-3} - 0.125 \times 10^{-6} + 0.625 \times 10^{-10} = 1.0004999 \tag{B-17b}$$

which equals the true value to the number of decimal points shown. Because the interval h is so small, the linear approximation would have been accurate to five decimal places.

B-4. PROBLEMS

B-1. Using the Taylor series expansion for cos(x) provided in Section B.2, evaluate the series starting at $x_0 = 0$ radian with $h = 0.1$ to $x = 1$ radian. Use one term, two terms, and three terms of the series. Evaluate the error in each value, and discuss the effect of incrementally adding the terms to the series. Compare your results with the true solution.

B-2. Using the Taylor series for sin(x) of Equation B-12, evaluate the series starting at $x_0 = 0$ radian with $h = 0.1$ to $x = 1$ radian. Use one term, two terms, and three terms of the series. Evaluate the error in each value, and discuss the effect of incrementally adding the terms to the series. Compare your results with the true solution.

B-3. Using the Taylor series expansion for e^x provided by Equation B-10, evaluate the series starting at $x_0 = 0$ with $h = 0.1$ to $x = 1$. Use one term, two terms, and three terms of the series. Evaluate the error in each value, and discuss the effect of incrementally adding the terms to the series. Compare your results with the true solution.

B-4. Develop a Taylor series expansion of the following polynomial:

$$f(x) = x^3 - 3x^2 + 5x + 10$$

Use $x = 2$ as the base (or starting) point and h as the increment. Evaluate the series for $h = 0.1, 0.2, 0.3, 0.4, 0.5, 0.6, 0.7, 0.8, 0.9$, and 1.0. Discuss the accuracy of the method as the terms of the Taylor series are added incrementally.

B-5. Develop a Taylor series expansion of the following polynomial:

$$f(x) = x^5 - 5x^4 + x^2 + 6$$

Use $x = 2$ as the base (or starting) point and h as the increment. Evaluate the series for h = 0.1, 0.2, 0.3, 0.4, 0.5, 0.6, 0.7, 0.8, 0.9, and 1.0. Discuss the accuracy of the method as the terms of the Taylor series are added incrementally.

B-6. Develop a Taylor series expansion of the following function:

$$f(x) = x^2 - 5x^{0.5} + 6$$

Use $x = 2$ as the base (or starting) point and h as the increment. Evaluate the series for h = 0.1, 0.2, 0.3, 0.4, 0.5, 0.6, 0.7, 0.8, 0.9, and 1.0. Discuss the accuracy of the method as the terms of the Taylor series are added incrementally.

B-7. Write a computer program or develop a spreadsheet to evaluate the Taylor series expansion for Problem B-1.

B-8. Write a computer program or develop a spreadsheet to evaluate the Taylor series expansion for Problem B-3.

B-9. Write a computer program or develop a spreadsheet to evaluate the Taylor series expansion for Problem B-5.

B-10. Write a computer program or develop a spreadsheet to evaluate the Taylor series expansion for Problem B-6.

Appendix C
Data for Simulation Projects

TABLE OF CONTENTS

C-1. STREAM EROSION STUDY

X_1	X_2	X_3	Y	X_1	X_2	X_3	Y
2.911	0.722	0.354	4.400	3.153	0.552	0.515	4.838
3.194	0.775	0.149	1.892	2.888	1.001	0.417	4.791
3.641	0.670	0.328	4.271	2.782	0.853	0.280	4.027
2.515	0.337	0.277	3.508	2.066	1.036	0.119	2.930
3.610	0.763	0.380	5.063	2.835	1.062	0.588	6.397
3.117	0.992	0.347	4.146	3.603	1.188	0.593	4.797
2.186	0.880	0.604	6.758	3.267	0.528	0.430	4.599
2.907	1.205	0.257	3.774	2.873	0.375	0.316	3.901
2.086	1.420	0.441	5.246	3.392	0.950	0.300	3.662
2.635	0.639	0.481	6.201	3.999	0.833	0.364	3.321
2.965	1.023	0.352	3.900	1.765	1.031	0.483	6.952
3.084	0.683	0.230	4.020	3.544	0.801	0.358	4.555
3.104	1.297	0.372	4.218	1.476	1.196	0.524	6.371
4.054	1.054	0.423	3.605	2.094	1.482	0.409	4.770
2.897	0.903	0.613	6.344	2.832	1.040	0.115	1.989
2.454	0.887	0.428	5.366	3.248	1.136	0.572	5.247
2.937	0.837	0.211	3.246	3.247	0.976	0.387	3.415
2.985	0.374	0.246	3.332	0.908	1.196	0.443	7.122
3.007	1.040	0.553	5.514	0.729	0.792	0.375	5.700
3.632	0.734	0.245	2.500	2.561	0.921	0.446	6.038
2.864	0.638	0.459	4.579	2.768	1.025	0.388	4.101
3.646	0.813	0.225	2.711	3.072	0.994	0.252	3.770
1.228	1.551	0.450	5.889	2.823	0.960	0.562	7.236
3.438	0.974	0.189	1.469	3.466	1.533	0.325	3.224
1.969	0.617	0.523	6.354	2.711	0.836	0.321	3.905
2.662	1.100	0.190	3.278	2.720	0.933	0.346	4.634
1.397	1.593	0.525	6.320	2.590	0.695	0.442	4.720
1.901	1.118	0.451	5.691	3.049	1.149	0.625	6.865
2.545	0.567	0.658	7.678	1.753	0.619	0.310	6.195
2.561	0.964	0.564	6.462	1.693	1.226	0.397	4.305
2.460	1.318	0.552	6.235	3.009	0.984	0.241	2.286

C-2. TRAFFIC ESTIMATION STUDY

X_1	X_2	X_3	Y	X_1	X_2	X_3	Y
20.009	6.330	10.223	1.776	23.235	4.170	7.156	1.616
21.739	2.327	5.661	1.479	14.258	7.398	12.071	1.500
24.471	2.170	7.509	0.858	18.487	2.936	6.098	0.788
17.589	1.851	6.237	0.542	10.761	4.131	6.923	0.912
24.285	5.109	7.806	1.627	13.840	2.186	4.882	0.783
21.270	5.476	8.034	1.289	17.776	6.699	8.775	1.469
15.582	6.016	9.108	1.384	17.875	4.310	6.234	1.281
19.987	3.575	6.452	1.103	17.254	4.445	9.435	1.171
14.970	6.717	12.142	1.256	21.493	4.356	7.367	0.987
18.327	4.375	7.073	1.081	19.874	4.069	6.643	1.282
20.342	3.391	6.071	1.440	19.225	1.425	3.873	0.571
21.068	4.160	7.764	1.175	14.848	4.227	6.173	1.135
21.192	2.976	5.792	1.556	19.547	3.478	5.234	0.956
26.998	3.671	9.176	1.455	24.242	2.931	7.135	1.173
19.925	3.566	11.634	1.432	22.188	3.929	6.513	1.175
17.219	4.905	7.845	1.489	19.780	8.150	11.272	1.743
20.172	3.410	8.665	1.233	22.949	2.850	4.857	1.458
20.465	3.715	7.291	1.217	26.663	2.809	8.262	1.167
20.599	3.115	9.182	1.094	13.010	5.570	8.999	1.086
24.416	1.209	5.719	0.721	23.880	6.388	8.616	1.887
19.724	3.643	6.821	0.797	11.243	4.672	5.357	1.011
24.505	4.954	7.716	1.640	15.016	2.526	6.983	0.919
9.726	4.313	7.450	0.975	19.528	4.797	7.932	1.176

C-3. WATER EVAPORATION STUDY

X_1	X_2	X_3	Y	X_1	X_2	X_3	Y
62.	32.	10.4	0.172	75.	58.	6.8	0.149
67.	56.	5.6	0.170	69.	64.	6.8	0.159
75.	35.	8.9	0.199	62.	53.	7.2	0.179
55.	74.	5.3	0.115	71.	68.	8.6	0.166
75.	58.	8.8	0.127	82.	42.	11.5	0.238
66.	49.	4.8	0.154	42.	50.	6.6	0.050
49.	48.	9.9	0.155	74.	57.	80.	0.180
62.	45.	4.8	0.157	37.	55.	5.6	0.036
48.	15.	9.9	0.179	48.	52.	8.1	0.125
57.	58.	8.8	0.094	61.	89.	6.4	0.141
63.	38.	120.	0.167	68.	44.	10.2	0.166
65.	32.	10.5	0.170	68.	51.	5.9	0.130
66.	43.	10.3	0.184	57.	35.	11.3	0.155
83.	90.	9.8	0.135	44.	65.	8.3	0.106
62.	53.	8.3	0.188	56.	31.	7.1	0.097
54.	30.	6.2	0.195	60.	70.	9.3	0.156
63.	61.	3.7	0.114	65.	54.	7.1	0.142
64.	93.	10.9	0.094	61.	49.	9.6	0.124
64.	63.	110.	0.136	72.	20.	6.9	0.190
75.	27.	8.5	0.170	59.	45.	10.7	0.132
61.	53.	6.7	0.103	59.	38.	8.1	0.142
75.	75.	6.5	0.176	57.	52.	6.8	0.145
42.	26.	7.4	0.107	95.	48.	7.3	0.210
72.	55.	9.3	0.202	42.	83.	8.2	0.092
46.	46.	7.4	0.104	61.	35.	8.5	0.177
58.	57.	9.9	0.171	64.	61.	7.4	0.157
35.	59.	8.1	0.063	39.	66.	6.7	0.050
44.	67.	3.6	0.092	42.	48.	6.8	0.129
56.	25.	10.6	0.163	56.	31.	5.5	0.159
56.	15.	7.8	0.197	47.	51.	8.1	0.102
54.	52.	8.7	0.109	41.	58.	6.9	0.134
67.	44.	7.6	0.190	55.	85.	4.7	0.139
82.	57.	8.7	0.215	48.	59.	7.2	0.052
60.	42.	7.8	0.097	87.	29.	8.8	0.241
47.	68.	5.7	0.060	79.	34.	6.7	0.175
61.	86.	10.5	0.140				

Index

C

D

E

F

H

I

J

K

L

M

N

Q

R

S

T

U

V

W

Y

Z